Glycosylation Demystified

Glycosylation Demystified

Edited by **Skylar Washington**

Callisto Reference

New York

Published by Callisto Reference,
106 Park Avenue, Suite 200,
New York, NY 10016, USA
www.callistoreference.com

Glycosylation Demystified
Edited by Skylar Washington

International Standard Book Number: 978-1-63239-365-4 (Hardback)

Printed in China.

Contents

Preface

I am honored to present to you this unique book which encompasses the most up-to-date data in the field. I was extremely pleased to get this opportunity of editing the work of experts from across the globe. I have also written papers in this field and researched the various aspects revolving around the progress of the discipline. I have tried to unify my knowledge along with that of stalwarts from every corner of the world, to produce a text which not only benefits the readers but also facilitates the growth of the field.

Glycosylation is generally described as a form of co-translational and post-translational modification. This book emphasizes on the concepts of glycobiology, and is a collective effort of a group of recognized researchers. It addresses some of the crucial topics in this domain, presenting a broad range of theoretical and practical topics in the subject of glycobiology. This book will be of great significance for scholars and researchers involved in drug glycoengineering and biomedical research.

Finally, I would like to thank all the contributing authors for their valuable time and contributions. This book would not have been possible without their efforts. I would also like to thank my friends and family for their constant support.

Editor

Section 1

Glycosylation and Structure

Chapter 1

Beyond the Sequon: Sites of N-Glycosylation

Benjamin Luke Schulz

Additional information is available at the end of the chapter

1. Introduction

Asparagine (N-) linked protein glycosylation is a common and essential post-translational modification of proteins in eukaryotes, archaea and some bacteria. It plays crucial roles in protein folding and in regulation of protein function. Although the general principles of N-glycosylation have been long known, the precise details governing whether a particular asparagine residue will be N-glycosylated or not are not well understood. This is of broad general importance in understanding the structure and function of the immense variety of N-glycoproteins in diverse biological systems. This chapter will review the current understanding of the mechanisms that determine how asparagine residues are selected for glycosylation by the enzyme oligosaccharyltransferase.

2. Overview of N-glycosylation in the endoplasmic reticulum

The initial steps in N-glycosylation take place in the lumen of the endoplasmic reticulum (ER). The enzyme oligosaccharyltransferase (OTase) catalyzes the key step in N-glycosylation, *en bloc* transfer of mature glycan from a lipid carrier to selected asparagine residues in nascent polypeptide chains (Kelleher & Gilmore, 2006). Glycan to be transferred to protein is synthesized by sequential addition of monosaccharides linked to a dolichol pyrophosphate lipid carrier (Burda & Aebi, 1999). This process is essentially linear, and in most organisms OTase specifically recognizes the final α-1,2-linked glucose, ensuring efficient transfer of only the mature $Glc_3Man_9GlcNAc_2$ glycan structure (Karaoglu *et al.*, 2001).

2.1. Oligosaccharyltransferase

The OTase enzyme is a multiprotein complex in most eukaryotes, and in yeast consists of 8 protein subunits (Ost1p, Ost2p, Ost3/6p, Ost4p, Ost5p, Swp1p, Wbp1p and Stt3p) (Kelleher & Gilmore, 2006). It is now clear that the Stt3p protein houses the catalytic site of OTase, while the accessory protein subunits of multiprotein complex OTases are required for complex stability, enzymatic regulation of OTase activity, substrate recognition and OTase enzyme localization (Mohorko *et al.*, 2011). OTase physically associates with the translocon (Shibatani *et al.*, 2005, Yan & Lennarz, 2005) and the ribosome (Harada *et al.*, 2009), and so has direct access to nascent polypeptides immediately as they enter the ER lumen (Dempski & Imperiali, 2002). Glycosylation of many asparagines is co-translocational, and occurs essentially as soon as they enter the ER lumen and can reach the OTase active site (Whitley *et al.*, 1996). Other sites are also glycosylated post-translocationally, with extended residence of protein in the ER lumen (Ruiz-Canada *et al.*, 2009). However, in all cases the protein substrate of OTase must be unfolded for glycosylation to occur.

2.2. Roles of N-glycans in protein folding

The key role of N-glycans on proteins in the ER is to assist in productive protein folding (Helenius & Aebi, 2004). By virtue of their hydrophilic bulk, N-glycans alter the overall biophysical properties of nascent polypeptides, increasing their solubility and constraining local polypeptide conformation (Wormald & Dwek, 1999). N-glycans can also function as signals for incomplete folding of particular domains of proteins, and so direct these to the ER resident thiol oxidoreductase ERp57 via the lectins calnexin and calreticulin (Oliver *et al.*, 1999). Timed trimming of N-glycans on glycoproteins in the ER lumen is also key for regulating retro-translocation of incorrectly folded glycoproteins to the cytoplasm for degradation (Aebi *et al.*, 2010).

3. The 'glycosylation sequon'

The key recognition factor for selection of asparagines for glycosylation by OTase is the 'glycosylation sequon'. This has been historically defined as Asn-Xaa-Ser/Thr (Xaa ≠ Pro). However, it has also long been clear that this is not an adequate predictor of glycosylation, as ~1/3rd of Asn in sequons in secreted proteins are not glycosylated. In addition to this, several examples of glycosylation of Asn residues not in sequons have been reported in recent years.

3.1. Definition of the sequon

The term 'sequon' was likely first used by Derek Marshall (Marshall, 1974) to describe the apparent three amino acid local sequence requirement for N-glycosylation. However, it was long recognized that the presence of a sequon was not sufficient for N-glycosylation to occur at a given Asn in portions of polypeptides entering the ER lumen. Nonetheless, the efficiency of glycosylation at a given asparagine is primarily determined by the flanking amino acids, with the primary factor increasing glycosylation being the presence of a threonine or serine at the +2 position. This has such a strong influence of the efficiency of glycosylation that it has been termed the 'glycosylation sequon' in recognition of its importance. However, the presence of a glycosylation sequon is neither necessary nor sufficient for an asparagine to be glycosylated.

3.2. The '+2' position: Thr, Ser, Cys, Etc

Whilst both Ser and Thr are accepted as amino acids at the +2 position in glycosylation sequons, they are not equal, as glycosylation of Asn-Xaa-Thr sequons is approximately 40 times efficient than of Asn-Xaa-Ser sequons (Kasturi *et al.*, 1995, Kasturi *et al.*, 1997). Far and away the majority of glycosylated asparagines are in traditional Asn-Xaa-Ser/Thr (Xaa≠Pro) sequons. However, several very well validated examples have been reported of asparagines *not* in sequons that are nonetheless efficiently glycosylated.

Several reports have been made of glycosylation at asparagines in the sequence Asn-Xaa-Cys. Human CD69 has such an Asn-Xaa-Cys glycosylation site (Vance et al, 1997). Human beta protein C is glycosylated at an Asn with cysteine at the +2 position (Miletich & Broze, 1990). Interestingly, the Cys in beta protein C is involved in a disulfide bond in the mature protein, and the formation of this disulfide competes directly with glycosylation at the preceding Asn. CHO-cell expressed recombinant human epidermal growth factor receptor (EGRF) also has such a glycosylation site (Sato *et al.*, 2000). Heterologous expression of an insect cathepsin B-like counter-defense protein in *Pichia pastoris* resulted in glycosylation at an asparagine in the sequence Asn-Xaa-Cys (Chi *et al.*, 2010). It is unclear if this site is also natively glycosylated. This shows that both mammalian and fungal OTase are capable of glycosylating selected Asn-Xaa-Cys sequences.

Several large-scale discovery projects for identification of N-glycosylation sites have been performed. The largest of these, from mouse, identified over 5000 putatively glycosylated asparagines (Zielinska *et al.*, 2010). While the vast majority of these were in conventional Asn-Xaa-Ser/Thr sequons, a small but significant number of Asn not in such sequons were identified as being glycosylated. Asn-Xaa-Cys sites represented 65/5052, and Asn-Xaa-Val 20/5052. It was also reported that Asn-Gly sites were modified. However, this result must be treated with extreme caution, given the propensity for non-catalyzed spontaneous deamidation (asparagine-aspartate conversion) is especially high at Asn-Gly sequences (Palmisano *et al.*, 2012, Robinson *et al.*, 2004)

It was proposed that the hydroxyl group of Ser/Thr amino acids at the +2 position was directly involved in catalysis, via the formation of an 'Asparagine turn' (Imperiali & Hendrickson, 1995). This proposal was certainly powerful, and could withstand the observation of rare Asn-Xaa-Cys glycosylation sequons with the relatively weak hydrogen bonding capacity of the cysteine sulfhydryl group. However, apparent glycosylation of Asn-Xaa-Val sequons could not be explained by this mechanism. Resolution of the role of the +2 amino acid in determining glycosylation needed to wait until an atomic resolution structure of OTase was available.

3.3. Further a field: The 'X' position and beyond

The amino acids immediately proximal to the glycosylated Asn also influence the efficiency of its glycosylation. Experimental manipulation of model proteins has shown that the +1 position of an Asn has a strong effect on its extent of glycosylation, with bulky hydrophobic or acidic amino acids strongly reducing glycosylation occupancy, and small, hydrophilic or basic amino acids giving high levels of modification (Shakin-Eshleman *et al.*, 1996). These results may be misleading, as glycosylation only occurs before protein folding, and so mutations which disrupt or slow local protein folding could make extrapolation of such results difficult. However, roughly this same overall pattern has also been observed in non-experimental comparisons of glycosylated and non-glycosylated Asn (Petrescu *et al.*, 2004). Interplay with the amino acid at the +2 position has also been shown to be important. Studies in a model glycoprotein showed that amino acid substitutions at the +1 position that reduced glycosylation efficiency with Ser at the +2 position were still completely modified if Thr was at the +1 position (Kasturi et al., 1997). The major difficulty in interpreting these results is that the amino acids in the vicinity of a glycosylated Asn residue influence both specific interactions with OTase *and* local protein folding, stability and dynamics. As it is clear that protein folding and glycosylation are intimately linked, separating these effects is difficult.

In addition to local sequence dependency, the position of an asparagine within its protein sequence also contributes to the extent or probability of glycosylation. For instance, probability and extent of glycosylation increases with increasing distance from the C-terminus of a protein. This has been measured both experimentally using manipulation of model proteins and by *in silico* surveys of large sets of experimentally characterized native glycoproteins (Bano-Polo *et al.*, 2011, Rao *et al.*, 2011). This effect is perhaps due to the increased relative protein folding or translocation rates towards the C-terminus.

3.4. The extended bacterial glycosylation sequon

The discovery of N-glycosylation systems in bacteria that are homologous to those in eukaryotes promised rapid progress in understanding the molecular basis for their specificity and activity, given their comparative simplicity and ease of manipulation (Szymanski *et al.*, 1999, Wacker *et al.*, 2002). Initially it was observed that the *C. jejuni* N-glycosylaiton system modifies Asn with very similar local sequence requirements to eukaryotic N-glycosylation sites, that is an Asn-Xaa-Ser/Thr sequon was required but not sufficient for glycosylation (Wacker et al., 2002, Nita-Lazar *et al.*, 2005). Later, it was found that an extended 'sequon' was needed for bacterial glycosylation, with the added requirement of an acidic residue at the -2 position: Asp/Glu-Xaa-Asn-Xaa-Ser/Thr (Xaa≠Pro) (Kowarik *et al.*, 2006b). Close homologues to the *C. jejuni* PglB OTase showed a less strict sequon (Schwarz *et al.*, 2011b). In either case, such an extended sequon was not sufficient for modification. A key defining factor determining glycosylation was that such a sequon was efficiently glycosylated in unfolded polypeptide or in flexible stretches of folded proteins (Kowarik *et al.*, 2006a). Thus, as in the eukaryotic system, flexible acceptor substrate was a key requirement for bacterial OTase.

3.5. Structural insights into the requirement for the glycosylation sequon

The high-resolution 3D crystal structure of the *Campylobacter lari* PglB OTase finally provided a structural basis for the requirement of a glycosylation sequon (Lizak *et al.*, 2011b). This structure was solved with co-crystallization of an acceptor peptide. The key pertinent feature of the structure was that the +2 position Thr was too far away from the Asn to be directly involved in catalysis. Instead, this Thr was hydrogen bonded with two tryptophans and the aspartate in the WWDYG motif conserved in all known OTase homologues. Thr also formed van der Waals interactions with Ile572 of PglB, which Ser at the +2 position could not form, explaining the preference for Thr over Ser in sequons. Proline at the +1 or -1 position would not have allowed this binding conformation, providing a structural basis for the requirement that proline not be present at these positions at glycosylation sites. The requirement of bacterial OTases for an acidic amino acid in the -2 position (Kowarik et al., 2006b) was also explained by formation of a salt bridge from this residue to Arg331 that is conserved in bacterial, but not eukaryotic, PglB/Stt3p OTases.

This structure of the PglB OTase provides clear evidence that the role of the glycosylation sequon is to increase the binding affinity of asparagines to the active site of OTase (Lizak et al., 2011b). Accessory subunits of multiprotein complex OTases in many eukaryotes have been shown to bind substrate polypeptide, perhaps contributing to increasing the binding affinity of specific Asn and leading to the short requirement of specific binding of an Asn-Xaa-Ser/Thr. In contrast, the single protein OTases such as the bacterial PglB may have evolved the requirement for an extended sequon in the absence of such additional binding by accessory OTase subunits.

3.6. The future of the sequon

How to best define the glycosylation 'sequon'? Many factors influence whether a particular asparagine is glycosylated, including: binding affinity of the region immediately proximal to the Asn to the polypeptide acceptor site of OTase; local folding, such as secondary structural elements, disulfide bond formation or hydrophobic collapse; the regulatory state of OTase, including the concentration and structure of lipid-linked oligosaccharide donor; protein expression rate, both global (rate of protein secretion saturates OTase catalytic ability) and local (position of Asn within the protein sequence); and the affect of glycosylation at an Asn on the total possibility of protein folding. (If glycosylation at a given Asn would not allow correct folding of the protein, such that the portion of nascent polypeptides that were glycosylated there would never correctly fold, then that Asn would appear to never be glycosylated. The converse is also true, that if glycosylation is strictly required at a particular Asn for correct protein folding, then that Asn will appear to always be glycosylated, even if most of the nascent polypeptide is not modified and degraded by the quality control systems of the ER.)

It is the combination of these factors that determines if a particular Asn reaches the threshold for modification by OTase. However, even the definition of this threshold is an analytical artefact, as it is increasingly apparent that most glycosylated Asn are only partially modified, with some portion ranging from a fraction of a percent to essentially all copies of a protein, actually glycosylated (Hülsmeier *et al.*, 2007, Sumer-Bayraktar *et al.*, 2011). This pattern seems to contrast with the general requirement of many proteins for N-glycosylation for correct and efficient protein folding (Helenius & Aebi, 2004). Two key factors probably explain this conundrum. Many proteins can fold correctly even without glycosylation at many sites, as long as a certain critical level of glycosylation is present, perhaps sufficient for ER-lectin chaperone recruitment to crucial protein domains, or overall biophysical solubility. Additionally, Asn residues are inherently likely to be present at the ends of secondary structural elements. This means that glycosylation at such sites is, in general, not likely to strongly disrupt protein folding.

In the end it appears that the descriptive beauty of the 'glycosylation sequon' is actually a dramatic simplification. However, the current state of knowledge is far from being able to quantify the 'glycosylatability' of a particular Asn. In place of this developing skill, the 'sequon' as it is traditionally defined is still a very accurate predictor of the possibility of glycosylation.

4. Oligosaccharyltransferase defines the sequon

The enzyme oligosaccharyltransferase (OTase) catalyses transfer of oligosaccharide from lipid to nascent polypeptide in the ER. However, while this enzyme shows a high degree of conservation between species with respect to the small scale reaction it catalyses, the immense range of different polypeptide substrates in various biological systems can be efficiently glycosylated because of co-evolution of these substrate proteins and the acceptor specificities of OTase. In turn, this evolutionary history determines whether a particular asparagine residue will be efficiently glycosylated in a given biological system. The OTase defines the 'sequon'.

4.1. OTase protein subunits

OTase consists of the catalytic protein subunit Stt3p/PglB with varying numbers of additional accessory subunits in different organisms (reviewed in (Kelleher & Gilmore, 2006, Mohorko *et al.*, 2011)). Comparison of the evolutionary tree of eukaryotes with the protein subunit composition of OTase implies that accessory protein subunits have been added sequentially during eukaryotic evolution, starting from an ancestral single protein Stt3p OTase enzyme. The functions of most accessory OTase subunits are not clearly defined, although roles in recognition and regulation of glycan and protein substrate have been proposed.

4.2. Single protein OTases

Some divergent eukaryotes such as *Trypanosoma, Leishmania* and *Giardia* have single subunit OTases, consisting of only a catalytic Stt3p protein. However, many species within these groups have multiple different Stt3p homologues. In all of the systems in which the functions of these homologues have been characterized it is apparent that this duplication is functionally important, as the different enzymes vary in their protein acceptor and/or glycan donor substrate specificities.

4.2.1. Single protein OTases in Trypanosoma brucei

Trypanosoma brucei, the causative agent of sleeping sickness, has a genome encoding three full-length Stt3p homologues. Several lines of evidence *in vivo* in *T. brucei,* in *in vitro* enzyme assays and in a yeast *ex vivo* system support a model in which these three enzymes transfer different glycan structures to selected sets of Asn residues – they have different specificities for both the glycan they transfer, and the asparagines they modify (Izquiedro *et al.*, 2012, Izquiedro *et al.*, 2009). With regard to the asparagine residues on proteins glycosylated by these homologues, heterologous expression of these proteins in *S. cerevisiae* lacking the yeast *STT3* gene and quantitative analysis of glycosylation site occupancy in cell wall glycoproteins showed that the two of these proteins that allowed survival had different protein substrate specificities. The TbStt3B enzyme efficiently glycoslyated Asn surrounded by basic residues, while the TbStt3C enzyme preferentially glycosylated Asn surrounded by acidic residues. These substrate specificities correlated with the presence of complementary residues near the active site of the TbStt3B (acidic) and TbStt3C (basic) enzymes. This suggests that these enzymes have alternate protein substrate specificities determined by ionic interactions between the peptide-binding site and protein substrates. This specificity can be viewed as a type of ill-defined 'extended sequon', similar to the requirement of bacterial OTase for the extended Asp/Glu-Xaa-Asn-Xaa-Ser/Thr sequon, but with less stringency to the precise location of the charged residues.

4.2.2. Single protein OTases in Leishmania major

The single subunit OTase enzymes of the related *Leishmania major* have also been studied *ex vivo (Nasab et al 2008)*. Heterologous expression of the four different *Leishmania major* STT3 protein homologues in *S. cerevisiae* showed that these proteins do not integrate into the yeast OTase complex, but are instead truly single subunit enzymes. Not all of these homologues were capable of allowing survival of *S. cerevisiae* in the absence of the yeast OTase activity, and those that did complement lack of yeast OTase activity showed different protein substrate specific activities – the enzymes showed differences in the glycosylation sites they glycosylated efficiently.

4.2.3. Role of OTase catalytic subunit homologues STT3A and STT3B

Even when present in multiprotein complexes, STT3 homologues have different activities. OTase complexes containing either of the homologous mammalian STT3A and STT3B proteins have different kinetic parameters (Kelleher *et al.*, 2003), and are also responsible for either co-translocational or post-translocational N-glycosylation (Ruiz-Canada et al., 2009), thereby glycosylating different protein substrates (Wilson & High, 2007). However, it is not clear if there is further definition of protein or glycan substrate specificity defined by the presence of Stt3A or Stt3B in an OTase complex.

4.3. Role of accessory OTase proteins

In organisms with multiprotein complex OTases, there are several lines of evidence that some of these additional non-catalytic subunits provide different protein substrate specificities and allow regulation of oligosaccharide substrate recognition and enzymatics.

4.3.1. *Role of accessory OTase proteins Ost3p and Ost6p*

The *S. cerevisaie* Ost3p and Ost6p OTase subunits are homologous proteins with the same topology of a thioredoxin-like N-terminal ER lumenal domain, followed by four transmembrane helices (Fetrow *et al.*, 2001, Schwarz *et al.*, 2005). These proteins are homologous to the mammalian proteins TUSC3 and MagT1 (Kelleher & Gilmore, 2006). Only one of these homologues is incorporated into a given OTase complex, meaning that there exist two isoforms of OTase in yeast, defined by the presence of either Ost3p or Ost6p (Schwarz et al., 2005, Spirig *et al.*, 2005). These different isoforms have been shown to have different protein substrate specific glycosylation efficiencies at the level of individual glycosylation sites (Karaoglu *et al.*, 1995, Knauer & Lehle, 1999, Schulz & Aebi, 2009). Some Asn residues require the Ost3p-OTase for efficient glycosylation, while other Asn residues require the Ost6p-OTase (Schulz & Aebi, 2009). The mechanistic basis for this difference is likely transient binding of stretches of nascent polypeptide by peptide-binding grooves in the ER lumenal domains of Ost3p and Ost6p (Schulz *et al.*, 2009). *In vitro* assays have shown that Ost3p and Ost6p can transiently non-covalently bind peptides complementary to the characteristics of their peptide-binding grooves (Jamaluddin *et al.*, 2011, Schulz et al., 2009). As the amino acids forming these grooves are different in Ost3p and Ost6p, it is proposed that they would tend to bind different stretches of nascent polypeptide, thereby increasing the efficiency of glycosylation at distinct sets of glycosylation sites. The peptide-binding specificity of yeast Ost6p is for short stretches of aliphatic amino acids, with additional affinity provided by the presence of neighbouring acidic residues (Jamaluddin et al., 2011). Such sequences are complementary to the peptide binding groove of Ost6p, as revealed by its 3D crystal structure, which shows a groove with a hydrophobic base, and lined by neutral and basic amino acids (Schulz et al., 2009). The dimensions of the groove are appropriate for binding a ~4-5 amino acid stretch of extended polypeptide, or an amphipathic alpha helix. Ost3p also binds hydrophobic stretches of polypeptide, but with a distinct amino acid characteristic specificity to Ost6p (Jamaluddin et al., 2011). It has also been proposed that the oxidoreductase activity of the thioredoxin-like ER lumenal domain of Ost3p and Ost6p could form mixed disulfides with cysteines in nascent polypeptides (Schulz et al., 2009). This would serve to tether nascent polypeptide close to the active site of OTase, and also efficiently inhibit oxidative protein folding. This transient binding of cysteines and hydrophobic stretches of nascent polypeptide is an optimal strategy for inhibiting local protein folding, as such stretches would be normally internal to folded protein domains. This fits with the requirement of the catalytic site of OTase for unfolded or flexible protein acceptor substrate.

4.3.2. *Role of accessory OTase protein ribophorin I / Ost1p*

Mammalian Ribophorin I (Ost1p in yeast) is required for efficient glycosylation of selected membrane proteins. Ribophorin I physically associates with selected membrane proteins after insertion into the ER membrane (Wilson *et al.*, 2005). This interaction with these selected substrate proteins was also shown to be required for their efficient glycosylation by OTase (Wilson & High, 2007). The interaction between selected membrane proteins and ribophorin I is direct, but the precise mechanisms of the interaction are not clear (Wilson *et al.*, 2008). It is possible that ribophorin I / Ost1p function in a conceptually similar way to

Ost3/6p, in transiently tethering substrate protein close to the catalytic site of OTase to allow efficient glycosylation of a defined subset of glycosylation sites or glycoproteins.

4.3.3. Additional known accessory OTase proteins

An integral membrane protein with homology to the integral membrane domain of Ost3p and Ost6p has been identified in mammalian cells. This protein, DC2 or OSTC, is required for glycosylation of specific substrate glycoproteins (Wilson & High, 2007). A further protein, Keratinocyte-associated protein 2 (KCP2), has been shown biochemically to be a subunit of the mammalian OTase (Sanyal & Menon, 2010, Roboti & High, 2012), and to be required for glycosylation of some proteins (Wilson & High, 2007).

4.3.4. Putative accessory OTase protein presenilin 1

A direct link between site-specific glycosylation and Alzheimer's disease has been made, through the Presenilin-1 protein (Lee *et al.*, 2010). N-glycosylation of the vaculoar ATPase subunit V0a1 is mediated by selective binding of the Alzheimer's disease related protein presenilin-1 (PS1) to unglyclosylated V0a1 and OTase. V0a1 glycosylation is required for ER-lysosome trafficking, and so lack of PS1 causes deficiencies in lysosomal acidification and proteolysis during autophagy. It is not clear if PS1 is a truly protein-specific enhancer of glycosylation, or if it interacts with additional substrate glycoproteins to enhance their glycosylation.

4.3.5. How many OTase subunits are there?

Have all OTase subunits been identified? Most known OTase subunit proteins have been identified in the yeast *S. cerevisiae* through genetic screens, and so it is likely that this set is complete. These proteins are identifiable in other eukaryotes including animals and plants. However, the presence of additional subunits cannot be easily predicted. As described above, several such additional subunits have been identified biochemically in recent years in the mammalian OTase enzyme.

It is possible that other, less tightly bound or lowly expressed proteins are yet to be identified. It is also possible that sequential addition of accessory proteins to the OTase complex has proceeded divergently in different eukaryotic lineages. This would mean that biochemical analyses, rather than genomic comparisons, would be necessary to identify any additional OTase complexes in for example the plant or protozoan OTase. Any such additional subunits would likely have diverse additional roles in regulation of OTase core activity.

5. Analytical approaches to determine site-specific glycosylation occupancy

5.1. Glycosylation site identification and occupancy

A goal of understanding the function of OTase in diverse biological systems is to enable accurate prediction of whether a particular Asn will be efficiently glycosylated. However, such prediction depends on a complete understanding of how OTase interacts with substrate polypeptides in each biological system, and as such is probably a very difficult problem. In addition to the diversity of OTase subunit proteins, OTase activity may also be subject to regulation. In the absence of accurate prediction tools, analytical identification and quantification of glycosylation occupancy is therefore necessary for accurate characterisation of the glycosylation status of a protein. In addition, it is not sufficient to identify that a site is glycosylated, as an Asn can be identified as 'glycosylated' in enrichment experiments, but may actually only be modified at a very low occupancy. The physiological relevance of glycosylation at such sites is therefore questionable. The converse of this is also true, as it appears that with sensitive analytical detection some or even most glycosylation sites are not completely occupied (there exists a small but significant proportion of proteins that are not glycosylated at that particular site) (Hülsmeier *et al.*, 2007). Analytical methods should therefore consider the proportion of a particular Asn that is glycosylated, for instance using LC-MS approaches that can compare the abundance of glycosylated and non-glycosylated versions of the same peptide (Schulz & Aebi, 2009). Although these methods are not in general absolutely quantitative, they can provide relative quantification and a first step towards characterization of the site-specific extent of glycosylation.

5.2. Western blotting for measuring glycosylation occupancy

Numerous studies have made use of Western blotting with antibodies recognizing a specific protein of interest to gauge glycosylation occupancy. However, Western blotting is inherently limited to analysis of proteins for which specific antisera are available, and is constrained to low-throughput assays. Western blotting can also only identify protein-wide glycosylation occupancy, and cannot distinguish between partial glycosylation at different Asn residues on the same protein. Mass spectrometry can overcome both of these key difficulties, as it is a general analysis tool that can be used for site-specific analysis of protein glycosylation.

5.3. Glycoconjugate enrichment stragtegies

Detection of glycosylation at a specific site is the first step in its quantitative analysis (Schulz *et al.*, 2012). Enrichment of glycoproteins or glycopeptides is key to the success of high sensitivity detection of glycosylation sites. Various enrichment strategies can be employed depending on the biological system of interest, and the analytes of interest within that system. The physical properties of carbohydrates that distinguish them from protein can be used to enrich glycopeptides and glycoproteins. Typical enrichment strategies based on the physical properties of glycans include hydrophilic interaction chromatography (Mysling *et al.*, 2010, Gilar *et al.*, 2011, Christiansen *et al.*, 2010), phenyl boronic acid (Li *et al.*, 2000, Li *et al.*, 2001) and hydrazide (Zhang & Aebersold, 2006, Zhang *et al.*, 2003) attachment. A key mechanism mediating the functional roles of glycans in many biological systems is recognition of specific glycan structures by proteins, or lectins. The specificity of such lectins for defined glycan structures can be used to enrich particular subsets of glycopeptides or glycoproteins bearing those structures (Drake *et al.*, 2006, Zielinska et al., 2010).

5.4. Mass spectrometry for measuring glycosylation occupancy

To obtain quantitative or semi-quantitative measurement of the extent of glycosylation at that site subsequent comparison must be made with the unglycosylated form of the detected peptide. This can be done using comparison of ion intensities of the glycosylated and unglycosylated peptides. The unglycosylated form of the peptide will only be present in one form. However, as glycosylation generally results in a complex mixture of glycan structures at each glycosylation site, measurement of the abundance of the glycosylated form of a given site is not trivial. Some approaches have used detection of entire glycopeptides, although this approach generally requires more specialized and targeted LC-MS technologies (Sumer-Bayraktar *et al.*, 2011). Other approaches have focused on improving quantification of occupancy, and have discarded information on site-specific glycan structure by endoglycosidase treatment (Schulz & Aebi, 2009). For instance, PNGaseF cleaves N-glycans and converts previously glycosylated Asn to Asp, while EndoH leaves a single *N*-acetylglucosamine on previously glycosylated Asn residues. In both of these cases, there is a clear mass and retention shift readily detectable with modern LC-MS technologies, that can be used to differentiate and independently measure the glycosylated and unglycosylated forms of a given peptide.

5.5. Selected-reaction-monitoring mass spectrometry

Recent years have seen impressive success with targeted mass spectrometry approaches, using selected-reaction-monitoring (Lange *et al.*, 2008, Gallien *et al.*, 2011). N-glycosylation has been used as a useful tag to specifically enrich otherwise low abundant components of biological fluids (Stahl-Zeng *et al.*, 2007). Often this has been performed not out of direct interest in glycosylation per se, but because of the ubiquity of glycosylation, and its proven utility in biomarker discovery. However, some analyses have used this approach to specifically measure glycosylation occupancy, for instance in patients with congenital disorders of glycosylation (Hülsmeier *et al.*, 2007).

5.6. Future analytical directions

Use of tools such as those outlined above, in combination with experimental manipulation of growth conditions, N-glycan biosynthetic pathways, protein translation and translocation, and OTase function or composition, will allow identification of the regulation and roles of site-specific N-glycosylation occupancy at a systems level.

6. Is the 'glycosylation sequon' an example of convergent evolution? Insights into glycosylation site evolution

6.1. HMW-ABC glycosylation in non-typeable *Haemophilus influenzae*

A family of cytoplasmic bacterial enzymes have been recently described that catalyse an N-glycosylation reaction remarkably reminiscent of 'traditional' N-glycosylation. These enzymes are the HMW-C glycosyltransferase of non-typeable *Haemophilus influenzae* (NTHi), and related organisms primarily in the Pasteurellaceae. NTHi can be a commensal resident of the nasopharynx in humans, but in mixed cultures with *Streptococucus pneumoniae* and *Moraxella catarrhalis* is a primary cause of middle ear infections (otitis media) which in developed countries is the most common reason for children to visit doctors and for antibiotic prescriptions (Murphy *et al.*, 2009). Chronic otitis media is common in indigenous communities worldwide, and can lead to hearing loss with subsequent learning difficulties. NTHi is also associated with chronic obstructive pulmonary disease (COPD), a major burden on health care systems worldwide (Murphy, 2006). Understanding of the role of glycosylation of the outer membrane protein components in these organisms is likely of key importance in the development of effective vaccines against NTHi infection.

A key step in NTHi infection is adherence to the host epithelium. Surface exposed adhesin proteins mediate this adherence, with the high molecular weight (HMW) adhesin system being of key importance in many NTHi clinical isolates. HMW-C is a glycosyltransferase associated with this two-partner secretion system adhesin, encoded in the HMW-ABC locus. Two highly homologous loci are present in the ~80% of NTHi clinical isolates that encode this system, HMW1ABC and HMW2ABC respectively (St. Geme *et al.*, 1998, Ecevit *et al.*, 2004). HMW1A encodes an adhesin glycoprotein (Gross *et al.*, 2008, Grass *et al.*, 2003), which is secreted across the inner membrane via the Sec apparatus, and requires the outer membrane protein HMW1B for correct export across the outer membrane (St Geme & Yeo, 2009). HMW1C encodes a family 41 glycosyltransferase that glycosylates HMW1A (Grass *et al.*, 2010, Kawai *et al.*, 2011, Choi *et al.*, 2010). This glycosylation is required for stability, efficient folding and secretion of the HMW1A glycoprotein adhesin (Grass et al., 2003). In turn, the HMW1A adhesin is important for NTHi colonisation and pathogenesis (St Geme *et al.*, 1993, St Geme, 1994).

Similar to several other described bacterial protein glycosyltransferases, HMW1C glycosylates its HMW1A substrate protein in the cytoplasm, before secretion across the inner membrane (Fleckenstein *et al.*, 2006, Charbonneau *et al.*, 2012, Choi et al., 2010, Schwarz *et al.*, 2011a). Most of these other reported bacterial glycosyltransferases are O-glycosyltransferases, transferring nucleotide-activated monosaccharides to the hydroxyl groups of Ser or Thr. In contrast, HMW1C glycosylates Asn residues, with a strong tendency to glycosylate Asn within glycosylation sequons with the sequence Asn-Xaa-Ser/Thr (Gross et al., 2008).

6.2. HMW-C versus OTase: Unrelated enzymes, same sequon?

The HMW-C and OTase systems are not homologous, as traditional N-glycosylation as described above is catalysed by the integral membrane OTase, which transfers an oligosaccharide from a lipid linked carrier to nascent polypeptide in the lumen of the ER (or periplasm). In contrast, the HMW-C cytoplasmic system of some bacteria is catalysed by a soluble glycosyltransferase that transfers a nucleotide-activated monosaccharide to protein in the cytoplasm. However, it is striking that the bacterial cytoplasmic HMW-C enzymes have very similar site recognition to 'traditional' OTase enzymes: they efficiently glycosylate Asn in 'sequons' with Asn-Xaa-Ser/Thr (Xaa≠Pro), but are capable of glycosylating some selected asparagines lacking S/T at the +2 position (Choi et al., 2010, Grass et al., 2010, Schwarz et al., 2011a, Schwarz & Aebi, 2011). HMW-C enzymes also share the substrate requirement of OTase for unfolded protein, or flexible loops in folded protein (Schwarz et al., 2011a).

A high-resolution 3D crystal structure of an HMW-C enzyme from *Actinobacillus pleuropneumoniae* has been reported (Kawai et al., 2011), which is highly similar to HMW-C enzymes from NTHi. This structure was solved in the presence of acceptor peptide substrate, but electron density for the peptide was not visible in the structure. Nonetheless, the ability of HMW-C enzymes to glycosylate asparagines not in Asn-Xaa-Ser/Thr glycosylation sequons strongly suggests that this sequence is not directly involved in catalysis. It is likely that, as with OTase, this local sequence requirement is instead involved in increasing the affinity for acceptor polypeptide binding, and in substrate recognition.

This then raises the very curious observation that two non-homologous enzymatic systems for glycosylation of Asn have independently evolved essentially identical substrate recognition motifs. This suggests convergent evolution of enzyme-substrate interactions in these two systems, which would in turn imply that there is some functional benefit for site recognition of Ser/Thr at +2 amino acid residues of an Asparagine. It is tempting to speculate that this sequence may have evolved to balance the need for sufficient binding affinity of the polypeptide acceptor with the advantages of a general glycosylation system.

The selection pressure for OTase and HMW-C to require unfolded polypeptide substrate is not completely clear. However, this requirement is likely due to the benefit of glycoysylation in increasing both protein folding efficiency and the stability of folded proteins. Addition of glycans to already folded proteins can serve to increase their stability, potentially in a regulated manner (Yuzwa *et al.*, 2012). However, to be of assistance in protein *folding*, glycans must be transferred to proteins while they are unfolded or still in the process of folding. This would then lead to the requirement of binding of flexible polypeptide acceptor to OTase/HMW-C, which in turn limits recognition to stretches of amino acid sequence, rather than a more complex folded protein structural motif. Binding of a single Asn residue in a stretch of unfolded polypeptide would be the smallest possible recognition motif, but hydrogen bonding and van der Waals interactions would likely not be sufficient for binding of adequate affinity to allow efficient glycan transfer. A solution to allow increased affinity would be to increase the length of the recognition sequence, and apparently a 3 or 5 amino acid sequence is sufficient, as shown by the Asn-Xaa-Ser/Thr and Asp/Glu-Xaa-Asn-Xaa-Ser/Thr OTase glycosylation sequons of eukaryotes and bacteria. This is evidenced in the structure of bacterial PglB (STT3), in which most of the length of the extended sequon is involved in direct contacts with the OTase peptide binding site (Lizak *et al.*, 2011a).

6.3. Why is the sequon as it is?

Why then should Ser/Thr be part of a preferred glycosylation recognition sequence, and not any other amino acids? Perhaps part of the answer is that these hydroxyl-containing residues are typically surface exposed and are not charged. The hydrophilic nature of Ser and Thr means that they are generally not present internally in folded proteins, but are almost always surface exposed. As addition of a glycan in the hydrophobic core of a protein would be incompatible with correct protein folding, a hydrophilic recognition motif is necessary. Charged residues (His, Arg, Lys, Asp, Glu) would also be potential candidates for such a role, but here the generality of the neutral hydroxyl groups of Ser and Thr is perhaps important. Neutral hydrophilic residues such as Ser and Thr are compatible with almost any position on the surface of a folded protein. In contrast, charge-based attraction and repulsion is an important contributor to protein folding, stability and function. Point mutation to insert one of these charged amino acids on the surface of a protein is likely to disrupt the protein structure. Ser/Thr as an extended recognition sequence therefore likely provides the affinity and ubiquity necessary for evolution of OTase/HMW-C enzymes as general glycosylation enzymes capable of glycosylating multiple Asn residues in many different proteins.

7. Conclusion

The structural basis for the glycosylation sequon is now apparent. However, it is also clear that recognition and glycosylation of selected asparagine residues is subject to further control and regulation depending on variation within catalytic STT3 enzymes, and on the presence of accessory protein subunits of multiprotein OTase complexes. In order to understand the roles of these accessory proteins, it is however necessary for them to be completely identified. With recent years showing the identification and preliminary characterization of several novel accessory proteins of mammalian OTase, it is probable that additional subunits still remain to be discovered. Biochemical characterization of OTase complexes in other eukaryotes may well also present additional, non-homologous, accessory protein subunits. Further, OTase enzymatic activity is actively regulated, adding to the complexity of potential OTase function. Mass spectrometry-based future analytics for glycosylation analysis will enable phenotypic characterization of the site-specific activity of OTase in these varied biological circumstances. Such analysis will contribute to, and also benefit from, a complete quantitative understanding of the interplay between glycoprotein folding and N-glycosylation. Finally, understanding of the molecular mechanisms of N-glycosylation site selection is beginning to open the possibilities for co-engineering of glycosylation sites and OTases in synthetic biology approaches outside of natural evolutionary constraints, moving N-glycosylation beyond the sequon.

Author details

Benjamin Luke Schulz
The University of Queensland, School of Chemistry and Molecular Biosciences, St Lucia, QLD, Australia

Acknowledgement

The author acknowledges the support of NHMRC Career Development Fellowship APP1031542 and NHMRC Project Grant 631615.

8. References

Aebi, M., R. Bemasconi, S. Clerc & M. Molinari, (2010) N-glycan structures: recognition and processing in the ER. *Trends Biochem Sci* 35: 74-82.

Bano-Polo, M., F. Baldin, S. Tamborero, M. A. Marti-Renom & I. Mingarro, (2011) N-glycosylation efficiency is determined by the distance to the C-terminus and the amino acid preceding an Asn-Ser-Thr sequon. *Protein Sci* 20: 179-186.

Burda, P. & M. Aebi, (1999) The dolichol pathway of N-linked glycosylation. *Biochim Biophys Acta* 1426: 239-257.

Charbonneau, M. E., J. P. Cote, M. F. Haurat, B. Reiz, S. Crepin, F. Berthiaume, C. M. Dozois, M. F. Feldman & M. Mourez, (2012) A structural motif is the recognition site for a new family of bacterial protein O-glycosyltransferases. *Mol Microbiol.*

Chi, Y. H., Y. D. Koo, S. Y. Dai, J. E. Ahn, D. J. Yun, S. Y. Lee & K. Zhu-Salzman, (2010) N-glycosylation at non-canonical Asn-X-Cys sequence of an insect recombinant cathepsin B-like counter-defense protein. *Comp Biochem Physiol B Biochem Mol Biol* 156: 40-47.

Choi, K. J., S. Grass, S. Paek, J. W. St Geme, 3rd & H. J. Yeo, (2010) The Actinobacillus pleuropneumoniae HMW1C-like glycosyltransferase mediates N-linked glycosylation of the Haemophilus influenzae HMW1 adhesin. *PLoS One* 5: e15888.

Christiansen, M. N., D. Kolarich, H. Nevalainen, N. H. Packer & P. H. Jensen, (2010) Challenges of determining O-glycopeptide heterogeneity: a fungal glucanase model system. *Anal Chem* 82: 3500-3509.

Dempski, R. E. J. & B. Imperiali, (2002) Oligosaccharyl transferase: gatekeeper to the secretory pathway. *Curr Opin Chem Biol* 6: 844-850.

Drake, R. R., E. E. Schwegler, G. Malik, J. Diaz, T. Block, A. Mehta & O. J. Semmes, (2006) Lectin Capture Strategies Combined with Mass Spectrometry for the Discovery of Serum Glycoprotein Biomarkers. *Mol Cell Proteomics* 5: 1957-1967.

Ecevit, I. Z., K. W. McCrea, M. M. Pettigrew, A. Sen, C. F. Marrs & J. R. Gilsdorf, (2004) Prevalence of the hifBC, hmw1A, hmw2A, hmwC, and hia Genes in Haemophilus influenzae Isolates *J Clin Micro* 42: 3065-3072.

Fetrow, J. S., N. Siew, J. A. Di Gennaro, M. Martinez-Yamout, H. J. Dyson & J. Skolnick, (2001) Genomic-scale comparison of sequence- and structure-based methods of function prediction: does structure provide additional insight? *Protein Sci* 10: 1005-1014.

Fleckenstein, J. M., K. Roy, J. F. Fischer & M. Burkitt, (2006) Identification of a Two-Partner Secretion Locus of Enterotoxigenic Escherichia coli. *Infect. Immun.* 74: 2245-2258.

Gallien, S., E. Duriez & B. Domon, (2011) Selected reaction monitoring applied to proteomics. *J Mass Spectrom* 46: 298-312.

Gilar, M., Y. Q. Yu, J. Ahn, H. Xie, H. Han, W. Ying & X. Qian, (2011) Characterization of glycoprotein digests with hydrophilic interaction chromatography and mass spectrometry. *Anal Biochem* 417: 80-88.

Grass, S., A. Z. Buscher, W. E. Swords, M. A. Apicella, S. J. Barenkamp, N. Ozchlewski & J. W. St Geme, 3rd, (2003) The Haemophilus influenzae HMW1 adhesin is glycosylated in a process that requires HMW1C and phosphoglucomutase, an enzyme involved in lipooligosaccharide biosynthesis. *Molecular Microbiology* 48: 737-751.

Grass, S., C. F. Lichti, R. R. Townsend, J. Gross & J. W. St Geme, 3rd, (2010) The Haemophilus influenzae HMW1C protein Is a glycosyltransferase that transfers hexose residues to asparagine sites in the HMW1 adhesin *PLoS Pathogens* 6: e1000919.

Gross, J., S. Grass, A. E. Davis, P. Gilmore-Erdmann, R. R. Townsend & J. W. St Geme, 3rd, (2008) The Haemophilus influenzae HMW1 Adhesin Is a Glycoprotein with an Unusual N-Linked Carbohydrate Modification. *J Biol Chem* 283: 26010-26015.

Harada, Y., H. Li, H. Li & W. J. Lennarz, (2009) Oligosaccharyltransferase directly binds to ribosome at a location near the translocon-binding site. *Proc Natl Acad Sci U S A* 106: 6945-6949.

Helenius, A. & M. Aebi, (2004) Roles of N-linked glycans in the endoplasmic reticulum. *Annu Rev Biochem* 73: 1019-1049.

Hülsmeier, A. J., P. Paesold-Burda & T. Hennet, (2007) N-glycosylation site occupancy in serum glycoproteins using multiple reaction monitoring liquid chromatography mass spectrometry. *Mol Cell Proteomics* 6: 2132-2138.

Imperiali, B. & T. L. Hendrickson, (1995) Asparagine-linked glycosylation: specificity and function of oligosaccharyl transferase. *Bioorg Med Chem* 3: 1565-1578.

Izquiedro, L., A. Mehlert & M. A. Ferguson, (2012) The lipid-linked oligosaccharide donor specificities of Trypanosoma brucei oligosaccharyltransferases. *Glycobiology* 22: 696-703.

Izquiedro, L., B. L. Schulz, J. A. Rodrigues, M. L. Güther, J. B. Procter, G. J. Barton, M. Aebi & M. A. Ferguson, (2009) Distinct donor and acceptor specificities of Trypanosoma brucei oligosaccharyltransferases. *EMBO J* 28: 2650-2661.

Jamaluddin, M. F. B., U. M. Bailey, N. Y. J. Tan, A. P. Stark & B. L. Schulz, (2011) Polypeptide binding specificities of Saccharomyces cerevisiae oligosaccharyltransferase accessory proteins Ost3p and Ost6p. *Protein Sci* 20: 849-855.

Karaoglu, D., D. J. Kelleher & R. Gilmore, (1995) Functional characterization of Ost3p. Loss of the 34-kD subunit of the Saccharomyces cerevisiae oligosaccharyltransferase results in biased underglycosylation of acceptor substrates. *J Cell Biol* 130: 567-577.

Karaoglu, D., D. J. Kelleher & R. Gilmore, (2001) Allosteric regulation provides a molecular mechanism for preferential utilization of the fully assembled dolichol-linked oligosaccharide by the yeast oligosaccharyltransferase. *Biochemistry* 40: 12193-12206.

Kasturi, L., H. Chen & S. H. Shakin-Eshleman, (1997) Regulation of N-linked core glycosylation: use of a site-directed mutagenesis approach to identify Asn-Xaa-Ser/Thr sequons that are poor oligosaccharide acceptors. *Biochem J* 323: 415-419.

Kasturi, L., J. R. Eshleman, W. H. Wunner & S. H. Shakin-Eshleman, (1995) The Hydroxy Amino Acid in an Asn-X-Ser/Thr Sequon Can Influence N-Linked Core Glycosylation Efficiency and the Level of Expression of a Cell Surface Glycoprotein. *J. Biol. Chem.* 270: 14756-14761.

Kawai, F., S. Grass, Y. Kim, K. J. Choi, J. W. St Geme, 3rd & H. J. Yeo, (2011) Structural insights into the glycosyltransferase activity of the Actinobacillus pleuropneumoniae HMW1C-like protein. *J Biol Chem* 286: 38546-38557.

Kelleher, D. J. & R. Gilmore, (2006) An evolving view of the eukaryotic oligosaccharyltransferase. *Glycobiology* 16: 47R-62R.

Kelleher, D. J., D. Karaoglu, E. C. Mandon & R. Gilmore, (2003) Oligosaccharyltransferase isoforms that contain different catalytic STT3 subunits have distinct enzymatic properties. *Mol Cell* 12: 101-111.

Knauer, R. & L. Lehle, (1999) The oligosaccharyltransferase complex from Saccharomyces cerevisiae. Isolation of the OST6 gene, its synthetic interaction with OST3, and analysis of the native complex. *J Biol Chem* 274: 17249-17256.

Kowarik, M., S. Numao, M. F. Feldman, B. L. Schulz, N. Callewaert, E. Kiermaier, I. Catrein & M. Aebi, (2006a) N-linked glycosylation of folded proteins by the bacterial oligosaccharyltransferase. *Science* 314: 1148-1150.

Kowarik, M., N. M. Young, S. Numao, B. L. Schulz, I. Hug, N. Callewaert, D. C. Mills, D. C. Watson, M. Hernandez, J. F. Kelly, M. Wacker & M. Aebi, (2006b) Definition of the bacterial N-glycosylation site consensus sequence. *The EMBO Journal* 25: 1957-1966.

Lange, V., P. Picotti, B. Domon & R. Aebersold, (2008) Selected reaction monitoring for quantitative proteomics: a tutorial. *Mol Syst Biol* 4: 222.

Lee, J. H., W. H. Yu, A. Kumar, S. Lee, P. S. Mohan, C. M. Peterhoff, D. M. Wolfe, M. Martinez-Vicente, A. C. Massey, G. Sovak, Y. Uchiyama, D. Westaway, A. M. Cuervo & R. A. Nixon, (2010) Lysosomal Proteolysis and Autophagy Require Presenilin 1 and Are Disrupted by Alzheimer-Related PS1 Mutations. *Cell* 141: 1146-1158.

Li, Y., E. L. Larsson, H. Jungvid, I. Y. Galaev & B. Mattiasson, (2000) Affinity chromatography of neoglycoproteins. *Bioseparation* 9: 315-323.

Li, Y., U. Pfüller, E. L. Larsson, H. Jungvid, I. Y. Galaev & B. Mattiasson, (2001) Separation of mistletoe lectins based on the degree of glycosylation using boronate affinity chromatography. *J Chromatogr A* 925: 115-121.

Lizak, C., Y. Y. Fan, T. C. Weber & M. Aebi, (2011a) N-Linked Glycosylation of Antibody Fragments in Escherichia coli. *Bioconjug Chem.*

Lizak, C., S. Gerber, S. Numao, M. Aebi & K. P. Locher, (2011b) X-ray structure of a bacterial oligosaccharyltransferase. *Nature* 474: 350-355.

Marshall, R. D., (1974) The nature and metabolism of the carbohydrate-peptide linkages of glycoproteins. *Biochem Soc Symp* 40: 17-26.

Miletich, J. P. & G. J. J. Broze, (1990) Beta protein C is not glycosylated at asparagine 329. The rate of translation may influence the frequency of usage at asparagine-X-cysteine sites. *J Biol Chem* 265: 11397-11404.

Mohorko, E., R. Glockshuber & M. Aebi, (2011) Oligosaccharyltransferase: the central enzyme of N-linked protein glycosylation. *J Inherit Metab Dis* 34: 869-878.

Murphy, T. F., (2006) The role of bacteria in airway inflammation in exacerbations of chronic obstructive pulmonary disease. *Curr Opin Infect Dis* 19: 225-230.

Murphy, T. F., L. O. Bakaletz & P. R. Smeesters, (2009) Microbial interactions in the respiratory tract. *Pediatr Infect Dis J* 28: S121-126.

Mysling, S., G. Palmisano, P. Højrup & M. Thaysen-Andersen, (2010) Utilizing ion-pairing hydrophilic interaction chromatography solid phase extraction for efficient glycopeptide enrichment in glycoproteomics. *Anal Chem* 82: 5598-5609.

Nasab, F. P., B. L. Schulz, F. Gamarro, A. J. Parodi & M. Aebi, (2008) All in One: Leishmania major STT3 proteins substitute for the whole oligosaccharyltransferase complex in Saccharomyces cerevisiae. Mol Biol Cell 19: 3758-3768

Nita-Lazar, A., M. Wacker, B. Schegg, S. Amber & M. Aebi, (2005) The N-X-S/T consensus sequence is required but not sufficient for bacterial N-linked protein glycosylation. *Glycobiology* 15: 361-367.

Oliver, J. D., H. L. Roderick, D. H. Llewellyn & S. High, (1999) ERp57 functions as a subunit of specific complexes formed with the ER lectins calreticulin and calnexin. *Mol Biol Cell* 10: 2573-2582.

Palmisano, G., M. N. Melo-Braga, K. Engholm-Keller, B. L. Parker & M. R. Larsen, (2012) Chemical deamidation: a common pitfall in large-scale N-linked glycoproteomic mass spectrometry-based analyses. *J Proteome Res* 11: 1949-1957.

Petrescu, A. J., A. L. Milac, S. M. Petrescu, R. A. Dwek & M. R. Wormald, (2004) Statistical analysis of the protein environment of N-glycosylation sites: implications for occupancy, structure, and folding. *Glycobiology* 14: 103-114.

Rao, R. S., O. T. Buus & B. Wollenweber, (2011) Distribution of N-glycosylation sequons in proteins: how apart are they? *Comput Biol Chem* 35: 57-61.

Robinson, N. E., Z. W. Robinson, B. R. Robinson, A. L. Robinson, J. A. Robinson, M. L. Robinson & A. B. Robinson, (2004) Structure-dependent nonenzymatic deamidation of glutaminyl and asparaginyl pentapeptides. *J Pept Res* 63: 426-436.

Roboti, P. & S. High, (2012) Keratinocyte-associated protein 2 is a bona fide subunit of the mammalian oligosaccharyltransferase. *J Cell Sci* 125: 220-232.

Ruiz-Canada, C., D. J. Kelleher & R. Gilmore, (2009) Cotranslational and Posttranslational N-Glycosylation of Polypeptides by Distinct Mammalian OST Isoforms. *Cell* 136: 272-283.

Sanyal, S. & A. K. Menon, (2010) Stereoselective transbilayer translocation of mannosyl phosphoryl dolichol by an endoplasmic reticulum flippase *Proc Natl Acad Sci U S A*.

Sato, C., J. H. Kim, Y. Abe, K. Saito, S. Yokoyama & D. Kohda, (2000) Characterization of the N-oligosaccharides attached to the atypical Asn-X-Cys sequence of recombinant human epidermal growth factor receptor. *J Biochem* 127: 65-72.

Schulz, B. L. & M. Aebi, (2009) Analysis of Glycosylation Site Occupancy Reveals a Role for Ost3p and Ost6p in Site-specific N-Glycosylation Efficiency. *Mol Cell Proteomics* 8: 357-364.

Schulz, B. L., C. U. Stirnimann, J. P. A. Grimshaw, M. S. Brozzo, F. Fritsch, E. Mohorko, G. Capitani, R. Glockshuber, M. G. Grütter & M. Aebi, (2009) Oxidoreductase activity of oligosaccharyltransferase subunits Ost3p and Ost6p defines site-specific glycosylation efficiency. *Proc Natl Acad Sci U S A* 106: 11061-11066.

Schulz, B. L., J. C. White & C. Punyadeera, (2012) Saliva proteome research: current status and future outlook. *Crit Rev Biotech.*

Schwarz, F. & M. Aebi, (2011) Mechanisms and principles of N-linked protein glycosylation. *Curr Opin Struc Biol* 21: 576-582.

Schwarz, F., Y. Y. Fan, M. Schubert & M. Aebi, (2011a) Cytoplasmic N-glycosyltransferase of Actinobacillus pleuropneumoniae is an inverting enzyme and recognizes the NX(S/T) consensus sequence. *J Biol Chem* 286: 35267-35274.

Schwarz, F., C. Lizak, Y. Y. Fan, S. Fleurkens, M. Kowarik & M. Aebi, (2011b) Relaxed acceptor site specificity of bacterial oligosaccharyltransferase in vivo. *Glycobiology* 21: 45-54.

Schwarz, M., M. Knauer & L. Lehle, (2005) Yeast oligosaccharyltransferase consists of two functionally distinct sub-complexes, specified by either the Ost3p or Ost6p subunit. *FEBS Lett* 579: 6564-6568.

Shakin-Eshleman, S. H., S. L. Spitalnik & L. Kasturi, (1996) The amino acid at the X position of an Asn-X-Ser sequon is an important determinant of N-linked core-glycosylation efficiency. *J Biol Chem* 271: 6363-6366.

Shibatani, T., L. L. David, A. L. McCormack, K. Frueh & W. R. Skach, (2005) Proteomic analysis of mammalian oligosaccharyltransferase reveals multiple subcomplexes that contain Sec61, TRAP, and two potential new subunits. *Biochemistry* 44: 5982-5992.

Spirig, U., D. Bodmer, M. Wacker, P. Burda & M. Aebi, (2005) The 3.4-kDa Ost4 protein is required for the assembly of two distinct oligosaccharyltransferase complexes in yeast. *Glycobiology* 15: 1396-1406.

St Geme, J. W., 3rd, (1994) The HMW1 adhesin of nontypeable Haemophilus influenzae recognizes sialylated glycoprotein receptors on cultured human epitheial cells. *Infect Immun* 62: 3881-3889.

St Geme, J. W., 3rd, S. Falkow & S. J. Barenkamp, (1993) High-molecular-weight proteins of nontypeable Haemophilus influenzae mediate attachment to human epithelial cels. *Proc Natl Acad Sci U S A* 90: 2875-2879.

St Geme, J. W., 3rd & H.-J. Yeo, (2009) A prototype two-partner secretion pathway: the Haemophilus influenzae HMW1 and HMW2 adhesin systems. *Trends in Microbiology* 17: 355-360.

St. Geme, J. W., III, V. V. Kumar, D. Cutter & S. J. Barenkamp, (1998) Prevalence and Distribution of the hmw and hia Genes and the HMW and Hia Adhesins among Genetically Diverse Strains of Nontypeable Haemophilus influenzae. *Infect. Immun.* 66: 364-368.

Stahl-Zeng, J., V. Lange, R. Ossola, K. Eckhardt, W. Krek, R. Aebersold & B. Domon, (2007) High sensitivity detection of plasma proteins by multiple reaction monitoring of N-glycosites. *Mol Cell Proteomics* 6: 1809-1817.

Sumer-Bayraktar, Z., D. Kolarich, M. P. Campbell, S. Ail, N. H. Packer & M. Thaysen-Andersen, (2011) N-glycans modulate the function of human corticosteroid-binding globulin. *Mol Cell Proteomics* 10: M111.009100.

Szymanski, C. M., R. Yao, C. P. Ewing, T. J. Trust & P. Guerry, (1999) Evidence for a system of general protein glycosylation in Campylobacter jejuni. *Molecular Microbiology* 32: 1022-1030.

Vance, B. A., W. Wu, R. K. Ribaudo, D. M. Segal & K. P. Kearse, (1997) Multiple dimeric forms of human CD69 result from differential addition of N-glycans to typical (Asn-X-Ser/Thr) and atypical (Asn-X-cys) glycosylation motifs. *J Biol Chem* 272: 23117-23122.

Wacker, M., D. Linton, P. G. Hitchen, M. Nita-Lazar, S. M. Haslam, S. J. North, M. Panico, H. R. Morris, A. Dell, B. W. Wren & M. Aebi, (2002) N-Linked Glycosylation in Campylobacter jejuni and Its Functional Transfer into E. coli. *Science* 298: 1790-1793.

Whitley, P., I. Nilsson & G. von Heijne, (1996) A Nascent Secretory Protein May Traverse the Ribosome/Endoplasmic Reticulum Translocase Complex as an Extended Chain. *J Biol Chem* 271: 6241-6244.

Wilson, C. M. & S. High, (2007) Ribophorin I acts as a substrate-specific facilitator of N-glycosylation. *J Cell Sci* 120: 648-657.

Wilson, C. M., C. Kraft, C. Duggan, N. Ismail, S. G. Crawshaw & S. High, (2005) Ribophorin I associates with a subset of membrane proteins after their integration at the sec61 translocon. *J Biol Chem* 280: 4195-4206.

Wilson, C. M., Q. Roebuck & S. High, (2008) Ribophorin I regulates substrate delivery to the oligosaccharyltransferase core. *Proceedings of the National Academy of Sciences* 105: 9534-9539.

Wormald, M. R. & R. A. Dwek, (1999) Glycoproteins: glycan presentation and protein-fold stability. *Structure* 7: R155-160.

Yan, A. & W. J. Lennarz, (2005) Two oligosaccharyl transferase complexes exist in yeast and associate with two different translocons. *Glycobiology* 15: 1407-1415.

Yuzwa, S. A., X. Shan, M. S. Macauley, T. Clark, Y. Skorobogatko, K. Vosseller & D. J. Vocadlo, (2012) Increasing O-GlcNAc slows neurodegeneration and stabilizes tau against aggregation. *Nat Chem Biol* 8: 393-399.

Zhang, H. & R. Aebersold, (2006) Isolation of glycoproteins and identification of their N-linked glycosylation sites. *Methods Mol Biol* 328: 177-185.

Zhang, H., X. J. Li, D. B. Martin & R. Aebersold, (2003) Identification and quantification of N-linked glycoproteins using hydrazide chemistry, stable isotope labeling and mass spectrometry. *Nat Biotechnol* 21: 660-666.

Zielinska, D. F., F. Gnad, J. R. Wisniewski & M. Mann, (2010) Precision mapping of an in vivo N-glycoproteome reveals rigid topological and sequence constraints. *Cell* 141: 897-907.

Chapter 2

Structural Biology of Glycoproteins

Joanne E. Nettleship

Additional information is available at the end of the chapter

1. Introduction

In recent years, the field of structural biology has seen many advances in technology for the production of recombinant proteins, mainly led by the high-throughput techniques of the structural genomics community. These technologies have largely focussed on expression using *Escherichia coli* with around 88 % of the protein chains in the Protein Data Bank (PDB) in 2010 being produced using this bacterium [1]. However, many proteins require post-translational modification for correct folding and/or activity. By far the most common modification is glycosylation and it has been estimated that over half of all human proteins including many membrane proteins are glycosylated [2-3]. Therefore, the production of correctly glycosylated proteins for functional and structural studies requires the use of eukaryotic systems [4].

Glycoproteins represent a unique challenge to the structural biologist due to the size and heterogeneity of the oligosaccharide chains. Glycans can constitute from 1 % to over 80 % of the total protein mass [5] with variation in the type and number of sugars attached to a glycosylation site and also the occupancy of a site. The heterogeneity introduced by glycosylation can hinder crystallization of a glycoprotein [4] thus limiting the use of X-ray crystallography, one of the major techniques using in protein structure solution.

The following chapter reviews options for the production of glycoproteins from different types of expression system, both prokaryotic and eukaryotic, along with methods for addressing glycan heterogeneity in order to produce homogeneous glycoprotein samples which are amenable for structural studies. The use of such homogeneous glycoprotein samples in both crystallization and nuclear magnetic resonance (NMR) experiments are discussed.

2. Glycosylation in mammalian cells

In eukaryotes, glycans are added to the polypeptide chain in the endoplamic reticulum (ER) and the Golgi as the protein is secreted (see Figure 1). There are two types of glycosylation, one involves linkage of an oligosaccahride to an asparagine residue and the second involves linkage to a serine or threonine residue referred to as N- and O-glycosylation respectively.

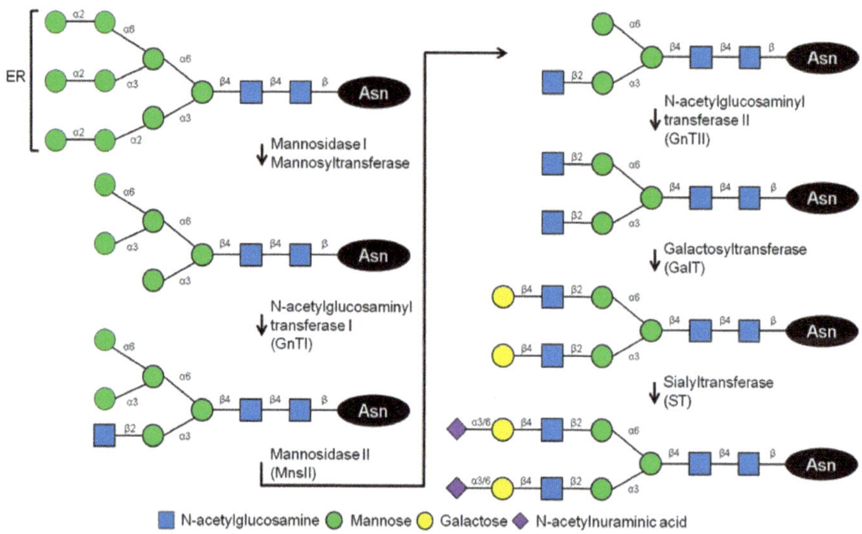

Figure 1. Simplified representation of the N-glycosylation pathway in humans with the initial core $Man_9GlcNAc_2$ sugar being formed in the endoplasmic reticulum (ER) and further modifications taking place in the Golgi.

N-glycosylation occurs co-translationally in the ER at the consensus sequon Asn-X-Ser/Thr/Cys-X where X is any amino acid except proline. The exact sequence of the sequon has a bearing on the occupancy of the glycosylation site. N-glycosylation is often essential for the expression and folding of a glycoprotein [6], with the initial glycan formed in the ER being further modified and decorated in the Golgi apparatus. This elaboration of the glycan core leads to a large number of possible structures which are classified as high mannose, complex or hybrid (Figure 2).

O-glycosylation (See Figure 3) usually occurs in regions containing large numbers of sequential serine, threonine and proline residues which are known as mucin domains. These regions show little secondary structure and are therefore often excluded from structural studies as proteins containing such disorder are unlikely to crystallize. O-glycosylation occurring outside mucin regions is difficult to predict accurately and so is usually only detected after production of a protein. As O-glycosylation occurs in the Golgi, it has little bearing on the early stages of protein folding and therefore sites found to be O-glycosylated can be engineered out of the protein by site-directed mutagenesis of the acceptor Ser/Thr residues.

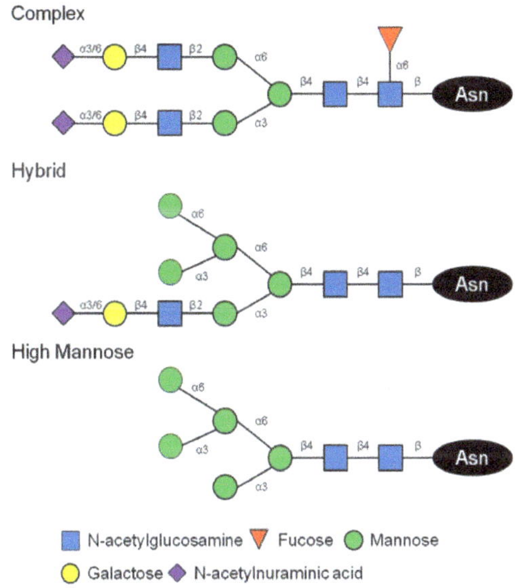

Figure 2. Diagrammatic representation of the major classes of N-linked sugars produced by mammalian cells.

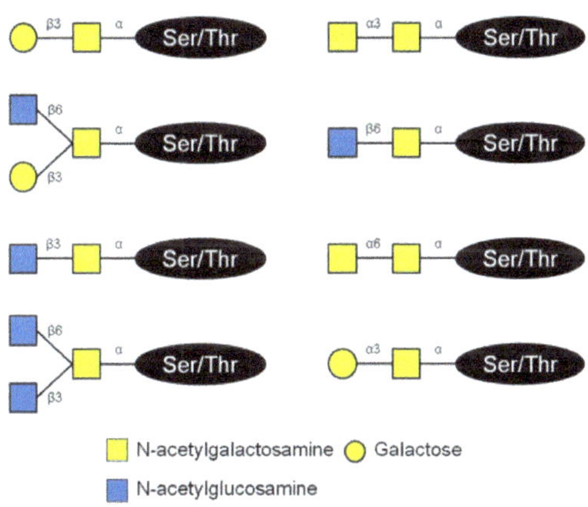

Figure 3. Representation of the core structures of the common O-glycans produced by mammalian cells.

3. Expression systems

For the correct production and folding of glycoproteins, the expression host needs to post-translationally modify the protein chain by adding sugars at the glycosylation sites. Therefore, production of glycoproteins is most often performed using eukaryotic cells, although the aglycosylated protein can be expressed as inclusion bodies in *E. coli* and refolded and also produced using cell-free systems. Recent advances in glycosylation pathway engineering have resulted in both *E. coli* [7] and cell-free systems [8] which are capable of introducing N-glycosylation onto a protein. These methods are discussed in the relevant sections below.

In addition to finding an appropriate system for the over-expression of the target glycoprotein, the ability of this system to incorporate selenomethioine into the protein must be assessed. Selenomethioine labelling enables phasing of X-ray diffraction data by multi-wavelength anomalous diffraction (MAD) [9]. Proteins expressed using *E. coli* routinely incorporate 100 % selenomethionine, whereas incorporation is more variable for proteins produced in higher eukaryotic systems.

3.1. *Escherichia coli*

E. coli is an attractive host as it is fast, simple to use, robust and cost-effective and therefore remains the dominant host for the production of recombinant proteins. In terms of glycoprotein production, the protein is usually expressed in inclusion bodies as bacteria do not have either the endoplasmic reticulum or the Golgi apparatus needed to secrete and post-translationally modify glycoproteins. As the protein is produced in inclusion bodies, refolding is needed to produce samples for structural biology – a process which can be time consuming and inefficient. However this method has proved useful for a number of products, for instance Watson *et al.* used *E. coli* to produce crystals of the extracellular domain of human UL16-binding protein ULBP1 [10]. Here the authors investigated various refolding strategies before finding the optimum protocol of slow dilution of guanidine-solubilised protein in solution containing arginine, ethylenediaminetetraacetic acid, reduced and oxidized glutathione and phenylmethyl-sulphonyl fluoride over 48 hours. In addition, recently the crystal structures of the murine class I major histocompatibility complex H-2Kb (PDB entry 3ROL) [11] and human β-secretase I with bound inhibitors (PDB entry 3S7M and 3S7L) [12] and the NMR structure of the sterile alpha motifs of EphA2 and SHIP2 (PDB entry 2KSO) [13] have solved using production in *E. coli* followed by refolding.

Another approach is to express the glycoprotein in the periplasm of *E. coli* using a bacterial signal sequence such as OmpA. For example, human α1-microglubulin was produced in the periplasm of *E. coli* and used to obtain the crystal structure to 2.3 Å resolution which revealed a potential heme binding site (PDB entry 3QKG) [14].

Recently an engineered eukaryotic protein glycosylation pathway has been inserted into *E. coli* resulting in cells capable of producing glycoproteins with $Man_3GlcNAc_2$ sugars attached [7]. Four glycosyltransfereases from *Saccharomyces cerevisae*, the uridine diphosphate-N-acetylglucosamine transferases Alg13 and Alg14 and the mannosyltransferases Alg1 and Alg2, are used to generate the glycan. The glycan is then transferred onto an N-glycosylation site using the oligosaccharyltransferase PglB from *Campylobacter jejuni*. Valderrama-Rincon *et al.* tested production of three eukaryotic glycoproteins; the Fc domain of human IgG1, bovine RNaseA and the placental variant of human growth hormone, and detected expression of glycosylated proteins [7]. Although currently only ~1 % of the expressed proteins were found to be glycosylated, this technology represents a huge potential for the cost-effective production of glycoproteins with a defined glycosylation pattern.

3.2. Insect cells

Glycoproteins produced by insect cells, such as *Spodoptera frugiperda, Trichoplusia ni* and *Drosophila melanogaster,* express products with glycans which are oligomannose and paucimannose in nature and mainly of the form α1-6 fucosylated $Man_3GlcNAc_2$ (Figure 4A) [15]. This compact, relatively homogeneous glycoform is compatible with protein crystallization and a number of structures have been determined of glycoproteins produced using insect cells. For example, the crystal structure of FcγRIIa was solved to 1.5 Å resolution using baculovirus infected *S. frugiperda* (Sf) 21 cells (PDB entry 3RY4) and also its structure bound to human IgG1-Fc was resolved to 3.8 Å (PDB entry 3RY6) [16].

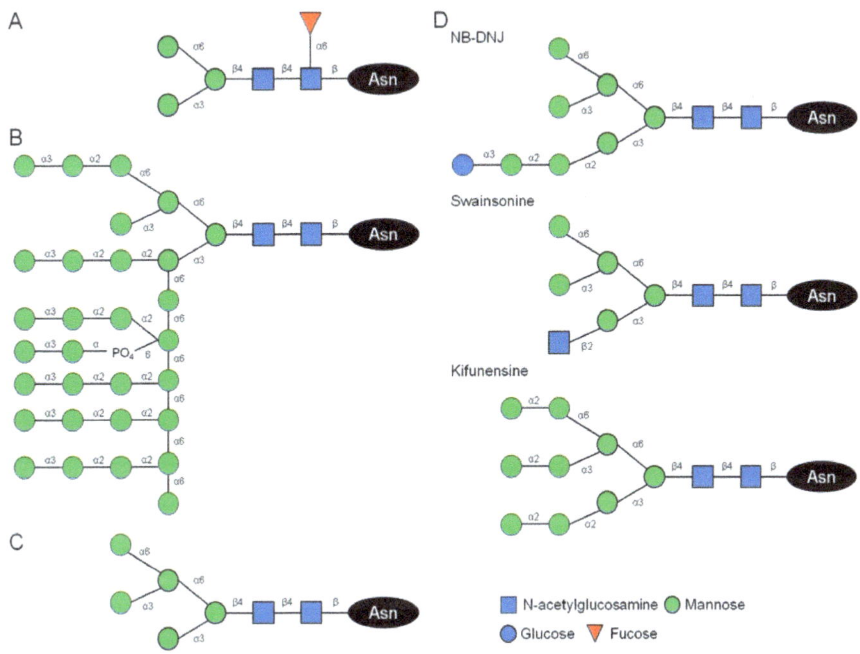

Figure 4. Representation of the N-glycans attached to proteins produced by (A) insect cells; (B) the yeast *Saccharomyces cerevisiae* (selected yeast oligosaccharide); (C) mutant CHO and HEK cell lines; and (D) mammalian cell lines in the presence of small molecule inhibitors.

The glycoforms produced by insect cells can be trimmed by treatment with endoglycosidases, for example endoglycosidase (endo) H or endo F1 and endo D will remove oligomannose and paucimannose sugars respectively leaving one GlcNAc residue on N-glcyosylation sites. In addition, endo F2 can be used in combination with endo F3 to cleave oligomannose and biantennary complex sugars and core fucosylated bi and triantennary complex gycans. Using this de-glycosylation strategy Fan *et al.* reported the successful crystallization and structure determination of a complex of human follicle stimulating hormone with its receptor (PDB entry 1XWD). Here crystals of fully glycosylated complex diffracted to 9 Å whereas 2.9 Å resolution was obtained after de-glycosylation with endo F2 and endo F3 [17]. Also, the structure of HIV-1 envelope glycoprotein gp120 was solved to 2.2 Å using de-glycosylation with endo H and endo D (PDB entry 1G9N) [18]. Although less common, tunicamycin has been used during production to block all glycosylation in order to promote crystallization. For example, evasin-1 structure was solved to 1.63 Å in its non-glycosylated form and to 2.7 Å in its glycosylated form (PDB entries 3FPR and 3FPT respectively) [19].

Selenomethionine labelling in insect cell systems can be difficult due to the toxicity of selenomethionine and the long incubation times needed as late baculovirus promoters such as polyhedron or P10 are normally used [1]. The levels of incorporation for secreted glycoproteins is higher than for intracellular proteins as unlabelled protein is removed during the media exchange process [20]. For example, 85 % selenomethionine incorporation was achieved for envelope glycoprotein D from HSV1 [21] and 76 % for palmitoyl protein thioesterase 1 (PDB entry 1EI9 and 1EH5) [22]. In 2007, Cronin *et al.* addressed this problem of variability in levels of incorporation and developed a method which consistently gave 70-75 % selenomethionine incorporation with a yield of around 20 % of the unlabelled protein [23].

3.3. Yeast cells

Yeast have been used for the production of human glycoproteins, with the most popular expression hosts being *Saccharomyces cerevisiae* and *Pichia pastoris* [24-25]. Recombinant proteins are secreted and post-translationally glycosylated to give glycoforms that are sensitive to endo H or endo F1 treatment and are therefore suitable for crystallization (Figure 4B). For example, de-glycosylation of protein produced in *P. pastoris* was used to prepare the G-protein coupled receptor (GPCR) human β_2 adrenergic receptor for crystallization [26] and production of gastric intrinsic factor with cobalamin (vitamin B12) bound (IF-Cbl) in *P. pastoris* was followed by complex formation with the cubilin IF-Cbl-binding-region and endo H treatment prior to crystallization of the complex to give a 3.3 Å resolution crystal structure (PDB entry 3KQ4) [27].

There are examples in the literature where retaining the N-glycans on the glycosylated protein produced in yeast is important, such as the of human and mouse glutaminyl cyclases (QC), enzymes linked with Alzheimer's disease, (PDB entry 3SI0) [28]. The structure of human QC had already been solved using protein produced in *E. coli* (PDB entry 2AFO) [29]. However, the glycosylated structure was shown to be more stable and to contain a disulphide bond not present in the non-glycosylated structure. The reduced stability of the aglycosylated human QC was associated with ready loss of the catalytic zinc ion [28].

Production of selenomethionine labelled proteins is possible in yeast with incorporation levels of around 50 % routinely reported for both *P. pastoris* and *S. cerevisiae* [20]. However higher selenomethionine incorporation has been reported. For instance, incorporation of ~98 % selenomethionine was reported for production of the carboxy-terminal fragment of the RNA-dependent RNA polymerase from *Neurospora crassa*, QDE-1 using *S. cerevisiae* [30]. This labelled protein was used to generate crystals which diffracted to 3.2 Å and allowed phasing of the native data giving a 2.3 Å resolution crystal structure (PDB entry 2J7N). More recently Malkowski *et al.* described generation of a selenomethionine-resistant strain of *S. cerevisiae* by blocking the S-adenosylmethionine synthesis pathway [31]. This strain was used to produce tryptophanyl-tRNA synthetase with >95 % selenomethionine incorporation which led to the determination of the structure of this yeast enzyme. [31]. Using the strain to produce and label heterologous proteins has so far not been reported.

3.4. Mammalian cells

Mammalian cells are a popular choice for the expression of glycoproteins, particularly for the production of human proteins. These cells have the correct machinery to fold and post-translationally modify glycoproteins. The two main cell lines used are human embryonic kidney (HEK) 293 cells and Chinese hamster ovary (CHO) cells, both of which are readily transfected with polythyleneimine (PEI), calcium phosphate or commercial lipids giving expression in 60-80 % of the cells [32-33]. There are two main variants of HEK 293 cells, (i) 293T which expresses the SV40 large-T antigen and (ii) 293E which expresses the Epstein-Barr virus (EBV) nuclear antigen 1 (EBNA1). Plasmids containing the SV40 or EBV origins of replication are amplified within these variants, giving more copies of the plasmid per cell and therefore a higher levels of protein expression [34]. An alternative method of introducing genes into mammalian cells is to use viral-mediated transduction such as the BacMam system [35], which has been shown to give milligram quantities of protein for structural studies [36].

Mammalian cells can be grown in either attached or suspension culture and automation of the culturing processes is possible in both cases [37-38]. Structural studies have been performed on proteins produced both by transient transfection, formation of a stable cell line, and using stable pools. Transient transfection is attractive due to the short timeframe between transfection and purified protein. Glycoprotein yields for non-antibodies of up to 40 mg/L have been obtained from attached HEK cells [39] and 36 mg/L from HEK cells in suspension [40], with yields of 27 mg/L from CHO cells in suspension [40].

Formation of a stable cell line producing a glycoprotein of interest often gives higher yields than utilizing transient transfection, however establishment of a stable cell line can take 2-6 months [41]. Automation can give a three-fold increase in throughput, although the timeline is the same [42]. An advantage of the system is that once the stable cell line has been established, production of the glycoprotein is fast and robust.

Use of stable transfection pools for expression are becoming increasingly popular as the protein yield is usually higher than for transient transfection but the timeline is shorter than for making a stable cell line. Post-transfection, the cells are sorted, often using a fluorescence marker, in order to enrich the pool for high producers. Using such a method, highly productive pools of cells can be obtained in 3 weeks, though yields may decline over time in culture as the transfection cells are not clonal. Using this method, expression levels of monoclonal antibodies from 100 mg/L to 1 g/L can be achieved in 2 months post transfection using CHO cells [43].

Selenomethionine labelling in mammalian cells is not as efficient as in *E. coli*, however levels of up to 90 % have been reported for stable HEK cell lines [44] and 78 % for transient glycoprotein production [1]. Generally cells are grown for a period of about 12 hours in media lacking methionine in order to deplete the intracellular methionine pools before the addition of selenomethionine. Keeping the concentration of selenomethionine used to 60 mg/L or below is important due to its toxicity which will lead to a lower yield of glycoprotein.

Unlike insect and fungi, mammalian cells produce glycoproteins with complex oligosaccharide chains (Figure 2) which are very heterogeneous and therefore not readily amenable to crystallization. Complete removal of N-linked glycans can be achieved by treatment of a purified glycoprotein with Peptide N-glycosidase (PNGase) F thus aiding crystallization [45]. In practice, two problems are encountered which limit this approach. Firstly, incomplete removal of glycans leading to partial de-glycosylation of the product; and secondly, insolubility due to protein aggregation following removal of all the sugars. Alternatively, glycosylation can be completely blocked in cells by treatment with tunicamycin. However, if glycosylation is required for proper folding and/or solubility in situ, then this approach will compromise the synthesis of the product. Two methods have been developed to get round these problems both of which depend upon manipulating N-glycosylation during glycoprotein biosynthesis by blocking the action of processing enzymes using either chemical inhibitors or null mutant cell lines.

3.4.1. Inhibitors

Three inhibitors have been used in the production of glycoproteins to manipulate the glycosylation pathway: N-butyldeoxynojirimycin (NB-DNJ), swainsonine and kifunensine. NB-DNJ inhibits α-glucosidase, thus blocking the early stages of N-glycan processing and giving products which contain high mannose or hybrid type sugars (Figure 4A). NB-DNJ has mainly being used in combination with the mutant CHO cell line, CHO Lec3.2.8.1 (Section 3.4.2), for instance in the crystallization of human costimulatory molecule B7-1 which is important in human immune response which gave crystals diffracting to 2.7 Å after treatment with endo H [46]. Swainsonine blocks α-mannosidase II resulting in high mannose or hybid type sugars shown in Figure 4D. Kifunensine strongly inhibits α-mannosidase I activity resulting in sugars of the form $Man_9GlcNAc_2$ (Figure 4D) [4]. Treatment of cells with the any of the above inhibitors results in relatively simple and chemically uniform glycoforms. Further, the high mannose and hybrid glycans resulting from the use of these drugs are cleavable using endoglycosidase (endo) H or endo F1 to leave one GlcNAc residue attached to the N-glycosylation site.

In practice, kifunensine is the most commonly used inhibitor in the production of glycoproteins for structural studies as the resulting glycans are the most homogeneous. In fact, the structures of glycoproteins can be solved following kifunensine treatment, without the use of endoglycosidase (for example, PDB entry 2WAH [47]), however more commonly the sugars are trimmed with endo H before crystallization studies. For example, Bishop *et al.* used transient transfection of HEK 293T cells in the presence of kifunensine to produce a number of glycoproteins involved in human Hedgehog signalling pathway. These were treated with endo H before crystallization and co-crystallization to give structures of hedgehog interaction protein ectodomain (Hhip) and Desert hedgehog in isolation and Hhip in complex with Desert hedgehog and sonic hedgehog (PDB entries 2WFT, 2WFR, 2WFQ, 2WG3, 2WFX and 2WG4) [48]. Production in the presence of kifunensine followed by endo H treatment has also been used with a stable CHO cell line in the crystallization of the extracellular region of cytotoxic T-lymphocyte antigen 4 (CTLA-4) giving crystals which diffracted to 1.8 Å [49].

3.4.2. Mutant cell lines

CHO and HEK cell lines have been generated with mutations in their glycosylation pathways which disrupt the action of GlcNAc transferase I (GnTI) [50-51]. CHO Lec3.2.1 contains four mutations which are in the *Gne, Slc35a1, Slc35a2* and *Mgat1* open reading frames leading to lower activity of various glycosylation enzymes including GnTI [50]. Use of CHO Lec3.2.8.1 leads to proteins mainly containing high mannose glycans which are sensitive to removal of all but the last GlcNAc residue with endo H or endo F1 (Figure 4C). However, the mutations in CHO Lec3.2.8.1 do not completely inhibit the formation of complex glycans [50] with, in some cases, as little as 10 % of the glycoprotein produced being endoglycosidase sensitive [4]. Structural studies have been carried out using CHO Lec3.2.8.1 both in the absence and presence of NB-DNJ which has a complementary inhibition effect to the mutant cell line giving endoglycosidase sensitive sugars (For examples see [46, 52-54]).

The GnTI-deficient cell line, HEK 293 $GnTI^-$ (also known as HEK 293S) produces glycoproteins with high mannose glycans (Figure 4C) which are endo H and endo F1 sensitive. Use of HEK 293 $GnTI^-$ cells gives product containing a very uniform glycosylation pattern of the form $Man_5GlcNAc_2$ with only traces of other glycan patterns being detected [51, 55]. This cell line has proved popular, recently facilitating structure solution of NetrinG1 and NetrinG2 in complex with their respective ligands at 3.25 Å and 2.6 Å resolution respectively (PDB entries 3ZYJ and 3ZYJ) [56]; the human glutamate receptor GluR2 amino terminal domain at 1.8 Å resolution (PDB entries 2WJW and 2WJX) [57]; and the orphan domain of the membrane glycoprotein endoglin using small angle X-ray scattering (SAXS) [58].

3.5. Other

Although not currently in mainstream use for the production of glycoproteins, two other expression systems are worth mentioning, namely cell-based production in the protozoan *Leishmania tarentolae* and cell-free synthesis using coupled transcription-translation systems.

The structure of human Cu/Zn superoxide dismutase has been recently published and represents the first structure solved using *L. tarentolae* as the expression host (PDB entry 3KH3) [59]. The advantages of the *L. tarentolae* system (commercialized by Jena Biosciences as LEXSY) are that it is simple to use, gives high yields and is inexpensive compared with the higher eukaryotes. *L. tarentolae* has eukaryotic folding and post-translational machinery and has been used in a proof of principle experiment to express glycosylated human erythropoietin [60]. These data show that use of *L. tarentolae* as a host system is applicable to the expression of glycoproteins for structural studies.

Cell-free synthesis systems have been used for structural studies for a number of years and are commonly based on lysates from *E. coli*, wheat germ and rabbit reticulocytes [61]. Glycoproteins can be produced by supplementing lysates with microsomal fractions [62], or using extracts of eukaryotes such as insect cells [63], hybridomas [64] and mammalian cells [65], although the yields are often poor. These systems give glycosylation patterns native to the host which in the case of insect cells can be modified using endoglycosidases and in the case of mammalian cells can be blocked using inhibitors (Section 3.4.1). Recently both the *E. coli* lysate cell free synthesis system and the PURE system for *in vitro* translation using purified components of *E. coli* [66] have been adapted for production of glycoproteins [8]. In this article, Guarino and DeLisa used the protein glycosylation locus from *Campylobater jejuni* to supplement both systems and produce glycoprotein with the GlcGalNAc$_5$Bac (where Bac represents bacillosamine) glycosylation pattern of *C. jejuni* (Figure 5) [8]. Cell-free systems have yet to yield a glycosylated protein for which a structure has been deposited in the PDB. However the ability to micro-engineer the components of the glycosylation pathway in an "open system" is an interesting and useful addition to the structural biologist's toolbox.

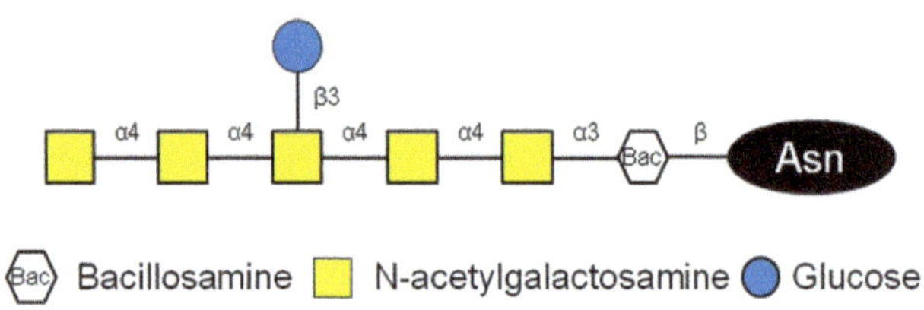

Figure 5. Representation of the glycosylation pattern added to glycoproteins using the cell-free synthesis system of Guarino and DeLisa [8].

4. Variable occupancy of glycosylation sites

In addition to the types of oligosaccharide chain attached to a glycosylation site, i.e. the glycoform, heterogeneity occurs in the occupancy of a glycosylation site. The occupancy of an N-glycosylation site depends upon the recognition sequon along with the structure of the protein in proximity to the glycosylation site. Some N-glycosylation sites have very low occupancy as a result of the local sequence composition, secondary structure and also distance to the C-terminus [67-69]. Bioinformatics programmes such as NetNGlyc and NetOGlyc (http://www.cbs.dtu.dk/services/NetNGlyc/ and http://www.cbs.dtu.dk/services /NetOGlyc/) [70-71] can predict occupancy of N-glycosylation and O-glycosylation sites respectively; however these are sequence based predictions and so need to be verified experimentally. After experimental identification of glycosylation site occupancy, variably occupied sites can be removed. Removal of variably occupied glycosylation sites should not affect the activity of a glycoprotein as these glycans are only present in a proportion of the sample and therefore must not be important for folding and solubility.

4.1. Identification methods

Analysis of glycosylation site occupancy is usually carried out using mass spectrometry. For a review of recent developments in glycoprotein analysis and glycomics see Zaia [72]. In the most straightforward experiment, the occupancy of an N-glycosylation site can be determined using a combination of tryptic digest and PNGase F treatment, followed by analysis of the resulting peptide [73]. The peptide contains asparagine at the proposed site if no glycan is present and, due to the action of PNGase F, an aspartic acid if sugars are attached. As glycopeptides are not readily ionized within a mass spectrometer, they can be enriched in the sample using hydrophilic interaction liquid chromatography (HILIC), thus increasing the likelihood of a glycosylation site-containing peptide being detected [74-75]. Mapping of glycoproteins, including both analysis of the glycosylation site and analysis of the glycans themselves, has been miniaturized and automated, such as methods using functionalized magnetic nanoparticles [76] and integrating chromatography steps [77] which allow for rapid acquisition of results.

Determination of O-glycosylation site occupancy is more difficult than for N-glycoslation sites as there is no endoglycosidase which universally releases O-linked glycans in the same way that PNGase F does for N-linked glycans. However, Halfinger *et al.* have developed a robust protocol using partial digestion with exoglycosidase and β-elimination using a mild alkylamine base followed by Michael addition and analysis using collision induced dissociation (CID) MS/MS [78].

Variable occupancy of glycosylation sites has also been detected by mutation studies. Here all the potential sites are mutated and the resulting protein analysed for activity. For example, Garman *et al.* analysed human IgE-FcεRIα and identified three glycosylation sites, N74, N135 and T142, which were not essential for protein folding, secretion and activity [79]. These sites were later confirmed by mass spectrometric analysis to be variably occupied [74]. However, mapping glycosylation occupancy by systematic mutation studies is labour intensive and is likely to result in many mutants that do not express or fold correctly as essential glycosylation sites have been removed.

4.2. Removal of glycosylation sites

Glycosylation sites found to be non-essential are often removed using site directed mutagenesis of the asparagine codon to a glutamine codon as this amino acid is similar in size and charge, but is not an attachment site for a glycan chain.

Formation of an N174Q mutant version of human ephrinA2 was essential in obtaining crystals of the EphA4:ephrinA2 complex which resulted in the structure being solved to 2.3 Å resolution (PDB entry 2WO3) [80]. In contrast, previous attempts at crystallization using wild type ephrinA2 with EphA4 were not successful. Although the N174 was shown to be invariably glycosylated, this site is not conserved across the ephrin family, (Nettleship, Bowden, unpublished results), and therefore was presumed to be non-essential.

In the case of human alpha-N-acetylgalactosaminidase, which has five N-glycosylation sites, mutation of N201 to glutamine led to crystals giving diffraction to 1.9 Å resolution (PDB entry 3H53) as opposed to the 8 Å resolution data collected using wild type glycoprotein crystals [81].

5. Structural studies

5.1. Crystallization and X-ray crystallography

Glycoproteins crystallize in a variety of conditions and trials are usually set up using the same screens as non-glycosylated proteins; such as those found in kits sold by Hampton Research, Molecular Dimensions and Emerald Biosystems. A number of reviews have addressed the problems connected with crystallization of glycoproteins such as the increase in surface entropy associated with large post-translational modifications and the microheterogenity of glycans [82-83]. However, it is to be noted that the presence of glycans can be an advantage for crystallization as they can form essential intermolecular contacts in crystal lattices [82]. Indeed glycoprotein crystallization has a success rate of around 50 % which is comparable to that for non-glycosylated proteins [82]. Methods around glycoprotein crystallization have developed to include automation and miniaturization using microscale crystallization techniques with as little as 65 μg of protein sample [45].

The strategy which is unique to the crystallization of glycoproteins is manipulation of the glycoform as described above. Such manipulations affect the propensity of a glycosylated protein to crystallize and also the quality of the resulting crystals. As shown in Figure 6, the type of oligosaccharide attached to the protein affects crystal morphology. Interestingly, in the case of human IgE-FcεRIα, the $Man_5GlcNAc_2$ glycoform gave better diffracting crystals than the shorter $Man_1GlcNAc_2$ (Figure 6E and 6D respectively) and therefore setting up crystallization trials with more than one glycoform can be advantageous.

X-ray crystal structures of glycoproteins often only show the initial GlcNAc residue even if more sugar residues are known to be attached to the protein because the glycan chains are flexible and so electron density for further sugar residues is not present. In rare cases, more of the oligosaccharide chain is resolved due to it either forming crystal contacts or interacting with the polypeptide chain. For example Figure 7 shows that the oligosaccharide chain attached to N297 of human IgG1-Fc could be fully resolved (PDB entry 2WAH) [47].

In solving a glycoprotein structure, it is important that the carbohydrate chains are built into the relevant electron density taking into account the restrictions on sugar conformation, including torsion angles, and linkages found in nature [87-88]. Recently, Crispin *et al.* have described the biosynthesis of glycans with emphasis on how these pathways lead to the glycosidic linkages found in glycoprotein structures [88]. If the production host gave sugars of a defined composition, for example HEK 293 $GnTI^-$ gives sugars of the form $Man_5GlcNAc_2$, modelling the glycans is relatively straightforward. For other cases, there are a number of databases where the structures of known glycans are deposited, with GlycomeDB (www.glycome-db.org) being the most comprehensive and unified carbohydrate database containing information from all public carbohydrate databases and the PDB [89-90]. Such databases are a useful resource for checking on the possible oligosaccharides and linkages likely to be attached to a glycoprotein produced using a given host cell. For building a three-dimensional model of an oligosaccharide chain, bioinformatics resources such as Sweet-II (www.glycosciences.de/spec/sweet2) [91] and Glycam Biomolecule Builder (glycam.ccrc.uga.edu/CCRC/biombuilder/biomb_index.jsp) [92] are available. After modelling the glycan, Glycam Biomolecule Builder attaches the sugar chain to the protein structure at the glycosylation site, with glycans generated by Sweet-II being attached using the glyProt software (www.glycosciences.de/modeling/glyprot/php/main.php) [93]. Upon deposition of the glycoprotein structure into the PDB, pdb-care (PDB carbohydrate residue check) can help with problems found within glycan structures [94-95]. Overall, use of these web-based tools allows for the correct building of sugar chains into glycoprotein structures when electron density is available.

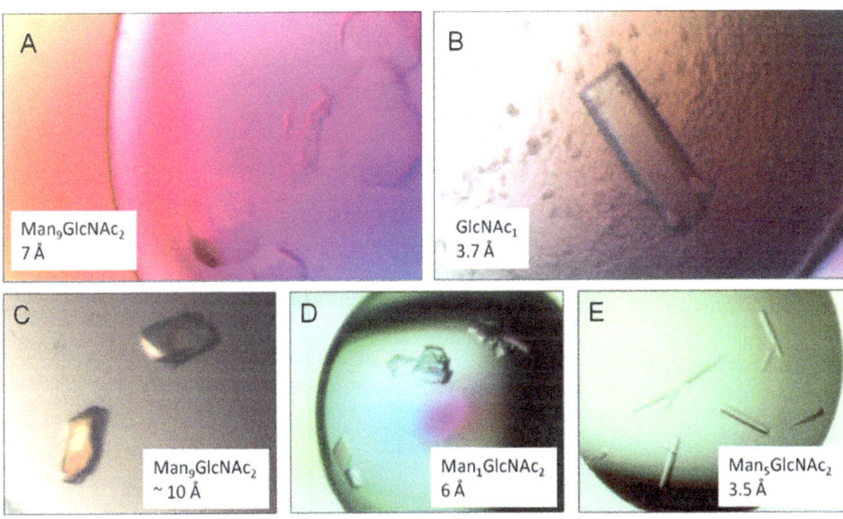

Figure 6. Crystals of human programmed cell death 1-ligand 1 formed from glycoprotein secreted by HEK 293T cells (A) in the presence of kifunensine and (B) after treatment with endo F1. Diffraction was to 7 Å and 3.7 Å respectively. Crystals of human IgE-FcεRIα (N74A, N135A, T142A) from protein produced using HEK 293T cells (C) in the presence of kifunensine; (D) after treatment with jack bean mannosidase and (E) produced by HEK 293 GnTI- cells. Diffraction was around 10 Å, to 6 Å and to 3.5 Å respectively. In each case the insert shows the major glycosylation state of the protein upon crystallization and the resolution of diffraction data collected. (Human programmed cell death 1-ligand 1 - PDB entry 3BIS [84] and human IgE-FcεRIα - PDB entry 2Y7Q [85])

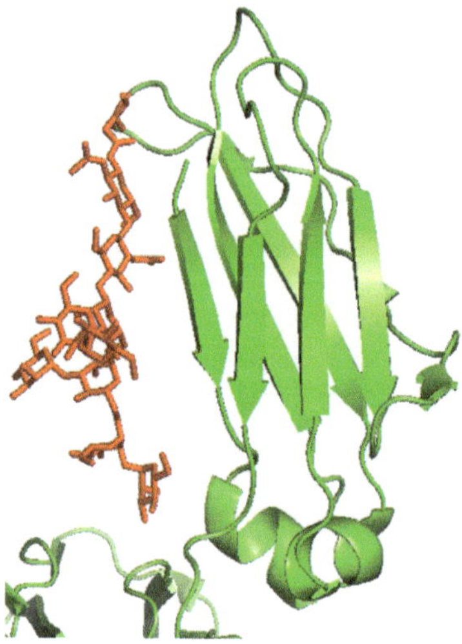

Figure 7. Cartoon representation of the human IgG1-Fc glycan attached to N297 shown as sticks in red with the protein backbone shown in green. The glycoprotein was produced in HEK 293T cells in the presence of kifunensine, giving sugars of the form $Man_9GlcNAc_2$. (The diagram was produced using PyMOL [86] and PDB entry 2WAH [47]).

5.2. NMR spectroscopy

Nuclear magnetic resonance (NMR) spectroscopy can be used to study protein structure, but the technique faces some obstacles in terms of its use for glycoproteins, namely the ^{1}H chemical shift overlap between carbohydrate and protein signal [96] and the difficulty of obtaining sufficient yield of labelled protein from eukaryotic systems. These challenges have led to the majority of glycoproteins structures solved by NMR being of the aglycosylated protein chains expressed in *E. coli.*

Labelling with ^{15}N or both ^{15}N and ^{13}C has been demonstrated in mammalian cells [97], insect cells [98] and yeast [99]. Such eukaryotic expression systems have been used to solve a few recombinant glycoprotein structures containing glycans such as fragments of human fibronectin (PBD entry 1E8B), human thrombomadulin (PBD entry 1DQB) and the extracellular domain of human cytotoxic T lymphocylte-associated protein (CTLA)-4 (PDB entry 1AH1) [100-102].

The problem of signal overlap between proteins and carbohydrates has been tackled by Slynko *et al.* who added unlabelled glycan chains using in vitro N-glycosylation technology [103] to AcrA from *C. jejuni* produced using *E. coli* with ^{15}N and ^{13}C labelling (PDB entry 2K32) [96]. This allowed a NOESY (Nuclear Overhauser effect spectroscopy) experiment to be performed without confusion between protein and glycan resonances which often cause difficulty in structure determination of unlabelled or ^{15}N-labelled glycoproteins [96].

In NMR structures, the protein chains and attached glycans are in solution and so are able to move around. Unlike in crystal structures, solution NMR allows the flexibility of both the oligosaccharide chain and the protein chain around a glycosylation site to be observed. In the case of AcrA (PDB entry 2K32), the flexibility was seen to be in the α-helical loop region containing the N42 glycosylation site (Figure 8A), with the heptasaccharide glycan having a well defined rod-like structure [96]. Human CTLA4 (PDB entry 1AH1) contains two N-glycosylation sites, N78 and N111 which are occupied with partially deglycosylated glycans of the form $Man_1GlcNAc_2$ (Figure 8B). The glycan attached to N78 interacts extensively with the side chains of a nearby β-sheet and is therefore ordered, whereas N111 had limited interaction with the protein chain and so only the initial GlcNAc residue is well defined [102].

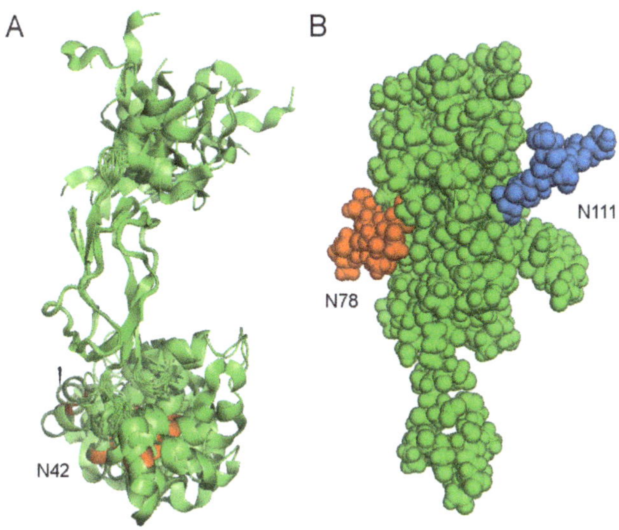

Figure 8. (A) Cartoon representation of the NMR ensemble structure of AcrA showing the N42 glycosylation site (in red) in the flexible lower domain of the protein. (B) A space filling model of CTLA4 with the glycans attached to N78 (in red) showing interaction with the nearby β-sheet and the N111 sugar chain (in blue) being remote from the polypeptide chain. (Diagrams were produced using PyMOL [86] and PDB entries 2K32 [96] and 1AH1 [102] respectively).

6. Conclusions and future prospects

This chapter has reviewed some of the recent developments in the field of glycoprotein structural biology. Increasingly, the expression system of choice for the production of recombinant glycoproteins is the mammalian cell, and in particular HEK in the presence of kifunensine which modifies glycan processing. This enables treatment of the purified glycoprotein with endoglycosidase to reduce the sugar "load" to a single GlcNAc at each N-linked attachment site. Experience shows that by preparing glycoproteins with different glycoforms the chances of obtaining diffraction quality crystals are significantly increased. Combining this approach with the inclusion of novel additives in the crystallization experiment, for example "smart materials", such as the molecularly imprinted polymers (MIPs) recently reported by Sarkidakis *et al.* [104], adds another dimension to crystallization.

Future developments in simpler low cost expression technology as exemplified by *Leishmania tarentolae* system will also have an impact on the structural biology of glycoproteins. It is anticipated that the number of structures solved by X-ray crystallography and NMR for this important class of proteins will rapidly increase in the next few years.

Author details

Joanne E. Nettleship
OPPF-UK, Research Complex at Harwell, R92 Rutherford Appleton Laboratories, Harwell Oxford, Didcot, UK

Division of Structural Biology, Oxford University, Wellcome Trust Centre for Human Genetics, Roosevelt Drive, Oxford, UK

Acknowledgement

The author would like to thank Ray Owens and Max Crispin for critical reading of the manuscript. The OPPF-UK is funded by the UK Medical Research Council and Biotechnology and Biological Sciences Research Council.

7. References

[1] Nettleship JE, Assenberg R, Diprose JM, Rahman-Huq N, Owens RJ. Recent advances in the production of proteins in insect and mammalian cells for structural biology. J Struct Biol. 2010 Oct;172(1):55-65.
[2] Apweiler R, Hermjakob H, Sharon N. On the frequency of protein glycosylation, as deduced from analysis of the SWISS-PROT database. Biochim Biophys Acta. 1999 Dec 6;1473(1):4-8.
[3] Zafar S, Nasir A, Bokhari H. Computational analysis reveals abundance of potential glycoproteins in Archaea, Bacteria and Eukarya. Bioinformation. 2011;6(9):352-5.
[4] Davis SJ, Crispin M. Solutions to the Glycosylation Problem for Low- and High-Throughput Structural Glycoproteomics. In: Owens RJ, Nettleship JE, editors. Functional and Structural Proteomics of Glycoproteins2011. p. 127-58.
[5] Varki A, Sharon N. Historical Background and Overview. 2009.
[6] Parodi AJ. Protein glucosylation and its role in protein folding. Annu Rev Biochem. 2000;69:69-93.
[7] Valderrama-Rincon JD, Fisher AC, Merritt JH, Fan YY, Reading CA, Chhiba K, et al. An engineered eukaryotic protein glycosylation pathway in Escherichia coli. Nat Chem Biol. 2012;8(5):434-6.

[8] Guarino C, Delisa MP. A prokaryote-based cell-free translation system that efficiently synthesizes glycoproteins. Glycobiology. 2011 Nov 8;8:8.
[9] Hendrickson WA, Horton JR, LeMaster DM. Selenomethionyl proteins produced for analysis by multiwavelength anomalous diffraction (MAD): a vehicle for direct determination of three-dimensional structure. Embo J. 1990 May;9(5):1665-72.
[10] Watson AA, Christou CM, O'Callaghan CA. Expression, purification and crystallization of the human UL16-binding protein ULBP1. Protein Expr Purif. 2011 Sep;79(1):44-8.
[11] Madhurantakam C, Duru AD, Sandalova T, Webb JR, Achour A. Inflammation-Associated Nitrotyrosination Affects TCR Recognition through Reduced Stability and Alteration of the Molecular Surface of the MHC Complex. PLoS One. 2012;7(3):e32805.
[12] Malamas MS, Erdei J, Gunawan I, Barnes K, Hui Y, Johnson M, et al. New pyrazolyl and thienyl aminohydantoins as potent BACE1 inhibitors: exploring the S2' region. Bioorg Med Chem Lett. 2011 Sep 15;21(18):5164-70.
[13] Lee HJ, Hota PK, Chugha P, Guo H, Miao H, Zhang L, et al. NMR structure of a heterodimeric SAM:SAM complex: characterization and manipulation of EphA2 binding reveal new cellular functions of SHIP2. Structure. 2012 Jan 11;20(1):41-55.
[14] Meining W, Skerra A. The crystal structure of human alpha1-microglobulin reveals a potential haem-binding site. Biochem J. 2012 Apr 19;19:19.
[15] Harrison RL, Jarvis DL. Protein N-glycosylation in the baculovirus-insect cell expression system and engineering of insect cells to produce "mammalianized" recombinant glycoproteins. Adv Virus Res. 2006;68:159-91.
[16] Ramsland PA, Farrugia W, Bradford TM, Sardjono CT, Esparon S, Trist HM, et al. Structural basis for Fc gammaRIIa recognition of human IgG and formation of inflammatory signaling complexes. J Immunol. 2011 Sep 15;187(6):3208-17.
[17] Fan QR, Hendrickson WA. Structure of human follicle-stimulating hormone in complex with its receptor. Nature. 2005 Jan 20;433(7023):269-77.
[18] Kwong PD, Wyatt R, Desjardins E, Robinson J, Culp JS, Hellmig BD, et al. Probability analysis of variational crystallization and its application to gp120, the exterior envelope glycoprotein of type 1 human immunodeficiency virus (HIV-1). J Biol Chem. 1999 Feb 12;274(7):4115-23.
[19] Dias JM, Losberger C, Deruaz M, Power CA, Proudfoot AE, Shaw JP. Structural basis of chemokine sequestration by a tick chemokine binding protein: the crystal structure of the complex between Evasin-1 and CCL3. PLoS One. 2009;4(12):e8514.
[20] Walden H. Selenium incorporation using recombinant techniques. Acta Crystallogr D Biol Crystallogr. 2010 Apr;66(Pt 4):352-7.
[21] Carfi A, Gong H, Lou H, Willis SH, Cohen GH, Eisenberg RJ, et al. Crystallization and preliminary diffraction studies of the ectodomain of the envelope glycoprotein D from herpes simplex virus 1 alone and in complex with the ectodomain of the human receptor HveA. Acta Crystallogr D Biol Crystallogr. 2002 May;58(Pt 5):836-8.
[22] Bellizzi JJ, 3rd, Widom J, Kemp C, Lu JY, Das AK, Hofmann SL, et al. The crystal structure of palmitoyl protein thioesterase 1 and the molecular basis of infantile neuronal ceroid lipofuscinosis. Proc Natl Acad Sci U S A. 2000 Apr 25;97(9):4573-8.
[23] Cronin CN, Lim KB, Rogers J. Production of selenomethionyl-derivatized proteins in baculovirus-infected insect cells. Protein Sci. 2007 Sep;16(9):2023-9.

[24] Mattanovich D, Branduardi P, Dato L, Gasser B, Sauer M, Porro D. Recombinant protein production in yeasts. Methods Mol Biol. 2012;824:329-58.
[25] Damasceno LM, Huang CJ, Batt CA. Protein secretion in Pichia pastoris and advances in protein production. Appl Microbiol Biotechnol. 2012 Jan;93(1):31-9.
[26] Noguchi S, Satow Y. Purification of human beta2-adrenergic receptor expressed in methylotrophic yeast Pichia pastoris. J Biochem. 2006 Dec;140(6):799-804.
[27] Andersen CB, Madsen M, Storm T, Moestrup SK, Andersen GR. Structural basis for receptor recognition of vitamin-B(12)-intrinsic factor complexes. Nature. 2010 Mar 18;464(7287):445-8.
[28] Ruiz-Carrillo D, Koch B, Parthier C, Wermann M, Dambe T, Buchholz M, et al. Structures of glycosylated mammalian glutaminyl cyclases reveal conformational variability near the active center. Biochemistry. 2011 Jul 19;50(28):6280-8.
[29] Huang KF, Liu YL, Cheng WJ, Ko TP, Wang AH. Crystal structures of human glutaminyl cyclase, an enzyme responsible for protein N-terminal pyroglutamate formation. Proc Natl Acad Sci U S A. 2005 Sep 13;102(37):13117-22.
[30] Laurila MR, Salgado PS, Makeyev EV, Nettelship J, Stuart DI, Grimes JM, et al. Gene silencing pathway RNA-dependent RNA polymerase of Neurospora crassa: yeast expression and crystallization of selenomethionated QDE-1 protein. J Struct Biol. 2005 Jan;149(1):111-5.
[31] Malkowski MG, Quartley E, Friedman AE, Babulski J, Kon Y, Wolfley J, et al. Blocking S-adenosylmethionine synthesis in yeast allows selenomethionine incorporation and multiwavelength anomalous dispersion phasing. Proc Natl Acad Sci U S A. 2007 Apr 17;104(16):6678-83.
[32] Huh SH, Do HJ, Lim HY, Kim DK, Choi SJ, Song H, et al. Optimization of 25 kDa linear polyethylenimine for efficient gene delivery. Biologicals. 2007 Jun;35(3):165-71.
[33] Martin-Montanez E, Lopez-Tellez JF, Acevedo MJ, Pavia J, Khan ZU. Efficiency of gene transfection reagents in NG108-15, SH-SY5Y and CHO-K1 cell lines. Methods Find Exp Clin Pharmacol. 2010 Jun;32(5):291-7.
[34] Van Craenenbroeck K, Vanhoenacker P, Haegeman G. Episomal vectors for gene expression in mammalian cells. Eur J Biochem. 2000 Sep;267(18):5665-78.
[35] Jardin BA, Elias CB, Prakash S. Expression of a secreted protein in mammalian cells using baculovirus particles. Methods Mol Biol. 2012;801:41-63.
[36] Dukkipati A, Park HH, Waghray D, Fischer S, Garcia KC. BacMam system for high-level expression of recombinant soluble and membrane glycoproteins for structural studies. Protein Expr Purif. 2008 Dec;62(2):160-70.
[37] Zhao Y, Bishop B, Clay JE, Lu W, Jones M, Daenke S, et al. Automation of large scale transient protein expression in mammalian cells. J Struct Biol. 2011 Aug;175(2):209-15.
[38] Gonzalez R, Jennings LL, Knuth M, Orth AP, Klock HE, Ou W, et al. Screening the mammalian extracellular proteome for regulators of embryonic human stem cell pluripotency. Proc Natl Acad Sci U S A. 2010 Feb 23;107(8):3552-7.
[39] Aricescu AR, Lu W, Jones EY. A time- and cost-efficient system for high-level protein production in mammalian cells. Acta Crystallogr D Biol Crystallogr. 2006 Oct;62(Pt 10):1243-50.
[40] Bollin F, Dechavanne V, Chevalet L. Design of Experiment in CHO and HEK transient transfection condition optimization. Protein Expr Purif. 2011 Jul;78(1):61-8.

[41] Wurm FM. Production of recombinant protein therapeutics in cultivated mammalian cells. Nat Biotechnol. 2004 Nov;22(11):1393-8.
[42] Lindgren K, Salmen A, Lundgren M, Bylund L, Ebler A, Faldt E, et al. Automation of cell line development. Cytotechnology. 2009 Jan;59(1):1-10.
[43] Ye J, Alvin K, Latif H, Hsu A, Parikh V, Whitmer T, et al. Rapid protein production using CHO stable transfection pools. Biotechnol Prog. 2010 Sep-Oct;26(5):1431-7.
[44] Barton WA, Tzvetkova-Robev D, Erdjument-Bromage H, Tempst P, Nikolov DB. Highly efficient selenomethionine labeling of recombinant proteins produced in mammalian cells. Protein Sci. 2006 Aug;15(8):2008-13.
[45] Lee JE, Fusco ML, Saphire EO. An efficient platform for screening expression and crystallization of glycoproteins produced in human cells. Nat Protoc. 2009;4(4):592-604.
[46] Davis SJ, Ikemizu S, Collins AV, Fennelly JA, Harlos K, Jones EY, et al. Crystallization and functional analysis of a soluble deglycosylated form of the human costimulatory molecule B7-1. Acta Crystallogr D Biol Crystallogr. 2001 Apr;57(Pt 4):605-8.
[47] Crispin M, Bowden TA, Coles CH, Harlos K, Aricescu AR, Harvey DJ, et al. Carbohydrate and domain architecture of an immature antibody glycoform exhibiting enhanced effector functions. J Mol Biol. 2009 Apr 17;387(5):1061-6.
[48] Bishop B, Aricescu AR, Harlos K, O'Callaghan CA, Jones EY, Siebold C. Structural insights into hedgehog ligand sequestration by the human hedgehog-interacting protein HHIP. Nat Struct Mol Biol. 2009 Jul;16(7):698-703.
[49] Yu C, Crispin M, Sonnen AF, Harvey DJ, Chang VT, Evans EJ, et al. Use of the alpha-mannosidase I inhibitor kifunensine allows the crystallization of apo CTLA-4 homodimer produced in long-term cultures of Chinese hamster ovary cells. Acta Crystallogr Sect F Struct Biol Cryst Commun. 2011 Jul 1;67(Pt 7):785-9.
[50] North SJ, Huang HH, Sundaram S, Jang-Lee J, Etienne AT, Trollope A, et al. Glycomics profiling of Chinese hamster ovary cell glycosylation mutants reveals N-glycans of a novel size and complexity. J Biol Chem. 2010 Feb 19;285(8):5759-75.
[51] Chang VT, Crispin M, Aricescu AR, Harvey DJ, Nettleship JE, Fennelly JA, et al. Glycoprotein structural genomics: solving the glycosylation problem. Structure. 2007 Mar;15(3):267-73.
[52] Wilke S, Groebe L, Maffenbeier V, Jager V, Gossen M, Josewski J, et al. Streamlining homogeneous glycoprotein production for biophysical and structural applications by targeted cell line development. PLoS One. 2011;6(12):e27829.
[53] Fischer PB, Karlsson GB, Butters TD, Dwek RA, Platt FM. N-butyldeoxynojirimycin-mediated inhibition of human immunodeficiency virus entry correlates with changes in antibody recognition of the V1/V2 region of gp120. J Virol. 1996 Oct;70(10):7143-52.
[54] Butters TD, Sparks LM, Harlos K, Ikemizu S, Stuart DI, Jones EY, et al. Effects of N-butyldeoxynojirimycin and the Lec3.2.8.1 mutant phenotype on N-glycan processing in Chinese hamster ovary cells: application to glycoprotein crystallization. Protein Sci. 1999 Aug;8(8):1696-701.
[55] Crispin M, Harvey DJ, Chang VT, Yu C, Aricescu AR, Jones EY, et al. Inhibition of hybrid- and complex-type glycosylation reveals the presence of the GlcNAc transferase I-independent fucosylation pathway. Glycobiology. 2006 Aug;16(8):748-56.

[56] Seiradake E, Coles CH, Perestenko PV, Harlos K, McIlhinney RA, Aricescu AR, et al. Structural basis for cell surface patterning through NetrinG-NGL interactions. Embo J. 2011 Nov 2;30(21):4479-88.
[57] Clayton A, Siebold C, Gilbert RJ, Sutton GC, Harlos K, McIlhinney RA, et al. Crystal structure of the GluR2 amino-terminal domain provides insights into the architecture and assembly of ionotropic glutamate receptors. J Mol Biol. 2009 Oct 9;392(5):1125-32.
[58] Alt A, Miguel-Romero L, Donderis J, Aristorena M, Blanco FJ, Round A, et al. Structural and functional insights into endoglin ligand recognition and binding. PLoS One. 2012;7(2):e29948.
[59] Gazdag EM, Cirstea IC, Breitling R, Lukes J, Blankenfeldt W, Alexandrov K. Purification and crystallization of human Cu/Zn superoxide dismutase recombinantly produced in the protozoan Leishmania tarentolae. Acta Crystallogr Sect F Struct Biol Cryst Commun. 2010 Aug 1;66(Pt 8):871-7.
[60] Breitling R, Klingner S, Callewaert N, Pietrucha R, Geyer A, Ehrlich G, et al. Non-pathogenic trypanosomatid protozoa as a platform for protein research and production. Protein Expr Purif. 2002 Jul;25(2):209-18.
[61] Katzen F, Chang G, Kudlicki W. The past, present and future of cell-free protein synthesis. Trends Biotechnol. 2005 Mar;23(3):150-6.
[62] Walter P, Blobel G. Preparation of microsomal membranes for cotranslational protein translocation. Methods Enzymol. 1983;96:84-93.
[63] Tarui H, Murata M, Tani I, Imanishi S, Nishikawa S, Hara T. Establishment and characterization of cell-free translation/glycosylation in insect cell (Spodoptera frugiperda 21) extract prepared with high pressure treatment. Appl Microbiol Biotechnol. 2001 May;55(4):446-53.
[64] Mikami S, Kobayashi T, Yokoyama S, Imataka H. A hybridoma-based in vitro translation system that efficiently synthesizes glycoproteins. J Biotechnol. 2006 Dec 15;127(1):65-78.
[65] Shibutani M, Kim E, Lazarovici P, Oshima M, Guroff G. Preparation of a cell-free translation system from PC12 cell. Neurochem Res. 1996 Jul;21(7):801-7.
[66] Ohashi H, Kanamori T, Shimizu Y, Ueda T. A highly controllable reconstituted cell-free system--a breakthrough in protein synthesis research. Curr Pharm Biotechnol. 2010 Apr;11(3):267-71.
[67] Rao RS, Bernd W. Do N-glycoproteins have preference for specific sequons? Bioinformation. 2010;5(5):208-12.
[68] Petrescu AJ, Milac AL, Petrescu SM, Dwek RA, Wormald MR. Statistical analysis of the protein environment of N-glycosylation sites: implications for occupancy, structure, and folding. Glycobiology. 2004 Feb;14(2):103-14.
[69] Bano-Polo M, Baldin F, Tamborero S, Marti-Renom MA, Mingarro I. N-glycosylation efficiency is determined by the distance to the C-terminus and the amino acid preceding an Asn-Ser-Thr sequon. Protein Sci. 2011 Jan;20(1):179-86.
[70] Gupta R, Brunak S. Prediction of glycosylation across the human proteome and the correlation to protein function. Pac Symp Biocomput. 2002:310-22.
[71] Julenius K, Molgaard A, Gupta R, Brunak S. Prediction, conservation analysis, and structural characterization of mammalian mucin-type O-glycosylation sites. Glycobiology. 2005 Feb;15(2):153-64.

[72] Zaia J. Mass spectrometry and glycomics. Omics. 2010 Aug;14(4):401-18.
[73] Nettleship JE, Aplin R, Aricescu AR, Evans EJ, Davis SJ, Crispin M, et al. Analysis of variable N-glycosylation site occupancy in glycoproteins by liquid chromatography electrospray ionization mass spectrometry. Anal Biochem. 2007 Feb 1;361(1):149-51.
[74] Nettleship JE. Hydrophilic Interaction Liquid Chromatography in the Characterization of Glycoproteins. In: Wang PG, He W, editors. Hydrophilic Interaction Liquid Chromatography (HILIC) and Advanced Applications: CRC Press; 2011. p. 523-50.
[75] Hagglund P, Bunkenborg J, Elortza F, Jensen ON, Roepstorff P. A new strategy for identification of N-glycosylated proteins and unambiguous assignment of their glycosylation sites using HILIC enrichment and partial deglycosylation. J Proteome Res. 2004 May-Jun;3(3):556-66.
[76] Kuo CW, Wu IL, Hsiao HH, Khoo KH. Rapid glycopeptide enrichment and N-glycosylation site mapping strategies based on amine-functionalized magnetic nanoparticles. Anal Bioanal Chem. 2012 Jan 29;29:29.
[77] Qu Y, Xia S, Yuan H, Wu Q, Li M, Zou L, et al. Integrated sample pretreatment system for N-linked glycosylation site profiling with combination of hydrophilic interaction chromatography and PNGase F immobilized enzymatic reactor via a strong cation exchange precolumn. Anal Chem. 2011 Oct 1;83(19):7457-63.
[78] Halfinger B, Sarg B, Lindner HH. Evaluation of non-reductive beta-elimination/Michael addition for glycosylation site determination in mucin-like O-glycopeptides. Electrophoresis. 2011 Dec;32(24):3546-53.
[79] Garman SC, Wurzburg BA, Tarchevskaya SS, Kinet JP, Jardetzky TS. Structure of the Fc fragment of human IgE bound to its high-affinity receptor Fc epsilonRI alpha. Nature. 2000 Jul 20;406(6793):259-66.
[80] Bowden TA, Aricescu AR, Nettleship JE, Siebold C, Rahman-Huq N, Owens RJ, et al. Structural plasticity of eph receptor A4 facilitates cross-class ephrin signaling. Structure. 2009 Oct 14;17(10):1386-97.
[81] Clark NE, Garman SC. The 1.9 a structure of human alpha-N-acetylgalactosaminidase: The molecular basis of Schindler and Kanzaki diseases. J Mol Biol. 2009 Oct 23;393(2):435-47.
[82] Mesters JR, Hilgenfeld R. Protein Glycosylation, Sweet to Crystal Growth? Cryst Growth Des. 2007;7(11):2251-3.
[83] Stura EA, Nemerow GR, Wilson IA. Strategies in the crystallization of glycoproteins and protein complexes. J Cryst Growth. 1992;122(1–4):273-85.
[84] Lin DY, Tanaka Y, Iwasaki M, Gittis AG, Su HP, Mikami B, et al. The PD-1/PD-L1 complex resembles the antigen-binding Fv domains of antibodies and T cell receptors. Proc Natl Acad Sci U S A. 2008 Feb 26;105(8):3011-6.
[85] Holdom MD, Davies AM, Nettleship JE, Bagby SC, Dhaliwal B, Girardi E, et al. Conformational changes in IgE contribute to its uniquely slow dissociation rate from receptor FcvarepsilonRI. Nat Struct Mol Biol. 2011 May;18(5):571-6.
[86] Schrödinger LLC. The PyMOL molecular graphics system. 1.3 ed.
[87] Crispin M, Stuart DI, Jones EY. Building meaningful models of glycoproteins. Nat Struct Mol Biol. 2007 May;14(5):354; discussion -5.
[88] Crispin M, Scanlan CN, Bowden TA. Structure and Biosynthesis of Glycoprotein Carbohydrates. In: Moo-Young M, editor. Comprehensive Biotechnology. 2 ed: Elsevier; 2011. p. 73-90.

[89] Ranzinger R, Maaβ K, Lütteke T. Bioinformatics Databases and Applications Available for Glycobiology and Glycomics. In: Owens RJ, Nettleship JE, editors. Functional and Structural Proteomics of Glycoproteins2011. p. 59-90.
[90] Ranzinger R, Herget S, von der Lieth CW, Frank M. GlycomeDB--a unified database for carbohydrate structures. Nucleic Acids Res. 2011 Jan;39(Database issue):D373-6.
[91] Bohne A, Lang E, von der Lieth CW. SWEET - WWW-based rapid 3D construction of oligo- and polysaccharides. Bioinformatics. 1999 Sep;15(9):767-8.
[92] Woods Group. GLYCAM Web. Complex Carbohydrate Research Center, University of Georgia, Athens, GA 2005-2012.
[93] Bohne-Lang A, von der Lieth CW. GlyProt: in silico glycosylation of proteins. Nucleic Acids Res. 2005 Jul 1;33(Web Server issue):W214-9.
[94] Lutteke T, von der Lieth CW. pdb-care (PDB carbohydrate residue check): a program to support annotation of complex carbohydrate structures in PDB files. BMC Bioinformatics. 2004 Jun 4;5(69):69.
[95] Read RJ, Adams PD, Arendall WB, 3rd, Brunger AT, Emsley P, Joosten RP, et al. A new generation of crystallographic validation tools for the protein data bank. Structure. 2011 Oct 12;19(10):1395-412.
[96] Slynko V, Schubert M, Numao S, Kowarik M, Aebi M, Allain FH. NMR structure determination of a segmentally labeled glycoprotein using in vitro glycosylation. J Am Chem Soc. 2009 Jan 28;131(3):1274-81.
[97] Egorova-Zachernyuk TA, Bosman GJ, Degrip WJ. Uniform stable-isotope labeling in mammalian cells: formulation of a cost-effective culture medium. Appl Microbiol Biotechnol. 2011 Jan;89(2):397-406.
[98] Saxena K, Dutta A, Klein-Seetharaman J, Schwalbe H. Isotope labeling in insect cells. Methods Mol Biol. 2012;831:37-54.
[99] Pickford AR, O'Leary JM. Isotopic labeling of recombinant proteins from the methylotrophic yeast Pichia pastoris. Methods Mol Biol. 2004;278:17-33.
[100] Pickford AR, Smith SP, Staunton D, Boyd J, Campbell ID. The hairpin structure of the (6)F1(1)F2(2)F2 fragment from human fibronectin enhances gelatin binding. Embo J. 2001 Apr 2;20(7):1519-29.
[101] Wood MJ, Sampoli Benitez BA, Komives EA. Solution structure of the smallest cofactor-active fragment of thrombomodulin. Nat Struct Biol. 2000 Mar;7(3):200-4.
[102] Metzler WJ, Bajorath J, Fenderson W, Shaw SY, Constantine KL, Naemura J, et al. Solution structure of human CTLA-4 and delineation of a CD80/CD86 binding site conserved in CD28. Nat Struct Biol. 1997 Jul;4(7):527-31.
[103] Kowarik M, Numao S, Feldman MF, Schulz BL, Callewaert N, Kiermaier E, et al. N-linked glycosylation of folded proteins by the bacterial oligosaccharyltransferase. Science. 2006 Nov 17;314(5802):1148-50.
[104] Saridakis E, Khurshid S, Govada L, Phan Q, Hawkins D, Crichlow GV, et al. Protein crystallization facilitated by molecularly imprinted polymers. Proc Natl Acad Sci U S A. 2011 Jul 5;108(27):11081-6.

Chapter 3

The Structural Assessment of Glycosylation Sites Database - SAGS – An Overall View on N-Glycosylation

Marius D. Surleac, Laurentiu N. Spiridon, Robi Tacutu, Adina L. Milac, Stefana M. Petrescu and Andrei-J Petrescu

Additional information is available at the end of the chapter

1. Introduction

The expansion of high-throughput technologies have led over the past decade to an unprecedented increase of the pace of data accumulation in biological science at molecular level. However this increase was not even - as while over 2000 genomes are already completed (Pagani *et al*, 2012) the vast majority of encoded proteins have not yet an assigned 3D structure and the functions of these proteins are still under investigation. While next-generation massive parallel sequencers are heading toward a 1 genome per day threshold, with significant impact on medical and environmental research (Sastre 2011; Shokralla *et al* 2012), the overall number of known protein structures remains well below 10^5 as their determination still relies on techniques such as crystallography and NMR which are time consuming, expensive and do not always work, despite the efforts made to automate the experimental flow in crystallization factories (Stuart *et al*, 2006) and the emergence of over 20 consortiums of Structural Genomics (Chandonia & Brenner, 2006) aiming at large scale 3D protein structure solving in Europe, United States and Japan.

Similarly in glycomics and glycoproteomics the gap between compositional and structural information has also increased. By combining techniques such as liquid chromatography, capillary electrophoresis and mass spectrometry significant advances were made in the last couple of years related to glycan profiling and the assessment of glycan heterogeneity and site occupancy (Liu *et al*, 2011; Mittermayr *et al*, 2011; North *et al*, 2010; Song *et al*, 2011; Zaia, 2010, Zauner *et al*, 2010). For example by pairing glycosylation site-specific stable isotope tagging of lectin affinity-captured N-linked glycopeptides with mass spectrometry a number of 1465 N-glycosylated sites were identified on 829 proteins expressed in Caenorhabditis elegans (Kaji *et al*, 2007). More recently, using a "filter aided sample preparation" (FASP) method in which glycopeptides are enriched by binding to lectins on the top of a filter, it was possible to map using mass spectrometry 6367 N-glycosylation sites on 2352 proteins in four mouse tissues and blood plasma (Zielinska *et al* 2010). By contrast the pace of data accumulation in structural glycobiology remained very low as less than 4% of the structural files deposited in the protein data bank (PDB) contain oligosaccharide structures.

In this context bioinformatics became crucial in the endeavor of interpreting the raw experimental data, in depositing, ordering and annotating the massive amount of accumulated information and also, when combined with molecular modeling, in getting more structural insights were experimental data is still unavailable.

For example to assist the interpretation of MS data in Glycobiology tools with different aims and flavors such as GlycoWorkbench, SysBioWare or SimGlycan were lately developed. GlycoWorkbench resulted from the EUROCarbDB initiative and was designed to assist the manual interpretation of MS data, hence this aims to evaluate user proposed structures by matching their theoretical list of fragment masses against the list of peaks derived from the MS spectrum (Ceroni *et al*, 2008). On the other hand SysBioWare performs isotopic grouping of detected peaks after de-noising and wavelet analysis and allows compositional assignment according to the tuned building block library (Vakhrushev *et al*, 2009); while SimGlycan predicts the glycan structure using a MS/MS database searching technique, and also facilitates novel glycans prediction by drawing a glycan and mapping it onto an experimental spectrum to check the degree of proximity between the theoretical and the experimental glycans (Apte & Meitei, 2010). Similarly other software platforms were developed for modeling the oligosaccharide composition starting from HPLC data - such as autoGU (Campbel *et al*, 2008) or from NMR Spectra - such as CASPER (Loss *et al*, 2006) or CCPN (Stevens *et al*, 2011). For depositing, ordering and annotating data in Glycobiology, a large number of web portals and frameworks appeared in the last couple of years such as KEGG (Hashimoto *et al*, 2006), CFG (Raman *et al*, 2006), GLYCOSCIENCES.de (Lütteke *et al*, 2006), RINGS (Akune *et al*, 2010) or UniCarbKB (Campbell *et al*, 2011). Finally, related to glycan and glycoprotein modeling we would mention here the results of Rob Woods group (Woods & Tessier, 2010) and the tools such as SWEET-II or GlyProt developed at GLYCOSCIENCES.de.

Structural information on glycans and glycoproteins is scarce. Intrinsic flexibility make oligosaccharides to be seldom resolved by crystallography as either they simply do not crystallize or they generate local disorder in glycoprotein crystals that make them unidentifiable in the electron density. In addition, the presence of glycoforms and glycan conformational heterogeneity usually prevent glycoprotein crystallization (Chang *et al*, 2007). Consequently structural glycoinformatics get relatively low attention and only recently a number of resources and services were developed such as GlycoCD (Lütteke, 2006), SAGS (Petrescu *et al*, 2006) or Glycoconjugate Data Bank (Nakahara *et al*, 2008).

We will concentrate here on SAGS - the Structural Assessment of Glycosylation Site Database (web: http://sags.biochim.ro) that contains information on N-linked oligosaccharides and structural properties of the protein core around N-glycosylation sites, and was used over the past decade to fulfill statistical analysis on structural aspects of protein glycosylation.

2. Redundant and non-redundant sets in SAGS for structural analysis

Historically SAGS was initiated around 1997 aiming to gather information related to the structure of N- and O-linked oligosaccharides starting from crystallographic data, in order to derive from this the general conformational rules governing the glycosidic linkages for use in glycoprotein model refinement (Wormald *et al*, 2002; Paduraru *et al*, 2006; Cioaca *et al*, 2011). Compared to NMR or theoretical methods, crystallography has the advantage that delivers a complete deterministic model of the oligosaccharide structure from the experimental data, yet this is static in nature and gives no clue on the flexibility of glycosidic linkages - as can be inferred from some NMR or theoretical models. However in ergodic conditions the statistical ensemble of the overall set of static structures found in protein data bank, PDB, is expected to be equivalent to the configurational space sampled during dynamics by any given glycan and thus it gives information on the flexibility of glycosidic linkages. This is why the overall nonredundant set of glycan structures found in PDB was used to gauge various glycosidic linkage configurations (Petrescu *et al*, 1999, Wormald *et al*, 2002).

Gradually the focus of the database has extended to cover also the assessment of structural properties of the protein core around N-glycosylation sites. These are critical in understanding the function of protein glycosylation in general, while in particular we were interested to see if from an overall view on glycosylation patterns some lessons could be retrieve related to site occupancy and GP folding which was our main interest at that stage. Whether N-glycans are used only to anchor glycoproteins to the quality control system during their folding and processing by the plethora of lectins, chaperones and enzymes (Petrescu *et al*, 1997, Zapun *et al*, 1997, Petrescu *et al*, 2000, Branza-Nichita *et al*, 2000), or do have a more direct effect onto the structure of the protein core is a fundamental problem in which a survey of the location of N-glycosylation sites in proteins is critical.

Nevertheless, investigating properties related to the (glyco)protein core implies a different approach from that used in glycosidic linkage evaluation. The overall set of glycoproteins found in PDB comprise in many cases multiple copies of the same protein or very close homologues crystallized or co-crystallized in various conditions, for instance there are no less than 50 copies of various variants of Hemagglutinin in PDB. While atomic level parameters such as the dihedral angles of various glycosidic linkages at a given site may vary drastically from crystal to crystal (Fig. 1) higher level properties such as the glycosylation site occupancy or accessibility, the sequence and secondary structure upstream and downstream a given site do not change so much and have to be counted only once in the statistics as otherwise multiple copies induce a bias into the statistics. To overcome this bias SAGS was provided with a system of clustering glycoproteins in general, according to their fold, but also the individual N-glycosylation sites.

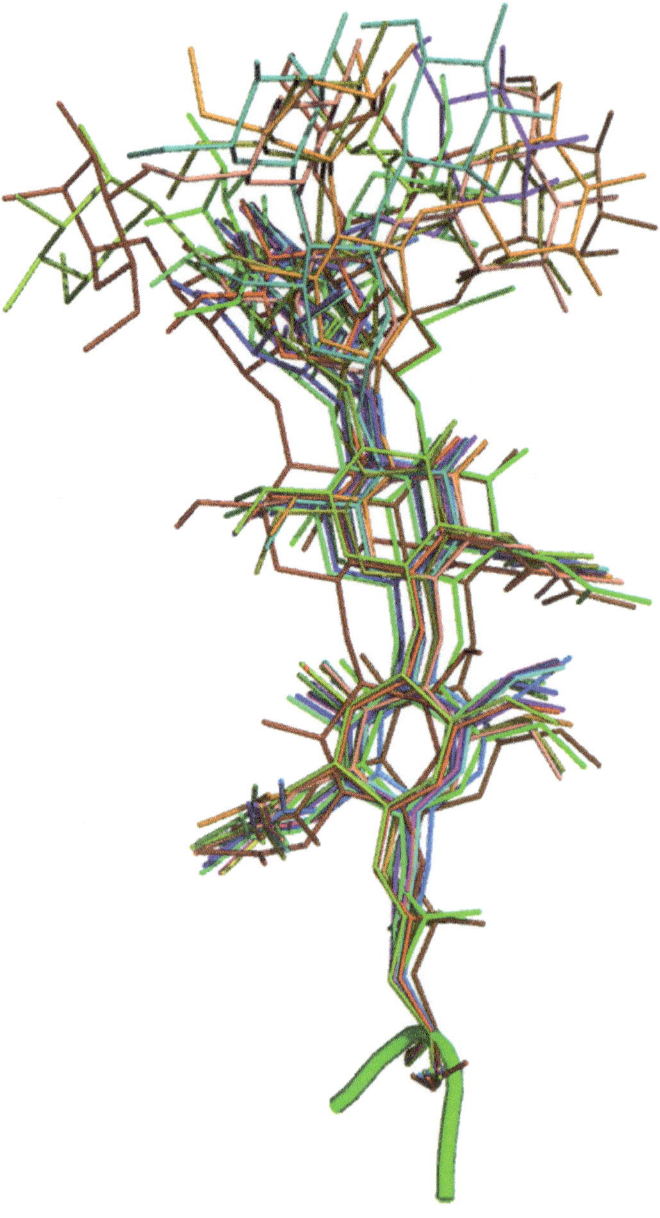

Figure 1. Structure variability of the glycan linked at N42 in all 19 crystal structures of the alpha subunit of the high affinity IG ε-receptor showing SAGS cluster ID: b.060.0040.ETT.h.12.

No simple N-site clustering system based entirely on either the sequence or the 3D structure was found to be adequate. For example very different sequences may adopt identical configurations, equally identical sequences may be found in significantly different configurations (Fig. 2). Tests have shown that an optimal separation between groups of similar N-sites is obtained by a two step procedure. First the sequence between -15 to +15 around a new N-site is compared to that of the existing clusters and in the case the identity is less than 60% the site is considered a new entry in SAGS. Secondly, if the identity is higher than 60% but lower than 90% a structural superposition onto the existing sites is further performed and again the site is retained as a new entry if the RMS to the rest of the sites is higher than 2Å, otherwise it is clustered to the closest group. Hence SAGS was automated to cluster groups of N-glycosylation sites that have over 60% sequence similarity and their structure is in less than 2Å from each other. Representatives for each cluster are further selected for representation and statistical analysis of higher scale properties based on non-redundant sets. Selection of the cluster representative relies on two criteria. First, sites with the highest number of glycosidic units are selected and secondly at equal glycan size the site in the crystal structure at the highest resolution is retained.

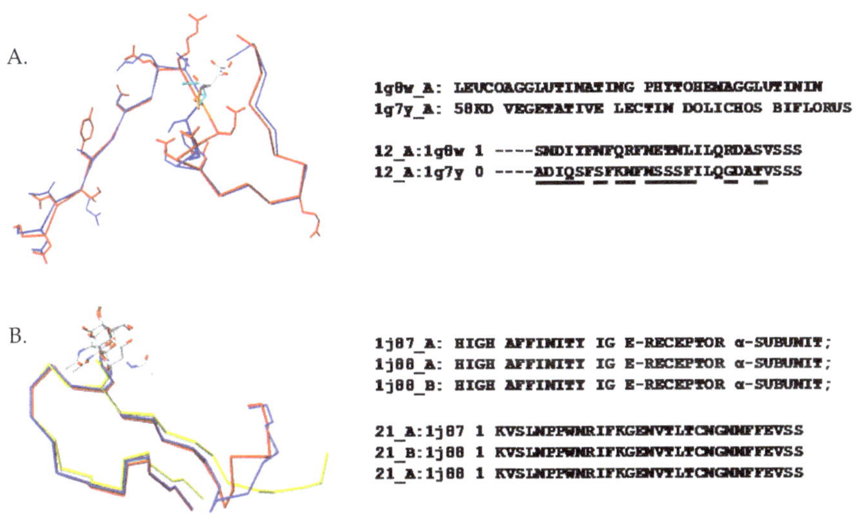

Figure 2. (A) Very different sequences can adopt almost identical structures.
12_A_1g8w - SAGS cluster ID: b.060.0120.CCG.h.53
12_A_1g7y - SAGS cluster ID: b.060.0120.CCS.m.55
(B) The same sequence can be found in different main chain configurations in various crystals.
SAGS cluster ID: b.060.0040.CCE.h.87

Each cluster is described by an identifier that captures the main structural properties of the group such that closely related groups to be close into the list. The identifier describes the class, architecture and topology of the glycosylated protein according the CATH taxonomy

(Greene *et al*, 2007), it then incorporates information on the secondary structure (ss) in positions N-1, N and N+1 based on DSSP (Kabsch & Sander, 1983) followed by a three state accessibility quantification: l, m, h (low, medium and high) and a counter. As CATH involves expert assessment of the protein fold and it is not automatically updated while SAGS is now fully automated we have had to introduce five states describing the class: a, b, c, d or x - corresponding to all alpha (a), all beta (b), alpha/beta (c), few secondary (d) and not yet assigned (x). An example is given in Table 1:

c.020.0020.HHH.m.36:							
class	architecture	topology	N-1_ss	N_ss	N+1_ss	accessibility	counter
alpha/beta	*barrel*	*Tim Barrel*	*Helix*	*Helix*	*Helix*	*medium*	*36*

Table 1. Example for the structural significance of an identifier in SAGS

Introduction of glycoprotein and N-site non-redundant sets with their identifiers was particularly instrumental not only in ordering the data, with consequences on a more accurate statistical assessment of various structural aspects of protein N-glycosylation, but also in generating and overall map of N-glycosylation onto the protein fold taxonomy. Even if ~40% of recently accumulated structures are yet to be assigned by CATH it is still striking to note that ~50% of all assigned glycoproteins in SAGS are of 'sandwich' type architecture. In particular, over 35% of all assigned structures are β-sandwiches which is two times more than their occurrence into the overall set of known protein structures present in CATH (Fig. 3). Therefore sandwich type architecture seems particularly fit to accommodate N-glycosylation either due to its prevalence in the secretome either due to its stability or other features favoring N-glycosylation. Further investigation is needed to address questions related to this striking prevalence.

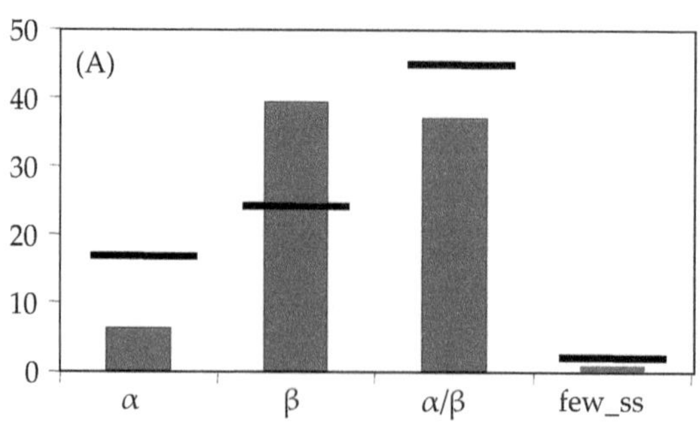

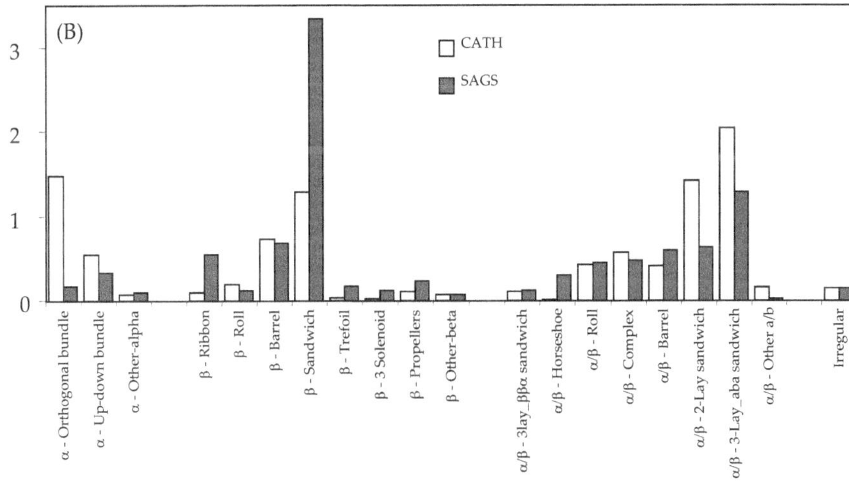

Figure 3. Occurrence of N-glycosylation into the four CATH classes (A) and more resolved architectures (B).

3. Evolution of SAGS and the robustness of the structural assessments over time

During the years SAGS increased steadily. For instance the first results on glycosidic linkage statistics rely on a set of only 639 representatives (Petrescu *et al* 1999); to date this set increased more than an order of magnitude to 9848.

In 2003, the analysis based on a non-redundant set of 386 entries revealed a number of patterns related to the properties of the sequence, secondary structure, shape and composition of the surface around occupied site with consequence on better understanding N-site occupancy and the relation between glycosylation and folding (Petrescu *et al* 2004). As the set was quite small it was important to assess how robust these findings would stand with data accumulation over the time. This was one of the main reasons that led to further developing SAGS and fully automate it for the foreseeable increase of structural calculations. Table 2 shows the evolution of N-glycosylation sites included in SAGS over the past decade.

	2003	2005	2007	2008	2010	2011	2012
PDB	23596	34140	47814	54778	70066	78157	82679
PDBs with N-glycans	711	1027	1436	1716	2006	2359	2605

	2003	2005	2007	2008	2010	2011	2012
% from PDB	*3.0*	*3.0*	*3.0*	*3.1*	*2.8*	*3.0*	*3.2*
glycosylated polypeptides	763	1443	2546	3010	3574	4349	4842
potential N-sites (sequons)	2219	5043	9157	11184	13895	16979	19164
occupied N-sites	1362	3013	5774	6985	8425	10142	11267
occupation [%]	*61*	*60*	*63*	*62*	*61*	*60*	*59*
non-redundant sequons	626	1055	1497	1825	2150	2580	2910
non-redundant N-sites	386	622	990	1184	1389	1666	1853
occupation [%]	*62*	*59*	*66*	*65*	*65*	*65*	*64*

Table 2. The evolution of the PDB data processed in SAGS over the past decade

As can be seen since 2003 the amount of data increased more than four fold. Interestingly, the main patterns proved surprisingly consistent over the time while new, more refined patterns shaped up as a result of the significant accumulation of data.

For example the ratio between NxT and NxS occupied sequons remained consistently 65% vs 35%. Similarly the occurrence of aromatic aminoacids in position N-2 and N-1 remained consistently in the range 15÷17% - namely 2-3 standard deviations (σ) over the expected percentage of ~10%; while the occurrence of acidic aminoacids in the same positions is >2σ lower than the 11% expected value, namely 5÷6%. Similar to the initial set the main deviations from the expected values remained located close to the N-site, in the region N-2 ÷ N+3. For the 2012 set these are shown in Fig. 4.

As the set of occupied sites increased over 4 fold more refined correlations are now possible to investigate. For instance when looking more closely by sequons types it is striking to see that in NxS sequons the occurrences of aromatic aminoacids raise to 19% in position N-2 (i.e. 4σ) and for acidic aminoacids these fall below 4% (i.e. 3σ) suggesting that the 'signature' for occupation was enhanced to compensate the presence of serine in position N+2. Similarly more subtle correlations between occurrence and accessibility shape up in occupied sites, e.g. in low accessibility sites aromatics are highly preferred in position N-1 (>3.6σ) while in highly accessible sites aromatics are far more frequent in position N-2 (>3.2σ).

Over the years the secondary structure statistics has also varied only marginally in SAGS, with a slight decrease of Turns, Bends and Coils in favor of β-structures that rise at glycosylation site (position N) from 20% to 27% (Fig. 5A). On the other hand these slight variations do not affect in any way the rate of change measured as the probability of having one type of structure in a given position followed by a different type for the next amino acid, suggesting that glycosylation remains strongly correlated with changes in secondary structure that make N-sites frequently landmarks for starting or ending regular secondary structure landmarks (Fig. 5B).

Turning now to the properties of the protein surface around N-glycosylation sites some of them remained remarkably stable while new other emerged during the analysis of the new 5 fold larger set of nonredundant entries in SAGS. For example the proportion of N-sites deeply buried in the surface remained surprisingly high as in the old, reduced set. Figure 6A shows the change in accessibility to a water molecule probe (1.5Å) when the glycan is taken into account at occupied sites. Here again, over 15% of the N-sites see their Asn side-chain accessibility reduced to less than 5%, showing that the glycan's first NAcGlc unit acts as part of the protein surface and completely obstructs the access of water to the Asn side-chain. Also very robust proved to be the distribution of contacts between the first two NAcGlc glycan units and the aminoacids brought close in space by the folding process. The percentage of contacts with aromatic aminoacids increased slightly to more than 3 fold than expected by chance (Fig. 6B). On the other hand as the non-redundant set and number of documented contacts increased significantly during time new details emerged from their analysis. For instance there are now 450 N-sites, from 1853, in which were identified 592 very close glycan-protein contacts, laying within less than 3Å. Interestingly over 30% of such hydrogen-bond contacts are formed with acidic aminoacids brought close in space by the folding process, which is twice the percentage expected by chance.

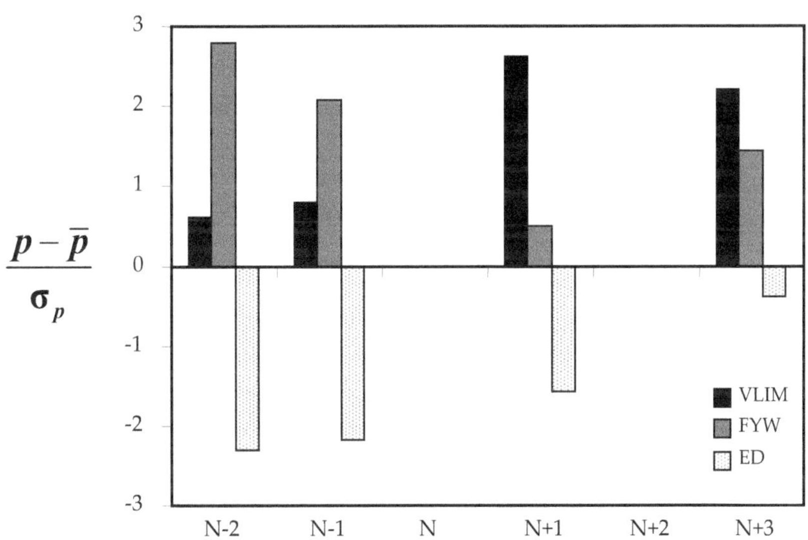

Figure 4. The main deviations from the expected occurrence around occupied N-glycosylation involve hydrophobic (VLIM), aromatic (F,Y,W) and acidic (E,D) aminoacids. These are shown here measured in standard deviations from the expected values.

A

B

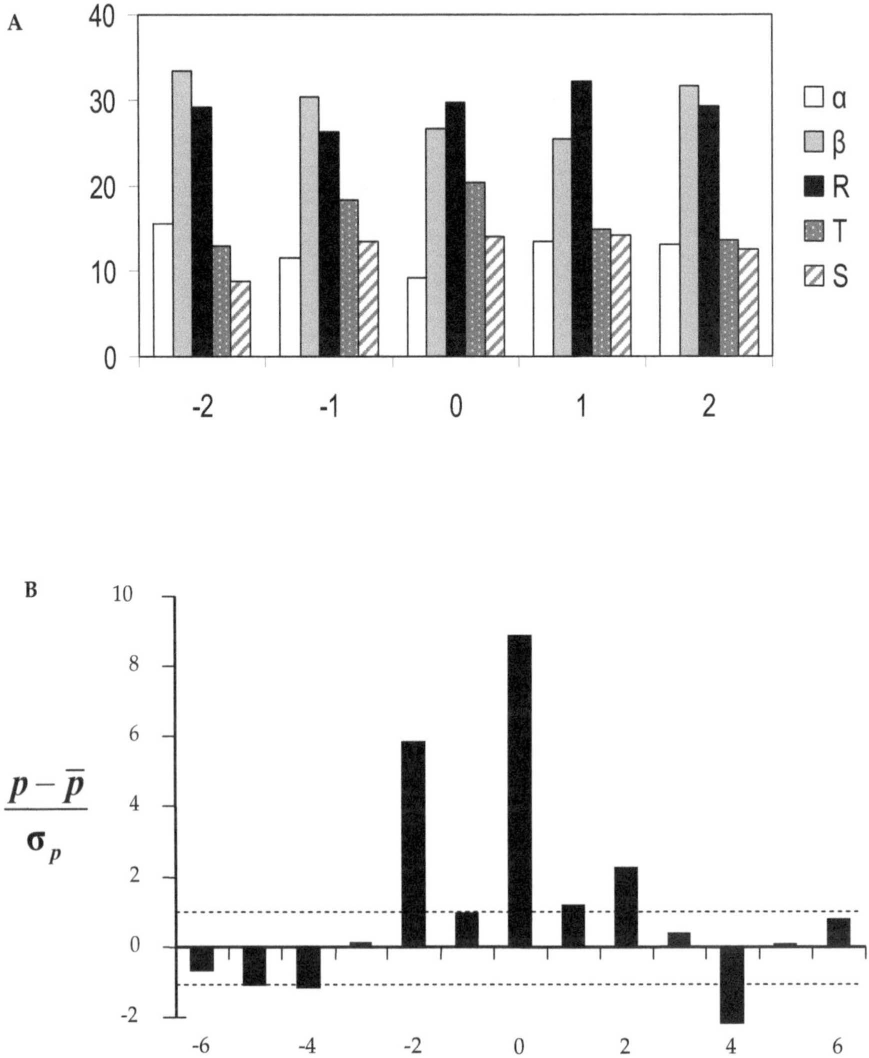

Figure 5. (A) Secondary structure distribution in positions N-2 to N+1.
(B) Rate of secondary structure change measured as the probability of having one type of structure in a given position followed by a different type for the next amino acid.

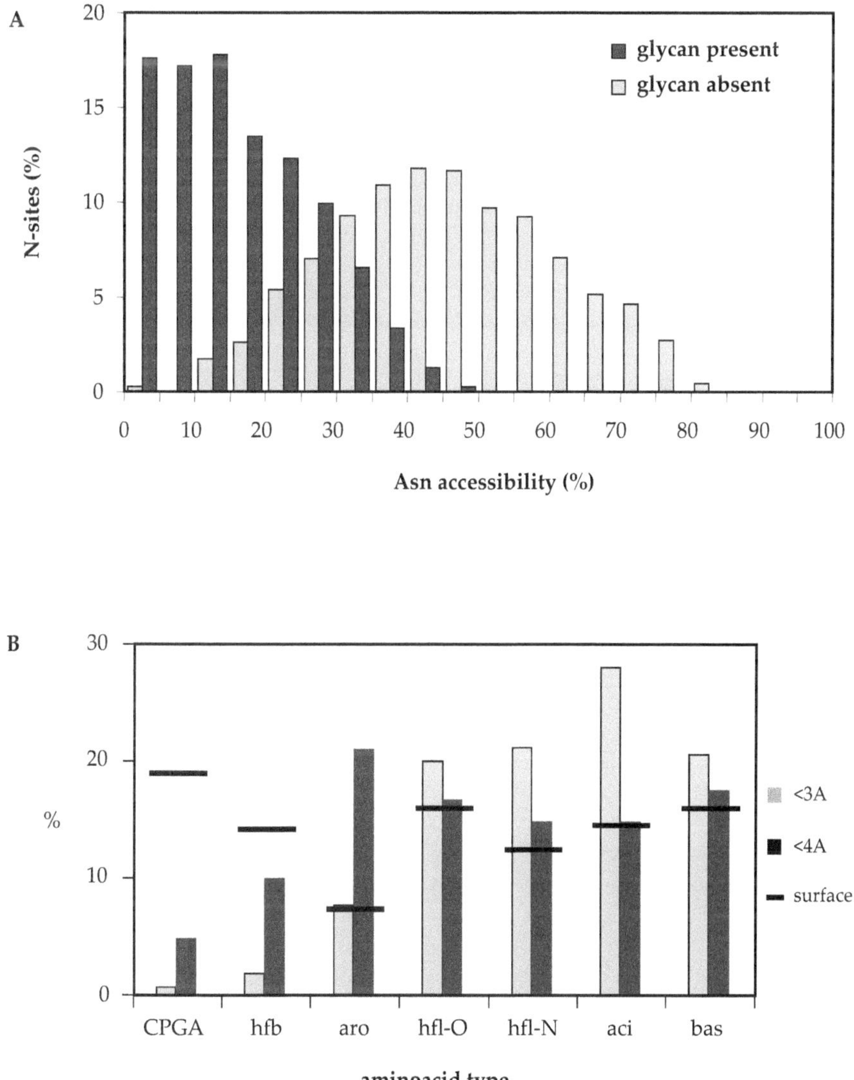

Figure 6. (A) Change of Asn side-chain accessibility over glycosylation.
(B) Percentage of glycan-protein contacts induced by folding.

The SAGS increase over time led also to a 10 fold increase of documented structures of glycans larger than the manosidic core. In the non-redundant set there are now available 29 complex glycan structures and 28 high mannose structures which makes now possible to investigate statistically not only the relation between N-glycosylation and folding but also if and how the structure of the protein core influence glycan processing into the Golgi.

These examples suggest that the analysis of ever increasing number of SAGS entries will soon make possible to develop more refined tools for glycoprotein modeling and predicting site occupancy based on the assessment of multiple sequence properties.

4. Significance and cross-validation of some results derived from SAGS analysis

Statistical analysis on SAGS suggests the existence of several sequence signatures that favor occupation. It is hence interesting to see if these correlate in any way with some different statistics performed on other data sets. Of particular importance is to use an as large as possible data set of sequences with known subcellular localization. On this line we found most useful LOCATE which curates all documented proteins sequences from mouse and human (Sprenger *et al*, 2008) and CYGD that comprise the documented localization in yeast (Güldener *et al*, 2005).

As cytosolic proteins are not subject to glycosylation, aminoacid occurrences is expected to be consistent with an independent probabilistic model while in the secretome or transmembrane proteins evolutionary pressures related to glycosylation, if any, are expected to influence the occurrences by inducing positive or negative correlations. Calculations are strikingly consistent with the statistical data from SAGS. Indeed in cytosolic proteins in all cases joint probabilities are within less than 5% from an independent probabilistic model. On the other hand within the secretome and membrane proteins significant correlations are shaping up. For example, in both secretome and TM proteins the occurrence of threonine in NxT sequences is 50% ÷ 60% higher than expected from independent probabilities in mouse and human respectively. Interestingly in yeast the deviation from independency is only 15%. Results on the occurrences of aromatic aminoacids in position N-2 and N-1 are even more striking. They reach an >80% increase in human secretome while in mouse and TM proteins the increase is >60%. In addition sequences of type Aro-N-x-S occur 70% (human) ÷ 80% (mouse) more frequently in the secretome than in the cytosol while for those of type Aro-N-x-T the difference in occurrence peaks even higher levels: 120% ÷ 150% in mouse and human respectively. In yeast these correlations still exists but again they are far less significant as they do not exceed 20-30%. Similar positive correlations shape up for hydrophobic bulky aminoacids (VLIM) in position N+1. These are on average 40% ÷ 60% more frequent than expected from independent probabilities in the secretome and 30%÷90% more frequent when compared to N(VLIM)S/T stretches in cytosolic proteins of mouse and humans. By contrast in yeast correlations are again dumped to less than 15%.

Hence an evolutionary pressure enhancing the occurrence of sequences known from SAGS to favoring occupation is obvious in the secretome and TM proteins. Besides this is far more significant in higher eukaryotes than in yeast.

N-site occupation is a direct consequence of the co- and post-translational interaction of the nascent polypeptide with the oligosaccharyltransferase (OST). Hence the yield of N-glycosylation at any given site depends on both the structure of OST and the sequence and local structure of the polypeptide at the time the transfer takes place.

In Eukaryotes OST is a multimeric complex located in the endoplasmic reticulum (ER), partly buried within the membrane of this compartment. Eukaryal OST transfers en-block a $Glc_3Man_9GlcNAc_2$ (G3M9) oligosaccharide from a dolichol pyrophosphate carrier to the side-chain amide nitrogen of only those asparagines located in sequons. The OST complex is composed from seven subunits of which two: the catalytic subunit and the thioredoxin-like subunit, come each in two isoforms: Stt3A/Stt3B and N33-Tusc3/IAP respectively. This results in four possible forms of the complex which are presumed to be optimized for co- vs. post-translational events and/or different types of substrates (Mohorko *et al*, 2011). In lower eukaryotes OST is a single-polypeptide membrane protein consisting alone in a Stt3 homologue (Maita *et al*, 2010) and similar monomeric Stt3 homologues performing OST functions were even identified in Bacteria and Archaea (Wacker *et al* 2002, Calo *et al*, 2010, Dell *et al*, 2010). In this diversity it is hence expected that peculiarities of each OST dictate the occupation rules, as indicated by the increasing number of studies reporting glycosylation at N-sites breaking the general accepted NxS/T sequon rule (Schwarz *et al*, 2011). Even in higher eukaryotes were the occupation control by OST is highly sophisticated deviations from the generally accepted NxS/T rules were recently reported (Schwarz & Aebi, 2011). Interestingly even the occurrence of N-glycosylation at an NPS site was recently identified in SAGS (ID: x.000.0000.ECC.m.77), in viral envelope GP160 (pdb code: 2qad-A362, Huang *et al*, 2007). This brings the absence of proline in position N+1 into the realm of statistical rare events, rather than a compulsory rule - as previously considered.

The mechanism by which OST transfers the G3M9 glycan to the asparagine is not yet fully understood but most likely this involves a nucleophilic attack to the amide nitrogen of the Asn side-chain. Historically there were two mechanisms proposed for this transfer both of which imply the involvement of β-OH group of the serine/threonine found in position N+2 of the sequon (NxS/T). However significant distance constraints have to be fulfilled in order that such an interaction between the amide of the Asn side-chain and the OH group of Ser/Thr to take place. These are satisfied only in a local geometry of the main chain consistent with β-turn or Asx turn configurations.

On this line SAGS provided a good opportunity to assess the proportion of sites in folded glycoproteins that actually fulfill the constraints imposed by the historic models of N-glycosylation. Both the distribution of N-sites in secondary structures and the distance measurements between the ND2 atom of Asn and the OG atom of Ser/Thr have shown that less than 20% of the N-sites found in SAGS are consistent with the two mechanisms. This indicates that either a different mechanism involving the nucleophilic attack provided by

OST is actually in place, or the polypeptide adopts transiently, at each site, a β-turn configuration before folding (Petrescu *et al,* 2006). Yet this later possibility is highly improbable in a process in which both the initial unfolded state and the final native state are locally far more extended than a β-turn, at over 80% of the N-sites.

It is only recently that the inference on the existence of a mechanism involving the intervention of OST in the nucleophilic attack was substantiated by the work performed at ETH in Zurich by the groups of Markus Aebi and Kaspar Locher (Lizak *et al,* 2011). From their seminal work on a bacterial homologue of OST, the PglB protein from *Campylobacter lari* - which was co-crystallized with an acceptor polypeptide - it results that the nucleophilic attack is facilitated by the aspartic acid D56 and glutamic acid E319 of Pg1B and not by the threonine in position N+2 of the glycosylated polypeptide. The crystal structure shows also, in agreement with what is an expected state along the folding pathway, that the polypeptide complexed to Pg1B is in fact in an extended configuration. The structure also indicates that the threonine/serine in position N+2 lays at the interface between the periplasmic and TM domain of Pg1B and that its side chain β-OH group is anchored by hydrogen bonds to the WWD motif 462-464 of Pg1B. In addition threonine (N+2) based complexes are further stabilized by hydrophobic interactions with I572 and V575 of Pg1B, as their side-chains are in contact with threonine gamma carbon (CG2) which in addition prevents free rotation of its side chain. Conversely these stabilization factors lack in serine (N+2) based complexes due to the absence of the gamma carbon. The additional stability induced by CG2 methyl group might explain the higher transfer yield to threonine as compared to serine sequons, shown by SAGS.

As seen many aspects of the statistics preformed on SAGS have a firm structural basis and the details of higher eukaryote OST structure, which are not yet known, will probably explain the other aspects of occupation signatures.

Author details

Marius D. Surleac, Laurentiu N. Spiridon, Robi Tacutu,
Adina L. Milac, Stefana M. Petrescu and Andrei-J Petrescu*
Institute of Biochemistry of the Romanian Academy, Bucharest, Romania

Acknowledgement

This work was supported by the Romanian Academy Research Plan, CNCS grant PCE-ID-3-0342/2011 and the POSDRU/89/1.5/S/60746 Program.

5. References

Akune Y, Hosoda M, Kaiya S, Shinmachi D, Aoki-Kinoshita KF. *The RINGS resource for glycome informatics analysis and data mining on the Web. OMICS.*,14(4), 475-486 (2010)

* Corresponding Author

Apte A, Meitei NS. *Bioinformatics in glycomics: glycan characterization with mass spectrometric data using SimGlycan. Methods Mol Biol.* 600, 269-281 (2010)

Branza-Nichita N, Petrescu A-J, Negroiu G, Dwek RA, Petrescu SM, *N-glycosylation processing and glycoprotein folding-lessons from the tyrosinase- related proteins, Chem. Rev.*, 100, 4697-4711 (2000)

Calo D, Kaminski L, Eichler J, *Protein glycosylation in Archaea: sweet and extreme, Glycobiology,* 20(9), 1065–1076 (2010).

Campbell MP, Hayes CA, Struwe WB, Wilkins MR, Aoki-Kinoshita KF, Harvey DJ, Rudd PM, Kolarich D, Lisacek F, Karlsson NG, Packer NH. *UniCarbKB: putting the pieces together for glycomics research. Proteomics.* 11(21), 4117-21 (2011)

Campbell MP, Royle L, Radcliffe CM, Dwek RA, Rudd PM. *GlycoBase and autoGU: tools for HPLC-based glycan analysis. Bioinformatics,* 24, 1214–1216 (2008)

Ceroni A, Maass K, Geyer H, Geyer R, Dell A, Haslam SM. *GlycoWorkbench: a tool for the computer-assisted annotation of mass spectra of glycans. J Proteome Res.* 7(4):1650-1659 (2008)

Chandonia JM, Brenner SE. "*The impact of structural genomics: expectations and outcomes*", *Science* 311, 347-351 (2006)

Chang VT, Crispin M, Aricescu AR, Harvey DJ, Nettleship JE, Fennelly JA, Yu C, Boles KS, Evans EJ, Stuart DI, Dwek RA, Jones EY, Owens RJ, Davis SJ. *Glycoprotein structural genomics: solving the glycosylation problem.* Structure, 15(3), 267-273 (2007).

Cioaca D, Ghenea S, Spiridon LN, Marin M, Petrescu AJ, Petrescu SM. *C-terminus glycans with critical functional role in the maturation of secretory glycoproteins. PLoS One.* 6(5):e19979, 2011.

Dell A, Galadari A, Sastre F, Hitchen P. *Similarities and differences in the glycosylation mechanisms in prokaryotes and eukaryotes. Int J Microbiol.* 2010,148178 (2010)

Greene LH, Lewis TE, Addou S, Cuff A, Dallman T, Dibley M, Redfern O, Pearl F, Nambudiry R, Reid A, Sillitoe I, Yeats C, Thornton JM, Orengo CA, *The CATH domain structure database: new protocols and classification levels give a more comprehensive resource for exploring evolution. Nucleic Acids Research* 35, D291-D297 (2007).

Güldener U, Münsterkötter M, Kastenmüller G, Strack N, van Helden J, Lemer C, Richelles J, Wodak SJ, Garcia-Martinez J, Perez-Ortin JE, Michael H, Kaps A, Talla E, Dujon B, Andre B, Souciet JL, De Montigny J, Bon E, Gaillardin C, Mewes HW. *CYGD: the Comprehensive Yeast Genome Database. Nucleic Acids Res,* 33, D364-D368 (2005)

Hashimoto K, Goto S, Kawano S, Aoki-Kinoshita KF, Ueda N, Hamajima M, Kawasaki T, Kanehisa M. *KEGG as a glycome informatics resource. Glycobiology.* 16(5), 63R-70R (2006)

Huang CC, Lam SN, Acharya P, Tang M, Xiang SH, Hussan SS, Stanfield RL, Robinson J, Sodroski J, Wilson IA, Wyatt R, Bewley CA, Kwong PD. *Structures of the CCR5 N terminus and of a tyrosine-sulfated antibody with HIV-1 gp120 and CD4. Science.* 317(5846), 1930-1934 (2007)

Kabsch W, Sander C, *Dictionary of protein secondary structure: pattern recognition of hydrogen-bonded and geometrical features. Biopolymers.* 22, 2577-2637 (1983).

Kaji H, Kamiie J, Kawakami H, Kido K, Yamauchi Y, Shinkawa T, Taoka M, Takahashi N, Isobe T. *Proteomics reveals N-linked glycoprotein diversity in Caenorhabditis elegans and suggests an atypical translocation mechanism for integral membrane proteins. Mol Cell Proteomics.* 6(12), 2100-2109 (2007)

Liu L, Telford JE, Knezevic A, Rudd PM. *High-throughput glycoanalytical technology for systems glycobiology. Biochem Soc Trans.* 38(5), 1374-1377 (2010).

Loss A, Stenutz R, Schwarzer E, von der Lieth CW. *GlyNest and CASPER: two independent approaches to estimate 1H and 13C NMR shifts of glycans available through a common web-interface. Nucleic Acids Res.* 34:W733-W737 (2006)

Lütteke T, Bohne-Lang A, Loss A, Goetz T, Frank M, von der Lieth CW. *GLYCOSCIENCES.de: an Internet portal to support glycomics and glycobiology research. Glycobiology.* 16(5):, 71R-81R (2006)

Lizak C, Gerber S, Numao S, Aebi M, Locher KP. *X-ray structure of a bacterial oligosaccharyl-transferase. Nature* 474(7351), 350-355 (2011)

Maita N, Nyirenda J, Igura M, Kamishikiryo J, Kohda D. *Comparative structural biology of eubacterial and archaeal oligosaccharyltransferases. J Biol Chem.* 285(7), 4941-4950 (2010)

Mittermayr S, Bones J, Doherty M, Guttman A, Rudd PM. *Multiplexed analytical glycomics: rapid and confident IgG N-glycan structural elucidation. J Proteome Res.* 10(8), 3820-3829 (2011).

Mohorko E, Glockshuber R, Aebi M. *Oligosaccharyltransferase: the central enzyme of N-linked protein glycosylation. J Inherit Metab Dis.* 34(4), 869-878 (2011)

Nakahara T, Hashimoto R, Nakagawa H, Monde K, Miura N, Nishimura S. *Glycoconjugate Data Bank: Structures--an annotated glycan structure database and N-glycan primary structure verification service. Nucleic Acids Res.* 36, D368-71 (2008).

North SJ, Hitchen PG, Haslam SM, Dell A. *Mass spectrometry in the analysis of N-linked and O-linked glycans. Curr Opin Struct Biol.* 19(5), 498-506 (2009).

Pagani I, Liolios K, Jansson J, Chen IM, Smirnova T, Nosrat B, Markowitz VM, Kyrpides NC. *The Genomes OnLine Database (GOLD) v.4: status of genomic and metagenomic projects and their associated metadata. Nucleic Acids Res.*, 40, D571-9 (2012)

Paduraru C, Spiridon L, Yuan W, Bricard G, Valencia X, Porcelli S, Besra G, Petrescu SM, Petrescu A-J, Cresswell P. "*An N-linked glycan modulates the interaction between the CD1d heavy chain and beta 2-microglobulin.*", *J Biol Chem.*, 281(52), 40369-78 (2006)

Petrescu A-J, Butters TD, Petrescu SM, Platt FM, Dwek RA, Wormald MR, *The solution NMR structure of Glc3Man9 unit in Glc3Man7GlcNAc2, Embo J.* 16, 4302-4310 (1997)

Petrescu A-J, Wormald MR, Dwek RA."*Structural aspects of glycomes with a focus on N-glycosylation and glycoprotein folding.*", *Curr Opin Struct Biol.* 16(5): 600-607 (2006)

Petrescu A-J, Milac A-L, Petrescu SM, Dwek RA, Wormald M.R."*Statistical analysis of the protein core around N-glycosylation sites. Implications on occupancy, folding and function*", *Glycobiology*, 14: 103-114 (2004)

Petrescu A-J, Petrescu S.M., Dwek R.A., Wormald M.R., "*A Statistical Analysis of N- and O-glycan linkages from crystallographyc data*" *Glycobiology*, 9, 343-352 (1999)

Petrescu SM, Branza-Nichita N, Negroiu G, Petrescu A-J, Dwek RA, *Tyrosinase and glycoprotein folding: roles of chaperones that recognize glycans, Biochemistry*, 39, 5229-5237 (2000)

Raman R, Venkataraman M, Ramakrishnan S, Lang W, Raguram S, Sasisekharan R *Advancing glycomics: Implementation strategies at the consortium for functional glycomics. Glycobiology* 16, 82R–90R (2006)

Sastre L., *New DNA sequencing technologies open a promising era for cancer research and treatment. Clin Transl Oncol.* 3(5):301-6 (2011).

Schwarz F, Aebi M. *Mechanisms and principles of N-linked protein glycosylation. Curr Opin Struct Biol.* 21(5), 576-82 (2011).

Schwarz F, Lizak C, Fan YY, Fleurkens S, Kowarik M, Aebi M. *Relaxed acceptor site specificity of bacterial oligosaccharyltransferase in vivo. Glycobiology.* 21(1), 45-54 (2011).

Shokralla S, Spall JL, Gibson JF, Hajibabaei M. Next-generation sequencing technologies for environmental DNA research. Mol Ecol. 21(8):1794-805 (2012).

Song W, Henquet MG, Mentink RA, van Dijk AJ, Cordewener JH, Bosch D, America AH, van der Krol AR. *N-glycoproteomics in plants: perspectives and challenges. J Proteomics.* 74(8), 1463-1474 (2011).

Sprenger J, Lynn Fink J, Karunaratne S, Hanson K, Hamilton NA, Teasdale RD, *LOCATE: a mammalian protein subcellular localization database. Nucleic Acids Res.* 36, D230-D233 (2008).

Stevens TJ, Fogh RH, Boucher W, Higman VA, Eisenmenger F, Bardiaux B, van Rossum BJ, Oschkinat H, Laue ED. *A software framework for analysing solid-state MAS NMR data. J Biomol NMR.* 51(4), 437-47 (2011)

Stuart DI, Jones EY, Wilson KS, Daenke S "*SPINE: Structural Proteomics IN Europe - the best of both worlds*", *Acta Cryst*, D62, 10 (2006).

Vakhrushev SY, Dadimov D, Peter-Katalinić *Software platform for high-throughput glycomics. J.Anal Chem.* 81(9), 3252-3260 (2009)

Wacker M, Linton D, Hitchen PG, Nita-Lazar M, Haslam SM, North SJ, Panico M, Morris HR, Dell A, Wren BW, Aebi M. *N-linked glycosylation in Campylobacter jejuni and its functional transfer into E. coli. Science.* 298(5599):1790-1793 (2002).

Woods RJ, Tessier MB. *Computational glycoscience: characterizing the spatial and temporal properties of glycans and glycan-protein complexes. Curr Opin Struct Biol.* 20(5), 575-83 (2010)

Wormald M, Petrescu A-J, Pao Y-L, Glythero A, Elliot T, Dwek RA, "*Conformational Studies of Oligosacharides and Glycopeptides: Complementarity of NMR, X-Ray Crystallogrphy and Molecular Modelling*", *Chem.Rev.*, 102, 371-387 (2002)

Zaia J. *Mass spectrometry and glycomics. OMICS* 14(4), 401-418 (2010).

Zapun A, Petrescu SM, Rudd PM, Dwek RA, Thomas DY, Bergeron JJM, *Conformation - independent binding of monoglucosylated ribonuclease B to calnexin, Cell*, 88, 29-38 (1997)

Zauner G, Deelder AM, Wuhrer M. *Recent advances in hydrophilic interaction liquid chromatography (HILIC) for structural glycomics. Electrophoresis.* 32(24), 3456-3466. (2011).

Zielinska DF, Gnad F, Wisniewski JR, Mann M, *Precision Mapping of an In Vivo N-Glycoproteome Reveals Rigid Topological and Sequence Constraints, Cell* 141, 897–907 (2010).

Chapter 4

Unravelling Glycobiology by NMR Spectroscopy

Vitor H. Pomin

Additional information is available at the end of the chapter

1. Introduction

1.1. NMR spectroscopy: Approaching glycobiology

Generally speaking, nuclear magnetic resonance (NMR) spectroscopy seems the most powerful technique in current use for structural analysis of biomolecules. Four Nobel prizes have been awarded so far due to discoveries related to NMR: 1952 (Physics) to Felix Bloch and Edward Mills Purcell for explanations of the physical properties of nuclei under magnetization; 1991 (Chemistry) to Richard Ernest for the development of the principles for the multidimensionality in NMR spectroscopy; 2002 (Chemistry) to Kurt Wüthrich for applying NMR in structural determination of biomolecules, mainly proteins; and 2003 (Medicine) for Paul Lauterbur and Peter Mansfield to the discoveries concerning the use of magnetic resonance imaging in medical diagnostics. The significant boom of NMR spectroscopy in structural biology however dates from the beginning of 80´s, mainly due to the implementation of two-dimensional techniques associated with advances in instrumentation and *in vivo* or *in vitro* methods for making suitable samples for NMR analysis (especially labeling procedures of NMR-active nuclei) [1].

Particularly, we could generalize that in the following two decades, proteins and nucleic acids were the primary biomolecule types in NMR studies. This was somewhat related to the usual 3D structures these molecules may present in solution. This consequently facilitates the detection of valuable spatial contacts between residues that are displaced away in a polymeric chain. The achievements by NMR in studies of proteins and nucleic acids were also significant to push the genome and proteome projects. Although, carbohydrates were also analyzed by NMR at this period, the association between glycobiology and NMR was somewhat neglected. For sure, this has happened because of the general idea of high structural complexity of carbohydrates combined with their high-order dynamic properties (high molecular flexibility). The conception of an unclear or absent three-dimensional states for carbohydrate molecules was one of the greatest reasons to this limited attention.

However, persistent research groups worldwide have been proving otherwise and showed to the community interesting NMR results concerning glycans. These results have helped to unravel the biological roles of glycans. Above all, since that time, glycobiology turned to be a fertile field for NMR spectroscopy. Due to the current glycomics′ boom, the high number of vital biological roles triggered by carbohydrate molecules, and the advances in NMR methods, NMR and glycobiology are now a very promising combination. Hence, in this chapter, we want to illustrate this new trend. For this, we cover the basis of the major solution NMR methods in glycobiology, taking examples from our experiments or from published works. We offer to the readership the adequate conclusions of these NMR data and moreover, what they represent for explaining the physicochemical mechanisms of the glycans addressed.

2. The essential high-throughput NMR methods exploited in glycobiology

NMR spectroscopy has a special position compared to the other analytical techniques once it provides a variety of experiment types to be performed as many as structural features to investigate in a biologically relevant macromolecule. Moreover, the NMR studies can be performed in solution which can mimic the physiological conditions. The major disadvantages in application of NMR spectroscopy are that of sensitivity and limitation in high-molecular mass systems. In order to overcome respectively these two obstacles, methods of *in vitro* or *in vivo* isotopic labeling with specific magnetically active nuclear spins (principally ^{13}C and ^{15}N) [2,3] and new pulse sequences to specifically reduce losses from transverse relaxation mechanisms [4-6] were developed along the years. These methods, established primarily to proteins and nucleic acids [4-9], are now being used to study carbohydrates and glycoconjugates as well [10,11].

Among many NMR techniques to be applied in glycobiology, we shall explain the most basic and utilized ones, such as 1) information from different NMR-active isotopes (^{1}H, ^{13}C, ^{15}N); 2) cross-peak assignments through multi-dimensional NMR spectra for tabulating chemical shift values or structural characterization purposes; 3) dynamic studies by measurements of longitudinal or spin-lattice (T_1) and transversal or spin-spin (T_2) relaxation rates (R_1 and R_2, respectively, in ms to s), and of line-widths or line broadening (in Hz) in certain NMR timescale; 4) chemical shift changes (in ppm) that may reflect conformational changes in carbohydrates due to temperature variations, or in carbohydrate-binding proteins in the presence of ligants where localized chemical shift perturbation is induced by presence of intermolecular contacts; 5) measurements of constants of either scalar (*J*) or dipolar (*D*) couplings (in Hz) and their contribution to the averaged conformational shape of carbohydrates or oligosaccharides, and 6) NOE-based though space contacts for determination of sugar preferential conformational shapes in solution or in studies of intermolecular complexes, especially those of carbohydrate-protein interactions. The NMR data presented herein is discussed with the primary objective to help the readership in NMR data interpretation and thus conclusions about the physicochemical properties of certain glycans in solution.

2.1. The NMR-active isotopes mostly used in carbohydrate studies

The three isotopes with magnetically active nuclei spin-½ mostly studied in glycobiology NMR are ^{1}H, ^{13}C, and ^{15}N. Each one has its own magnetic relative susceptibility (Table 1), its own precessional frequency (Larmor frequencies at 800 MHz for ^{1}H, for example, in Table 1), relaxation properties, and principally differential atomic positions within a molecule. These differenced localizations ultimately provide useful information concerning structural features, in the atomic perspective, and dynamic properties, of specific regions or sites within a molecule. In dynamic studies, localized motions as well as global motions of the analyzed molecules can also be evaluated, depending on the atom or group of atoms that have to be examined. It is worth mentioning that the relative NMR receptivity (dependent on the isotopomeric abundance, relaxation and magnetogyric ratio) of these three biomolecularly abundant NMR-active isotopes is quite different, which leads to different NMR sensitivity. While ^{1}H has its value set as 1.0, ^{13}C and ^{15}N have their values as $1.76x10^{-4}$, and $3.85x10^{-6}$ (Table 1).

Nuclide	Spin	Natural abundance	Gyromagnetic ratio γ [10^{7} rad T^{-1} s^{-1}]	NMR Frequency (at 18.8 Tesla)	Relative receptivity
Proton (^{1}H)	½	99.985	26.7522	799.734 (1)	1.00
Carbon-12 (^{12}C)	0	98.9	-	-	-
Carbon-13 (^{13}C)	½	1.108	6.7283	201.133 (1/3.976)	$6.73x10^{-7}$
Nitrogen-14 (^{14}N)	1	99.63	1.9338	57.820 (1/13.831)	$1.00x10^{-3}$
Nitrogen-15 (^{15}N)	½	0.37	-2.7126	81.093 (1/9.861)	$3.85x10^{-6}$

Table 1. Some nuclei properties important for NMR detection.

In principle, one would say it would be one of the greatest luck for NMR spectroscopists, the existence of biomolecules, including carbohydrates, rich in sensitive and active isotopes for NMR studies (Table 1). This is true if we consider the essential magnetic properties and abundance of hydrogen atoms in such molecules. Coincidently or not, in proteins, nucleic acids, lipids, and not differently in carbohydrates, hydrogen is not the most abundant atom, but it also fundamentally participates directly (by physical contacts) in many of the biological reactions, either through hydrogen-bond networks in protonated states during binding events, or even indirectly because of its absence in deprotonated states during pH-changes occurred during physiological reactions. Therefore, one would guess that ^{1}H-atoms would be the most used isotope in NMR analysis, and this is exactly what happens not only for glycomics but also for lipidomics, proteomics, and genomics. The changes and profiles of

^{1}H-signal distribution are informative in terms of molecular states, reactions, and dynamics, as exemplified by the simple mutarotation changes of the anomeric hydrogens of reducing terminal sugars in solution as a function of time (Figure 1). For example, only a simple structural change at the C2 of the glucose (Glc)-derived monosaccharides is enough to cause kinetic alterations of the ^{1}H-anomerics equilibrium in solution. The intensities of the ^{1}H-peaks properly indicate the proportion of these protons within a molecule, and therefore they are extremely useful to estimate the atoms and the percentage of each enantiomeric form (percentages in Figure 1). Actually, just after mass spectrometry (MS) techniques, measurements based on integrals of ^{1}H-signals seem to be the most reliable quantitative procedure to determine the number of atoms within a molecule, as well as their conformational states.

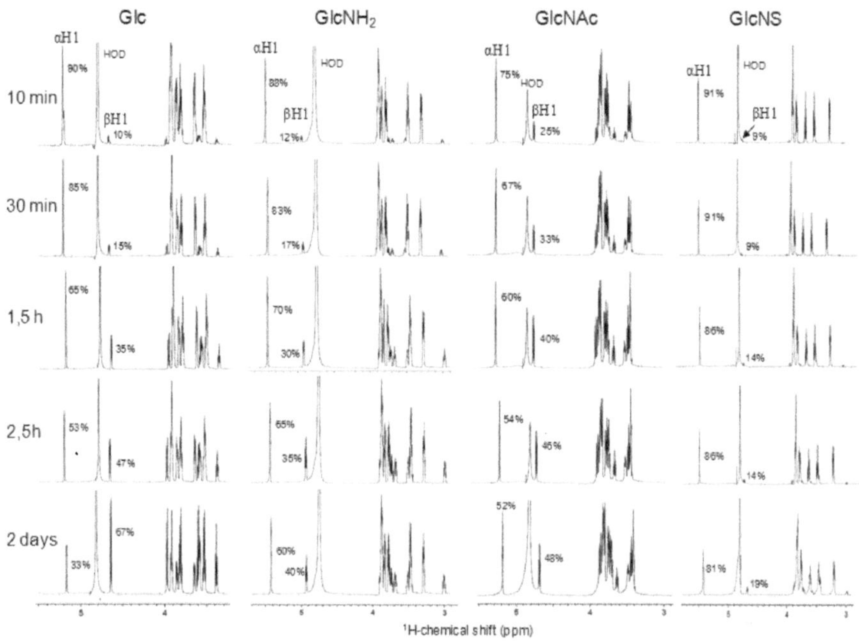

Figure 1. 1D ^{1}H-NMR spectra of glucose (Glc), and its derivatives: glucosamine ($GlcNH_2$), *N*-acetylglucosamine (GlcNAc), and N-sulfated glucosamine (GlcNS), all in ~100% D_2O-solution. The spectra were recorded in a Varian 800 MHz spectrometer at 25° C and at different time courses after dissolution, as indicated in the left of the panel. The percentage of the α- and β-^{1}H-anomerics changes as a function of time at different rates accordingly to the monosaccharide type, solvent and temperature. The water (HOD) and anomeric signals are indicated at the top of the panel.

The profile of ^{1}H-signals in 1D ^{1}H-NMR spectra offers a general fingerprint of the biomolecule in a solution such as the degree of pureness crucial for carbohydrates with biomedical purposes. However, due to the large density of protons in biomolecules, including carbohydrates, sometimes the ^{1}H-NMR profile is overwhelming because of the many condensed, overlapped and thus unresolved signals. Therefore, the use of other nuclei such as ^{13}C and ^{15}N, either coupled or uncoupled to ^{1}H, becomes valuable and complementary in the analyses, resulting important conclusions with respect the physicochemical properties of carbohydrates. For example, the intensity of ^{13}C-signals of anomerics can also be diagnostic for the changing rates of the α- and β-states during mutarotation in solution, as demonstrated below. The directly observed ^{13}C-signals are frequently much sharper than ^{1}H-signals due to its differential relaxation property. The thinner ^{13}C-signals in 1D NMR profile result usually more well-resolved peaks, and through the unidimensional scale, less superimpositions should occur. This is exactly what happens in structural analyses of algal sulfated polysaccharides with high degrees of structural heterogeneities [12-14]. ^{1}H-coupled ^{13}C-signals are very diagnostic of the methine (CH), methylene (CH_2) and methyl (CH_3) groups highly abundant in carbohydrate molecules, and thus quite useful in NMR structural glycobiology. In case of carbohydrates or glycosylated molecules that contain amino sugars, like the glycosaminoglycans (GAGs) that bear hexosamines with ^{15}N-natural abundant amino groups, the ^{15}N-related NMR signals although few in number are well-resolved and still quite useful for structural diagnosis [10]. Hence, hetero-atoms are very useful in solution NMR structural glycobiology through multi-dimensional heteronuclear experiments, or in direct observe experiments, as discussed next to describe the assignments of glucose and the novel method for characterizing GAG molecules through ^{15}N-atoms.

2.2. Peak assignments through multi-dimensional NMR spectra of glycans

Although ^{1}H-NMR spectra are quite informative in terms of the general view of the molecule and diagnostic for the presence of contamination, fully protonated molecules with high-molecular weights, and with a certain degree of heterogeneity, result in very complex ^{1}H-NMR spectra. This can be seen by the presence of many unresolved peaks from polysaccharides with no clear pattern of structural regularity. In general, in glycan analysis, except the ^{1}H-signals belonging to the anomerics that resonate at the most downfield region of the spectra (usually located somewhere between 4.5 and 6.0 ppm), all other ring protons from the most monosaccharide types resonate very squeezedly between 3.0 and 4.5 ppm. This can be seen through the 1D ^{1}H-NMR spectrum in solution of the simple and most common monosaccharide, the glucopyranose (Glc*p*) (Figure 2B) after reaching the anomeric equilibrium in solution (Figure 1). The low chemical shift dispersion (also called chemical shift degeneracy) typical of carbohydrates makes assignments of individual peaks hard of full accomplishment, and thus the acquisition of 2D homonuclear NMR spectra becomes essential for further structural ^{1}H-assignments. The combination of some 2D homonuclear ^{1}H-based spectra together with the 1D ^{1}H-NMR facilitates resonance assignments as exemplified with the acquisition and annotation of the correlation spectroscopy (COSY) and total correlation spectroscopy (TOCSY) spectra of glucose (Figure 3). This procedure allows the proper full assignment of all chemical shifts that belong to non-exchangeable protons (in black in structures of Figure 2A).

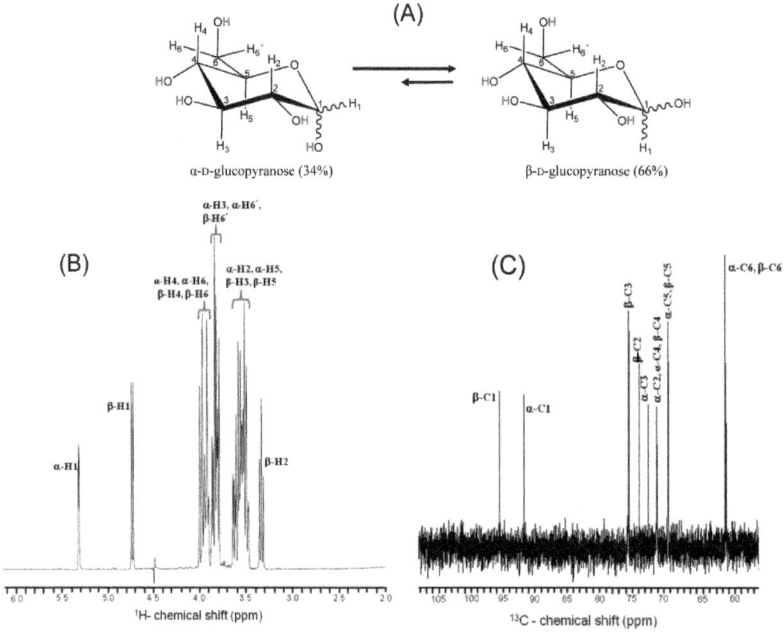

Figure 2. (**A**) Structural representation of D-glucose and its anomeric configurations (α- and β-forms) in aqueous solution. The numbers of the hydrogen and carbon atoms are indicated by digits, and the solvent exchangeable protons are red-colored. (**B**) ^{1}H- and (**C**) ^{13}C-NMR 1D spectra, Bruker 400 MHz and 37° C of D-glucose in solution after waiting to reach the anomeric equilibrium (2 days as shown at Figure 1). The signals are labeled according to the atom numbers and anomeric forms designated in the molecules represented on panel A.

Observing either the 1D ^{1}H-NMR spectrum (Figure 2B), or the 2D ^{1}H-based NMR spectra (Figure 3A, and 3B), it is clear that the anomeric ^{1}H-signals at the downfield region of the suppressed water (HOD) peak at ~4.50 ppm are well-resolved, and thus they may serve as good starting-points to trace connectivities using 2D-NMR spectra. Usually, α-^{1}H-anomeric signals resonate more downfield than β-ones in the majority of the carbohydrate units. In the case of glucose, the α-^{1}H signals resonate exactly at 5.32 ppm, while the β-^{1}H signals resonates exactly at 4.74 ppm (Table 2).

Since COSY experiments create ^{1}H-^{1}H cross-peaks within $^{3}J_{H(n)\text{-}H(n+1)}$ (scalar couplings of protons of three-bond distances) like those observed for $^{3}J_{H1\text{-}H2}$, $^{3}J_{H2\text{-}H3}$, $^{3}J_{H3\text{-}H4}$, (Figure 3A), most of the ^{1}H-chemical shifts (in ppm) for each type of anomeric form of Glc can be assigned (Table 2). The TOCSY spectrum, which gives the full displacement of all *J*-connected protons (spin systems), can be used either to reinforce the assignments made using COSY or to allow additional assignments which were missing or unclear on the COSY spectrum. Once all ^{1}H-chemical shifts are properly obtained, the 1D ^{1}H-NMR spectrum (Figure 2B) can be completely assigned accordingly, and a table of ^{1}H-chemical shifts can be filled up. Furthermore, through a carbon-related experiment such as the heteronuclear single quantum coherence (HSQC) spectrum, all ^{1}H-linked carbons signals can be ultimately assigned through the correlation of the one-bonded ^{1}H-^{13}C J-couplings such as $^1J_{H1\text{-}C1}$, $^1J_{H2\text{-}C2}$, $^1J_{H3\text{-}C3}$, and so on. Therefore, all ^{13}C-chemical shifts are finally obtained from previously assigned ^{1}H-chemical shifts. And the directly observed ^{13}C-NMR spectrum (Figure 2C) can be ultimately assigned as well.

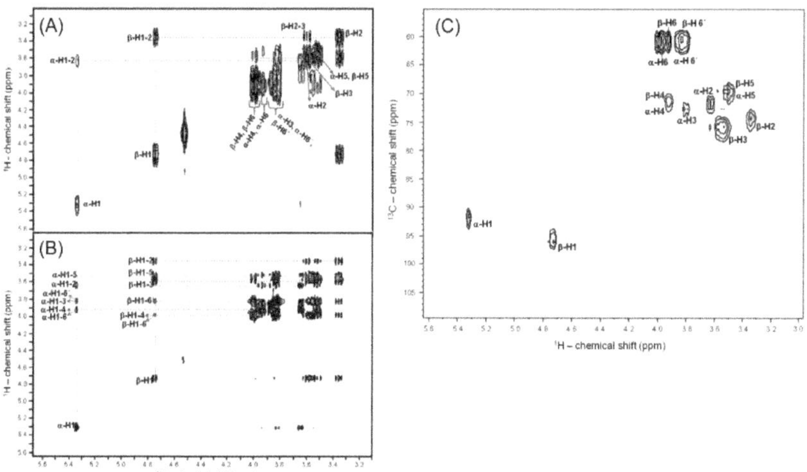

Figure 3. (**A**) COSY, (**B**) TOCSY and (**C**) HSQC spectra, Bruker 400 MHz at 37° C of D-glucose in 100% D_2O-solution after waiting to reach the anomeric equilibrium. The signals are labeled according to the atoms in the molecule (Figure 1A).

α-D-Glc*p*	β-D-Glc*p*
α-H1 – 5.32	β-H1 – 4.74
α-H2 – 3.63	β-H2 – 3.37
α-H3 – 3.83	β-H3 – 3.60
α-H4 – 3.92	β-H4 – 3.92
α-H5 – 3.50	β-H5 – 3.5
H6/6′ - 3.91/3.82	H6/6′ - 3.91/3.82
α-C1 – 91.4	β-C1 – 95.9
α-C2 – 71.8	β-C2 – 74.1
α-C3 – 72.4	β-C3 – 75.8
α-C4 – 71.2	β-C4 – 71.2
α-C5 – 69.8	β-C5 – 69.8
α-C6 – 60.3	β-C6 – 60.3

*Obtained from assignments of NMR spectra at Figures 2B, 2C, and 3.

Table 2. ^{1}H- and ^{13}C-chemical shifts* of both α- and β-D-Glc*p* residues.

The downside in collecting spectra containing heteronuclear-filters (either ^{13}C-, or ^{15}N-) even with acquisition through ^{1}H, such as in the HSQC experiment (Figure 3C) is that it is more time-consuming than those based exclusively on ^{1}H-atoms. This is because of the sensitivity and abundance of the heteronuclear isotopes, which are much lower than those of proton (see NMR receptivity in Table 1). This can be proved by different signal-to-noise ratio clearly seen by the baseline widths comparatively observed in the 1D ^{1}H- and ^{13}C-NMR spectra (Figure 2B vs 2C). In ^{13}C-directely observed NMR spectra, the peaks are usually thinner compared to ^{1}H-signals (Figure 2B vs 2C). This is, in turn, due to different relaxation properties of the ^{13}C-nucleus. The longitudinal or T_1 relaxation of carbons is much faster than protons, which result a considerably sharper peak in 1D NMR (Figure 2B vs 2C).

Due to the lower relative receptivity of heteronucleus, sometimes isotopic labeling techniques are necessary to develop 2D NMR analysis for certain carbohydrates. This is crucial, especially for those cases where the amount of material is a limitation, such as glycans isolated from cell cultures. Recently, we developed an *in vivo* method to spin-label cellular GAG molecules with nitrogen-15 [10] to allow observation and consequently get structural characterization via ^{15}N-chemical shift. This is the third isotope type used in NMR analysis in glycosaminoglycanomics. Our method was based on cell cultures using media enriched in ^{15}N-labeled side-chain glutamine [10]. The side-chain amino group of this amino acid is responsible to donate the NH_2 group to the amino sugar during hexosamine biosynthesis in the cytosol. Through this *in vivo* labeling method, ^{15}N-HSQC spectra became recordable for a couple of few hundred micrograms of GAG isolated from cellular sources [10]. On the other hand, as standard GAGs can be readily available from suppliers, the HSQC spectra using ^{15}N-isotope at natural abundance proved to be successfully recordable at 15 mg/ml concentration-samples within just a couple of hours of acquisition. ^{15}N-HSQC spectrum has been proved to be very straightforward in structural characterization of GAGs in solution (Figure 4).

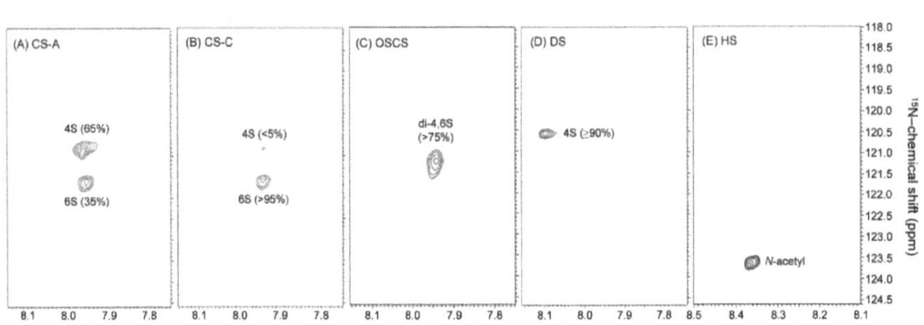

Figure 4. ^{15}N-HSQC spectra from Varian 800MHz at 25° C of different GAG standards: (A) bovine tracheal chondroitin sulfate A (CS-A), (B) shark cartilage chondroitin sulfate-C (CS-C), oversulfated chondroitin sulfate (OSCS), (D) dermatan sulfate (DS), or chondroitin sulfate-B, and (E) heparan sulfate (HS), dissolved in 50 mM acetate buffer 12.5% D_2O (pH 4.5) 0.1 % sodium azide, at a final 15 mg/mL concentration. Reproduced from [10].

Note that through ^{15}N-HSQC spectroscopy, all GAG standards show just the ^{1}H-^{15}N cross-peak related to their differential hexosaminyl units, in a very simple way that allows rapid structural assignments. Although the NMR peaks are few, these resonances are still quite useful for structural determination in glycosaminoglycanomics. All signals from different standards are characteristic, and resonate with distinct ^{1}H- or ^{15}N-chemical shifts (Figure 4) [10]. Structural features such as the hexosaminyl type (galactosamines at upfield ^{1}H- and ^{15}N-chemical shifts, Figures 4A-D, as opposed to glucosamines at a more downfield ^{1}H- and ^{15}N-chemical shifts, Figure 4E), sulfation pattern (4-sulfation at the upfield ^{15}N-chemical shifts, Figures 4A, B and D, as opposed to more downfield ^{15}N-chemical shifts of 4,6-di-sulfated units, Figure 4C, and further more downfield ^{15}N-chemical shift of 6-sulfated units, Figure 4A, and B), and adjacent uronic acid type (glucuronic acid at upfield ^{1}H-chemical shifts, Figures 4A-C, as opposed to iduronic acid at the more downfield ^{1}H-chemical shifts, Figure 4D) can be easily determined through this method [10]. These studies concerning the use of ^{15}N-NMR for structural characterization in GAG molecules turned out to be quite valuable also in predicting the anomericities in GAG-derived oligosaccharides as well as the sulfation patterns. For more details read reference 10 about this topic.

One downside of the ^{15}N-NMR application on structural studies of GAGs is the very fast exchange rates of protons from sulfamate groups ($NHSO_3^-$) that are quite abundant in glucosamines of certain GAG types, such as heparan sulfates and heparins. Although the ^{1}H-^{15}N cross-peaks from ^{15}N-HSQC spectra of residual N-acetylated glucosamines (GlcNAc) can be fairly used to quantify the amounts and types of uronic acid units, to which GlcNAc are linked (Figure 5), the remaining amounts of uronic acids linked to N-sulfated glucosamines (GlcNS) can be missed under normal conditions. Alternative ways to force the protonated states in sulfamate groups is the use of controlled samples at narrow pH range (7.0-8.0) [15], or to slow down the fast ^{1}H-amide exchange rates recording experiments at very low-temperatures, as detailed next with preliminary results using commercially available sodiated GlcNS as a molecular model to mimic the composing N-sulfated amino sugars in heparins and heparan sulfates.

Very low temperatures such as 3°C (20% acetone is added to avoid freezing) can slow down considerably the solvent exchange of the labile protons. All protons, inclusively the exchangeable ones from sulfamate or hydroxyl groups of GlcNS, become thus detectable by ^{1}H-based NMR spectroscopy (Figure 6A). This can be exemplified with the 1D spectrum of GlcNS at Figure 6A. The ^{1}H-peaks can be assigned through spin-systems from TOCSY spectrum using this same hydrated low-temperature condition (Figure 7A), like the same way undertaken and explained for Glc (Figure 2). The α- and β-anomers with ^{1}H-chemical shifts respectively at ~ 5.42 and ~ 4.68 ppm (Figures 1 and 6A) serve as starting point for tracing connectivities by cross-peaks in TOCSY spectrum also recorded in water-rich solution (Figure 7A). They have $^{3}J_{1H1\text{-}1H2}$ respectively around ~ 3.17 and ~ 2.98 ppm (Figure 7A). These H2 resonances have consequently further connections with ~ 5.33 and ~ 5.86 ppm, respectively for α- and β-anomerics. The latter ones are amide protons with chemical shifts determined for further assignments in ^{15}N-HSQC spectrum. These signals were not observed in the protonated amide condition (data not shown). Using the hydrated and cold condition, the ^{15}N-HSQC spectrum (Figure 7B) turns now to be useful in detection of the protonated state of sulfamate groups, and the ^{1}H-^{15}N cross-peaks of each NH are at the expected chemical shifts as seen by TOCSY spectrum (Figure 7B). Note that in the ^{15}N-HSQC spectrum, the amides anomerics showed swapped ^{1}H-chemical shifts, since the α-form of GlcNS is the predominant population in solution (Figure 1).

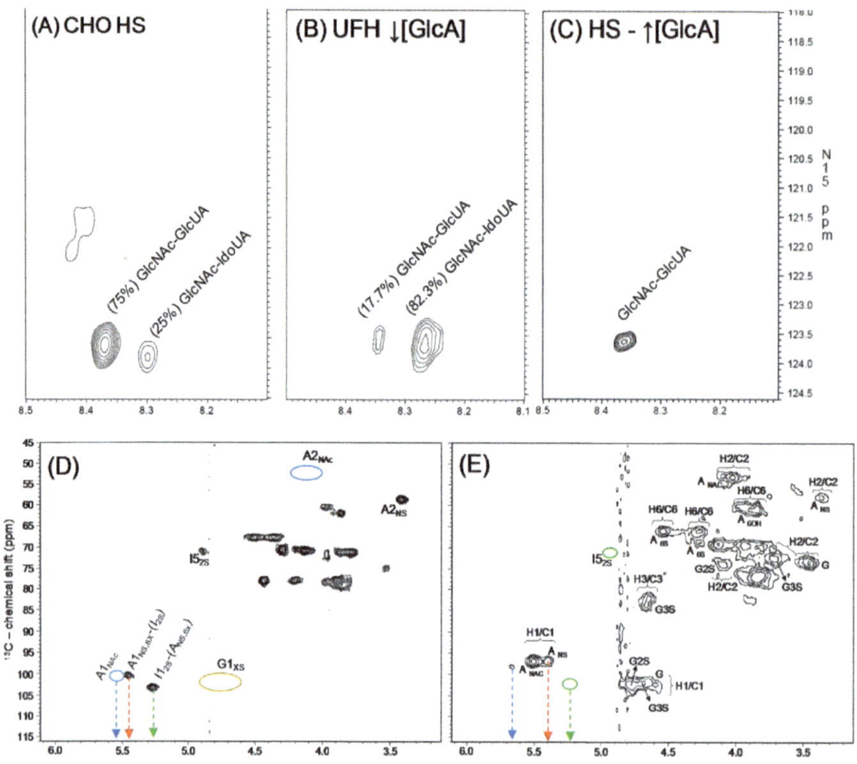

Figure 5. ^{15}N- (A-C) and ^{13}C- (D, E) HSQC spectra of heparan sulfate (HS) (A, C, and E) and unfractionated heparin (UFH) (B, D), recorded at Varian 800 MHz (A-C) and Bruker 400 MHz (D, E). The HS in panel A was from cultures of chinese hamster ovarian (CHO) cells [10], the UFH in panel B was from a pharmaceutical supplier, and the HS in panel C was extracted from the bivalve *Nodipecten nodosus* [16]. Note that the (B) UFH and (C) HS have respectively low and high amounts of glucuronic acids (GlcUA), as demonstrated by the tiny or missing color-coded peaks in their respective ^{13}C-HSQC spectra of panels D and E. In these panels, the letters used are A for glucosamine, I for iduronic acid, G for glucuronic acid, NAc for N-acetyl, NS for N-sulfation, 6X for 6-(un)substituted, 2S for 2-sulfation, XS for 2- and/or 3-sulfation, and 3S for 3-sulfation. The number right after the capital letters designate the ring position, for example, $I5_{2S}$ means a cross-peak of the ^{1}H-^{13}C pair at the 5-position of a 2-sulfated iduronic acid ring. Hence, the colors indicate blue for N-acetyl glucosamines, red for N-sulfated glucosamines, green for iduronic acids, and yellow for glucuronic acids.

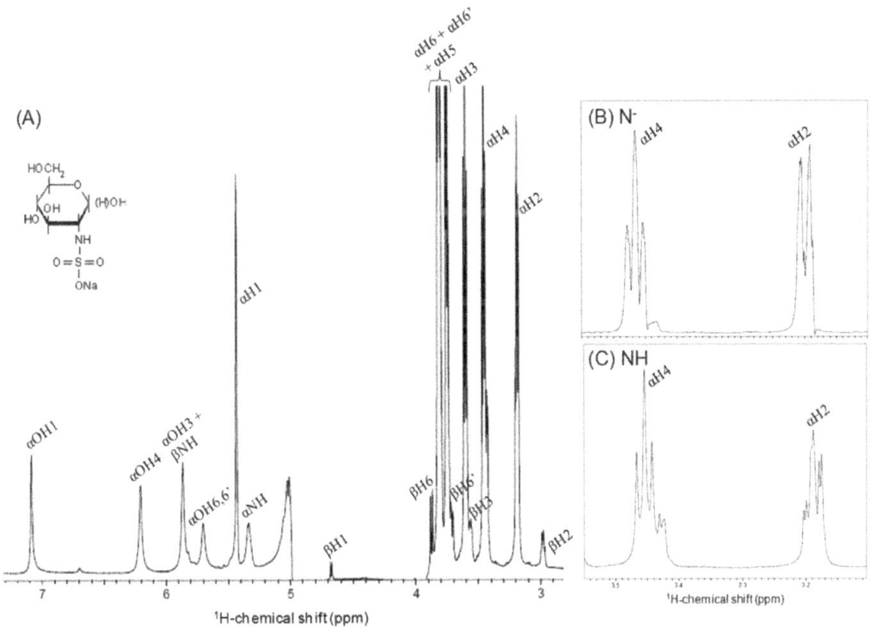

Figure 6. 1D ^{1}H-NMR expansions 2.8-7.3 ppm (A) and 3.10-3.55 ppm (B, C) of 5 mg/mL GlcNS in 10:20:70% D_2O:Acetone:H_2O (pH 4.5) at 3°C (A, C) and at 15°C (B) in spectrometer Varian 800 MHz. Note that in panel B, the resonance of αH2 splits in a dublet of dublets due to $^3J_{1H2-1H1}$ and $^3J_{1H2-H3}$, while in panel C there are a triplet of dublets for the split of αH2 due to $^3J_{1H2-1H1}$, $^3J_{1H2-H3}$, and now the $^3J_{1H2-HN}$ since in the former case the amide is deprotonated because of the higher temperature, but in case of panel C, the amide is protonated because of lower temperature. The structure of the GlcNS in its protonated and sodiated state is shown on the top left side of the panel A.

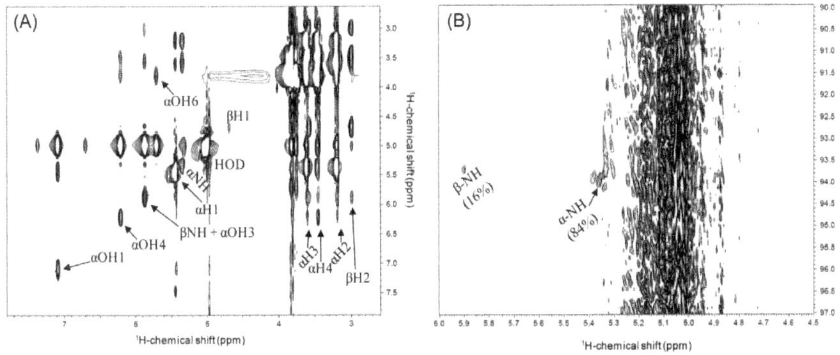

Figure 7. (A) TOCSY and (B) ^{15}N-HSQC spectra of GlcNS, at 3°C, in a Varian 800 MHz. The exchangeable protons from hydroxyl and amide groups are accordingly labeled. The population percentage of the anomeric forms are indicated in panel B. The α-anomers of GlcNS are more populated as seen in Figure 1.

2.3. Examining flexibility and dynamic properties of glycans by relaxation rates measurements

In radio-frequency pulsed NMR experiments, the magnetization aligned with the static magnetic field (B_0) is titled away from the longitudinal *Z*-axis (parallel to the static magnetization), and place into the *X,Y*-transverse plan. The fundamental property that brings back the magnetization to the *Z*-plane, restoring therefore the equilibrium during this modern pulsed NMR experiments is the spin relaxation phenomena [17]. The relaxation mechanisms are particularly dependent of the molecular motions of either internal localized sites or the overall tumbling of the molecule in solution. Therefore, relaxation measurements are very useful for studying molecular dynamic processes on a fast time scale. The longitudinal or spin-lattice relaxation process is given by the relaxation time T_1, and this is a measure of how fast the longitudinal magnetization goes back to its original state before the pulses. The transversal or spin-spin relaxation effect denoted by the time course T_2, measures how fast the Free Induction Decay (FID) dissipates its magnitude from the *X,Y*-transverse plane. Several mechanisms are known to influence the T_2 relaxation time, such as the dipolar interactions like dipole-dipole and dipole-chemical shift anisotropy [17]. The relaxation rates (R_1 and R_2) measurements (in ms to s) are the inverse of T_1 and T_2 respectively. As discussed below using some published data from the literature, these rates and thus relaxation times, combined with measurements of line-widths or line broadening (in Hz) during NMR experimentation, can be successfully used for determine dynamic properties of either free glycans in solution as glycan domains of glycoconjugates.

In a recent study about structural dynamics of the saccharidic portion of immunoglobulin G (IgG) using NMR spin-relaxation [18], the authors have proved that both glycan branches at the Fc fragment (Figure 8) are accessible and dynamic in solution. This motion, and glycosylated rates (Figure 8) are responsible to participate during the cellular responses of the adaptive immune systems, mainly modulating the health balance between hyposensitivity to foreign particles versus hypersensitivity to auto-antigens. Again, the spin labeling technique as discussed in section 2.2 was crucial to advance this work. In this reference, the N-glycan at Asparagine-297 was enzymatically remodeled by sialytransferases and glycosidases to build up branches specifically labeled with ^{13}C-isotope in galactosyl units (Figure 8). This initial procedure of spin labeling indicated already an apparent accessibility of these branches to the remodeling activity of the enzymes, unlikely previous conceptions of a more static and internalized behavior of these branches. As discussed next, the spin relaxation measurements of these branches have ultimately confirmed these dynamic properties [18].

NMR resonance linewidths (half-height of the peak, in Hz), which are directly related to the transverse relaxation rates (R_2 values) are values indicative of dynamic properties. Narrow lines indicate decreased relaxation rates and thus increased rates of rotational motions. The linewidths of the $^{13}C2$-labeled galactose in the α1-6Man branch was over three-times that of the corresponding resonance from the α1-3Man branch (Table 3) [18]. This suggested that the α1-6Man branch is considerably more immobilized than the other antennary. At experiments undertaken at higher fields, the ^{13}C linewidths of the α1-6Man $^{13}C2$ resonance of galactose gave values more than three-times than the correspondent on the other branch. Interpretation using an isotropic model gave an effective correlation time of 37 ns, which is considerably longer than the hydrodynamic predictions for the Fc fragment, and the protein tumbling time (~20 ns). This suggested that although the α1-6Man branch is partially immobilized, there should be still some dynamic properties of this branch to explain the higher experimental value. Therefore a more complete set of relaxation time was measured at two different magnetic strengths (Table 3). A field-dependent R_2 values were obtained, with smaller relaxation at 14.0 T (600 MHz) as opposed to 21.1 T (900 MHz). All R values combined at the lower magnetic field were not consistent with prediction based on isotropic tumbling model and have suggested that if the relaxation were solely dipole in origin, R_2 would be even less at this field strength [18]. Both the field dependence and these inconsistencies have suggested contribution from chemical exchange contributions to R_2 relaxation mechanisms that originated from the α1-6Man branch sampling multiple conformational states on the microsecond to millisecond timescale and therefore, modulating chemical shift in this process. At least, some of these states have to have substantial internal motion to raise the rotational R_1 value ($R_{1\rho}$).

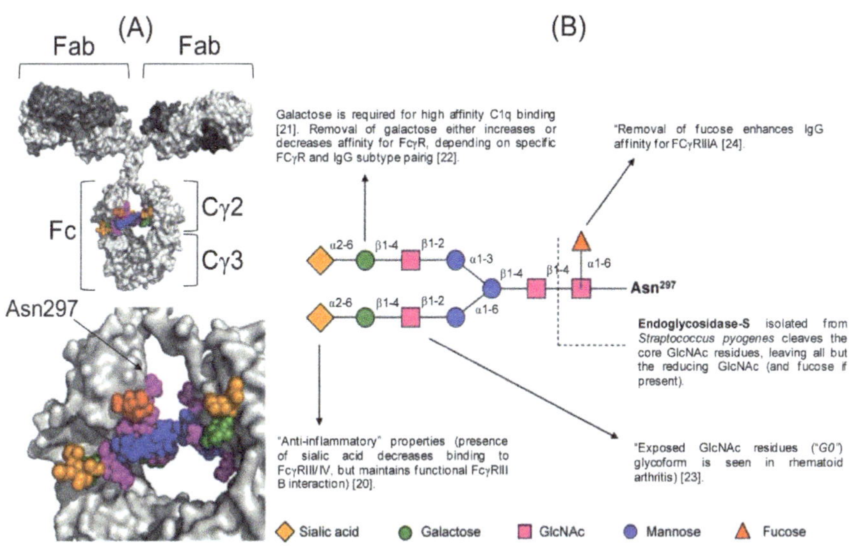

Figure 8. The conserved N-linked carbohydrate domain of IgG is attached to Asn-297 of the Cγ2 domains of each IgG heavy chain. Data modified from [19]. A range of other glycoforms (full or partial unsialylated and/or ungalactosilated) also exists on serum IgG [19]. These other glycans are based mostly on the biantennary structure but without some or all of the sialic acid or galactose residues from the non-reducing terminus (or fucose residues from the core). The presence of a particular carbohydrate maintains a defined structure of the Fc region that leads to specific consequences for the Fc function. Some of the key glycan-mediated interactions with receptors, for example, are indicated and properly referenced [20-24]. Heavy and light chains are colour-coded in light and dark grey respectively.

	α1-3Man-linked	α1-6Man-linked
	Relaxation rates	Relaxation rates
900 MHz	**Galactose ^{13}C6**	**Galactose ^{13}C6**
R_1	2.8 ±0.1 s^{-1}	3.1 ±0.4 s^{-1}
R_2	48.4 ±5.5 s^{-1}	59.7 ±9.2 s^{-1}
900 MHz	**Galactose ^{13}C2**	**Galactose ^{13}C2**
R_1	1.2 ±0.1 s^{-1}	1.2 ±0.2 s^{-1}
R_{1p}	20.1 ±1.0 s^{-1}	56.1 ±10 s^{-1}
$R_{2\ linewidths}$	40 s^{-1}	150 s^{-1}
600 MHz		
R_{1p}	14.7 ±1.1 s^{-1}	48.0 ±5.4 s^{-1}
$R_{2\ linewidths}$	28 s^{-1}	107 s^{-1}
R_2	29 s^{-1}	ND

ND, not determined owing to low signal-to-noise ratio.
* Reproduced from [18].

Table 3. ^{13}C relaxation measurements of galactose residues from Fc isotopic labeled at either C2 or C6 positions.

In order to probe the existence of chemical exchange contribution, relaxation dispersion experiments can be employed, in which transverse relaxation rates (R_2) can be measured using Carr-Purcell-Meiboom-Gill (CPMG) pulse sequences [25]. Through this NMR experiment type, resonances that experience multiple chemical environments on a timescale near to ν_{CPMG} (variable pulsing rates due to varying delays between the forming 180° pulses of CPMG pulse sequences) must become more intense as a function of increasing pulsing rates. Results from this experiment using both α1-3Man and α1-6Man branches of IgG Fc with ^{13}C-labeled galactose residues at 50° C, and at two different magnetic strengths (21.1 T, 900 MHz, and 18.8 T, 800 MHz) have shown this profile (Figure 9). At low pulsing rates, the relaxation measurements approach predictions using linewidths. At higher pulsing rates, chemical shifts are canceled out due to rapid chemical shift refocus, therefore chemical exchange contributions can be sorted out. Clearly, the R_2 values of α1-6Man branch decrease with pulsing rates, whereas the values for the other branch have kept constant (Figure 9A). The change in R_2 in the presence of chemical exchange between two states (A and B) was described using the equation $R_2(1/\tau_{cp}) = R_2^0 + \phi_{ex}/k_{ex}\ [1-^{2\tanh}(k_{ex}\tau_{cp/2})/k_{ex}\tau_{cp}]$, where $\phi_{ex} = P_A P_B \Delta\omega^2$, R_2^0 is the relaxation rate in the absence of chemical exchange, k_{ex} is the exchange rate, τ_{cp} is the delay between refocusing pulses, and $\Delta\omega^2$ is the frequency difference of states A and B in angular units [26]. Thought this equation, the contribution of chemical exchange to the observed R_2 was seen to depend on the rate of exchange, the time between pulses and the scaling factor ϕ_{ex}. Fitting this equation to the data that was obtained for the α1-6Man branch, an effective exchange constant (k_{ex}) of 5,300 ±1,700 s^{-1} was generated [18]. The scaling factor ϕ_{ex} in principle can yield information about the nature of the exchange process by describing populations of states and chemical shift changes on moving between states. The ϕ_{ex} at 21.1T and 18.8 T were respectively, 380,000 ±60,000 $rad^{-2}\ s^{-2}$, and 300,000 ±100,000 $rad^{-2}\ s^{-2}$. The R_2^0 for each magnetic field were respectively 26 ±13 s^{-1}, and 39 ±7 s^{-1} [18].

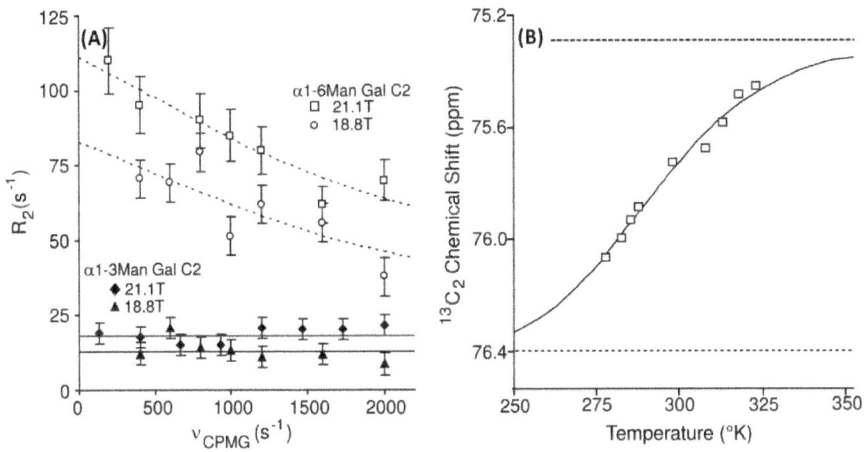

Figure 9. (A) Relaxation dispersion and (B) temperature-dependent chemical shift measurements show evidence of two states (A and B). (A) The relaxation dispersion experiments indicate motion at the micro-millisecond timescale for the α1-6Man branch but not for the α1-3Man branch. (B) The chemical shift of the α1-6Man branch-linked ^{13}C2-galactose approaches a saturation point at higher temperatures and permits the estimation of chemical shift values for two states. Chemical shift asymptotes for high-temperature (state A) and low-temperature (state B) are represented with discontinued lines. Reproduced from [18].

The effects of chemical exchange rates on R_2 can be also investigated using the rotational spin lattice-relaxation (R_{1p}) measurements. R_{1p} of both branches (Table 3) approach the rapid pulsing limit of R_2 in CPMG experiments (Figure 9), which represents the approaching to a non-exchange contribution to R_2. The large deviation seen for the α1-6Man branch is consistent with the considerable chemical exchange contribution. Taking all these relaxation-based data, two conformational states were observed for α1-6Man branch: one related to chemical exchange contribution within contact with the peptide chain, and another regardless the molecular contact resulted from low chemical exchange contributions [18]. Taking together the existence of these both states, a dynamic behavior for the α1-6Man branch has been proved, while the constant low chemical exchange contribution of α1-3Man branch has pointed towards a glycan segment more externalized from the Cγ2 domain (Figure 8). These conclusions will be sustained by other results about chemical shift changes [18], as described in the following item.

2.4. Applications and implications of chemical shifts in glycobiology

Chemical shift values (δ, in ppm) together with coupling constants comprise the major information extracted in NMR studies of proteins and nucleic acids [27], and thus are of great usefulness in glycobiology as well. Chemical shift values provide reliable information concerning the chemical environments in which a given atom (nucleus) molecularly experiments at certain timescale. If there were no other kinds of interactions in addition to the Zeeman interaction [28] during the NMR experimentation, all nuclei of a given molecule would lead to the same frequency in a 1D NMR spectrum. Zeeman states represent the two spin possibilities of a spin-half nucleus (usually defined as α- and β-spin states) under the static magnetic field (B_0) [29]. Due to changes on the chemical environment of the nucleus within the molecules chemical shifts may experience different values which will result different frequencies even though the same isotope and with same Larmor frequency. One of the major determinants in the chemical shifts is the electronic shielding generated by the electrons that surround the spin-half nuclei. Electrons at these surrounds have also angular momentum and thus magnetic moment. Differential conformational states may give different chemical shift values in a limited time-range. This can be explored to understand dynamic behaviors of glycans and to understand the chemical structures of biomolecules. Another application in the exploration of chemical shift changes is in glycan-involved intermolecular complexes, such as those of carbohydrate-binding proteins. The former phenomenon will be briefly explained through the dynamic properties of the α1-6Man branch of IgG Fc [18], continuing the same example taken at the previous section. The latter phenomena on concerning intermolecular complexes involving carbohydrate-binding proteins and glycans as common ligants will be discussed afterwards, taken some unpublished data about the binding properties of the chemokine RANTES complexed with chondroitin sulfate hexaccharides of well-defined sulfation patterns.

2.4.1. The dynamic behavior of glycans through chemical shift changes: the bound and unbound states of the N-glycan of IgG Fc

The chemical shifts of the Fc α1-6Man galactose residue were observed to be clearly temperature-dependent (Figure 9B). The ^{13}C2-resonance can show a significant displacement between 5-50°C with its ^{13}C-chemical shift trending toward a plateau of 75.3 ±0.2 ppm at the high-temperature limit (Figure 9B) [18]. This plateau value is quite consistent with chemical shift values from the free glycan in solution which implies that the glycan is far away from chemical environment promoted by the Fc amino acids (state A, Figure 10). Although Figure 9B showed a curve-fitting only for the C2 resonances, the ^{13}C6-chemical shift showed a similar trend and moved from a perturbed resonance position at lower temperatures towards a free glycan position at higher temperatures [18].

Through the use of the chemical shift values determined for state A in conjunction with the observed chemical shifts at 50°C obtained from CPMG relaxation dispersion data, the 0.8 value was extracted for the state A population (P_A) using the equation $P_A = 1/[\omega_0^2(\delta_X-\delta_A)^2/\varphi_{ex}]+1$. Then the P_A value was used to deconvolute φ_{ex} and derive a value for $\Delta\omega$ using the relationship of $\phi_{ex} = P_A P_B \Delta\omega^2$. The determined chemical shift for state B was therefore 76.4±0.5 ppm using the resultant value for $\Delta\omega$ of 1,500 ± 300 rads^{-1} along with the state A chemical shift. Moreover, the data obtained have indicated that populations of both states are nearly equal at 15°C and shift towards higher population of state A (the state with chemical shifts similar to free glycans in solution) at temperatures higher than 15°C. At body physiological temperature state A is approximately 70% populated.

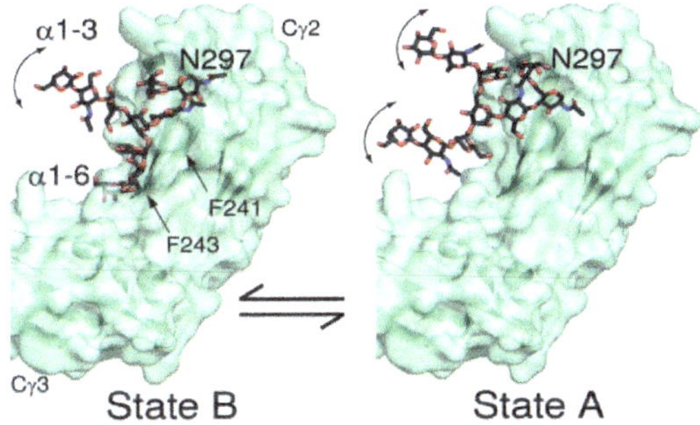

Figure 10. Models for Fc N-glycan dynamics and accessibility showing that exposed conformational states of branches are possible. While the α1-3Man branch is highly flexible out of the protein pocket, the α1-6Man branch is partially dynamic within two states: externalized A state *versus* internalized B state. Reproduced from [18].

2.4.2. *Unveiling binding sites of carbohydrate-binding proteins through chemical shift perturbation: The chemokine CCL5/RANTES-chondroitin sulfate hexasaccharides complexes*

Chemokines are a family of small cytokines with chemotactic properties. They are small and soluble proteins (with a mass of 8-16 kDa), and are produced and released by a variety of cell types during the initial phase of host responses to injury, allergens, antigens, or invading microorganisms. Chemokines share a sequence homology and possess four cysteines in conserved locations. These form two disulfide linkages that are keys to their tertiary structure and stability (Figure 11). These cysteines are also involved in the nomenclature and classification of the chemokines. Among many chemokines, CCL5, also known as RANTES, is a 68-residue proinflammatory chemokine responsible to control migration and trigger activation of leukocytes during their trafficking to the inflamed sites. Both these biological activities of CCL5 have shown to be critically influenced by interactions with GAG in endothelial surface proteoglycans [30]. The immobilization of the CCL5 onto surface proteoglycans will attract leukocytes to the sites of injury, helping the rolling step which is heavily driven by selectin-mediated interactions; followed by the activation step of the leukocytes. This step is driven by the interaction of the CCL5 receptor on the leukocyte (CCR5) with the immobilized CCL5. GAGs in proteoglycans do not only regulate these two steps but also induce the oligomerization of CCL5 [31]. This GAG-induced oligomerization process is crucial to i) create a high local concentration of chemokine to optimize the chemoattraction of cell for sites of lesion, ii) increase the *in vivo* CCL5 half-life due to enhancement of protection from natural proteolysis, iii) promote resistance to physical disruptions caused by the blood flow in vessels, and iv) serve as storage resources for rapid mobilization of chemokines without biosynthesis [31]. Recently, the GAG-induced oligomeric structure of CCL5 was revealed by NMR, MS and small-angle X-ray scattering (SAXS) experiments [32]. Although, the heparin biding site of CCL5 has already been reported [33], comprising the segment ^{44}RKNR^{47}, the potential GAG binding sites of CCL5 are yet undocumented, especially considering chondroitin sulfate which is the most abundant GAG in the human body, and widely spread across surface proteoglycans of blood vessels.

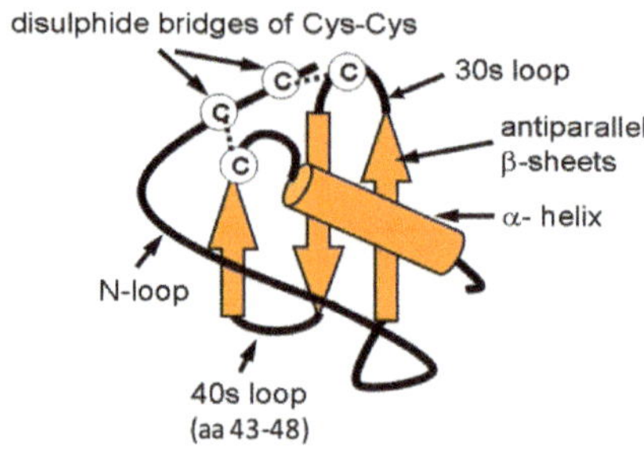

Figure 11. 3D structural model of the CC chemokines.

The ^{15}N-HSQC spectrum of the ^{15}N-labeled E66S CCL5 (Figure 12), which is mostly found as monomers in solution rather than the oligomerization propensity of the wild type [35], showed well-resolved amide signals that allowed easily a near-complete chemical shift assignment (Table 4). These initial assignments are crucial to further inform the chemical shift changes induced by the presence of increasing concentration of the ligants. The ligants used were two chondroitin sulfate hexasaccharides fully characterized by NMR [34]. They are named CS 6-6-4, and CS 4-4-4, which represent the following respective structures: ΔUA(β1→3)GalNAc6S(β1→4)GlcA(β1→3)GalNAc6S(β1→4)GlcA(β1→3)GalNAc4S-ol, and GlcA(β1→3)GalNAc4S(β1→4)GlcA(β1→3)GalNAc4S(β1→4)GlcA(β1→3)GalNAc4S-ol. The abbreviations are ΔUA for $\Delta^{4,5}$unsaturated uronic acid; GalNAc for N-acetyl galactosamine; GlcA for glucuronic acid; "S" for sulfation group, digits before "S" represent the ring position; and -ol stands for reduced sugars (open rings at the reducing-end terminal units) [34].

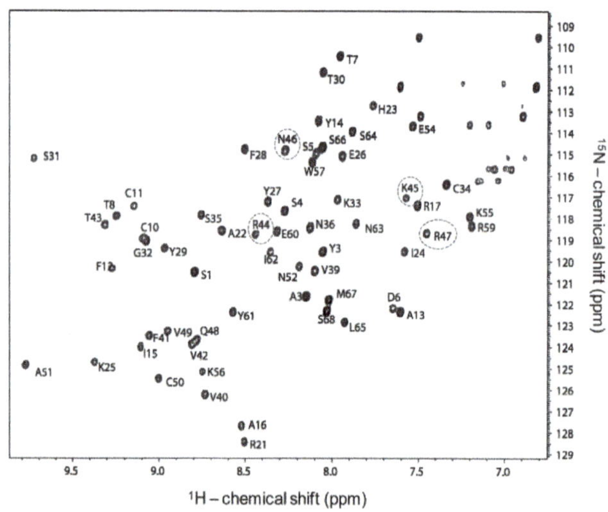

Figure 12. ^{15}N-HSQC spectrum of the ^{15}N-labeled E66S chemokine CCL5/RANTES (1-2 mM) in 150 mM sodium acetate buffer, pH 4.5. The amide signals are labeled accordingly to their respective amino acids. The signals circled in grey represent the heparin motif binding site (BBXB) [33].

Residue	^{15}N (ppm)	NH (ppm)	Residue	^{15}N (ppm)	NH (ppm)
S1	120.50	8.78	S35	117.88	8.74
P2	nd*	nd	N36	118.49	8.11
Y3	119.64	8.04	P37	nd	nd
S4	117.76	8.26	A38	121.67	8.14
S5	114.91	8.07	V49	120.49	8.08
D6	122.29	7.63	V50	126.32	8.72
T7	110.44	7.94	F41	123.53	9.04
T8	117.88	9.22	V42	124.07	8.81
P9	nd	nd	T43	118.28	9.31
C10	118.92	9.08	**R44**	118.69	8.43
C11	117.50	9.13	**K45**	117.01	7.57
F12	120.36	9.26	**N46**	114.91	8.24
A13	122.35	7.59	**R47**	118.63	7.43
Y14	113.41	8.07	Q48	123.72	8.78
I15	124.00	9.09	V49	123.30	8.92
A16	127.69	8.51	C50	125.46	8.99
R17	117.39	7.50	A51	124.84	9.76
P18	nd	nd	N52	120.24	8.16
L19	nd	nd	P53	nd	nd
P20	nd	nd	E54	113.87	7.52
R21	128.41	8.49	K55	117.89	7.19
A22	118.62	8.62	K56	127.69	8.51
H23	113.15	7.73	W57	115.25	8.08
I24	119.62	7.56	V58	122.48	5.61
K25	124.73	9.36	R59	118.49	7.17
E26	115.13	7.92	E60	118.62	8.29
Y27	117.14	8.35	Y61	122.35	8.54
F28	114.78	8.48	I62	119.62	8.34
Y29	119.40	8.96	N63	118.25	7.84
T30	111.25	8.03	S64	114.02	7.86
S31	115.27	9.71	L65	122.97	7.90
G32	119.08	9.06	S66	114.77	8.04
K33	117.13	7.95	M67	121.86	8.00
C34	116.51	7.32	S68	122.52	8.02

*not determined.

Table 4. ^{15}N and ^{1}H chemical shift of E66S at 297K, pH 4.5. The amino acids that form the heparin binding site motif (BBXB) are highlighted in bold.

As the concentration of the chondroitin sulfate ligants increases, either loss of signal intensity or chemical shift migration were observed on the ^{15}N-HSQC spectra of E66S CCL5 (Figure 13). In the case of the continuous titration using the ligant CS 6-6-4 (left-hand side spectrum of Figure 13), the signal intensity decreases proportionally, and no chemical shift migration was seen. This likely indicates a slow exchange rate between the on- and off-states of the complex, and the GAG-induced oligomerization was observed since large amounts of precipitates were visually formed on the bottom of the NMR tube during the titration experiment (data not shown). This precipitation phenomenon is one of the major reasons for the NMR signal intensity loss. Conversely, the titration using the ligant CS 4-4-4, no intensity loss was observed even using 10 equivalent molar of this ligant (blue spectrum), and clear chemical shift migration of certain peaks was observed (right-hand side spectrum of Figure 13). In the experiment using this ligant, the amino acids that showed the major chemical shift changes were S1, Y3, D6, T7, R17, K45, N46, R47, Q48, as indicated in the spectrum of Figure 13. The residues K45, N46, and R47 match perfectly with the heparin binding motif (Table 4) [33]. These residues that experience the largest chemical shift migration are highlighted on the structure of CCL5, represented in both cartoon and surface structural models (Figure 14). They are located at the pocket of the dimer consisted of the 40s loop, which in turn includes the heparin binding motif, together with the N-terminal (Figure 11).

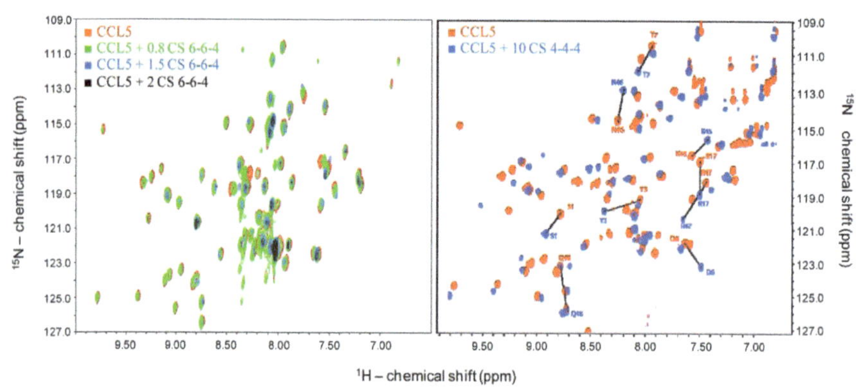

Figure 13. Superimposed ^{15}N-HSQC spectra from ligant titration on CCL5 in solution. The ligant types and relative concentration to the protein are indicated and color-coded as shown on the panels.

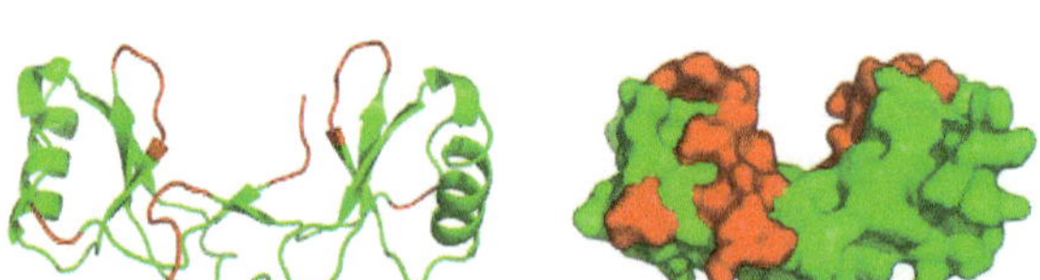

Figure 14. Cartoon (left) and surface (right) models of CCL5. In red are the amino acid S1, Y3, D6, T7, R17, K45, N46, R47, Q48 that have experienced the largest chemical shift migration induced by the presence of the ligant CS 4-4-4.

In order to understand the physiological condition of the GAG-induced oligomerization of CCL5 promoted by the CS 6-6-4, and the nature of the binding involved in the interaction between these two molecules, an experiment using increasing concentrations of sodium chloride was performed (Figure 15) at the 2-times molar excess of ligant. This was the condition that showed the huge precipitation and loss of signal intensity (black spectrum at left panel of Figure 13). It is clearly observed that at physiological salt concentration (150 mM NaCl) the low intensity of CCL5′s signals still indicate oligomerized states of the chemokine. However, this highly-complexed state is broken down at higher salt molar concentrations than the physiological condition as seen with the recovery of the signal intensity in the spectrum with 300-450 mM. This result indicates that only at higher salt concentrations than the physiological condition, the oligomerization of CCL5 is abrogated and the nature of the binding between the two molecules is just electrostatic. The return in loss of the signal intensity in the ^{15}N-HSQC spectra of CCL5 is just consequence of an instrumentation limitation due to the extreme high salt molarity that disturbs the magnetization sensitivity on the probe (the detection devise of the magnet). In synthesis, the CS 6-6-4 induces CCL5 aggregation whereas CS 4-4-4 just a little. This preliminary result indicates that the oligomerization of this chemokine is triggered by specific regions of the GAG chains in endothelial surface proteoglycans.

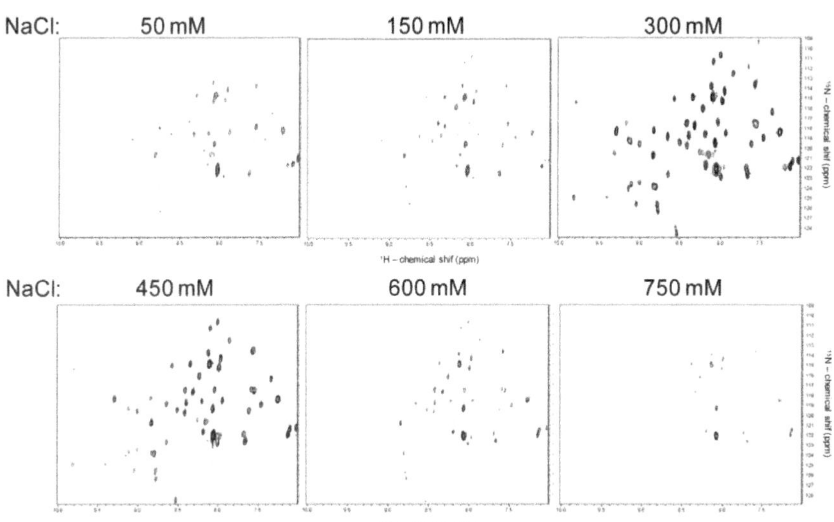

Figure 15. ^{15}N-HSQC spectra of CCL5 with 2 equivalent molar of CS 6-6-4 in increasing NaCl concentrations.

2.5. Scalar and dipolar coupling constants in glycobiology NMR

Besides chemical shifts, coupling constant values, either scalar (*J*) or dipolar (D) couplings, both measured in Hz, are another important set of information obtained from

NMR experiment to unveil structural shapes of molecules under certain NMR timescale. In glycobiology, scalar and dipolar couplings are fundamental NMR tools to respectively predict ring conformation, and overall conformational shape of oligosaccharides in solution.

2.5.1. *Scalar couplings in conformational analysis of sugar rings*

In general, three chemical bond proton-proton ($^3J_{H\text{-}H}$) scalar coupling constants are the most commonly used parameters in 3D conformational shape of sugar rings. This is because that $^3J_{H\text{-}H}$ depends directly upon the dihedral angle of the bonded atoms (Karplus-type relationship). Scalar coupling are usually easily seen by splitting resonances on 1D NMR spectra, but very often, J-resolved NMR experiments must be undertaken for obtaining such values. $^3J_{H\text{-}H}$ can primarily serve as diagnostic for the ring structural conformers of certain glycans. The usefulness of scalar coupling constants can be used by the rule that β-anomers usually exhibit larger $^3J_{H1\text{-}H2}$ values than α-^{1}H1-anomers (Table 5, and Figure 2B), which thus help structural assignment. Another example is the GlcNSO$_3^-$ addressed at item 2.2, Figure 6. At normal experimental conditions, N-sulfated glucosamine has its amide group unprotonated as observed by the doublets of doublet of the H2 resonance shown at panel 6B. This is a result from the $^3J_{H\text{-}H}$ coupling of the H2 with H1 and H3. However, at lower temperatures (panel 6C) the H2 resonance showed a splitting resonance of three doublets, meaning $^3J_{H2\text{-}H1}$, $^3J_{H2\text{-}H3}$, and now the protonated NH seen by the third $^3J_{H2\text{-}NH}$. Experimental $^3J_{H\text{-}H}$ values have been determined in many differently sulfated glucosamines, galactosamines and iduronic acids (Tables 5, and 6).

Through the *J*-values tabulated on Table 5, a sulfation pattern-dependent sugar ring conformation can be clearly noted, except for GalNAc units. And these vicinal coupling constants, for example in α-GlcN,6S is in agreement in various oligo- or polysaccharides containing such monosaccharide [36, 37, 47]. The $^3J_{H1\text{-}H2}$ values ranging from 3.6 to 3.8 Hz for GlcNS; GlcN,6S; GlcN,3,6S; GlcNAc,6S; GalNAc,4S; Glc2,3,4,6S together with the larger coupling values for the other intra-ring protons ($^3J_{H2\text{-}H3}$, $^3J_{H3\text{-}H4}$, $^3J_{H4\text{-}H5}$), somewhere between 8.8 and 11.1 Hz signifies that these monosaccharides are in their $^4C_1$ chair conformation of the pyranose ring, where the anomeric H1 is in the equatorial position as opposed to the other protons H2, H3, H4, and H5 are in their axial position (Figure 16) [46]. The $^3J_{H\text{-}H}$ values of GalNAc units are also in agreement with the $^4C_1$ chair conformation regardless the sulfation pattern (4- or 6-sulfated), and the $^3J_{H3\text{-}H4}$ is also compatible with the C4 epimerization of Glc. A different set of $^3J_{H\text{-}H}$ values implying different conformational equilibrium has been determined in hexuronic acids having the non-reducing end 4,5-unsaturation due to an eliminase action (Table 5) [41,43]. These values are consistent with the entire population in the half-chair $^1H_2$ conformation (Figure 16) in solution [41], and not to the $^2H_1$ half-chair conformation, which have quite different $^3J_{H\text{-}H}$ values (Table 5).

Residue type	$^3J_{H1-H2}$	$^3J_{H2-H3}$	$^3J_{H3-H4}$	$^3J_{H4-H5}$	Ref.
α-GlcNS	3.6	10.2	9.4	9.9	[36]
α-GlcN,6S	3.6	10.3	9.2	10.0	[37]
α-GlcN,3,6S	3.5	10.7	9.1	10.0	[37]
α-GlcNAc,6S	3.0	11.0	8.0	9.0	[38]
α-GalNAc,4S	3.5	11.0	2.5	nd	[38]
α-GalNAc,6S	4.0	11.0	3.0	nd	[38]
α-Glc2,3,4,6S	3.6	9.8	8.8	nd	[39]
β-Glc2,3,4,6S	5.7	5.2	6.8	6.0	[39]
β-GlcA2,3,4S	6.4	3.2	4.0	2.8	[39]
β-GlcA2,3S	5.9	nd*	1.5	1.5	[40]
α-IdoA (exp) (80:20% $^4C_1$:$^2S_0$)	3.8	8.4	7.6	7.6	[41]
α-IdoA $^4C_1$ (calc)	2.0	2.1	2.0	2.0	[41]
α-IdoA $^2S_0$ (calc)	4.3	11.1	5.4	5.4	[41]
α-IdoA (exp) (76:24% $^4C_1$:$^2S_0$)	2.4	4.8	3.4	2.5	[42]
α-IdoA $^4C_1$ (calc)	1.7	3.0	3.1	2.5	[42]
α-IdoA $^2S_0$ (calc)	4.8	10.5	3.4	2.5	[42]
ΔHexA (exp)	2.9	1.3	4.8	4.8	[41]
ΔHexA $^1H_2$ (calc)	3.0	1.5	4.1	4.1	[41]
ΔHexA $^2H_1$ (calc)	8.3	7.7	2.8	2.8	[41]
ΔHexA (exp)[a]	3.4	2.6	4.7	nd	[42]
ΔHexA (exp+calc)[b]	4.2	2.4	5.0	nd	[42]
ΔHexA $^1H_2$ (calc)	3.1	0.2	6.0	nd	[42]
ΔHexA $^2H_1$ (calc)	8.3	10.2	1.7	nd	[42]
Standard $^1C_4$	1.8	2.6	2.6	1.3	[43]
Standard $^2S_0$	6.3	9.5	9.3	2.8	[43]
Standard $^1H_2$	3.2	2.1	4.7	nd	[43]
Standard $^2H_1$	8.3	7.7	2.8	nd	[43]

[a] Experimental values only.

[b] Obtained by least square analysis of the theoretical and experimental values combined.

The abbreviations are: exp, experimental values; and calc, calculated values. GlcN, glucosamine; GalN, galactosamine; GlcA, glururonic acid; IdoA, iduronic acid; ΔhexA, 4,5-unsaturated hexuronic acid; nd, not determined.

Table 5. $^3J_{H-H}$ values (in Hz) in differently sulfated pyranosyl residues from various selected saccharides in solution NMR studies.

Compound number	$^3J_{H1\text{-}H2}$	$^3J_{H2\text{-}H3}$	$^3J_{H3\text{-}H4}$	$^3J_{H4\text{-}H5}$	$^1C_4$	$^4C_1$	$^2S_0$	Ref.
1	4.0	6.6	5.2	3.7	45%	29%	26%	[36]
2	1.9	3.7	3.7	2.2	87%	13%	-	[36]
3	1.8	3.3	3.4	2.2	90%	-	10%	[44]
4	4.9	6.9	6.4	4.2	38%	45%	17%	[45]
5	2.5	4.5	2.8	2.2	75%	-	25%	[46]
6	2.5	4.6	3.1	2.3	75%	-	25%	[47]
7	4.0	7.5	3.6	3.1	35%	-	65%	[44]
8	5.2	9.8	4.1	4.0	10%	-	90%	[47]
9	2.6	5.9	3.4	3.1	60%	-	40%	[44]

Table 6. $^3J_{H\text{-}H}$ values (in Hz) in composing IdoA units in various selected GAG molecules in solution NMR studies, and the abundance of their $^1C_4$, $^4C_1$, $^2S_0$, $^1H_2$, and $^2H_1$ conformers. These are all experimental values.

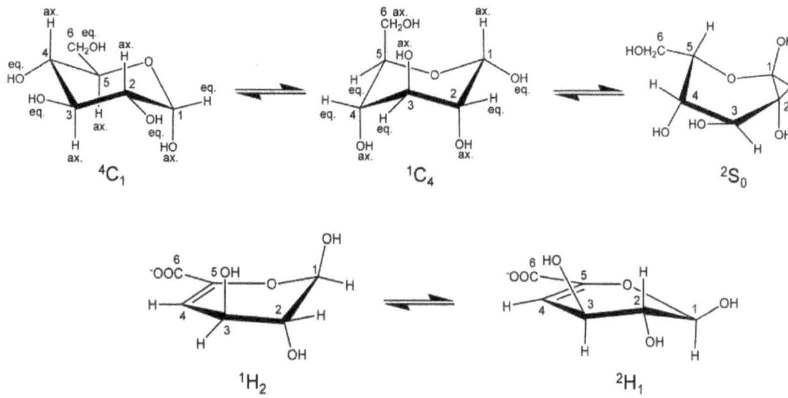

Figure 16. Sugar ring conformations for β-glucopyranosyl units as model for the nomenclatures $^4C_1$, $^1C_4$, $^2S_0$, $^1H_2$, and $^2H_1$. The latter two conformers have 4,5-unsaturation as a consequence of β-elimination from a eliminase reaction [41, 43]. The carbon numbers are indicated by digits in each conformer.

The conformation set observed for iduronic acid (IdoA) units (Tables 5 and 6), is complex and seems to be influenced not only by their sulfation patterns but also by the adjacent residues in which the IdoA units are linked to. For example, in case of IdoA units of a dermatan sulfate-tetrasaccharide studied at reference [41], the conformer population was estimated through Karplus relationship, and showed the 80/20 ratio for the chair $^4C_1$ and skew-boat $^2S_0$ conformations in solution. These IdoA units are linked to GalNAc units in dermatan sulfate, and are not sulfated. But when 2-sulfated or linked to GlcNAc units, as shown in other works presented at Table 6, different population ratios were observed, including the chair $^4C_1$ conformer together with the chair $^1C_4$, and the skew-boat $^2S_0$ conformers (Figure 16). In a heparin-derived tetrasaccharide, the chair $^1C_4$ and skew-boat $^2S_0$ conformer population ratios changed slightly for 76:24% (Table 5) [42]. The half-chair $^1H_2$ and $^2H_1$ populations of the unsaturated uronic acid in this heparin-derived tetrasaccharide changed to the 78:22% ratio [42] rather than 100% $^1H_2$ conformer population for Δ4,5HexA in a dermatan sulfate-derived tetrasaccharide [41] (Table 5) due to the presence of the adjacent GlcNAc unit rather than the GalNAc unit. Therefore, IdoA units can experience different ring conformer population ratios based on the neighboring units (Tables 5 and 6).

2.5.2. Dipolar couplings in conformational studies of oligosaccharides

Dipolar couplings (D) arise from through space spin-spin interactions and are strictly dependent on both inter-nuclear distance (r) and the angle (θ) between the magnetic field (B_0) and the inter-nuclear vector as described by the equation $D_{ij} = \xi_{ij}[(3\cos^2\theta - 1)/2](1/r^3)$, where ξ_{ij} is a constant that depends on the properties of the nuclei *i* and *j*. Dipolar coupling-based NMR data can provide both short-range (distance-dependent) as well as long-range (angular-dependent) structural information [48]. Since the inter-nuclear vector averages to zero in solution through isotropic tumbling, partially aligned molecules, using alignment media such as gels, liquid crystals or polyethylene glycol (PEG), can orient the molecules in solution forcing a partial anisotropic behavior, as the same time the resolution is preserved high in NMR spectra due to some residual fast tumbling in solution [49]. At this condition, the dipolar couplings do not average to zero, and splittings of dipolar coupled spin pairs become measurable [48].

Evidences of spatial structures of oligosaccharides are gradually increasing along the past few years with the advent of specific NMR methods and their application in glycobiology. Although NOE seemed the primary choice in conformational NMR studies, sometimes in the analysis of oligosaccharides the NOE-signal intensity that relies on the efficiency of polarization transfer between proton pairs may go to zero or close to zero because of the correlation time dependence cross-relaxation. In addition, the usual high-order flexibility of oligosaccharides, as opposed to a more rigid structural behavior of nucleic acids and proteins, may result NOE-contacts only from few conformers experienced within a short timescale. This ultimately lowers the efficacy of the NOE-based method in carbohydrate studies. Moreover, while NOE-based structural determination is basically based on through space inter-proton physical contacts, theoretically all types of nuclei pair combinations of spin vectors (^1H-^1H, ^{13}C-^1H, ^{15}N-^1H, ^{13}C-^{13}C) can be evaluated by the RDC method. Therefore, RDC-based measurements become the alternative NMR method in studies of oligosaccharides, not only on the conformational perspective but also in dynamic analysis. Using residual dipolar couplings (RDC), many carbohydrates have been characterized by the conformational perspective [42, 43, 48]. Here, we briefly revised the conformational and dynamical view of a heparin-tetrasaccharide [ΔUA2SO$_3^-$(1→4)-GlcNS,6S-(1→4)-IdoA2S-(1→4)-GlcNS,6S], notated as D-C-B-A, studied by RDC [42]. And in this study, a limited flexibility of the IdoA-composing unit of this tetrasaccharide was unexpectedly observed through the dynamic point-of-view.

Table 7 summarizes experimental $^1D_{H\text{-}H}$ and $^1D_{H\text{-}C}$ coupling constants of the heparin-derived tetrasaccharide aligned in PEG/hexanol neutral media. The former and the latter couplings were determined from COSY type spectra and ^1H-coupled non-refocused ^1H-^{13}C HSQC spectra. Five $^1D_{C\text{-}H}$ were used to calculate separately the order tensors via singular value decomposition method [50]. The theoretical RDC values were back calculated using REDCAT [51]. These calculations were performed for both $^1C_4$ and $^2S_0$ conformations of the B ring, the IdoA unit close to the non-reducing end. The comparison between the theoretical and experimental RDCs showed good agreement, proving high accuracy of the measured coupling constants. The better agreement was achieved with the $^1C_4$ conformation of the iduronic acid B unit [42].

Ring	One-bond ^{1}H-^{13}C couplings	*J* (Hz)	*J* + *D* (Hz)	RDC (Hz)	Three-bond ^{1}H-^{1}H couplings	*J* (Hz)	*J* + *D* (Hz)	RDC (Hz)
	C1-H1	172.2	168.8	-3.4	H1-H2	3.53	2.71	-0.82
	C2-H2	139.1	144.9	5.8	H2-H3	10.25	11.13	0.88
A	C3-H3	148.0	154.1	6.1	H3-H4	8.79	9.75	0.96
	C4-H4	147.2	152.6	5.4	H4-H5	9.85	10.75	0.9
	C5-H5	146.9	153.1	6.2	-	-	-	-
	C1-H1	174.1	178.9	4.8	H1-H2	2.42	3.97	1.55
	C2-H2	151.4	152.5	1.1	H2-H3	4.76	3.53	-1.22
B	C3-H3	151.5	148.9	-2.6	H3-H4	3.41	3.30	-0.11
	C4-H4	148.8	152.8	4.0	-	-	-	-
	C5-H5	146.1	145.7	-0.4	-	-	-	-
	C1-H1	172.6	177.7	5.1	H1-H2	3.67	2.46	-1.21
	C2-H2	138.7	135.6	-3.1	H2-H3	10.59	11.14	0.55
C	C3-H3	147.4	144.4	-3.0	H3-H4	9.04	8.43	-0.61
	C4-H4	147.2	142.4	-4.8	H4-H5	10.16	8.80	-1.36
	C5-H5	147.0	143.4	-3.6	-	-	-	-

Table 7. Scalar and dipolar coupling constants of the heparin-derived tetrasaccharide from the work [41].

In order to obtain the order tensor parameters, the analysis was divided into two sets depending on the conformation of the ring B, and thus, their diagonalized order tensors were inspected. The average Saupe matrix, mostly utilized to build up the molecular frame in RDC-based studies [48], of the $^1C_4$ form was found to have principle axis components: $S_{ZZ} = 1.44e^{-04}$, $S_{YY} = 1.36e^{-04}$, $S_{XX} = -0.70e^{-05}$ (generalized degree of order, GDO, = $1.62e^{-04}$), while in the $^2S_0$ form, the values $S_{ZZ} = 1.58e^{-04}$, S_{YY} $1.43e^{-04}$, and S_{XX} = -1;44e^{-05} (GDO = $1.74e^{-04}$) was given [42]. The GDO is a parameter indicative of molecular motion, and the large difference between GDO of two rigid fragments of a molecule indicates conformational flexibility [42, 48]. The two largest components of the tensors obtained for the heparin-derived tetrasaccharide [42], S_{ZZ}, and S_{YY}, differ by less than 10%, which caused their occasional swapping, seen in particular for the B ring in the chair $^1C_4$ conformer. Importantly, there is less than 10% difference between the S_{ZZ} (S_{YY}) values of the two groups of B conformers. In addition, the orientation of the principle axes of the two tensors was very similar. Figure 17 shows that S_{ZZ} is parallel to the long axis of the trisaccharide CBA of the heparin-derived tetrasaccharide [42]. The large S_{YY} component arises from the presence of bulky sulfate groups, and the fact that the D ring is positioned at an acute angle to the S_{ZZ}-axis. The smallest component, S_{XX}, is at least an order of magnitude smaller than the others. Although there is 100% difference between the two forms of ring B, this can easily be a result of small changes in the average orientation of some sulfate groups [42].

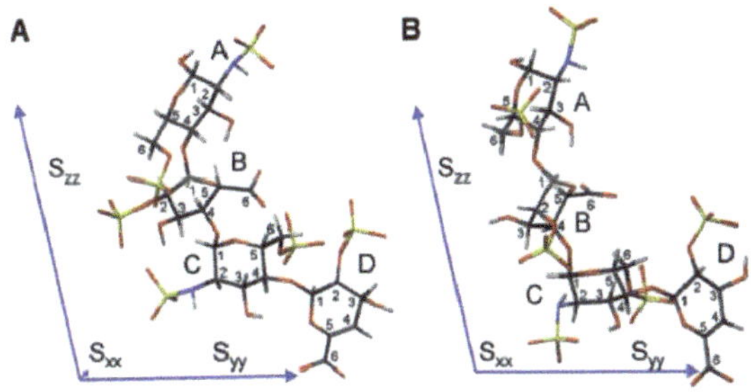

Figure 17. Orientation of the principle axis frame of the order tensor relative to the line-model structure of the heparin-derived tetrasaccharide: (A) ring B in its skew-boat $^2S_0$ conformation; and (B) ring B in its chair $^1C_4$ conformation. The relative lengths of the principle axis correspond to the principle order parameters. The rings are indicated by letters and the carbons by digits. White, hydrogen; dark grey, carbon; red, oxigen; blue, nitrogen; yellow, sulfur. Data reproduced from [42].

2.6. NOE in glycobiology NMR

The nuclear Overhauser effect (NOE)-based NMR studies have been considered the foundations of biomolecular NMR not only through the conformational perspective but also useful in deciphering the hydrogen-bond networks of intermolecular complexes containing glycans. This was the main technique applied by the Nobel Prize-awarded group of Kurt Wuthrich in structural studies of biomolecules by NMR spectroscopy. NOE-signals result from a relaxation phenomenon and provide a through-space distance between two nuclei (usually protons) from the same molecule (intra-molecular NOE) from the same residue (intra-residue NOE) or those induced in the presence of binding residue (inter-residue NOE), as well as from different molecules (inter-molecular NOEs). The latter is usually studied by the specific transferred NOE (trNOE)-signals [52]. Next, we will give few brief examples in glycobiology successfully using these two NOE-related NMR techniques.

2.6.1. The classical through space NOE-based structural and conformational studies

As already documented in item 2.5.1, IdoA units can experience in solution the chair $^1C_4$, and $^4C_1$ as well as the skew-boat $^2S_0$ conformations. The relative proportion of each conformer population varies as function of the adjacent units and attached substituents (if any) [41]. The contribution of each of these conformation populations can be determined by $^3J_{H\text{-}H}$, but also by intra-residue NOE-signals. In the work of Silipo et al. [41], the NOE contact between H2 and H5 of the IdoA unit, which can only be seen in the skew-boat $^2S_0$ conformation, supports the $^3J_{H\text{-}H}$-based results obtained from a study of a dermatan sulfate-derived tetrasaccharide (Table 5) [41].

Unlikely the previous example of intra-residue NOE-signal, in the work of Castro et al. [53], an intramolecular inter-residue NOE-signal has been assigned between the (1→3) glycosidic bonds of β-galactosyl units in a marine sulfated galactan. This NOE-signal was crucial to define the repetitive disaccharide unit of this new structure of sulfated glycan [53].

NOE-signals are the most important NMR information for conformational studies, including of carbohydrates. The existence of conformational structures of some oligosaccharides in solution can be seen by the examples of tri-, tetra-, and pentasaccharides of (1→2)-β-mannopyranosyl units studied by NOESY (nuclear Overhauser effect spectroscopy), and ROESY (rotational frame nuclear Overhauser effect spectroscopy) [54]. These both 2D NMR spectral types are capable to display cross-peaks from through space ^{1}H-^{1}H connections, however, with different intensities as a function of the molecular size or correlation-time. ROE-signals are always positive, ranging from ~40 to 65% intensity, while NOE-signals has larger range, from 50% positive peak intensity to 100% negative peak intensity, according to the molecular size and correlation-time [55]. Therefore, as NOE- and ROE- resonances depend on the size of the molecules, both experiments were recorded for the mannopyranan oligosaccharides [54]. Inter-proton distances were calculated based on the r^{-6} relationship between distance and NOE/ROE cross-peak intensity related to a signal reference from the same studied molecule with known parameters. In this example case, the intra-residue H1-H5 contact, set to 2.39 Å was used found on relevant crystal structure models [54], and the values obtained are displayed in Table 8.

Observed NOE- or ROE-contacts	**Inter-proton distances calculated (Å)**					
	trisaccharide[a]		tetrasaccharide[b]		pentasaccharide[a]	
Contacts across glycosidic linkage	observed	calculated	observed	calculated	observed	calculated
H2A-H1B	2.3	2.5	2.3	2.4	2.2	2.5
H2B-H1C	2.3	2.4	2.2	2.4	2.2	2.4
H2C-H1D	-	-	2.3	2.5	nq[c]	2.4
H2D-H1E	-	-	-	-	2.3	2.5
Contacts separated by one residue						
H4A-H1C	3.3	2.7	3.1	2.8	3.2	2.9
H4A-H2C	2.9	3.0	2.8	2.7	2.7	2.7
H4B-H1D	-	-	3.1	2.8	3.1	2.9
H4B-H2D	-	-	2.7	3.0	2.7	2.8
H4C-H1E	-	-			3.2	2.8
H4C-H2E	-	-			2.8	3.1
Contacts separated by two residues						
H4A-H1D	3.7	3.4	4.1	3.6	3.7	3.4
H4B-H1E	4.2	3.7	-	-	4.2	3.7

[a]ROE-contacts

[b]NOE-contacts

[c]not quantified because of overlap.

Table 8. NOE- or ROE-contacts for (1→2)-β-mannopyranosyl units tri-, tetra- and pentasaccharides. Oligosaccharides are labeled from the reducing terminus. Reproduced from [54].

All the inter-glycosidic contacts (H1-H2) between linked mannosyl units are assigned and with similar distance values (Table 8). The rising numbers of other spatial contacts in tri-, tetra- and pentasaccharide, point clearly towards defined 3D conformers of these molecules. The presence of contacts between H4 with H1 and H2 of residues located two units away in a chain, strikingly supports the presence of ordered structures of these oligosaccharides in solution [54].

2.6.2. Protein-bound carbohydrate conformations seen by transferred NOE

Transferred NOE (trNOE) experiments can reveal the bioactive conformations of protein-bound carbohydrates [56-59]. This is very important in glycobiology since the recognition of binding conformers of sugars to enzymes, antibodies and lectins are of great interest in glycobiology, especially for in carbohydrate-based drug design. trNOE is obtained from a regular NOESY experiment although at a protein-carbohydrate mixed samples that posses some dynamic exchange. The carbohydrate ligand must be in excess of the protein, and thus trNOE can be collected from the free ligant that experience the "on-state" upon physical contact with the protein. In complexes involving large molecules, cross-relaxation rates of the bound state (σ^B), which depend on the respective inter-proton distances, the spectrometer frequency, and the correlation time (molecular size) of the complex, have opposite sign compared to the free state (σ^F), and produce negative resonance. Therefore, the existence of binding formation can be proved by simple visual inspection since the small molecules, free ligants, would produce positive NOE signals. For a trNOE experiment, the following conditions must occur: 1) PROTEIN + $\text{SUGAR}_{(\text{excess})} \leftrightarrow$ PROTEIN.SUGAR, 2) K_a = [PROTEIN.SUGAR] / [PROTEIN] x [SUGAR], 3) $p_b\sigma^B > p_f\sigma^F$, and 4) $k_{-1} >> \sigma^B$, where K_a is the association constant, p_b and p_f are the fractions of bound and free sugar ligand, σ^B and σ^F are the cross-relaxation rates for the bound and unbound ligand states, and k_{-1} is the off-rate constant. Therefore, based on the fourth rule, the exchange reaction has to be fast on the relaxation timescale.

The use of trNOE can be illustrated by the work concerning the protein-bound conformational characterization of the mimetic ligand β-D-Glc*p*NAc-(1→2)-α-D-Man*p*-(1→6)-β-D-Glc*p*-*O*-Octyl, an acceptor of the enzyme N-acetylglucosaminyltransferase V (GnT-V) [59]. Figure 18 shows an expansion of the glycosyl region at NOESY spectra of the carbohydrate acceptor substrate, in the presence of the enzyme, and the enzyme only. Regardless the cross-peaks from the protein (Figure 18c), Figure 18b showed a number of cross-peaks not observed in the free acceptor (Figure 18a). These are trNOE that were used as distances constraints (Table 9) for modeling a GnT-V-bound conformation of the ligand (Figure 19). Simulated annealing was used to refine the structure of the bound-state acceptor using the NOE constraints listed in Table 9. From the 100 structures generated, an ensemble of the selected 10 lowest overall energy conformations-built structure is illustrated in Figure 19. With this example, trNOE-based NMR studies are shown to be quite efficient to determine, protein-bound conformational states of carbohydrates.

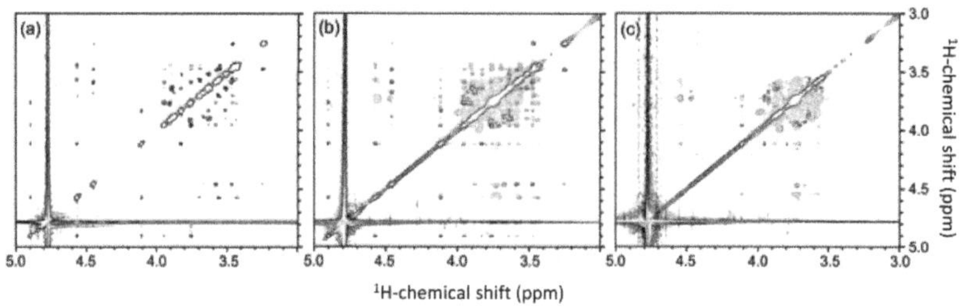

Figure 18. Expanded (carbohydrate region) NOESY spectra of (a) the acceptor, (b) the acceptor in the presence of GnT-V, and (c) GnT-V background. These allow assessment of conformational changes on binding to GnT-V. Reproduced from [59].

O-Octyl	Glc	Man′	GlcNAc′′	Distances (Å)
CH2-1b	H1	-	-	3.0
CH2-1a	H1	-	-	3.0
CH2-2	H1	-	-	3.5
-	H5-H6(S)	-	-	2.2
-	H1-H6(S)	-	-	3.3
-	H6(S)	H1′	-	2.5
-	H6(S)	H2′	-	2.9
-	H4	H3′	-	3.0
-	H5	H1′	-	3.5
-	H4	H1′	-	3.9
-	H6(S)	-	H1′′	3.0
-	H6(S)	-	NAc	3.6
-	-	H4-H6(S)	-	2.9
-	-	H2′	H1′′	2.6
-	-	H1′	H1′′	2.7
-	-	H2′	H5′′	3.5
-	-	H1′	NAc	3.5
-	-	H3′	H1′′	3.6
-	-	H6′(S)	NAc	3.6
		H2′	H4′′	3.7
		H1′	H5′′	4.1
		H2′	NAc	4.8
		H4′	NAc	4.9
			H5′′-H6′′(R)	2.5
			H5′′-H6′′(S)	2.5
			H4′′-H6′′(S)	3.1
			H1′′-NAc	4.0
			H2′′-NAc	4.5
			H3′′-NAc	

Table 9. ^1H-^1H distances determined from the trNOE signals from the acceptor in the presence of GnT-V. Data reproduced from [59].

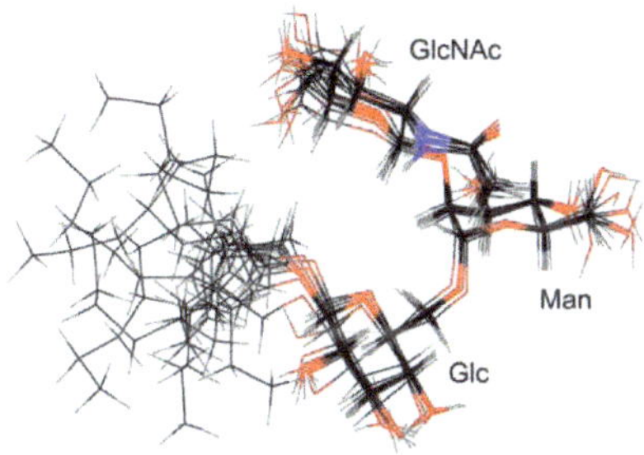

Figure 19. The 10 lowest energy structures from NOE restrained simulated annealing were aligned using the sugar ring atoms. Note the higher disorder at the *O*-octyl chain. Reproduced from [59].

3. Marked conclusions

Despite the recent association of NMR with glycobiology, relevant results have ultimately appeared. Although carbohydrates posses high-order degrees of flexibility, and usually great structural complexity; NMR methods still seem quite able to elucidate the main structural characteristics and dynamic behaviors of the majority of glycans. Since NMR spectroscopy is currently the most advanced and powerful structural technique, despite its sensitivity issue, its contribution to the glycobiology's progress, and thus to the current glycomics' is profound. New NMR methods have been adjusted just for carbohydrate analysis, inclusively specific isotopic labeling protocols to overcome the sensitivity problem. Proton, carbon-13, nitrogen-15 by either one- or multi-dimensional NMR experiments, chemical shifts, scalar coupling constants, dipolar coupling constants, and NOE-through space connections of free or protein-bound carbohydrates comprise the principle NMR spectroscopy methods for glycobiology. Many other NMR techniques, such as saturation transfer difference and modified pulse sequence destined just to address glycobiology-related problems do exist, although not covered in this chapter. The main idea of carbohydrates as just energetically or structurally involved-molecules, has falling apart as many other vital functions of glycans, mostly in signaling events, have been unraveled along the past few years. And NMR spectroscopy is making an outstanding contribution for these big discoveries.

Author details

Vitor H. Pomin
Program of Glycobiology, Institute of Medical Biochemistry, University Hospital Clementino Fraga Filho, Federal University of Rio de Janeiro, Rio de Janeiro, Brazil

Acknowledgement

The author acknowledges the Publisher for the kind invitation in contributing with this chapter and the support; Dr. John Glushka to the help on the GlcNS assignment; Dr.

Ana Paula Valente to the help on the ^{13}C-direct observe spectrum of Glc, and the accessibility to the Bruker 400 MHz spectrometer; and Prof. James H. Prestegard for all NMR background during my post-doctorate period, and the accessibility at the Varian 800 MHz spectrometer. All the data discussed in this chapter in which the Varian 800 MHz is indicated in figure legends, were recorded during the author´s post-doctorate period at Complex Carbohydrate Research Center, the University of Georgia, US under the partial financial support from the National Center for Research Resources of the National Institute of Health, RR005351, and from the post-doctoral fellowship (PDE #201019/2008-6) from Conselho Nacional de Desenvolvimento Científico e Tecnológico (CNPq), Brazil.

4. References

[1] Kurt Wüthrich (1986) NMR of Proteins and Nucleic Acids, Wiley Interscience, pg 1.

[2] LeMaster DM, Richards FM (1982) Preparative-scale isolation of isotopically labeled amino acids. Anal. biochem. 122: 238-247.

[3] Kainosho M, Tsuji T (1982) Assignment of the three methionyl carbonyl carbon resonances in Streptomyces subtilisin inhibitor by a carbon-13 and nitrogen-15 double-labeling technique. A new strategy for structural studies of proteins in solution. Biochem. 21: 6273-6279.

[4] Pervushin K, Riek R, Wider G, Wüthrich K (1997) Attenuated T2 relaxation by mutual cancellation of dipole-dipole coupling and chemical shift anisotropy indicates an avenue to NMR structures of very large biological macromolecules in solution. Proc. natl. acad. sci. USA. 94: 12366-12377.

[5] Salzmann M, Pervushin K, Wider G, Senn H, Wüthrich K (1998) TROSY in triple-resonance experiments: new perspectives for sequential NMR assignment of large proteins. Proc. natl. acad. sci. USA. 95: 13585-13590.

[6] Liu Y, Prestegard JH (2011) Multi-dimensional NMR without coherence transfer: minimizing losses in large systems. J. magn. reson. 212: 289-298.

[7] Pervushin K, Ono A, Fernández C, Szyperski T, Kainosho M, Wüthrich K (1998) NMR scalar couplings across Watson-Crick base pair hydrogen bonds in DNA observed by transverse relaxation-optimized spectroscopy. Proc. natl. acad. sci. USA. 95: 14147-14151.

[8] Riek R, Wider G, Pervushin K, Wüthrich K (1999) Polarization transfer by cross-correlated relaxation in solution NMR with very large molecules. Proc. natl. acad. sci. USA. 96: 4918-4923.

[9] Fiala R, Czernek J, Sklenář V (2000) Transverse relaxation optimized triple-resonance NMR experiments for nucleic acids. J. biomol. NMR. 16: 291-302.

[10] Pomin VH, Sharp JS, Li X, Wang L, Prestegard JH (2010) Characterization of glycosaminoglycans by 15N NMR spectroscopy and in vivo isotopic labeling. Anal. chem. 82: 4078-4088.

[11] Parella T, Nlis P (2005) Spin-edited 2D HSQC-TOCSY experiments for the measurement of homonuclear and heteronuclear coupling constants: application to carbohydrates and peptides. J. magn. reson. 176: 15-26.

[12] Zhang Q, Li N, Liu X, Zhao Z, Li Z, Xu Z (2004) The structure of a sulfated galactan from *Porphyra haitanensis* and its in vivo antioxidant activity. Carbohydr. res. 339: 105-111.
[13] Kolender AA, Matulewicz MC (2002) Sulfated polysaccharides from the red seaweed *Georgiella confluens*. Carbohydr. res. 337: 57-68.
[14] Liao ML, Chiovitti A, Munro SL, Craik DJ, Kraft GT, Bacic A (1996) Sulfated galactans from Australian specimens of the red alga Phacelocarpus peperocarpos (Gigartinales, Rhodophyta). Carbohydr. res. 296: 237-247.
[15] Langeslay DJ, Beni S, Larive CK (2011) Detection of the ^{1}H and ^{15}N NMR resonances of sulfamate groups on aqueous solution: a new tool for heparin and heparan sulfate characterization. Anal. chem. 83: 8006-8010.
[16] Gomes AM, Kozlowski EO, Pomin VH, de Barros CM, Zaganeli JL, Pavão MS (2010) Unique extracellular matrix heparan sulfate from the bivalve Nodipecten nodosus (Linnaeus, 1758) safely inhibits arterial thrombosis after photochemically induced endothelial lesion. J. biol. chem. 285: 7312-7323.
[17] Jiménez-Barbero J, Peters J (2003) NMR spectroscopy of glycoconjugates, Wiley-VCH, pg 4.
[18] Barb AW, Prestegard JH (2011) NMR analysis demonstrates immunoglobulin G N-glycans are accessible and dynamics. Nat. chem. biol. 7: 147-153.
[19] Scalan CN, Burton DR, Dwek RA (2008) Making antibodies safe. Proc. natl. acad. sci. USA. 105: 4081-4082.
[20] Kaneko Y, Nimmerjahn F, Ravetch JV (2006) Anti-Inflammatory Activity of Immunoglobulin G Resulting from Fc Sialylation. Science. 313: 670–673.
[21] Tsuchiya N, Endo T, Matsuta K, Yoshinoya S, Aikawa T, Kosuge E, Takeuchi F, Miyamoto T, Kobata A (1989) Effects of galactose depletion from oligosaccharide chains on immunological activities of human IgG. J. rheumatol. 16: 285–290.
[22] Nimmerjahn F, Anthony RM, Ravetch JV *(2007)* Agalactosylated IgG antibodies depend on cellular Fc receptors for *in vivo* activity. Proc. natl. acad. sci. USA. 104: 8433–8437.
[23] Malhotra R, Wormald MR, Rudd PM, Fischer PB, Dwek RA, Sim RB (1995) Glycosylation changes of IgG associated with rheumatoid arthritis can activate complement via the mannose-binding protein. Nat. med. 1: 237–243.
[24] Shields RL, Lai J, Keck R, O'Connell LY, Hong K, Meng YG, Weikert SH, Presta LG (2002) Lack of fucose on human IgG1 *N*-linked oligosaccharide improves binding to human FcγRIII and antibody-dependent cellular toxicity. J. biol. chem. 277: 26733–26740.
[25] Teng Q (2005) Structural biology – Practical NMR applications. Springer Science, pg 253-258.
[26] Cavanagh J (2007) Protein NMR Spectroscopy: Principles and Practice. 2nd ed., Academic Press, pg 711.
[27] Kurt Wüthrich (1986) NMR of Proteins and Nucleic Acids, Wiley Interscience, pg 1.
[28] Teng Q (2005) Structural biology – Practical NMR applications. Springer Science, pg 18.
[29] Teng Q (2005) Structural biology – Practical NMR applications. Springer Science, pg 3-4.

[30] Martin L, Blanpain C, Garnier P, Wittamer V, Parmentier M, Vita C (2001) Structural and Functional Analysis of the RANTES-Glycosaminoglycans Interactions. Biochem. 40: 6303-6318.

[31] Hoogewerf AJ, Kuschert GS, Proudfoot AE, Borlat F, Clark-Lewis J, Power CA, Wells TN (1997) Glycosaminoglycans mediate cell surface oligomerization of chemokines. Biochem. 36: 13570-13578.

[32] Wang X, Watson C, Sharp JS, Handel TM, Prestegard JH. (2011) Oligomeric structure of the chemokine CCL5/RANTES from NMR, MS, and SAXS data. Structure. 19: 1138-1148.

[33] Proudfoot AE, Fritchley S, Borlat F, Shaw JP, Vilbois F, Zwahlen C, Trkola A, Marchant D, Clapham PR, Wells TN. (2001) The BBXB motif of RANTES is the principal site for heparin binding and controls receptor selectivity. J. biol. chem. 276: 10620-10626.

[34] Pomin VH, Park Y, Huang R, Heiss C, Sharp JS, Azadi P, Prestegard JH (2012) Exploiting enzyme specificities in digestions of chondroitin sulfates A and C: production of well-defined hexasaccharides. Glycobiology, 22: 826-838.

[35] Czaplewski LG, McKeating J, Craven CJ, Higgins LD, Appay V, Brown A, Dudgeon T, Howard LA, Meyers T, Owen J, Palan SR, Tan P, Wilson G, Woods NR, Heyworth CM, Lord BI, Brotherton D, Christison R, Craig S, Cribbes S, Edwards RM, Evans SJ, Gilbert R, Morgan P, Randle E, Schofield N, Varley PG, Fisher J, Waltho JP, Hunter MG (1999) Identification of amino acid residues critical for aggregation of human CC chemokines macrophage inflammatory protein (MIP)-1 alpha, MIP-1 beta, and RANTES - Characterization of active disaggregated chemokine variants. J. biol. chem. 274: 16077-16084.

[36] van Boeckel CAA, van Aelst SF, Wagenaars GN, Mellema J-R, Paulsen H, Peters T, Pollex A, Sinnwell V (1987) Conformational analysis of synthetic heparin-like oligosaccharides containing α-L-idopyranosyluronic acid. Recl. trav. chim. pays-bas. 106: 19-29.

[37] Torri G, Casu B, Gatti G, Petitou M, Choay J, Jacquinet JC, Sinaÿ P (1985) Mono- and bidimensional 500 MHz 1H-NMR spectra of a synthetic pentasaccharide corresponding to the binding sequence of heparin to antithrombin-III: evidence for conformational peculiarity of the sulfated iduronate residue. Biochem. biophys. res. com. 128: 134-140.

[38] Yamada S, Yoshida K, Sugiura M, Sugahara K (1992) One- and two-dimensional 1H-NMR characterization of two series of sulfated disaccharides prepared from chondroitin sulfate and heparan sulfate/heparin by bacterial eliminase digestion. J. biochem. 112: 440-447.

[39] Wessel HP, Bartsch S (1995) Conformational flexibility in highly sulfated beta-D-glucopyranoside derivatives. Carbohydr. res. 274: 1-9.

[40] Maruyama T, Toida T, Imanari T, Yu G, Linhardt RJ.(1998) Conformational changes and anticoagulant activity of chondroitin sulfate following its O-sulfonation. Carbohydr. res. 306: 35-43.

[41] Silipo A, Zhang Z, Cañada FJ, Molinaro A, Linhardt RJ, Jiménez-Barbero J (2008) Conformational analysis of a dermatan sulfate-derived tetrasaccharide by NMR, molecular modeling, and residual dipolar coupling. Chembiochem. 9: 240-252.

[42] Jin L, Hricovíni M, Deakin JA, Lyon M, Uhrín D (2009) Residual dipolar coupling investigation of a heparin tetrasaccharide confirms the limited effect of flexibility of the iduronic acid on the molecular shape of heparin. Glycobiology. 19: 1185-1196.
[43] Mikhalov D, Linhardt RJ, Mayo KH (1997) NMR solution conformation of heparin-derived hexaccharide. Biochem. j. 328: 51-61.
[44] Ferro DR, Provasoli A, Ragazzi M, Torri G, Casu B, Gatti G, Jacquinet JC, Sinay P, Petitou M, Choay J (1986) Evidence for conformational equilibrium of the sulfated L-iduronate residue in heparin and in synthetic heparin mono- and oligo-saccharides: NMR and force-field studies. J. am. chem. soc. 108: 6773-6778.
[45] Foster MJ, Mulloy B (1993) Molecular dynamics study of iduronate ring conformation. Biopolymers. 33: 575-588.
[46] Hricovíni M, Guerrini M, Bisio A (1999) Structure of heparin-derived tetrasaccharide complexed to the plasma protein antithrombin derived from NOEs, J-couplings and chemical shifts. Eur. j. biochem. 261: 789-801.
[47] Jiménez-Barbero J, and Peters T (2003) NMR Spectroscopy of Glycoconjugates, Wiley-VCH, pg 197.
[48] Tian F, Al-Hashimi HM, Craighead JL, Prestegard JH (2001) Conformational analysis of a flexible oligosaccharide using residual dipolar couplings. J. am. chem. soc. 123: 485-492.
[49] Prestegard JH, Al-Hashimi HM, Tolman JR (2000) NMR structures of biomolecules using field oriented media and residual dipolar couplings. Q. rev. biophys. 33: 371-424.
[50] Losonczi JA, Andrec M, Fischer MW, Prestegard JH (1999) Order matrix analysis of residual dipolar couplings using singular value decomposition. J. magn. reson. 138: 334-342.
[51] Valafar H, Prestegard JH (2004) REDCAT: a residual dipolar coupling analysis tool. J. magn. reson. 167: 228-241.
[52] Willianson MP (2006) The transferred NOE. Modern magn. reson. 28: 1357-1362.
[53] Castro MO, Pomin VH, Santos LL, Vilela-Silva AC, Hirohashi N, Pol-Fachin L, Verli H, Mourão PA (2009) A unique 2-sulfated {beta}-galactan from the egg jelly of the sea urchin Glyptocidaris crenularis: conformation flexibility versus induction of the sperm acrosome reaction. J. biol. chem. 284: 18790-18800.
[54] Jiménez-Barbero J, and Peters T (2003) NMR Spectroscopy of Glycoconjugates, Wiley-VCH, pg 160-168.
[55] Otter A, Kotovych (1988) The solution conformation of the synthetic tubulin fragment Ac-tubulin-α(430–441)-amide based on two-dimensional ROESY experiments. Can. j. chem. 66: 1814-1820.
[56] Angulo J, Rademacher C, Biet T, Benie AJ, Blume A, Peters H, Palcic M, Parra F, Peters T (2006) NMR analysis of carbohydrate-protein interactions. Methods enzymol. 416: 12-30.
[57] Bevilacqua VL, Thomson DS, Prestegard JH (1990) Conformation of methyl beta-lactoside bound to the ricin B-chain: interpretation of transferred nuclear Overhauser effects facilitated by spin simulation and selective deuteration. Biochem. 29: 5529-5537.

[58] Bevilacqua VL, Kim Y, Prestegard JH (1992) Conformation of beta-methylmelibiose bound to the ricin B-chain as determined from transferred nuclear Overhauser effects. Biochem. 31: 9339-9349.

[59] Macnaughtan MA, Kamar M, Alvarez-Manilla G, Venot A, Glushka J, Pierce JM, Prestegard JH (2007) NMR structural characterization of substrates bound to N-acetylglucosaminyltransferase V. J. mol. biol. 366: 1266-1281.

Section 2

Systems Glycosylation

Chapter 5

Distribution, Characterization of Mycobacterial Glycolipids and Host Responses

Nagatoshi Fujiwara

Additional information is available at the end of the chapter

1. Introduction

Robert Koch discovered the acid-fast bacterium that is the pathogenic germ of tuberculosis (TB) in 1882 [1]. TB is a major public health problem in the world and the leading cause of death from a single infectious agent. The World Health Organization estimates that one third of the world's population is infected with *Mycobacterium tuberculosis*, and annually reports the global burden of the disease caused by TB. Over 8 million new cases and nearly 1.5 million deaths from TB occur each year, presenting a significant threat to the world health [2]. The pathogenicity of mycobacteria is related to their ability to evade being ingested by macrophages, produce latent infection, and induce delayed-type hypersensitivity lesions such as granulomas. Moreover, the global spread of multi- and extensively drug-resistant TB (MDR-TB and XDR-TB, respectively) and the number of immunocompromised hosts, including victims of the human immunodeficiency virus (HIV) epidemic, are important problems [3].

The mycobacteria include the TB-causative acid-fast bacteria that are widely pathogenic to humans: *M. tuberculosis, M. avium-intracellulare* complex (MAC), *M. leprae*, and *M. bovis*, the source of the only available TB vaccine, Bacillus Calmette Guérin (BCG). The acid-fast bacteria are rich in lipids, and many mycoloyl glycolipids, such as cord factor/trehalose-6,6'-dimycolate (TDM), phenolic glycolipid (PGL), sulfolipid (SL), glycopeptidolipid (GPL), phosphatidylinositol mannoside (PIM), lipomannan (LM), and lipoarabinomannan (LAM), are distributed in the cell wall [4-7]. Among them, glycolipids specific to the mycobacteria play a key role in the pathogenesis, because mycobacteria have large amounts of lipids that possess pleiotropic activities. The complex interaction between a range of mycobacterial components and the host cause the pathogenesis.

In this review, the distribution of major glycolipids in several *Mycobacteria* and structural analyses using mass spectrometry (MS) are described. MS and MS/MS are useful to analyze the glycosyl linkage, composition, and sequence of the sugar moiety. These results make it possible to discuss the heterogeneity and biosynthesis of glycolipids, and host responses to them. We hope that this review will promote better understanding of the structure-function relationships of glycolipids and open new avenues for prevention of infectious diseases.

2. Distribution of glycolipids and phospholipids in mycobacteria

The cell envelope surrounds the cytoplasm and is important for bacterial physiology and protection of microorganisms from their environment. Unlike other pathogenic microbes, the cell envelope of the acid-fast bacteria, including mycobacteria, is wax-like. The characteristic component is mycolic acids (MAs) which are α-branched β-hydroxy fatty acids (FAs), and have species-specific carbon-chain lengths and subclasses (α, methoxy, keto, dicarboxy, epoxy, etc.). The total lipid fraction was extracted with chloroform/methanol (3:1 and 2:1 v/v) and developed by two-dimensional thin-layer chromatography (TLC). Many glycolipids and phospholipids were detected, as shown in Fig. 1. Some glycolipids, such as cord factor/TDM and trehalose 6-monomycolate (TMM), exist ubiquitously in mycobacteria, and other glycolipids, including SLs and penta-acyl trehalose, exist only in virulent strains [8]. In general, the mycobacterial species have heterogeneous compositions and concentrations of glycolipids. According to current models of the mycobacterial cell envelope, which are ultimately based on an idea of Minnikin [9], arabinogalactan-attached MAs are believed to form one highly-arranged leaflet of a second membrane-like structure adjacent to the cell-wall skeleton (CWS). Complemented by other lipids forming the outer leaflet, this structure represents a very hydrophobic barrier that is responsible for the resistance to certain drugs [10]. The free glycoconjugates of MA, in particular TMM and cord factor/TDM, appear to be more difficult to localize, and occasionally they are described or depicted as distributed on the cell surface or buried more deeply in the cell envelope. A proposed structure of the mycobacterial cell envelope is shown in Fig. 2 [4, 11, 12].

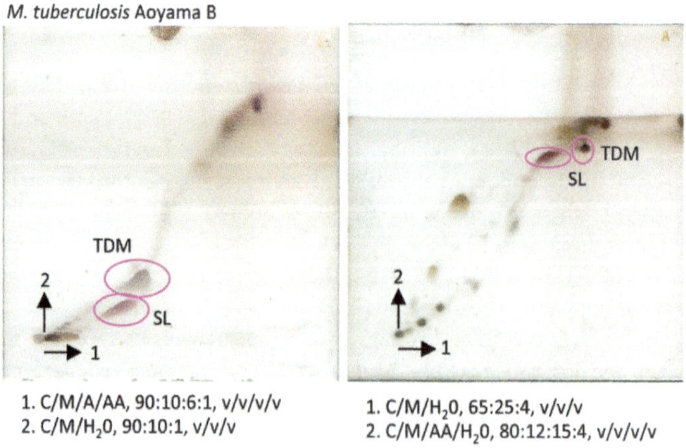

Figure 1. TLC of total lipids derived from *Mycobacterium tuberculosis* Aoyama B strain.

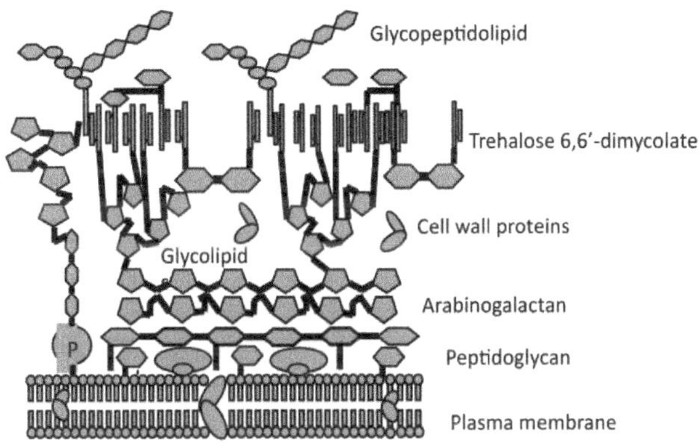

Figure 2. Schematic representation of mycobacterial cell envelope.

3. Host responses to mycobacterial glycolipids

Mycobacteria are intracellular pathogens. Mycobacterial infection has dual consequences for the host: development of inflammatory lesions and clearance of the pathogen. Genetic regulation and expression of cell-mediated immunity and delayed-type hypersensitivity play critical roles in the outcome. Cell-mediated immunity participates in host defense, whereas delayed-type hypersensitivity is involved in the development of granulomatous inflammation. After invasion to the host cell, the mycobacteria multiply inside macrophages. The mechanism of pathogenesis is focused on evading the host's killing system and induction of delayed-type hypersensitivity [13]. Certain cell wall glycolipids, such as cord factor/TDM, SL, and LAM [5-7], are involved in the host-pathogen interaction, and induce the primary host immune responses.

Peptides and proteins with a wide range of antigenic moieties are recognized by the host immune system using major histocompatibility complex (MHC) class I or II. The recognition of lipids is important for host defense against mycobacterial infection as well as other antigenic responses. Several bacterial lipid antigens are recognized by specific T cells. MA derived from mycobacteria is a lipid antigen that stimulates CD1b-restricted T cells [14, 15]. Other immunogenic mycobacterial glycolipids, glucose-6-monomycolate (GMM), PIM, and LAM, are available for loading onto CD1 molecules and recognized by specific T cells [16, 17].

4. Mycoloyl glycolipids

Acid-fast bacteria produce several mycolyl glycolipids that are composed of sugar and MA. The carbon-chain lengths and subclasses of MAs are species-specific. *Nocardia, Rhodococcus,* and *Diezia* species have short carbon-chain length MAs, and the subclass is mainly the α type. MAs of *Mycobacterium* species have long carbon-chains and are of various subclasses (α, methoxy, keto, dicarboxy, and epoxy types). In general, the average number of carbons in MAs increases from *Dietzia, Rhodococcus, Nocardia,* and *Gordona* to *Mycobacteria.* The MAs are extracted with *n*-hexane or chloroform after alkaline-hydrolysis of heat-killed bacteria, and the subclasses are detected by using TLC developed with methyl ester derivatives (Fig. 3). The molecular species of the MA subclasses are determined by matrix-assisted laser desorption/ionization time-of-flight mass spectrometry (MALDI-TOF MS) using an Ultraflex II (Bruker Daltonics, Billerica, MA, USA) with 10 mg/ml 2,5-dihydroxybenzoic acid (DHB) in chloroform-methanol (1:1, v/v) as a matrix, and analyzed in Reflectron mode with an accelerating voltage operating in positive mode at 20 kV (Fig. 4). The MAs with short carbon-chains can be analyzed by using gas-chromatography mass spectrometry GC/MS. The trimethylsilyl derivatives of MA methyl esters are volatile, and the GC/MS spectra show specific fragment patterns that are assigned to detailed structures (Fig. 5), although the MAs with long carbon chains are difficult to vaporize and detect by GC/MS [18-20]. In addition, MAs are analyzed by using high performance liquid chromatography (HPLC) [21]. The resolution of HPLC is poor compared to that of GC/MS. GMM, fructose-6-monomycolate (FMM), and mannose-6-monomycolate (MMM) are produced by acid-fast bacteria cultured with their respective carbon source: glucose, fructose, and mannose. The structures are composed of MAs attached to glucose, fructose, and mannose, respectively, at the C6 position [20, 22]. Recently, it was reported that *M. tuberculosis* produces GMM by transferring MA to glucose from cord factor/TDM, and thus evades the host immune system [23]. The molecular weights of mycoloyl glycolipids are measured by MALDI-TOF MS, and this information makes it possible to determine the combination of MA subclasses [24]. Cord factor/TDM is probably the most prominent and best-studied MA-containing compound of mycobacteria, and shows pleiotropic activities, including the development of granulomatous inflammation, anti-tumor immune response, and adjuvant effects based on the induction of proinflammatory and type 1 helper T cell (Th1)-related cytokines from host cells [25-27]. We demonstrated that administration of cord factor/TDM could induce delayed-type hypersensitivity-like lesions and foreign-body granulomas in euthymic and athymic mice, regardless of preimmunization with mycobacteria and cord factor/TDM. In fact, preimmunized mice challenged with cord factor/TDM developed more severe lesions than unimmunized mice. At the active lesion, CC-chemokines attracting monocytes, proinflammatory cytokines, and immunoregulatory cytokines were found. The inflammatory and cytokine responses were augmented in immunized mice challenged with cord factor/TDM. As a result, cord factor/TDM can induce both foreign body-type (nonimmune) and hypersensitivity-type (immune) granulomas by acting as a nonspecific irritant and T-cell-dependent antigen, and both nonimmune and immune mechanisms participate in granulomatous inflammation induced by mycobacterial infection [13, 28]. Moreover, rhodococcal cord factor/TDM induced milder granulomatous lesions than a mycobacterial one, implying that the proinflammatory responses induced by cord factor/TDM are in proportion to the carbon-chain length and subclasses of MAs in addition to carbon species [20, 29]. Recently, it was clarified that macrophage-inducible C-type lectin (mincle) is an essential receptor for cord factor/TDM [30]. Mincle is a receptor for sugar, and our previous result showed that the proinflammatory responses are MA-dependent. I expect the discovery of other unknown receptors for cord factor/TDM in addition to mincle.

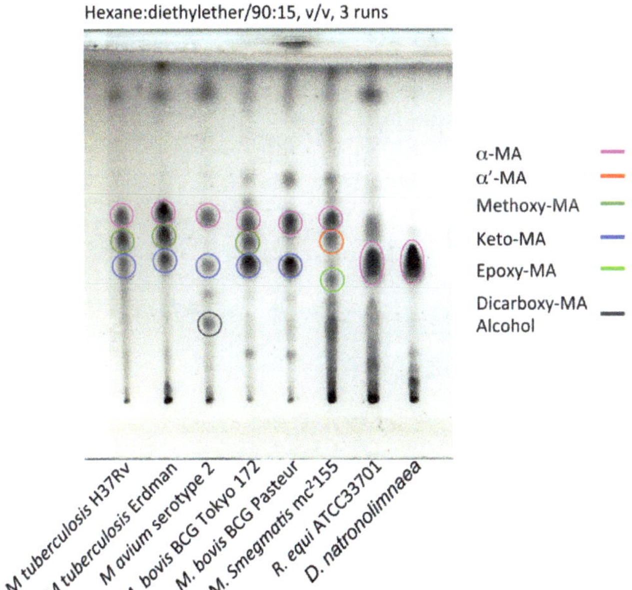

Figure 3. TLC of mycolic acid methyl esters derived from *Mycobacterium*, *Rhodococcus*, and *Dietzia* species.

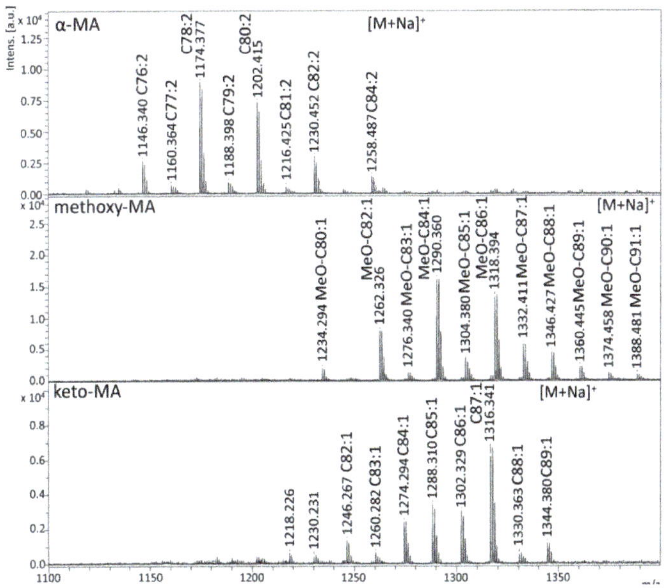

Figure 4. MALDI-TOF MS spectra of mycolic acid methyl ester subclasses from *M. tuberculosis* H37Rv.

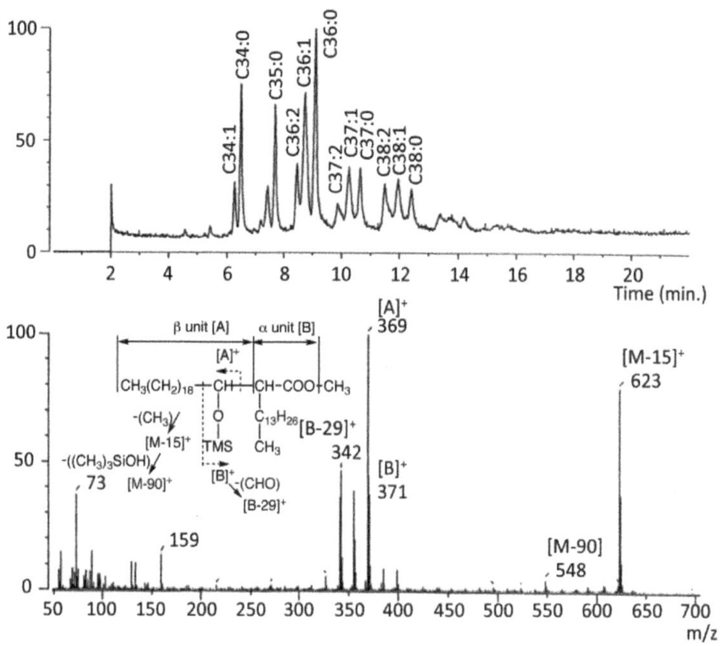

Figure 5. Total ion chromatogram and mass spectrum of mycolic acid trimethylsilyl derivatives from *R. equi*.

5. Glycopeptidolipid (GPL)

GPLs are produced by MAC, *M. scrofulaceuem*, *M. chelonae*, *M. fortuitum*, and *M. smegmatis*. Structurally, a GPL is composed of two parts, a common tetrapeptido-amino alcohol core and a serotype-specific oligosaccharide (OSE) elongated from 6-deoxy-talose (6-d-Tal). D-phenylalanine-D-*allo*-threonine-D-alanine-L-alaninol (D-Phe-D-*allo*-Thr-D-Ala-L-alaninol), which is modified with an amido-linked 3-hydroxy or 3-methoxy C26-C34 FA at the *N*-terminal of D-Phe; D-*allo*-Thr and terminal L-alaninol are further linked to a 6-d-Tal and 3,4-di-*O*-methyl rhamnose (3,4-di-*O*-Me-Rha), respectively [31]. This portion is common to all serotypes, and is called the serotype-nonspecific GPL (apolar GPL), which exhibits antigenicity [11]. Serotype-specific GPLs (polar GPLs) are further glycosylated with a variable haptenic OSE at 6-d-Tal. We determined the structures of the serotype 7, 13, and 16 GPLs and identified the gene clusters completing the OSE biosynthesis [32-34]. In addition, two methyltransferase genes of serotype 7- and 12-specific GPL biosynthesis were characterized [35]. The standard technique to classify MAC strains has employed serologic typing based on the OSE residue of the GPL. Recently, the biosynthetic pathway of various haptenic OSEs has been explored, and the genes encoding the pathways have been identified and characterized [36-38]. At present, 31 distinct serotype-specific polar

GPLs have been identified biochemically, and the complete structures of 17 GPLs are defined [12, 33]. The GPL present on the cell wall is considered to affect colony morphology. The MAC colony phenotype spontaneously changed from a smooth to a rough type, and this was due to a mutation lacking GPLs [39, 40]. The polar GPLs produced by MAC species are of particular interest because they are considered to be correlated with the physiology of the bacteria and the host responses to MAC infection, for example, colony morphology, sliding motility, biofilm formation, immune modulation, and virulence [40-43].

We have demonstrated the applicability of serodiagnosis of MAC pulmonary diseases using the GPL and GPL core antigens, and have also shown that the levels of GPL and GPL core antibodies reflect disease activity [44-46]. The GPLs are produced by MAC species but are absent in *M. tuberculosis*, making it possible to distinguish MAC from tuberculous mycobacteria [44, 47]. An anti-GPL antibody is produced in the sera of patients and the level reflects the extent of disease, which is useful in diagnosis and treatment [48, 49]. GPLs are one of the immunologically active molecules characteristic of MAC, and serotype-specific GPLs participate in the pathogenesis and immunomodulation in the host [50, 51]. It has been reported that the GPL core plays a role in suppression of mitogen-induced blastogenic response in spleen cells [52]. In addition, the immunomodulating activity of GPL on macrophage functions is serotype-dependent [53]. The serotype 4 GPL promotes phagocytosis and inhibits phagosome-lysosome (P-L) fusion, whereas the GPLs of serotypes 9 and 16 exhibit no effect on phagocytosis and P-L fusion. The serotype 8 GPL shows concomitant stimulation of both phagocytosis and P-L fusion. Because the GPL core, but not OSE, is common in all serotypes, the OSE of GPL may be involved in the mechanism of inhibition of P-L fusion, which is mediated through mannose receptors of macrophages [54]. The serotype 4 GPL inhibits the lymphoproliferative response to mitogens [51]. Thus, host responses to GPLs vary with the MAC serotype. It is reported that the uptake by and growth in macrophages of a MAC mutant with a gene in the GPL synthesis pathway inactivated by a transposon insertion were decreased [55]. The pathogenicity of GPL may comprise both a common peptide core and an OSE elongated from 6-d-Tal. The GPL is a pleiotropic molecule and participates in the pathogenesis of MAC disease. Elucidation of the structure-activity relationship of GPL is required for a better understanding of the pathogenesis.

6. Structural analysis of GPL

The polar GPL is species-specific in the portion of the OSE. MS is very useful to analyze the structure of OSE sequences and linkage positions in addition to total molecular weight. In this review, I describe detailed methods and results of structural analyses of some GPLs. The procedure for structural analysis of OSE is summarized in Fig. 6.

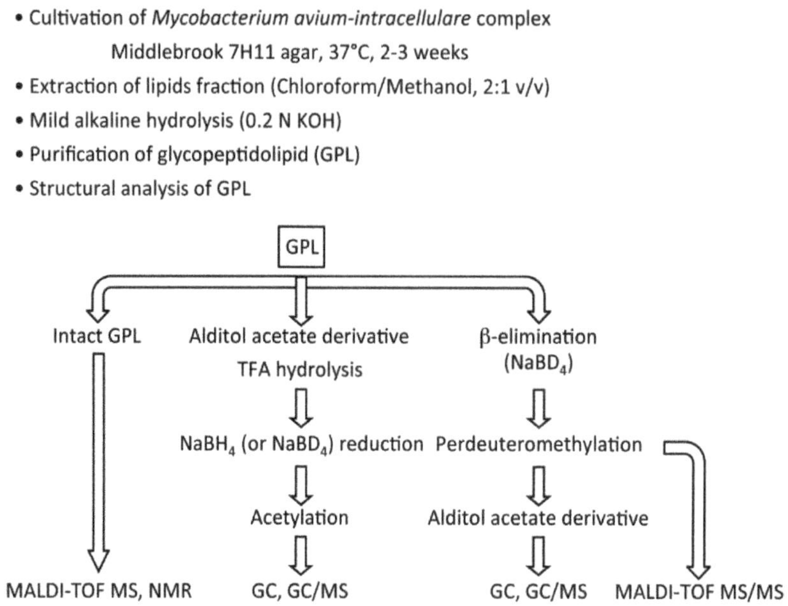

Figure 6. Procedure for structural analyses of OSE.

6.1. Preparation of GPL

MAC was grown on Middlebrook 7H11 agar (Difco Laboratories, Detroit, MI, USA) with 0.5% glycerol and 10% Middlebrook OADC enrichment (Difco) at 37°C for 2-3 weeks. Heat-killed bacteria were sonicated, and total lipids were extracted with chloroform-methanol (2:1, v/v). The total lipids were hydrolyzed with 0.2 N sodium hydroxide in methanol at 37°C for 2 h, followed by neutralization with 6 N hydrochloric acid. Alkaline-stable lipids were partitioned by a two-layer system with chloroform-methanol (2:1, v/v) and water. The organic phase was evaporated and precipitated with acetone to remove any acetone-insoluble components. The supernatant was partially purified with a Sep-Pak Silica Cartridge (Waters Corporation, Milford, MA, USA). The GPL was completely purified by preparative TLC of silicagel G (Uniplate; 20 × 20 cm, 250 μm; Analtech, Inc., Newark, DE, USA). The TLC plate was developed with chloroform-methanol-water (65:25:4 and 60:16:2, v/v), until a single spot was obtained.

6.2. Preparation of OSE moiety

β-Elimination of the GPL was performed with alkaline borohydride, and the OSE elongated from D-*allo*-Thr was released [34, 56]. The GPL was stirred in a solution of equal volumes of ethanol and 10 mg/ml sodium borodeuteride in 0.5 N sodium hydroxide at 60°C for 16 h. The reaction mixture was decationized with Dowex 50W X8 beads (Dow Chemical Company, Midland, MI, USA), and evaporated under nitrogen to remove boric acid. After partition into two layers of chloroform-methanol (2:1, v/v) and water, the upper aqueous phase was recovered and evaporated, and the OSE was purified as an oligoglycosyl alditol.

6.3. Molecular weight of intact GPL

The molecular species of the intact GPL was determined by MALDI-TOF MS. One μg of the GPL dissolved with chloroform-methanol (2:1,v/v) was applied to the target plate, and 1 μl of 10 mg/ml DHB in chloroform-methanol (1:1, v/v) was added as a matrix. The intact GPL was analyzed in the Reflectron mode with an accelerating voltage operating in positive mode at 20 kV [57]. The peak ions of intact GPL were detected in sodium adduct form, $[M+Na]^+$ as the main molecular-related ion; representative spectra of some GPLs are shown in Fig. 7. The mass numbers identified the proposed structures of each GPL.

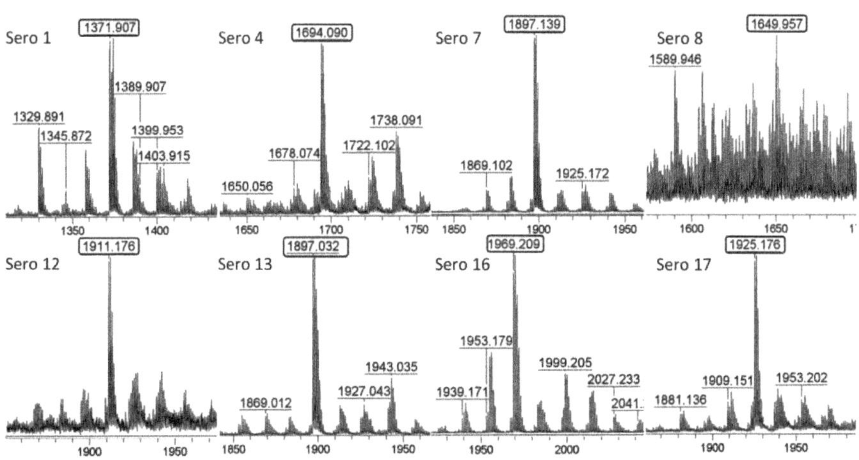

Figure 7. MALDI-TOF MS spectra of representative GPLs.

6.4. Glycosyl sequences of OSE.

The molecular weight of the OSE portion is measured by MALDI-TOF MS. The OSE and 10 mg/ml DHB were dissolved in ethanol-water (3:7, v/v) and applied to the target plate according to the method for intact GPL. To determine the glycosyl sequence of the OSE, MALDI-TOF MS/MS analysis of the oligoglycosyl alditol from the OSE was performed. The spectrum afforded the molecular ion $[M+Na]^+$, together with the characteristic mass increments in the series of glycosyloxonium ions formed on fragmentation at each glycosyl linkage from both terminal sugars to their opposites, respectively. It was shown the representative assignments of intact serotype 13 GPL and its OSE with the proposed structure in Fig. 8.

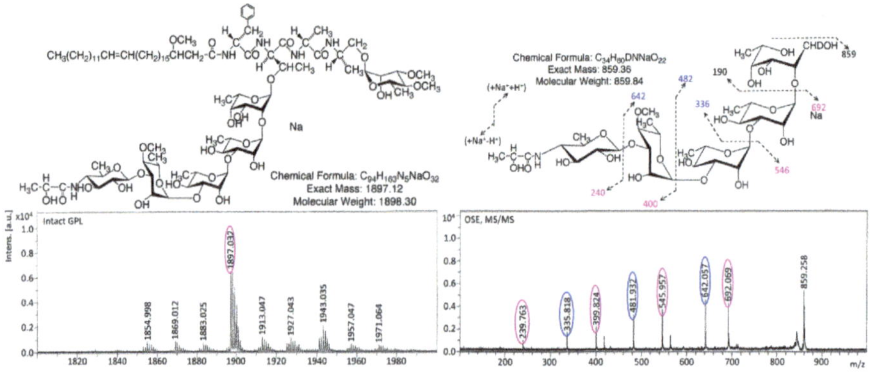

Figure 8. MALDI-TOF MS and MS/MS spectra of serotype 13 GPL. The molecular weight of the intact GPL was detected as m/z 1897 for $[M+Na]^+$ and fixed the proposed structure. The MS/MS spectra clearly show that the each ion was assigned to the fragmentation at each glycosyl linkage from both terminal sugars to their opposites, respectively.

6.5. GC/MS of carbohydrates

To determine the glycosyl composition and linkage position, GC/MS of partially methylated alditol acetate derivatives was performed. Perdeuteromethylation was conducted by the modified procedure of Hakomori [34, 58]. The OSE was dissolved with a mixture of dimethylsulfoxide and sodium hydroxide, followed by the addition of deuteromethyl iodide. After stirring at room temperature for 15 min, the reaction mixture was separated by a two-layer system of water and chloroform. The chloroform-containing perdeuteromethylated OSE layer was collected, washed twice with water, and evaporated completely. Partially deuteromethylated alditol acetate derivatives were prepared from perdeuteromethylated OSE by hydrolysis with 2 N trifluoroacetic acid at 120°C for 2 h, reduction with 10 mg/ml sodium borodeuteride at 25°C for 2 h, and acetylation with acetic anhydride at 100°C for 1 h [34, 59]. GC/MS was performed using a fused capillary column (SP-2380 and Equity-1; 30 m, 0.25 mm ID, Supelco, Bellefonte, PA). The linkage positions were acetylated, and no related positions were previously deuteromethylated. The linkage positions were assigned by the fragmentation patterns of partially deuteromethylated alditol acetate derivatives and retention time of the peaks. The representative fragmentation patterns of partially methylated alditol acetate derivatives are shown in Fig. 9.

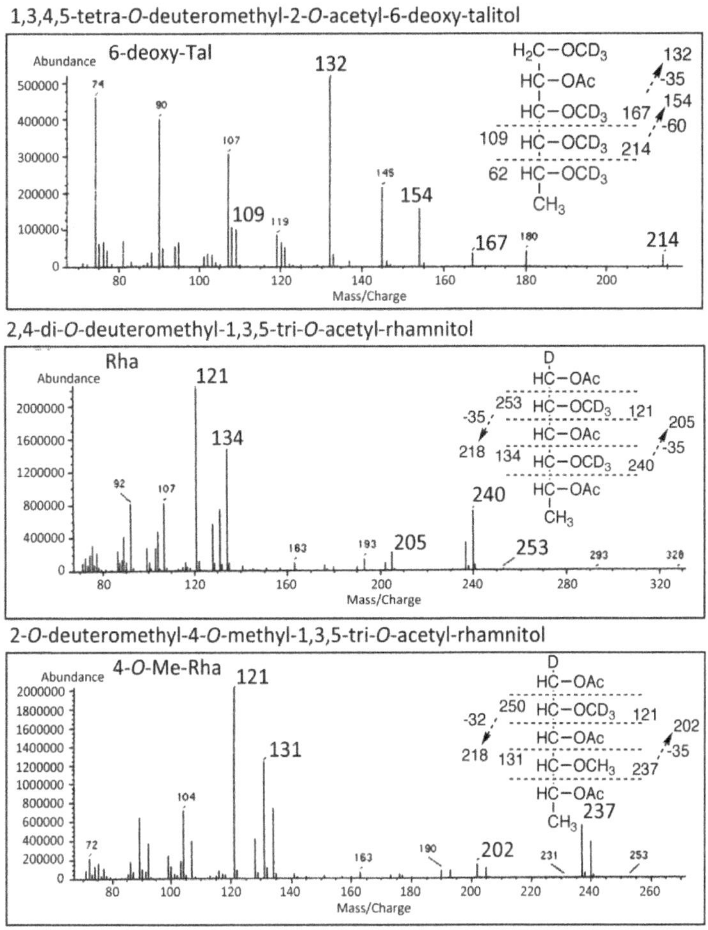

Figure 9. Representative mass spectrum of partially deuteromethylated alditol acetate derivatives.

6.6. Nuclear magnetic resonance (NMR) of GPL

The OSE was dissolved in deuterium oxide. To define the anomeric configurations of each glycosyl residue, ^{1}H and ^{13}C nuclear magnetic resonance (NMR) was employed. Both homonuclear correlation spectrometry (COSY) and ^{1}H-detected [^{1}H, ^{13}C] heteronuclear multiple-quantum correlation (HMQC) were performed as described previously [34, 56]. The ^{1}H NMR and ^{1}H-^{1}H homonuclear COSY analyses of the OSE derived from the GPL revealed distinct anomeric protons with corresponding H1-H2 cross-peaks in the low-field region, and the chemical shift and coupling constant are indicative of α-anomers and a β-hexosyl unit, respectively.

7. Biosynthesis gene of GPL

7.1. Isolation of cosmid clones carrying *rtfA* gene and sequence analysis

First, we constructed the *M. intracellulare* cosmid library. Genomic DNA of MAC was prepared by mechanical disruption of bacterial cells, which was accomplished by homogenizing a bacterial pellet with glass beads in phosphate-buffered saline, followed by phenol-chloroform extraction, and precipitation with ethanol. Genomic DNA fragments randomly sheared to 30–50 kb fragments during the extraction process were fractionated and electroeluted from agarose gels using Takara Recochip (Takara, Kyoto, Japan). These DNA fragments were rendered blunt-ended using T4 DNA polymerase and dNTPs, followed by ligation to dephosphorylated arms of pYUB412 (*Xba*I-*Eco*RV and *Eco*RV-*Xba*I). After *in vitro* packaging using Gigapack III Gold extracts (Stratagene, La Jolla, CA, USA), recombinant cosmids were introduced into the *E. coli* STBL2 [F–*mcrA D* (*mcrBC-hsdRMS-mrr*) *endA*1 *recA*1 *lon gyrA*96 *thi supE*44 *relA*1 l- D (*lac-proAB*)]. PCR was used to isolate cosmid clones carrying the rhamnosyltransferase (*rtfA*) gene with primers *rtfA*-F (5′-TTTTGGAGCGACGAGTTCATC-3′) and *rtfA*-R (5′-GTGTAGTTGACCACGCCGAC-3′). *RtfA* encodes an enzyme responsible for the transfer of Rha to 6-d-Tal in the OSE [38, 60]. We isolated the cosmid clones #49 (accession no. AB274811) and #253 (accession no. AB355138), which are responsible for the biosynthesis of GPL 7 and 16, respectively. The insert of a cosmid clone was sequenced using a BigDye Terminator v3.1 Cycle Sequencing Kit (Applied Biosystems, Foster City, CA, USA) and an ABI Prism 310 gene analyzer (Applied Biosystems). The putative function of each open reading frame (*orf*) was identified by similarity searches between the deduced amino acid sequences and known proteins using BLAST (http://www.ncbi.nlm.nih.gov/BLAST/) and FramePlot (http://www.nih.go.jp/~jun/cgi-bin/frameplot.pl), and the DNASIS computer program (Hitachi Software Engineering, Yokohama, Japan). The similarity of protein sequences of each ORF was compared to those of serotype 2, 4, 7, and 16 GPLs, and the genetic maps for GPL biosynthetic cluster are summarized in Fig. 10 [32, 36, 61]. The OSE of serotype 1 GPL produced by *M. avium* serotype 1 strain is composed of α-L-Rha-(1→2)-6-d-L-Tal, which is the core OSE of all serotypes [12]. *M. avium* serotype 1 strain (NF113) was transformed with pYUB412-cosmid clone #253 containing the serotype 16-specific gene cluster. The transformant produced a serotype 16 GPL with a different Rf value on TLC compared to serotype 1 GPL. The molecular weights of intact GPLs and the fragment patterns of their OSEs were completely equivalent to the serotype 16 GPL. Next, we identified some *orf* functions of #253. We hypothesized that *orf1, 16,* and *17* in #253 were correlated with the glycosyltransferase by the similarity of ORF sequences, in addition to *rtfA*. They were compared to the productive GPLs of serotype 1 transformants inserted with the combination of *orf1, 16,* and *17.* We determined that *orf1, 17,* and *16* were responsible for the elongation from 6-d-Tal-Rha to 6-d-Tal-Rha-Rha-Rha-Rha, and that these glycosyltransferases operated in this order, as shown in Fig. 11. After the elongation of an OSE, the OSE may be modified by aminotransferase, methyltransferase, and acyltransferase, and serotype 16 GPL may be completed.

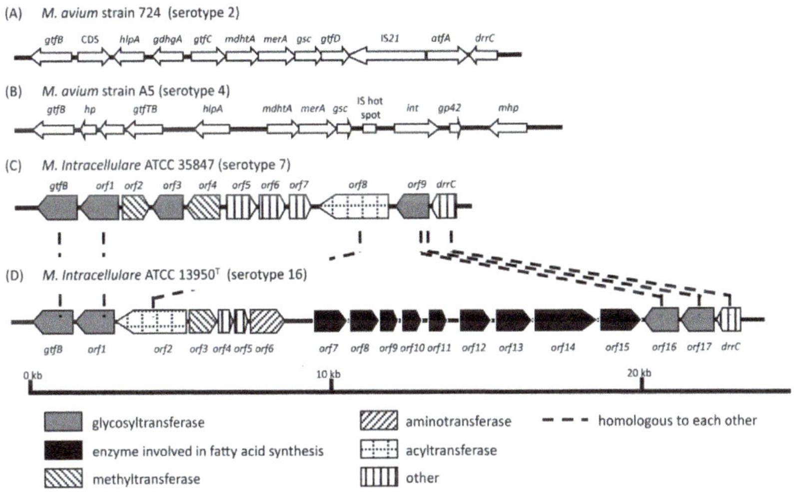

Figure 10. Comparison and overview of genetic maps of GPL biosynthetic cluster. (A), *M. avium* strain 724 (serotype 2, accession no. AF125999); (B), *M. avium* strain A5 (serotype 4, accession no. AY130970); (C), *M. intracellulare* ATCC 35847 (serotype 7, accession no. AB274811); (D), *M. intracellulare* ATCC 13950^T (serotype 16, accession no. AB355138).

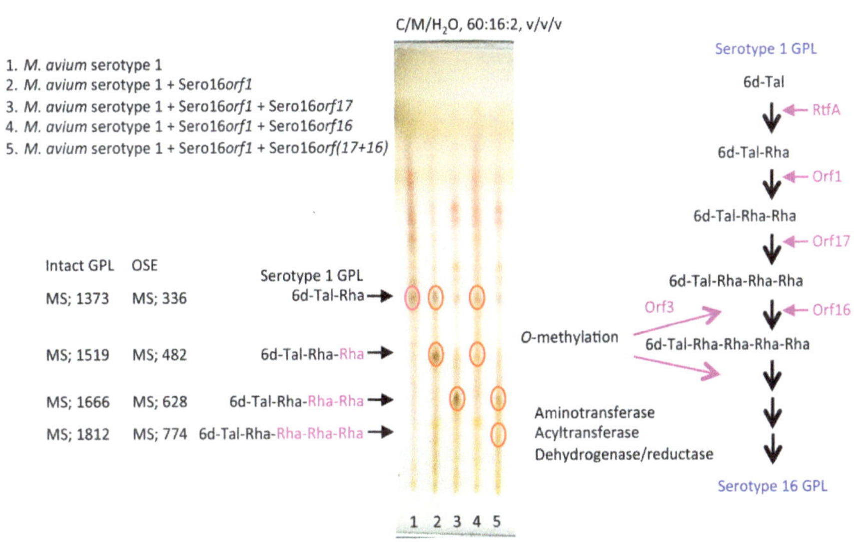

Figure 11. Proposed elongation of OSE in serotype 16 GPL.

7.2. Native conformation of GPL and host response

The native GPL was purified without alkaline treatment. The native GPLs were detected on TLC as several spots that expanded broadly and had different Rf values from that of the alkaline-treated GPL (Fig. 12). Alkaline treatment converged these spots into one spot. It was reported that the native GPLs were modified by several *O*-acetylations in the OSE portion and the alkaline treatment removed the acetylated groups [23, 33]. We are now analyzing in detail the positions and numbers of *O*-acetylations in the OSE by using MALDI-TOF MS/MS.

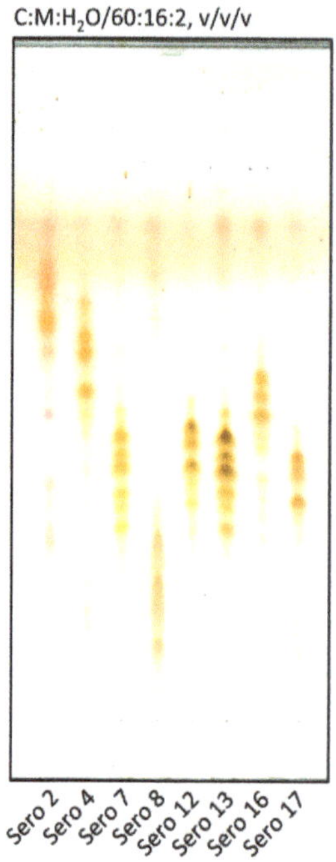

Figure 12. TLC of some native GPL fractions (non-alkali-treated).

The importance of Toll-like receptor (TLR)-mediated responses has been studied in tuberculous infections. Means et al. reported that *M. tuberculosis* activated both TLR2 and TLR4, whereas heat-killed *M. tuberculosis* and MAC activated only TLR2 [62]. It was observed that MyD88- and TLR2-deficient mice have increased susceptibility to MAC infection compared to TLR4-deficient and wild-type mice [63]. These lines of evidence suggest that TLRs are related to host recognition of the MAC components containing GPLs and affect MAC infections. To clarify the host recognitions of GPLs via TLRs, we stimulated HEK-blue-2, and -4 cells (InvivoGen, San Diego, CA, USA) with native and alkaline-treated GPLs. HEK-blue-2 and -4 cells are HEK293 cells stably transfected with multiple genes for recognition of TLR2 and TLR4 (including the co-receptors MD2 and CD14). The native GPL significantly activated HEK-blue-2 cells in a dose-dependent manner, but HEK-blue-4 cells did not respond (Fig. 13). The alkaline-treated GPL without *O*-acetylation did not activate either HEK-blue-2 or -4 cells. Re-acetylated alkaline-treated GPLs with *O*-acetyl groups substituted for all hydroxy groups of the OSE activated HEK-blue-2 cells, although the level of activation was lower than that of the native form. Moreover, we confirmed that only the native GPL stimulated mouse bone marrow-derived macrophages via TLR2 by using C57BL/6 and TLR2 knockout mice. Brennan and Goren first proposed that de-acetylated GPLs as alkaline-stable lipids, made it possible to classify serotyping [12, 31]. Schorey and colleagues clarified that serotype 1 and 2 GPLs can function as TLR2 agonists and promote macrophage activation in a TLR2 and MyD88-dependent pathway [64, 65]. They reported that the acetylated and methylated groups of GPLs were necessary for GPL-TLR2 interaction as a molecular requirement. Taken together with our results, native GPLs are a TLR2 agonist, and it may be important for GPL-TLR2 interaction to balance the hydrophobicity and hydrophilicity of the GPL molecules.

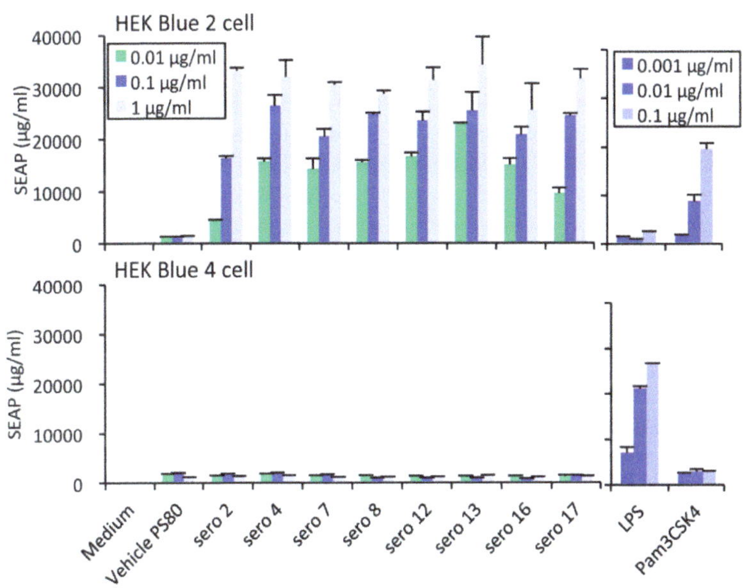

Figure 13. Native GPLs activate cells through TLR2.

8. Sulfolipid (SL)

Mycobacterial SLs are classified into three types by the number of attached short acyl chains. SL-1 is the predominant type, and SL-2 and -3 are intermediate types. The structure of SL-1 was identified as 2-palmitoyl(stearoyl)-3-phthioceranoyl-6,6′-bis-hydroxyphthioceranoyl-trehalose-2′-sulfate [66, 67]. The presence of SLs in mycobacteria is strain-specific. Several studies have demonstrated a significant relationship between SL-1 and virulence, such as the amount of SL-1 biosynthesis in virulent strains, and the inhibition of P-L fusion in macrophages by SL-1. SL-1 modulates superoxide release and secretion of interleukin (IL)-1β and tumor necrosis factor (TNF)-α by blocking activation of human macrophages and neutrophils [68, 69]. In contrast to these studies, it has been reported that *pks2* disruption and SL-1 deficiency do not significantly affect the replication, persistence, and pathogenicity of *M. tuberculosis* in mice, guinea pigs, or cultured macrophages [70, 71]. The pathogenicity of SL-1 is controversial. SL-1 does not induce granulomatous inflammation, but rather inhibits it and the release of TNF-α induced by cord factor/TDM [72]. SL-1 could contribute to virulence at an early stage of mycobacterial infection by counteracting the immunopotentiating effect of cord factor/TDM. On the other hand, it is reported that SL-3, 2-palmitoyl(stearoyl)-3-hydroxyphthioceranoyl-trehalose-2'-sulfate, is mainly recognized by CD1b-restricted T cells as a lipid antigen. The other tetraacylated and triacylated SLs (SL-1 and -2) were unable to stimulate diacylated SL (SL-3)-specific T cell clones, which implies that immunogenic SL-3 are not generated inside the antigen-presented cells [73]

SLs of *M. tuberculosis* have been implicated in the virulence of this organism by inhibiting the P-L fusion in macrophages, thus probably promoting the intracellular survival of *M. tuberculosis*, but the pathogenic role remains controversial. The gene, *pks2*, responsible for SL synthesis was identified and disrupted [74]. The *pks2* mutant defective in an early step of SL biosynthesis had no obvious growth defect in infected mice. By contrast, growth of a strain lacking MmpL8, a transporter of SL in *M. tuberculosis*, was highly attenuated in a mouse model of tuberculosis [70]. Although initial replication rates and containment levels of the MmpL8 mutant were identical, compared with the wild type, a significant attenuation of the mutant strain in time-to-death was observed. Early in infection, differential expression of cytokines and cytokine receptors revealed that the mutant strain less efficiently suppresses key indicators of a Th1-type immune response, suggesting an immunomodulatory role for SLs in the pathogenesis of tuberculosis [75].

Recently, Kummer et al. demonstrate that PapA2 and PapA1 are responsible for the sequential acylation of SL1 biosynthesis. Disruption of *papA2* and *papA1* in *M. tuberculosis* confirmed their essential role in SL-1 biosynthesis and their order of action. BALB/c mice infected by aerosol with wild-type, *ΔpapA2*, and *ΔpapA1* mutants showed no significant difference in the ability of the bacteria to grow or persist through the time to death of the mice. The loss of SL-1 did not appear to affect bacterial replication or trafficking. They suggested that the functions of SL-1 are specific to human infection [76].

9. Phenolic glycolipid (PGL)

The glycosylated phenolphthiocerol dimycocerosates (PDIM), so-called PGLs, are produced by a limited group of mycobacterial species, and most of them are pathogenic in humans [77]. PGL is distributed in *M. leprae* as a unique antigen, and inhibits the lymphoproliferative responses and suppresses monocyte oxidative responses [78-80]. It has also shown that disruption of PGL synthesis results in loss of the virulent phenotype without significantly affecting the bacterial load during disease in experimental models using mice and rabbits, and loss of PGL was found to correlate with an increase in the release of the pro-inflammatory cytokines *in vitro* [81, 82]. A PGL purified from *M. leprae* has been used as antigen in an enzyme-linked immunosorbent assay. Antibodies directed against the lipid were detected in sera of leprosy patients but not in sera from uninfected controls or patients infected with other mycobacteria, including *M. tuberculosis*. The antibody response distinguished between the *M. leprae* lipid and the structurally related PGL from *M. kansasii*. Similar to serodiagnosis of MAC disease using GPL, this assay based on PGL antigen has considerable potential as a specific serodiagnostic test for infection with *M. leprae* [83, 84].

We determined the structure of the PGL derived from BCG (Tokyo 172 strain). The PGL produced by BCG is a so-called mycoside B (PGL-BCG), and its sugar moiety is different from that of the PGL produced by *M. tuberculosis* (PGL-tb). The PGL-BCG has only a 2-*O*-Me-Rha branch elongated from the phenol moiety, although PGL-tb has it elongated to three sugar residues [85]. The composition of the PDIM in PGL is similar in both species. The MALDI-TOF MS spectrum of PGL-BCG showed m/z 1531 and other mass units at 14 Da intervals for $[M+Na]^+$ as molecule-related ions in positive mode (Fig. 14). In addition, the MS/MS spectrum showed fragment ion peaks m/z 1371 based on the elimination of methyl-deoxysugar; m/z 1135, 1093, based on the elimination of the C26:0, C29:0 FAs; m/z 696 based on the elimination of both C26:0 and C29:0 FAs; and m/z 535 for the phenol phthiocerol that eliminated methyl-deoxysugar and C26:0, C29:0 FAs (Fig. 14).

Reed et al. demonstrated that PGL-tb inhibits the innate immune response. Loss of PGL-tb was responsible for an increase in the release of TNF-α and IL-6 and IL-12 *in vitro*, and the PGL-tb-deficient mutant showed a phenotype with low virulence/pathogenicity [81]. The composition of the PDIM in PGL-BCG is similar to that in PGL-tb. Although purified PGL molecules by themselves had no effect on the activation of macrophages *in vitro*, we found that PGL suppressed the activation of murine bone marrow-derived macrophages elicited by total lipids. It is considered that the PGL may have a competitive inhibitory effect or mask the active site of other TLR2 agonistic lipid components, and decrease their activity.

10. Phosphatidylinositol mannoside (PIM), lipomannan (LM), and lipoarabinomannan (LAM)

PIMs and their multiglycosylated counterparts, LMs, and LAMs, are complex lipoglycans that are found ubiquitously in the envelopes of all mycobacterial species. Their structures originate from a phosphatidyl-*myo*-inositol (MPI) anchor, which is mannosylated to generate LM and further arabinosylated to give LAM [86]. The non-reducing termini of the arabinosyl side chains can be substituted by capping motifs, and LAMs are classified into three families. LAMs from slow growing mycobacteria bearing mannose caps, *i.e.* mono- or ($\alpha 1 \rightarrow 2$)-di- or trimannoside units, are designated as ManLAMs. In contrast, LAMs from fast growing mycobacteria capped by phospho-*myo*-inositol units and not capped at all, are termed PILAM and AraLAM, respectively [87]. LAMs and LMs exhibit a broad spectrum of

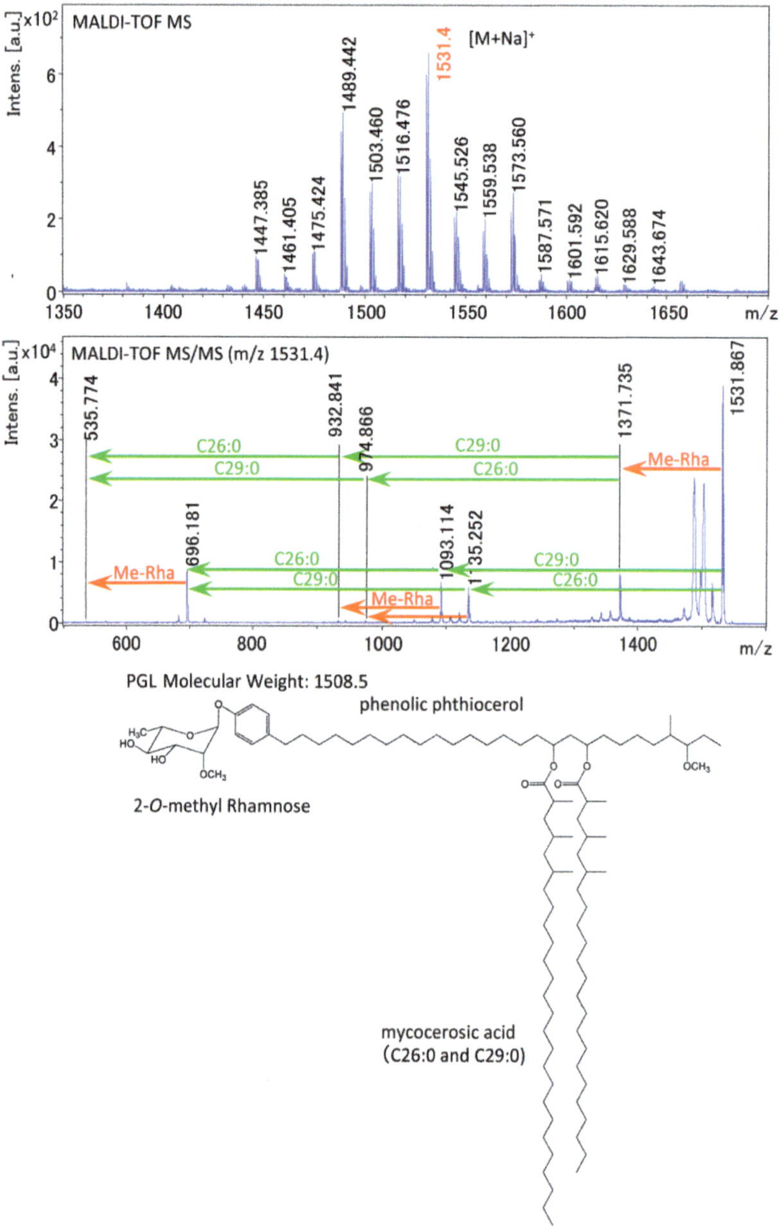

Figure 14. MALDI-TOF MS and MS/MS spectra of PGL derived from BCG Tokyo 172 strain, and proposed structure.

immunomodulatory activities, including the ability to modulate the production of macrophage-derived Th1 pro-inflammatory cytokines, most commonly TNF-α and IL-12. The ManLAM from *M. tuberculosis* is involved in the inhibition of phagosome maturation, apoptosis, and IFN-γ signaling in macrophages and IL-12 secretion of dendritic cells. ManLAMs contribute an immunosuppressive effect to the persistence of slow-growing mycobacteria in humans. In contrast, PILAMs are able to induce the release of a variety of proinflammatory cytokines. Recently it was reported that LMs from both pathogenic and nonpathogenic mycobacterial species, independent of their origin, are potent stimulators of TNF-α, IL-8, and IL-12, and activate macrophages via TLR2 [88-90]. The ManLAM/LM balance might be important to host immune responses against mycobacteria.

PIMs are composed of a phsophatidylinositol and mannose moieties. PIMs have 4–6 mannoses, the third and fourth of which are acylated in some cases. PIM_2 containing two mannoses, a main component of mycobacterial cell walls, is highly immunogenic. Several biological functions have been recently attributed to PIMs. PIM_2 was shown to recruit natural killer T cells, which play a role in the local granulomatous response [91, 92]. PIMs activate cells via TLR-2 [93]. PIM_6 as well as the ManLAMs from *M. leprae* and *M. tuberculosis* are presented by antigen-presenting cells in the context of CD1b [17]. The phosphatidylinositol moiety plays a central role in the process of binding PIMs and ManLAMs to CD1b proteins.

11. Conclusion

Mycobacterial glycolipids are pleiotropic molecules and play key roles in both microbes and hosts by acting as structural components of cell walls and immunologically active substances. A better understanding of mycobacterial glycolipids will help us develop novel diagnostics, therapeutics, and prophylactics for mycobacterial diseases.

Author details

Nagatoshi Fujiwara
Department of Bacteriology, Osaka City University Graduate School of Medicine, Japan

Acknowledgement

This work was supported by grants from the Ministry of Education, Culture, Sports, Science and Technology of Japan, and the Japan Health Sciences Foundation.

12. References

[1] Koch R, (1884) Aetiologie der Tuberkulose. Mitt. k. Gesundheitsamt. 2: 1-88.
[2] WHO, (2011) WHO report 2011 Global Tuberculosis Control. http://www.who.int/tb/publications/global_report/2011/gtbr11_full.pdf.

[3] Lee JH, Ammerman NC, Nolan S, Geiman DE, Lun S, Guo H, Bishai WR, (2012) Isoniazid resistance without a loss of fitness in *Mycobacterium tuberculosis*. Nat. Commun. 3: 753.

[4] Brennan PJ, Nikaido H, (1995) The envelope of mycobacteria. Annu. Rev. Biochem. 64: 29-63.

[5] Brennan PJ, (2003) Structure, function, and biogenesis of the cell wall of *Mycobacterium tuberculosis*. Tuberculosis (Edinb). 83: 91-97.

[6] Glickman MS, Jacobs WR, Jr., (2001) Microbial pathogenesis of *Mycobacterium tuberculosis*: dawn of a discipline. Cell. 104: 477-485.

[7] Russell DG, Mwandumba HC, Rhoades EE, (2002) *Mycobacterium* and the coat of many lipids. J. Cell. Biol. 158: 421-426.

[8] Fujiwara N, (1997) [Distribution of antigenic glycolipids among *Mycobacterium tuberculosis* strains and their contribution to virulence]. Kekkaku. 72: 193-205.

[9] Dmitriev BA, Ehlers S, Rietschel ET, Brennan PJ, (2000) Molecular mechanics of the mycobacterial cell wall: from horizontal layers to vertical scaffolds. Int. J. Med. Microbiol. 290: 251-258.

[10] Nikaido H, (2001) Preventing drug access to targets: cell surface permeability barriers and active efflux in bacteria. Semin. Cell. Dev. Biol. 12: 215-223.

[11] Aspinall GO, Chatterjee D, Brennan PJ, (1995) The variable surface glycolipids of mycobacteria: structures, synthesis of epitopes, and biological properties. Adv. Carbohydr. Chem. Biochem. 51: 169-242.

[12] Chatterjee D, Khoo KH, (2001) The surface glycopeptidolipids of mycobacteria: structures and biological properties. Cell Mol. Life Sci. 58: 2018-2042.

[13] Kobayashi K, Yoshida T, (1996) The immunopathogenesis of granulomatous inflammation induced by *Mycobacterium tuberculosis*. Methods. 9: 204-214.

[14] Beckman EM, Porcelli SA, Morita CT, Behar SM, Furlong ST, Brenner MB, (1994) Recognition of a lipid antigen by CD1-restricted alpha beta+ T cells. Nature. 372: 691-694.

[15] Moody DB, Porcelli SA, Porcelli SA, Modlin RL, (2003) Intracellular pathways of CD1 antigen presentation. Nat. Rev. Immunol. 3: 11-22.

[16] Moody DB, Guy MR, Grant E, Cheng TY, Brenner MB, Besra GS, Porcelli SA, (2000) CD1b-mediated T cell recognition of a glycolipid antigen generated from mycobacterial lipid and host carbohydrate during infection. J. Exp. Med. 192: 965-976.

[17] Sieling PA, Chatterjee D, Porcelli SA, Prigozy TI, Mazzaccaro RJ, Soriano T, Bloom BR, Brenner MB, Kronenberg M, Brennan PJ, et al., (1995) CD1-restricted T cell recognition of microbial lipoglycan antigens. Science. 269: 227-230.

[18] Kaneda K, Imaizumi S, Yano I, (1995) Distribution of C22-, C24- and C26-alpha-unit-containing mycolic acid homologues in mycobacteria. Microbiol. Immunol. 39: 563-570.

[19] Nishiuchi Y, Baba T, Yano I, (2000) Mycolic acids from *Rhodococcus, Gordonia*, and *Dietzia*. J. Microbiol. Methods. 40: 1-9.

[20] Ueda S, Fujiwara N, Naka T, Sakaguchi I, Ozeki Y, Yano I, Kasama T, Kobayashi K, (2001) Structure-activity relationship of mycoloyl glycolipids derived from *Rhodococcus* sp. 4306. Microb. Pathog. 30: 91-99.

[21] Butler WR, Guthertz LS, (2001) Mycolic acid analysis by high-performance liquid chromatography for identification of *Mycobacterium* species. Clin. Microbiol. Rev. 14: 704-726.

[22] Matsunaga I, Oka S, Fujiwara N, Yano I, (1996) Relationship between induction of macrophage chemotactic factors and formation of granulomas caused by mycoloyl glycolipids from *Rhodococcus ruber* (*Nocardia rubra*). J. Biochem. 120: 663-670.

[23] Matsunaga I, Naka T, Talekar RS, McConnell MJ, Katoh K, Nakao H, Otsuka A, Behar SM, Yano I, Moody DB, Sugita M, (2008) Mycolyltransferase-mediated glycolipid exchange in Mycobacteria. J. Biol. Chem. 283: 28835-28841.

[24] Fujita Y, Naka T, McNeil MR, Yano I, (2005) Intact molecular characterization of cord factor (trehalose 6,6'-dimycolate) from nine species of mycobacteria by MALDI-TOF mass spectrometry. Microbiology. 151: 3403-3416.

[25] Kaneda K, Sumi Y, Kurano F, Kato Y, Yano I, (1986) Granuloma formation and hemopoiesis induced by C36-48-mycolic acid-containing glycolipids from *Nocardia rubra*. Infect. Immun. 54: 869-875.

[26] Natsuhara Y, Oka S, Kaneda K, Kato Y, Yano I, (1990) Parallel antitumor, granuloma-forming and tumor-necrosis-factor-priming activities of mycoloyl glycolipids from *Nocardia* rubra that differ in carbohydrate moiety: structure-activity relationships. Cancer Immunol. Immunother. 31: 99-106.

[27] Yano I, Tomiyasu I, Kaneda K, Kato Y, Sumi Y, Kurano S, Sugimoto N, Sawai H, (1987) Isolation of mycolic acid-containing glycolipids in *Nocardia rubra* and their granuloma forming activity in mice. J. Pharmacobiodyn. 10: 113-123.

[28] Yamagami H, Matsumoto T, Fujiwara N, Arakawa T, Kaneda K, Yano I, Kobayashi K, (2001) Trehalose 6,6'-dimycolate (cord factor) of *Mycobacterium tuberculosis* induces foreign-body- and hypersensitivity-type granulomas in mice. Infect. Immun. 69: 810-815.

[29] Natsuhara Y, Yoshinaga J, Shogaki T, Sumi-Nishikawa Y, Kurano S, Kato Y, Kaneda K, Oka S, Yano I, (1990) Granuloma-forming activity and antitumor activity of newly isolated mycoloyl glycolipid from *Rhodococcus terrae* 70012 (Rt. GM-2). Microbiol. Immunol. 34: 45-53.

[30] Ishikawa E, Ishikawa T, Morita YS, Toyonaga K, Yamada H, Takeuchi O, Kinoshita T, Akira S, Yoshikai Y, Yamasaki S, (2009) Direct recognition of the mycobacterial glycolipid, trehalose dimycolate, by C-type lectin Mincle. J. Exp. Med. 206: 2879-2888.

[31] Brennan PJ, Goren MB, (1979) Structural studies on the type-specific antigens and lipids of the *Mycobacterium avium, Mycobacterium intracellulare, Mycobacterium scrofulaceum serocomplex, Mycobacterium intracellulare* serotype 9. J. Biol. Chem. 254: 4205-4211.

[32] Fujiwara N, Nakata N, Maeda S, Naka T, Doe M, Yano I, Kobayashi K, (2007) Structural characterization of a specific glycopeptidolipid containing a novel *N*-acyl-deoxy sugar from *Mycobacterium intracellulare* serotype 7 and genetic analysis of Its glycosylation pathway. J. Bacteriol. 189: 1099-1108.

[33] Naka T, Nakata N, Maeda S, Yamamoto R, Doe M, Mizuno S, Niki M, Kobayashi K, Ogura H, Makino M, Fujiwara N, (2011) Structure and host recognition of serotype 13 glycopeptidolipid from *Mycobacterium intracellulare*. J. Bacteriol. 193: 5766-5774.

[34] Fujiwara N, Nakata N, Naka T, Yano I, Doe M, Chatterjee D, McNeil M, Brennan PJ, Kobayashi K, Makino M, Matsumoto S, Ogura H, Maeda S, (2008) Structural analysis and biosynthesis gene cluster of an antigenic glycopeptidolipid from *Mycobacterium intracellulare*. J. Bacteriol. 190: 3613-3621.

[35] Nakata N, Fujiwara N, Naka T, Yano I, Kobayashi K, Maeda S, (2008) Identification and characterization of two novel methyltransferase genes that determine the serotype 12-specific structure of glycopeptidolipids of *Mycobacterium intracellulare*. J. Bacteriol. 190: 1064-1071.

[36] Eckstein TM, Belisle JT, Inamine JM, (2003) Proposed pathway for the biosynthesis of serovar-specific glycopeptidolipids in *Mycobacterium avium* serovar 2. Microbiology. 149: 2797-2807.

[37] Heidelberg T, Martin OR, (2004) Synthesis of the glycopeptidolipid of *Mycobacterium avium* Serovar 4: first example of a fully synthetic C-mycoside GPL. J. Org. Chem. 69: 2290-2301.

[38] Maslow JN, Irani VR, Lee SH, Eckstein TM, Inamine JM, Belisle JT, (2003) Biosynthetic specificity of the rhamnosyltransferase gene of *Mycobacterium avium* serovar 2 as determined by allelic exchange mutagenesis. Microbiology. 149: 3193-3202.

[39] Eckstein TM, Inamine JM, Lambert ML, Belisle JT, (2000) A genetic mechanism for deletion of the ser2 gene cluster and formation of rough morphological variants of *Mycobacterium avium*. J. Bacteriol. 182: 6177-6182.

[40] Howard ST, Rhoades E, Recht J, Pang X, Alsup A, Kolter R, Lyons CR, Byrd TF, (2006) Spontaneous reversion of *Mycobacterium abscessus* from a smooth to a rough morphotype is associated with reduced expression of glycopeptidolipid and reacquisition of an invasive phenotype. Microbiology. 152: 1581-1590.

[41] Belisle JT, Klaczkiewicz K, Brennan PJ, Jacobs WR, Jr., Inamine JM, (1993) Rough morphological variants of *Mycobacterium avium*. Characterization of genomic deletions resulting in the loss of glycopeptidolipid expression. J. Biol. Chem. 268: 10517-10523.

[42] Bhatnagar S, Schorey JS, (2007) Exosomes released from infected macrophages contain *Mycobacterium avium* glycopeptidolipids and are proinflammatory. J. Biol. Chem.

[43] Schorey JS, Sweet L, (2008) The mycobacterial glycopeptidolipids: structure, function, and their role in pathogenesis. Glycobiology. 18: 832-841.

[44] Enomoto K, Oka S, Fujiwara N, Okamoto T, Okuda Y, Maekura R, Kuroki T, Yano I, (1998) Rapid serodiagnosis of *Mycobacterium avium-intracellulare* complex infection by ELISA with cord factor (trehalose 6, 6'-dimycolate), and serotyping using the glycopeptidolipid antigen. Microbiol. Immunol. 42: 689-696.

[45] Kitada S, Maekura R, Toyoshima N, Fujiwara N, Yano I, Ogura T, Ito M, Kobayashi K, (2002) Serodiagnosis of pulmonary disease due to *Mycobacterium avium* complex with an enzyme immunoassay that uses a mixture of glycopeptidolipid antigens. Clin. Infect. Dis. 35: 1328-1335.

[46] Kitada S, Maekura R, Toyoshima N, Naka T, Fujiwara N, Kobayashi M, Yano I, Ito M, Kobayashi K, (2005) Use of glycopeptidolipid core antigen for serodiagnosis of *Mycobacterium avium* complex pulmonary disease in immunocompetent patients. Clin. Diagn. Lab. Immunol. 12: 44-51.

[47] Kitada S, Kobayashi K, Ichiyama S, Takakura S, Sakatani M, Suzuki K, Takashima T, Nagai T, Sakurabayashi I, Ito M, Maekura R, (2008) Serodiagnosis of *Mycobacterium avium*-complex pulmonary disease using an enzyme immunoassay kit. Am. J. Respir. Crit. Care Med. 177: 793-797.

[48] Maekura R, Okuda Y, Hirotani A, Kitada S, Hiraga T, Yoshimura K, Yano I, Kobayashi K, Ito M, (2005) Clinical and prognostic importance of serotyping *Mycobacterium avium-Mycobacterium intracellulare* complex isolates in human immunodeficiency virus-negative patients. J. Clin. Microbiol. 43: 3150-3158.

[49] Kitada S, Kobayashi K, Nishiuchi Y, Fushitani K, Yoshimura K, Tateishi Y, Miki K, Miki M, Hashimoto H, Motone M, Fujikawa T, Hiraga T, Maekura R, (2010) Serodiagnosis of pulmonary disease due to *Mycobacterium avium* complex proven by bronchial wash culture. Chest. 138: 236-237.

[50] Barrow WW, Davis TL, Wright EL, Labrousse V, Bachelet M, Rastogi N, (1995) Immunomodulatory spectrum of lipids associated with *Mycobacterium avium* serovar 8. Infect. Immun. 63: 126-133.

[51] Kano H, Doi T, Fujita Y, Takimoto H, Yano I, Kumazawa Y, (2005) Serotype-specific modulation of human monocyte functions by glycopeptidolipid (GPL) isolated from *Mycobacterium avium* complex. Biol. Pharm. Bull. 28: 335-339.

[52] Tassell SK, Pourshafie M, Wright EL, Richmond MG, Barrow WW, (1992) Modified lymphocyte response to mitogens induced by the lipopeptide fragment derived from *Mycobacterium avium* serovar-specific glycopeptidolipids. Infect. Immun. 60: 706-711.

[53] Takegaki Y, (2000) [Effect of serotype specific glycopeptidolipid (GPL) isolated from *Mycobacterium avium* complex (MAC) on phagocytosis and phagosome-lysosome fusion of human peripheral blood monocytes]. Kekkaku. 75: 9-18.

[54] Shimada K, Takimoto H, Yano I, Kumazawa Y, (2006) Involvement of mannose receptor in glycopeptidolipid-mediated inhibition of phagosome-lysosome fusion. Microbiol. Immunol. 50: 243-251.

[55] Danelishvili L, Wu M, Stang B, Harriff M, Cirillo S, Cirillo J, Bildfell R, Arbogast B, Bermudez LE, (2007) Identification of *Mycobacterium avium* pathogenicity island important for macrophage and amoeba infection. Proc. Natl. Acad. Sci. USA. 104: 11038-11043.

[56] Khoo KH, Chatterjee D, Dell A, Morris HR, Brennan PJ, Draper P, (1996) Novel *O*-methylated terminal glucuronic acid characterizes the polar glycopeptidolipids of *Mycobacterium habana* strain TMC 5135. J. Biol. Chem. 271: 12333-12342.

[57] Bhatt A, Fujiwara N, Bhatt K, Gurcha SS, Kremer L, Chen B, Chan J, Porcelli SA, Kobayashi K, Besra GS, Jacobs WR, Jr., (2007) Deletion of *kasB* in *Mycobacterium tuberculosis* causes loss of acid-fastness and subclinical latent tuberculosis in immunocompetent mice. Proc. Natl. Acad. Sci. USA. 104: 5157-5162.

[58] Hakomori S, (1964) A rapid permethylation of glycolipid, and polysaccharide catalyzed by methylsulfinyl carbanion in dimethyl sulfoxide. J. Biochem. (Tokyo). 55: 205-208.

[59] Khoo KH, Jarboe E, Barker A, Torrelles J, Kuo CW, Chatterjee D, (1999) Altered expression profile of the surface glycopeptidolipids in drug-resistant clinical isolates of *Mycobacterium avium* complex. J. Biol. Chem. 274: 9778-9785.

[60] Eckstein TM, Silbaq FS, Chatterjee D, Kelly NJ, Brennan PJ, Belisle JT, (1998) Identification and recombinant expression of a *Mycobacterium avium* rhamnosyltransferase gene (*rtfA*) involved in glycopeptidolipid biosynthesis. J. Bacteriol. 180: 5567-5573.
[61] Krzywinska E, Schorey JS, (2003) Characterization of genetic differences between *Mycobacterium avium* subsp. avium strains of diverse virulence with a focus on the glycopeptidolipid biosynthesis cluster. Vet. Microbiol. 91: 249-264.
[62] Means TK, Wang S, Lien E, Yoshimura A, Golenbock DT, Fenton MJ, (1999) Human toll-like receptors mediate cellular activation by *Mycobacterium tuberculosis*. J. Immunol. 163: 3920-3927.
[63] Feng CG, Scanga CA, Collazo-Custodio CM, Cheever AW, Hieny S, Caspar P, Sher A, (2003) Mice lacking myeloid differentiation factor 88 display profound defects in host resistance and immune responses to *Mycobacterium avium* infection not exhibited by Toll-like receptor 2 (TLR2)- and TLR4-deficient animals. J. Immunol. 171: 4758-4764.
[64] Sweet L, Schorey JS, (2006) Glycopeptidolipids from *Mycobacterium avium* promote macrophage activation in a TLR2- and MyD88-dependent manner. J. Leukoc. Biol. 80: 415-423.
[65] Sweet L, Zhang W, Torres-Fewell H, Serianni A, Boggess W, Schorey J, (2008) *Mycobacterium avium* glycopeptidolipids require specific acetylation and methylation patterns for signaling through toll-like receptor 2. J. Biol. Chem. 283: 33221-33231.
[66] Goren MB, (1970) Sulfolipid I of *Mycobacterium tuberculosis*, strain H37Rv. II. Structural studies. Biochim. Biophys. Acta. 210: 127-138.
[67] Mougous JD, Petzold CJ, Senaratne RH, Lee DH, Akey DL, Lin FL, Munchel SE, Pratt MR, Riley LW, Leary JA, Berger JM, Bertozzi CR, (2004) Identification, function and structure of the mycobacterial sulfotransferase that initiates sulfolipid-1 biosynthesis. Nat. Struct. Mol. Biol. 11: 721-729.
[68] Pabst MJ, Gross JM, Brozna JP, Goren MB, (1988) Inhibition of macrophage priming by sulfatide from *Mycobacterium tuberculosis*. J. Immunol. 140: 634-640.
[69] Zhang L, English D, Andersen BR, (1991) Activation of human neutrophils by *Mycobacterium tuberculosis*-derived sulfolipid-1. J. Immunol. 146: 2730-2736.
[70] Converse SE, Mougous JD, Leavell MD, Leary JA, Bertozzi CR, Cox JS, (2003) MmpL8 is required for sulfolipid-1 biosynthesis and *Mycobacterium tuberculosis* virulence. Proc. Natl. Acad. Sci. USA. 100: 6121-6126.
[71] Rousseau C, Turner OC, Rush E, Bordat Y, Sirakova TD, Kolattukudy PE, Ritter S, Orme IM, Gicquel B, Jackson M, (2003) Sulfolipid deficiency does not affect the virulence of *Mycobacterium tuberculosis* H37Rv in mice and guinea pigs. Infect. Immun. 71: 4684-4690.
[72] Okamoto Y, Fujita Y, Naka T, Hirai M, Tomiyasu I, Yano I, (2006) Mycobacterial sulfolipid shows a virulence by inhibiting cord factor induced granuloma formation and TNF-alpha release. Microb. Pathog. 40: 245-253.
[73] Gilleron M, Stenger S, Mazorra Z, Wittke F, Mariotti S, Bohmer G, Prandi J, Mori L, Puzo G, De Libero G, (2004) Diacylated sulfoglycolipids are novel mycobacterial

antigens stimulating CD1-restricted T cells during infection with *Mycobacterium tuberculosis*. J. Exp. Med. 199: 649-659.

[74] Sirakova TD, Thirumala AK, Dubey VS, Sprecher H, Kolattukudy PE, (2001) The *Mycobacterium tuberculosis pks2* gene encodes the synthase for the hepta- and octamethyl-branched fatty acids required for sulfolipid synthesis. J. Biol. Chem. 276: 16833-16839.

[75] Domenech P, Reed MB, Dowd CS, Manca C, Kaplan G, Barry CE, 3rd, (2004) The role of MmpL8 in sulfatide biogenesis and virulence of *Mycobacterium tuberculosis*. J. Biol. Chem. 279: 21257-21265.

[76] Kumar P, Schelle MW, Jain M, Lin FL, Petzold CJ, Leavell MD, Leary JA, Cox JS, Bertozzi CR, (2007) PapA1 and PapA2 are acyltransferases essential for the biosynthesis of the *Mycobacterium tuberculosis* virulence factor sulfolipid-1. Proc. Natl. Acad. Sci. USA. 104: 11221-11226.

[77] Daffe M, Laneelle MA, (1988) Distribution of phthiocerol diester, phenolic mycosides and related compounds in mycobacteria. J. Gen. Microbiol. 134: 2049-2055.

[78] Fournie JJ, Adams E, Mullins RJ, Basten A, (1989) Inhibition of human lymphoproliferative responses by mycobacterial phenolic glycolipids. Infect. Immun. 57: 3653-3659.

[79] Mehra V, Brennan PJ, Rada E, Convit J, Bloom BR, (1984) Lymphocyte suppression in leprosy induced by unique *M. leprae* glycolipid. Nature. 308: 194-196.

[80] Vachula M, Holzer TJ, Andersen BR, (1989) Suppression of monocyte oxidative esponse by phenolic glycolipid I of *Mycobacterium leprae*. J. Immunol. 142: 1696-1701.

[81] Reed MB, Domenech P, Manca C, Su H, Barczak AK, Kreiswirth BN, Kaplan G, Barry CE, 3rd, (2004) A glycolipid of hypervirulent tuberculosis strains that inhibits the innate mmune response. Nature. 431: 84-87.

[82] Tsenova L, Ellison E, Harbacheuski R, Moreira AL, Kurepina N, Reed MB, Mathema B, Barry CE, 3rd, Kaplan G, (2005) Virulence of selected *Mycobacterium tuberculosis* clinical isolates in the rabbit model of meningitis is dependent on phenolic glycolipid produced by the bacilli. J. Infect. Dis. 192: 98-106.

[83] Young DB, Buchanan TM, (1983) A serological test for leprosy with a glycolipid specific for *Mycobacterium leprae*. Science. 221: 1057-1059.

[84] zumi S, Fujiwara T, Ikeda M, Nishimura Y, Sugiyama K, Kawatsu K, (1990) Novel gelatin particle agglutination test for serodiagnosis of leprosy in the field. J. Clin. Microbiol. 28: 525-529.

[85] Simeone R, Leger M, Constant P, Malaga W, Marrakchi H, Daffe M, Guilhot C, Chalut C, (2010) Delineation of the roles of FadD22, FadD26 and FadD29 in the biosynthesis of phthiocerol dimycocerosates and related compounds in *Mycobacterium tuberculosis*. FEBS J. 277: 2715-2725.

[86] Hunter SW, Brennan PJ, (1990) Evidence for the presence of a phosphatidylinositol anchor on the lipoarabinomannan and lipomannan of *Mycobacterium tuberculosis*. J. Biol. Chem. 265: 9272-9279.

[87] Nigou J, Gilleron M, Puzo G, (2003) Lipoarabinomannans: from structure to biosynthesis. Biochimie. 85: 153-166.

[88] Dao DN, Kremer L, Guerardel Y, Molano A, Jacobs WR, Jr., Porcelli SA, Briken V, (2004) *Mycobacterium tuberculosis* lipomannan induces apoptosis and interleukin-12 production in macrophages. Infect. Immun. 72: 2067-2074.

[89] Quesniaux VJ, Nicolle DM, Torres D, Kremer L, Guerardel Y, Nigou J, Puzo G, Erard F, Ryffel B, (2004) Toll-like receptor 2 (TLR2)-dependent-positive and TLR2-independent-negative regulation of proinflammatory cytokines by mycobacterial lipomannans. J. Immunol. 172: 4425-4434.

[90] Vignal C, Guerardel Y, Kremer L, Masson M, Legrand D, Mazurier J, Elass E, (2003) Lipomannans, but not lipoarabinomannans, purified from *Mycobacterium chelonae* and *Mycobacterium kansasii* induce TNF-alpha and IL-8 secretion by a CD14-toll-like eceptor 2-dependent mechanism. J. Immunol. 171: 2014-2023.

[91] Apostolou I, Takahama Y, Belmant C, Kawano T, Huerre M, Marchal G, Cui J, Taniguchi M, Nakauchi H, Fournie JJ, Kourilsky P, Gachelin G, (1999) Murine natural killer T(NKT) cells [correction of natural killer cells] contribute to the granulomatous eaction caused by mycobacterial cell walls. Proc. Natl. Acad. Sci. USA. 96: 5141-5146.

[92] Gilleron M, Ronet C, Mempel M, Monsarrat B, Gachelin G, Puzo G, (2001) Acylation state of the phosphatidylinositol mannosides from *Mycobacterium bovis* bacillus Calmette Guerin and ability to induce granuloma and recruit natural killer T cells. J. Biol. Chem. 276: 34896-34904.

[93] Jones BW, Means TK, Heldwein KA, Keen MA, Hill PJ, Belisle JT, Fenton MJ, (2001) Different Toll-like receptor agonists induce distinct macrophage responses. J. Leukoc. Biol. 69: 1036-1044.

Chapter 6

Flagellar Glycosylation: Current Advances

Jumpei Hayakawa and Morio Ishizuka

Additional information is available at the end of the chapter

1. Introduction

In this chapter, we present the current advances in flagellar glycosylation. Glycosylation is well-known as one of the most frequent posttranslational protein modification. Glycosylation is well studied in eukaryotes as the superficial and secretory proteins are mostly glycosylated in the eukaryotic cell. Protein glycosylation was considered to be a eukaryotic organism specific modification for many years. However, reports of bacterial glycosylation have increased since the discovery of surface layer glycosylation on the cell envelope in archaea and hyperthermophiles in the mid-1970's (Mescher & Strominger, 1976; Sleytr, 1975; Sleytr & Thorne, 1976).

1.1. Protein glycosylation

Protein glycosylation is largely classified as N-linked or O-linked while C-mannosylation is rarely identified (Furmanek & Hofsteenge, 2000). Glycan structures are enzymatically transferred to amino acid residues where they can covalently conjugate via the amino group of asparagine residues (N-glycosylation) or the hydroxyl group of serine or threonine residues (O-glycosylation). Both linkage types are distributed in eukaryotes and prokaryotes. The N-linkage glycosylation pathway is characterized in all three domains of life (eukarya, archaea, and bacteria) (Calo et al., 2010; Haeuptle & Hennet, 2009; Szymanski & Wren, 2005; Weerapana & Imperiali, 2006). The carbohydrate chain is synthesized at the membrane (endoplasmic reticulum (ER) in eukarya, or on the cytoplasmic side of the plasma membrane in archaea and bacteria) via a specific glycosyltransferase which transfers a nucleotide-activated sugar precursor onto the lipid carrier (dolichol-phosphate in eukarya and archaea, or undecaprenyl-phosphate in bacteria). The synthesized carbohydrate chain (oligosaccharide) is flipped across the membrane using a specific flippase, and is transferred to the asparagine residue in the nascent protein en bloc by an oligosaccharyltransferase (OST), which is composed of nine subunits in eukarya. Archaeal and bacterial OST are encoded by the aglB (Abu-Qarn et al., 2007) or pglB (Wacker et al., 2002) gene to yield a single protein, respectively. In eukarya and archaea, the asparagine residues of the N-linkage glycosylation site are conserved in a sequon, the N-X-S/T motif (where X is any amino acid except proline). Recently, the bacterial N-linkage sequon was characterized from *Campylobacter jejuni* (Kowarik et al., 2006). The two N terminal extended residues, D/E-X1-N-X2-S/T (where X1 and X2 are any amino acid except proline), were required for the bacterial OST recognition of the glycosylation site.

In eukaryote, glycoproteins typically possess a pentasaccharide core carbohydrate structures which consist of $(Man)_2$-Man-GlcNAc-GlcNAc with one or more glycan chain (N-glycosylation) and di-, or trisaccharide core structure based on GalNAc attached to serine or threonine (O-glycosylation). However, recent reports on *C. jejuni, Haemophilus influenzae,* and *Desulfovibrio desulfuricans* N-glycosylations indicate that there is no conserved core structure in bacteria (Gross et al., 2008; Ielmini & Feldman, 2011; Young et al., 2002).

Thus, in prokaryotes many different glycoprotein structures have been observed that display much more variation than those observed in eukaryotes.

Glycoproteins have many biological functions: one example is recognition and adhesion among cells (Varki, 1993). The interactions between cells are mediated by the glycan structures on the cell surfaces. Therefore, the different glycan moieties on the cell surfaces serve as markers for cell recognition events, and modifications of the glycan structures can render several biological functions to the protein in eukarya. In recent years, accumulating studies for glycosylated bacterial proteins indicate that glycan structures mainly participate in the virulence of the mucosal pathogen (Szymanski & Wren, 2005). Most bacterial glycoproteins appear to be associated with the surface of the organism as in pili or flagella. Flagellin is one of the extensively studied glycosylated bacterial proteins and it is suggested that the flagellin glycosylation is responsible for their virulence, adherence, filament assembly, and filament stability (Arora et al., 2005; Goon et al., 2002; Szymanski et al., 2002; Taguchi et al., 2008).

1.2. Flagellar structures

Most bacterial species swim by means of rotating flagella that are powered by the monovalent cation (H^+ or Na^+) influx. Many bacteria have extracellular flagellar structures and the pattern of flagellar arrangement is an identification tool in bacteria. A variety of flagella structures and swimming patterns have been discussed in previous works (Armitage & Macna, 1987; Bardy et al., 2003; Charon & Goldstein, 2002; Macnab, 1977, McCarter, 2001, 2004; Shigematsu et al., 2005). There are classified in-to four flagella arrangements as follows: multi-flagella are randomly distributed on the overall cell surface (peritrichous; e.g. *Escherichia coli, Salmonella typhimurium, Bacillus subtilis*), several flagella are present at one end of the cell (lophotrichous; e.g. *Pseudomonas fluorescens*), a single polar flagella is projected from the cell end (monotrichous; e.g. *Vibrio cholerae, Rhodobacter sphaeroides*), and a single flagellum is present at each pole of the bacterium (amphitrichous; e.g. *Spirillum serpens*). Although bacteria possess different flagella arrangement, a basic flagella structure is common among many bacterial species. The study of bacterial flagella has been intensively investigated in *Escherichia coli* and *Salmonella typhimurium* by using genetic and biochemical approaches. A typical flagella structure is shown in Figure 1. About 50 genes, which are related to the bacterial flagellar assembly, have been identified. More than 20 distinct proteins make up the flagella structure and it consist of three main parts: a basal body, a hook, and a filament. The basal body consists of four rings (L, P, MS, and C-ring) and rod which is located just above the MS-ring and connected to the proximal region of the hook. However, the L and P ring are not observed in gram-positive bacterial species, because of the difference in the cell wall architecture (DePamphilis & Adler, 1971; Francis et al., 1995). The C-ring which is a part of flagella rotor consists of three proteins, FliG, FliM, and FliN (in gram-positive bacteria FliY correspond to FliN). In particular, FliG is the most directly involved in the rotation of the flagella motor among the C-ring proteins, as it interact with the motor complex (MotA/B or PomA/B) component of the force-generating unit in the flagellar motor. The motor complex acts as the stator and the rotation energy is generated by the monovalent cation influx from the periplasmic space across the inner membrane. The C-ring is also called the "switch complex" as it can switch the direction of flagellar motor (Irikura et al., 1993; Kihara et al., 1996; Sockett et al., 1992). The switching event is caused by the binding of phosphorylated CheY (chemotaxis related gene) to the FliM of switch complex, and clockwise (CW)/counterclockwise (CCW) switching enable the bacterial cell to change swimming direction (Mathews et al., 1998; Sockett et al., 1992).

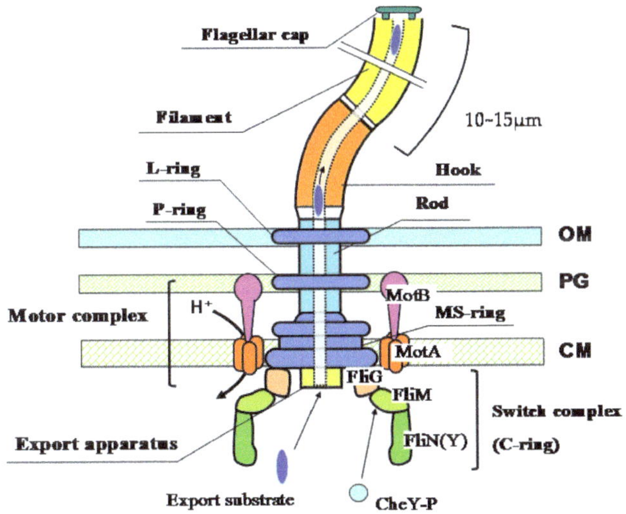

Figure 1. The model structure of the flagellar motor in gram-negative bacteria.

The motor consists of the Mot complex (MotA/B) and rotor (MS- and C-ring). The L- and P-ring do not exist in the gram-positive bacterial flagellar structure. The Mot complex is supposed to function as the force-generating unit via proton conduction while the C-ring functions as the switch. The phosphorylated form of CheY (CheY-P) interacts with FliM and promotes CW rotation. When CheY-P is not bound to it, the motor rotates CCW. Flagellin subunits are transported from the cytoplasm and are delivered into a central channel in the basal body–hook filament structure (the diameter of the central channel is only 2 nm). OM, Outer membrane; PG, peptidoglycan layer; CM, cytoplasmic membrane (Irikura et al., 1993; Mathews et al., 1998).

The most impressive structure of the bacterial flagella motor is an extracellular long helical filament. In general, the flagella filament is composed of 20,000 ~ 30.000 subunits of a single protein called flagellin, and it reaches to more than 10 μm in length (Namba & Vonderviszt, 1997). The flagellar specific export apparatus is located on the inside of the C-ring, and most of the flagellar components proteins are translocated across the cytoplasmic membrane by this apparatus, and then the proteins are diffused in the narrow nascent lumen structure and self-assemble at the distal end of the flagellar structure (Aizawa, 1996).

Flagellar-based motility is also common to archaea, but its structural features are quite distinct from bacteria. Archaeal flagella are closely related to bacterial type IV pili in their structure and assembly and the origin of bacterial flagella is considered to be a type III secretion system (Bardy et al., 2004). In bacteria, the flagellar filament is composed of a single flagellin subunit, in contrast, two or more distinct flagellin subunits are require for production of the flagellar filament. The other notable differences include that archaeal flagella rotation is powered by ATP, the flagellin subunit has a signal peptide which is cleaved by a specific peptidase for the secret matured flagellin subunit from the cell, and the flagella is grown from the proximal end of the cell surface by the addition of subunits to the base.

1.3. Flagellin glycosylations

Although flagella structures from both eubacteria and archaea are different, flagellin glycosylation is reported in both organisms. Eubacterial flagellin glycosylations are classified as either N- or O-linkage in a single subunit, and to date, most reports are about O-linkages. The O-linked glycan positions of bacterial flagellin proteins appear to be limited to the central region of the primary flagellin structure. The amino acid sequence alignment indicates that flagellin proteins are well conserved in the N- and C-terminal regions, while the central region is highly variable (Beatson et al., 2006). Although the intensively studied peritrichous flagella from *Salmonella typhimurium* do not have a glycosylated flagellin, a complete atomic model of its flagellin protein was resolved by excellent X-ray crystallography analysis and electron cryomicroscopic observation of the intact flagella filament (Samatey et al., 2001; Yonekura et al., 2003, 2005). These structural data revealed that the *S. typhimurium* flagellin protein consists of four major domains, D0, D1, D2, and D3 (Figure 2). The N- and C-terminal regions of flagellin correspond to D0 and D1, which are composed mainly of α-helices, and they form the core part of a flagellar filament. D0 has two α-helices (ND0 and CD0), whereas D1 has two long α-helices (ND1a and CD1), one short α-helix (ND1b) and one short β-sheet. These domains are positioned in the filament core. Gugolya et al. suggested that the alpha-helical terminal regions that correspond to the D0 domains are important for the coiled-coil model of flagellar filament formation (Gugolya et al., 2003). The central variable region of flagellin forms the outside surface-exposed domain (D2 and D3) in the assembled filament. Studies of the variable region have focused on its role in H antigenicity, the effect of deletions on filament formation and motility, and the insertion of foreign peptides for extracellular display on bacterial flagella (Malapaka et al., 2007; Reid et al., 1999; Westerlund-Wikström, 2000; Woodset al., 2007; Yoshioka et al., 1995). Thus, flagellin glycan structures are restricted in the D2 and D3 domains (a few glycans were located at the D2 proximal end of the D1 domain), and it is considered that these glycan moieties are exposed to environmental conditions (Figure 3).

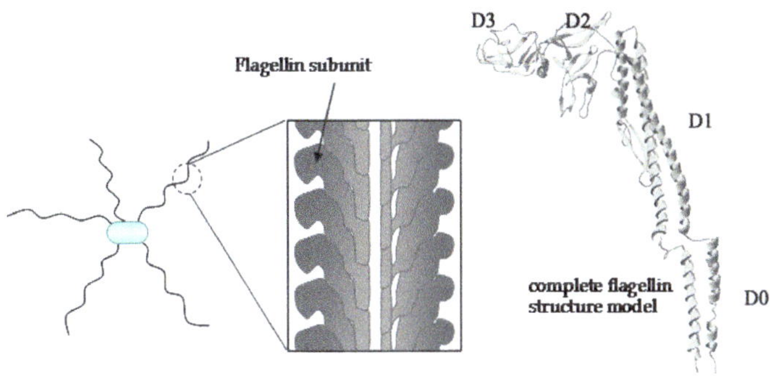

Figure 2. Bacterial flagellar filament.

Flagellar filament structure and complete flagellin 3D model of *Salmonella typhimurium* (Samatey et al., 2001; Yonekura et al., 2003, 2005). A flagellar filament consists of substantial amount of flagellin subunit and takes on a tubular structure.

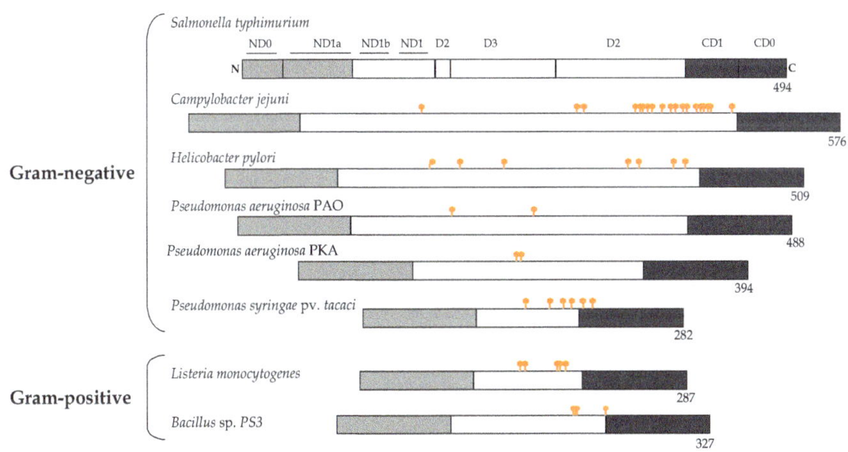

Glycosylated flagellins are schematically aligned with the *S. typhimurium* flagellin domain. Light and dark grey boxes indicate 100 highly conserved terminal amino acid residues which constitute the alpha-helical structures. The corresponding amino acid size of the flagellin proteins is indicated at the C-terminal. Positions of glycosylation on each flagellin are indicated ( ).

Figure 3. O-linked glycosylation sites of bacterial flagellin.

In contrast with eubacteria, three archaeal flagellin glycosylations are reported as N-linkage (Chaban et al., 2007; Voisin et al., 2005; Wieland & Sumper, 1985). Work on the most extensively studied flagellin from *Methanococcus voltae* are demonstrated that the glycan attached positions are not limited to anywhere specific in the flagellin primary structure and the N-linked asparagine residues seem to follow the classic eukaryotic type consensus sequon (N-X-S/T) rather than the recently identified bacterial N-linkage sequence (D/E-X_1-N-X_2-S/T).

2. Flagellar glycosylation

There have been many reports on flagella glycosylation since it was discovered about 20 years ago. Flagellin glycosylation is mainly found in gram-negative pathogenic bacterial species, and has been identified in about 30 microorganism strains including the archaea and gram-positive species (Logan, 2006). The distribution of flagellin glycosylation among several species is shown in Table 1, and a gene cluster which is potentially involved in the posttranslational modification of flagellin glycosylation is shown in Figure 4.

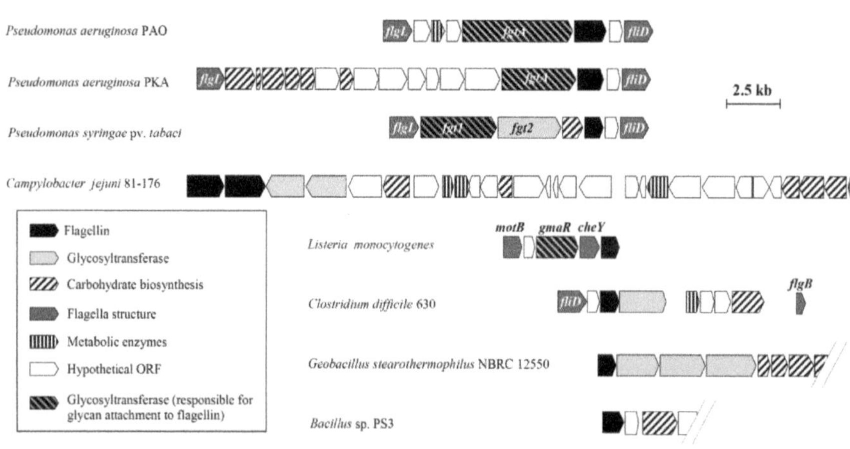

Figure 4. Organization of the glycosylation island located around the flagellin gene.

The location of glycosylation islands was not restricted to directly upstream or downstream of the flagellin gene, and the component genes were highly-diverse. A glycosyltransferase, which is responsible for the glycan attachment to a flagellin protein is usually included in the proximal glycosylation island of a flagellin gene, whereas, it was not identified to date in this region in *C. jejuni*.

Organism	linkage type	Glycan characterization	Function	GTase[*5] (OST)	Reference
Gram-negative					
Aeromonas caviae Sch3	O	Pse5Ac7Ac	Motility	n.d.	Tabei et al., (2009)
Aeromonas caviae UU 51	O	Pse5Ac7Ac8Ac	n.d.	n.d.	Schirm et al., (2005)
Aeromonas hydrophila AH3	n.d.[*1]	PAS[*2]	n.d.	n.d.	Rabaan et al., (2001)
Agrobacterium tumefaciens C58C1	n.d.	PAS	n.d.	n.d.	Deakin wt al., (1999)
Azospirillum brasilense Sp7[*2]	O	β-elimination[*3]	Adsorptio n	n.d.	Moens et al., (1995)
Campylobacter jejuni 81-176	O	Pse5Ac7Ac, PseAm, Pse8OAc PseAmGlnAc	Assembly	n.d.	Thibault et al., (2001) Schirm et al., (2005)
Campylobacter jejuni 11168	O	Leg5Am7Ac, Leg5AmNMe7Ac	n.d.	n.d.	Zampronio et al., (2011)
Campylobacter coli VC167	O	Pse5Ac7Ac, PseAm, Pse/PseAm-deoxypentose, Leg5Ac7Ac, Leg5Am7Ac, Leg5AmNMe7Ac	Assembly	n.d.	Logan et al., (2002)
Caulobacter crescentus CB15	n.d.	SDS-PAGE[*4]	Adherenc e Stability	n.d.	Faulds et al., (2011) Johnson et al., (1983)
Helicobacter pylori	O	Pse5Ac7Ac	Assembly	n.d.	Schirm et al., (2003)
Helicobacter felis CS1	n.d.	SDS-PAGE	n.d.	n.d.	Josenhans et al., (1999)
Pseudomonas aeruginosa PAK	O	Rha-(2-7 variable oligosaccharide chain)-deoxyhexosamine (dhexN)-deoxyhexose (dHex)	Virulence	FgtA	Schirm et al., (2004a)
Pseudomonas aeruginosa PAO1	O	dHex(PO_4)-192Da	Virulence	FgtA	Verma et al., (2006)
Pseudomonas aeruginosa JJ692	O	Rhamnose	n.d.	n.d.	Schirm et al., (2004a)
Pseudomonas syringae pv. *tabaci*	O	β-D-Quip4N(3-hydroxy-1-oxobutyl) 2Me-(1-3)-α-L-Rhap-(1-2)-α-L-Rhap	Virulence Stability	Fgt1	Takeuchi et al., (2007)
Burkholderia pseudomallei	O	291 Da carbohydrate	Motility	n.d.	Scott et al., (2011)
Burkholderia thailandensis	O	acetylated hexuronic acid	n.d.	n.d.	Scott et al., (2011)
Acidovorax avenae	O	n.d.	epitope recognitio n in rice	Fgt	Che et al., (2000) Hirai et al., (2011)
Borrelia burgdorferi	N	SNA, GNA lectin binding	n.d.	n.d.	Ge et al., (1998)
Serpulina hyodysenteriae	N	SDS-PAGE	n.d.	n.d.	Li et al., (1993)
Spirochaeta aurantia	n.d.	Lectin binding	n.d.	n.d.	Brahamsha & Greenberg, (1988)
Treponema pallidum	n.d.	PAS	n.d.	n.d.	Wyss, (1998)
Shewanella oneidensis	n.d.	n.d.	Motility	n.d.	Wu et al., (2011)
Marine magnetotactic ovoid bacterium MO-1	n.d.	n.d.	filament lubricant?	n.d.	Zhang et al., (2012)
Xanthomonas oryzae pv. *Oryzae*	O	n.d.	Virulence	*rbfCxoo*	Sun et al., (2009)

Organism	linkage type	Glycan characterization	Function	GTase*5 (OST)	Reference
Thermus thermophilus HB8*4	N	N-glycosydase F sensitive	n.d.	n.d.	Papaneophytou et al., (2012)
Clostridium tyrobutyricum	n.d.	β-elimination	n.d.	n.d.	Bedouet et al., (1998)
Clostridium acetobutylicum	O	PAS	n.d.	n.d.	Lyristis et al., (2000)
Clostridium difficile	O	HexNAc	Assembly	n.d.	Twine et al., (2009)
Clostridium botulinum	O	αLeg5GluNMe7Ac	n.d.	n.d.	Twine et al., (2008)
Butyrivibrio fibrisolvens	n.d.	PAS	n.d.	n.d.	Kalmokoff et al., (2000)
Listeria monocytogenes	O	N-acetylglucosamine (GlcNAc)	n.d.	GmaR	Shen et al., (2006)
Geobacillus stearothermophilus	O	PAS, β-elimination	Assembly	n.d.	Hayakawa et al., (2009a)
Bacillus sp. PS3	O	PAS, β-elimination	Assembly	n.d.	Hayakawa et al., (2009a)
Archaea					
Halobacterium salinarum	N	Glc(4-1)GlcASO4(4-1)GlcASO4 (4-1)GlcASO4	Stability	n.d.	Wieland et al., (1985)
Methanococcus voltae	N	β-ManpNAcA6Thr-(1-4)-β-GlcpNAc3NAcA-(1-3)-β-GlcpNAc	Assembly	AglA (OST)	Voisin et al., (2005)
Methanococcus maripaludis	N	Sug-4-β-ManNAc3NAmA6Thr-4-β-GlcNAc3NAcA-3-β-GalNAc	Motility	AglB (OST)	VanDyke et al., (2009)

*1 not determined; *2 Periodic acid stain; Carbohydrate specific staining method; *3 Removal of Ser/Thr binding carbohydrate structure; *4 Protein anomalous migration on SDS-PAGE; *5 Responsible for glycan attachment to flagellin.

Table 1. Flagellin glycosylations

2.1. Gram-negative

2.1.1. *Pseudomonas* spp.

Pseudomonas spp. are ubiquitous in nature and frequently isolated as opportunistic pathogen of both plant and animal. *P. aeruginosa* has a single polar flagellum which is classified by its type of flagella filament (type-a and type-b) by flagellin subunit size, amino acid sequence, and antigenicity (Allison et al., 1985; Lanyi, 1970). Both types of *P. aeruginosa* flagellin are known to contain O-linked glycosylated proteins. *P. aeruginosa* PAK and *P. aeruginosa* JJ692 produce glycosylated type-a flagellin protein, which is modified with a rhamnose (Rha) residue based on the glycan attached at two sites of each flagellin monomer (Schirm et al., 2004a). The glycan structure of PAK flagellin is a complex oligosaccharide which is composed of Rha-(2-7 variable oligosaccharide chain)-deoxyhexosamine (dhexN)-deoxyhexose (dHex), whereas JJ692 flagellin has only a single Rha glycosylation on both glycosylation site. The glycan form of PAO type-b flagellin is simpler than that of the PAK type-a flagellin, as it has a dHex linked sugar containing a phosphate moiety at two sites of the flagellin monomer (Verma et al., 2006). With regards to the *Pseudomonas* spp. of plant pathogens, flagellin glycosylation from *Pseudomonas syringae* pv. *glycinea, Pseudomonas syringae* pv. *tomato,* and *Pseudomonas syringae* pv. *tabaci* 6605 have been identified (Taguchi et al, 2003, and Takeuchi et al., 2003). The structural characterization of the flagellin protein from *P. syringae* pv. *tabaci* 6605 revealed that six sites of the flagellin were modified with a novel trisaccharide, which was composed of two rhamnosyl (Rha) residues and one modified 4-amino-4,6-dideoxyglucosyl (Qui4N; trivial name, viosamine; Vio) residue, β-D-Quip4N(3-hydroxy-1-oxobutyl)2Me-(1-3)-α-L-Rhap-(1-2)-α-L-Rhap (Takeuchi et al., 2007). The flagella glycosylation island of these *Pseudomonas* spp. have been identified and located in the upstream region of their flagellin gene (Arora et al., 2001; Taguchi et al., 2006) (Figure 4). The PAK glycosylation island is composed of 14 ORFs (~16 kb) containing putative carbohydrate synthesis related genes and glycosyltransferase (*orfN*). In contrast, both the PAO and *syringae* pv. *tabaci* glycosylation island are more simple (only 4 and 3 genes, respectively), and a putative glycosyltransferase is encoded for each. Functional analysis of these glycosyltransferases demonstrated that they are essential for the addition of glycan structure to flagellin protein (Schirm et al., 2004a; Verma et al., 2006; Taguchi et al., 2006). Recently, a flagellin glycan biosynthesis gene cluster was newly identified from *P. syringae* pv. *tabaci* (Nguyen et al., 2009). The gene cluster is related to viosamine biosynthesis (viosamine island) and these genes are homologous to a part of the PAK glycosylation island *(orfA-E* and *orfG*) (Chiku et al, 2011). Mutagenesis analysis of glycosyltransferases and flagellin subunits demonstrated that flagellin glycosylation of PAK and PAO was not require for flagella biosynthesis and motility, but a remarkable reduction of virulence was observed upon mutation (Montie et al., 1982; Arora et al., 2005). Whereas, in *P. syringae* pv. *tabaci,* loss of flagellin glycosylation reduce not only virulence but also motility, in addition, mutations which resulted in a loss of glycosylation showed differences in the bundle formation of flagella, i.e. flagella bundles on the wild-type cell were loose, and in contrast mutant filaments seemed to be tightly interacting with each other. These results indicated that glycosylation stabilizes the filament structure and lubricates the rotation of the bundle (Taguchi et al, 2008, 2010). A similar conclusion was drawn for the glycan function of flagellin in the marine magnetotactic ovoid bacterium MO-1. Flagella bundles of MO-1 were enclosed with in sheaths structure and its glycosylation was required for smooth swimming (Lefèvre et al., 2010). Flagellin proteins were also glycosylated and each flagella bundle consisted of seven individual flagella filament, which were organized in a hexagon with a seventh in the middle. Considering the compact arrangement of the seven flagella filaments in the bundle, flagellin glycosylation might function as a lubricant (Zhang et al., 2012). Recently, an *fgt2* inactivation mutant from the biosurfactant producing species *P. syringae* pv. *syringae* B728a demonstrate upregulation of the late-stage flagellar genes (class IV), and increase surfactant production (Burch et al., 2012; Xu et al., 2012). The authors suggested that over-production of the biosurfactant helps smooth cell migration and minimize flagella breakage on sticky surfaces, such as a leaf surface.

2.1.2. *Campylobacter* spp.

Campylobacter jejuni have polar flagellum at one or both ends of the cell. The flagellin proteins are extensively O-glycosylated with structural analogues of the nine-carbon sugar pseudaminic acid (Pse), legionaminic acid (Leg), and their derivatives. Flagellin glycosylation is well characterized in three species of *Campylobacter*, i.e. *Campylobacter jejuni* 81-176, *C. jejuni* NCTC 11168, and *C. coli* VC167. Flagellin modification of *Campylobacter* species were identified at 19 serine or threonine residues, 16 sites, and at least 4 sites, in *C. jejuni* 81-176, *C. coli* VC167, and *C. jejuni* NCTC 11168, respectively (Thibault et al., 2001; Logan et al., 2002; Zampronio et al., 2011). The flagellin molecular weight from *C. jejuni 81-176* is predicted by its amino-acid sequence to be 59.5 kDa, however flagellin from this strain is actually approximately 65 kDa. The additional 10% mass is attributed to attachment of substantial glycan structure (Thibault et al., 2001). In strain 81-176, the probable flagellin glycosylation related genes are largely involved in pseudaminic acid and acetamidino pseudaminic acid biosynthesis (Pse family), and lie downstream of the flagellin gene (about 27 kb) (Guerry et al., 2006). Similarly, Pse family glycosylation islands were identified in both *C. coli* VC167 and *C. jejuni* NCTC 11168, and in addition, in *C. coli* VC167 other flagellin modification genes which are involved in the synthesis of legionaminic acid and its derivatives (ptm family) were identified (McNally et al., 2007). However, the glycosyltransferase which catalyzes the addition of sugar residues to the protein backbone has not been identified. A mutation in the first step Pse biosynthesis gene leads to intercellular accumulation of unglycosylated flagellin protein (Goon et al., 2003). The biological roles of *Campylobacter* flagellin glycosylation have been mentioned in many reports. In 2007, an excellent review of the functions of flagellin glycosylation from *Campylobacter* species was published (Guerry, 2007).

2.1.3. *Helicobacter pylori*

Helicobacter pylori is a human gastric pathogen associated with gastric and duodenal ulcers as well as gastric cancer. Flagella and motility are important for colonization onto the mucosal of the human stomach. Two distinct flagellin subunits (FlaA and FlaB) were identified as glycosylated, and their glycan structures were characterized in a similar manner to that of *C. jejuni*, with Pse5Ac7Ac found at seven sites on FlaA and ten sites on FlaB, in addition flagellin glycosylation is required for functional filament assembly (Schirm et al., 2003, 2005; Josenhans et al., 2002). The mutagenesis analysis of four genes (HP0178, HP0326A, HP0326B, and HP0114) previously reported to be involved in flagellar glycosylation and polysaccharide biosynthesis demonstrated a non-motile phenotype with no structural flagella filament and only minor amounts of flagellin protein (Schirm et al., 2003). In contrast, inactivation of HP0518 resulted in altered motility and an increased level of flagellin glycosylation (Asakura et al., 2010). Complementation of a *H. pylori* HP0518 mutant and a recombinant HP0158 protein assay demonstrated the decreased glycosylation level of *H. pylori* flagellin *in vivo* and *in vitro* suggesting that HP0518 functions in the deglycosylation of flagellin. The *H. pylori* HP0518 mutant showed an increased colonization capability for the gastric tissues of mice. These results indicate that HP0518 is involved in the deglycosylation of flagellin, thereby regulating pathogen motility.

2.1.4. *Burkholderia* spp.

Burkholderia pseudomallei is also known as *Pseudomonas pseudomallei* and is important as a human and animal pathogen. In contrast, *Burkholderia thailandensis* is closely related to *B. pseudomallei* but is a nonpathogenic bacterium. Top-down and bottom-up mass spectrometry (MS) analyses of both flagellin proteins identified that there were posttranslationally modified with novel glycans (Scott et al., 2011). MS analysis of the flagellin carbohydrate moiety suggested that *B. pseudomallei* flagellin was modified with a glycan with a mass of 291 Da, while *B. thailandensis* flagellin protein was modified with related glycans with a mass of 300 or 342 Da which included an acetylated hexuronic acid. A mutagenesis analysis of the lipopolysaccharide (LPS) O-antigen biosynthetic cluster demonstrated that it was important for flagellin glycosylation and motility in *B. pseudomallei.*

2.2. Gram-positive

2.2.1. *Clostridium*

Clostridium spp., a gram-positive spore-forming anaerobic bacterium, is an emerging opportunistic pathogen towards humans and plants, and includes *Clostridium botulinum, Clostridium difficile,* and *Clostridium glumae.* The genus *Clostridium* provides the most examples of gram-positive bacterium flagellin glycosylation which has been known since the discovery of *Clostridium tyrobutyricum* (Bédouet et al., 1998, Arnold et al., 1998). Structural characterization of the carbohydrate moiety from *C. botulinum* flagellin has been achieved, and it was shown to be composed of the Leg derivative, 7-acetamido-5-(N-methyl-glutam-4-yl)-amino-3,5,7,9-tetradeoxy-d-glycero-α-d-galacto-nonulosonic acid (αLeg5GluNMe7Ac) (Twine et al., 2008). For the *C. botulinum* strain Langeland, a bioinformatic analysis of the flagella glycosylation island was completed between *flgB* and *fliD* as a large gene cluster (~48 kb), many of which appeared to be involved in carbohydrate biosynthesis (Sebaihia et al., 2007). This glycosylation island could be divided into two regions, a variable region which was located immediately downstream of the flagellin gene, and a subsequent conserved region. The carbohydrate biosynthesis genes, which are significantly related to the legionaminic acid biosynthesis genes (ptm family) in *Campylobacter coli,* were encoded in the variable region, whereas the conserved region also encoded the carbohydrate biosynthesis genes (McNally et al., 2007). The *C. botulinum* strain Langeland was found to have homologous proteins to the capsular biosynthetic proteins from *Streptococcus agalactia,* including those derived from a second set of the sialic acid biosynthetic genes, *neuA* and *neuB.*

2.2.2. *Listeria monocytogenes*

Listeria monocytogenes is a gram-positive bacterium responsible for listeriosis, and *Listeria* species are found throughout the food-processing environment. Flagellin subunits are covalently modified by monomeric β-O-linked N-acetylglucosamine (GlcNAc) residues at three to six sites per subunit (Schirm et al., 2004b). The functional consequence of flagellin glycosylation in *L. monocytogenes* was investigated by modification of the O-GlcNAc transferase (Lmo0688 renamed to GmaR), which is located just upstream of the flagellin gene (Shen et al., 2006). An in-frame deletion mutant of lmo0688 (Δ688) resulted in a nonmotile bacteria similar to what was observed for *Campylobacter* species and *Helicobacter pylori* (Josenhans et al., 2002), but this phenotype differed from that reported for gram-negative species, as it was caused by a loss of flagellin expression. The point mutation analysis of the functional residues involved in the glycosyltransferase activity demonstrated full flagellin expression (without glycosylation) and motility. The authors concluded that GmaR is a bifunctional glycosyltransferase. However, glycosylation of flagellin is not required for any flagella functions and it remains to be determined what role glycosylation of the flagellin protein plays in *Listeria monocytogenes.*

2.2.3. *Thermophilic Bacillus* spp.

Thermophilic *Bacillus* species have been isolated from deepest sea mud, hot springs, and soil, and produce multiple peritrichous flagella. These thermophiles belong to genus *Geobacillus* and are not considered to be pathogens regardless of their flagellin glycosylation. In recent years, O-linked flagellin glycosylation was reported in two thermophilic *Bacillus* species, *Geobacillus stearothermophilus* NBRC 12550 and *Bacillus* sp. PS3 (Hayakawa et al., 2009a). These flagellin glycosylations were confirmed by PAS staining and beta-elimination. The analysis of the modification sites indicated that glycan structures were attached to at least 4 sites of the flagellin monomer in *Bacillus* sp. PS3, but the structural detail of the carbohydrate chains and total number of the modification sites is currently unknown. Although it was a partial sequence, the probable glycosylation islands from both thermophilic bacterial species were confirmed downstream of these flagellin genes (J. Hayakawa and M. Ishizuka, unpublished data). In *G. stearothermophilus*, a dTDP-L-rhamnose biosynthesis gene cluster (rml operon) was also identified immediately after GTases, which is highly homologous to the glycan biosynthesis genes of the S-layer glycoprotein from a closely related *G. stearothermophilus* strain (*G. stearothermophilus* NRS2004/3a) (Novotny et al., 2004 and Steine et al., 2007). The heterologous gene expression of these flagellin in a *Bacillus subtilis* flagellin deficient mutant demonstrated that unglycosylated flagellin proteins were intracellularly accumulated and phenotypically paralyzed (Hayakawa et al., 2009a), however amino acid substitutions could restore functional filament assembly and motility (Hayakawa et al., 2009b. described below). These results supported the proposal that flagellin glycosylation is important for filament assembly. However, the carbohydrate structure and more detail of the biological functions remain to be elucidated.

2.3. Archaea

Archaeal flagellin glycosylation was first identified in *Halobacterium salinarum* (Wieland et al., 1985). Its flagellin subunit was glycosylated with sulfated glucuronic acid which is the same type as the cell surface S-layer glycoprotein. The detailed structural characterization of flagellin attached carbohydrate was accomplished for *Methanococcus voltae* (Voisin et al.,

2005). *M. voltae* flagellin proteins were modified with a novel trisaccharide, β-ManpNAcA6Thr-(1-4)-β-GlcpNAc3NAcA-(1-3)-β-GlcpNAc, N-linked to Asn. In addition, the peptide containing the N-linked sequence motif of the flagellin protein was Asn-X-Ser/Thr, which is identical to that observed for S-layer protein glycosylation. Recently, a tetrasaccharide glycan which was N-linked to the flagellin subunits in *M. maripaludis* was also characterized, with a reported structure of Sug-4-β-ManNAc3NAmA6Thr-4-β-GlcNAc3NAcA-3-β-GalNAc, where Sug is a (5S)-2-acetamido-2,4-dideoxy-5-O-methyl-α-l-erythro-hexos-5-ulo-1,5-pyranose, representing the first example of a naturally occurring diglycoside of an aldulose (Kelly et al., 2009, Jones et al., 2012). A deletion mutant analysis of three glycosyltransferases and an oligosaccharyltransferase (Stt3p homologue) from *M. maripaludis* revealed that these genes were responsible for flagellin glycosylation supported by the fact that glycan reduced flagellins were not assembled into the flagella filament (VanDyke et al., 2009). The structural and genetic analysis of archaeal flagellin glycosylation is frequently linked with S-layer protein glycosylation, and the reader is referred to a recent detailed review (Jarrell et al., 2010).

3. Glycosylation pathway

The complete pathway of bacterial flagellin glycosylation is still not clarified. There are two reviews which provide an overview of the O-linked flagellin glycosylation pathway (Logan et al., 2006; Nothaft & Szymanski, 2010). Bacterial flagella assembly occurs at the distal end of the basal body. The nascent flagellin protein is transported across the cytoplasmic membrane by a type three secretion system, and then proceeds through the narrow central channel of the flagella structure. Finally, the flagellin subunit associates with the tip of the filament structure which is elongated and reaches a length of about ten micrometers. In contrast to the archaeal flagellin export pathway, bacterial flagellin protein is not exposed outside of the inner membrane containing the periplasmic space until assembled into the filament. In other words, if flagellin glycosylation occurred extracellularly, it must be achieved far away from the cell. Therefore, it is reasonable to assume that the flagellin glycosylation machinery is located in the vicinity of the flagella basal body. Recently, the *C. jejuni* O-linked flagellin glycosylation machinery was localized at the pole of the cell along with the flagella (Ewing et al., 2009). Three genes involved in pseudaminic acid biosynthesis (PseC, which is the enzyme involved in the second step of PseAc synthesis, PseE, the putative PseAc transferase, and PseD, the putative PseAm transferase) were labeled with GFP fusion and expressed in *C. jejuni* 81-176. The fluorescent microscopic observation demonstrated that some, but not all, of the enzymatic glycosylation machinery was localized at the poles of the cells, consistent with a possible association with the flagellar basal body/export apparatus. Further study indicated that O-linked glycan biosynthesis could be reconstructed *in vitro* (Schoenhofen et al., 2009). The flagellin monomers from *Campylobacter* species are predominantly glycosylated with pseudaminic acid (Pse) and legionaminic acid (Leg). The precursors of these glycans are utilized in the form of CMP-activated sugars (CMP-Pse, CMP-Leg, and their derivatives), and they are added to the serine or threonine residues of flagellin by a specific glycosyltransferase (Note that the glycosyltransferases responsible for O-glycan attachment to flagellin have yet to be identified). The eleven candidates of glycan biosynthetic enzymes (PtmF, PtmA, PgmL, PtmE, GlmU, LegB, LegC, LegH, LegG, LegI, and LegF) from *Campylobacter jejuni* have been individually purified and characterized. It was confirmed that Leg and its CMP-activated form were synthesized from fructose-6-phosphate. The authors also suggested that O-linked glycan biosynthesis was involved in the synthesis of the N-linked glycan.

4. Amino acid substitutions of flagellin protein

Many attempts have been carried out to obtain insight into the significance of flagellin glycosylation. One of the most visible experiments is the disruption of glycosyltransferase activity which allows the evaluation of the flagella assembly, filament morphology, motility, and virulence (see above). In this section, we focus on the effects of amino acid substitution in glycosylated flagellin proteins.

4.1. Influence of loss of glycosylation to the motility and virulence

4.1.1. Campylobacter jejuni 81-176

The major flagellin of *Campylobacter jejuni* 81-176, FlaA, has been shown to be glycosylated at 19 serine or threonine residues, and this glycosylation is required for flagellar filament formation (Thibault et al., 2001; Goon et al., 2003). Mutants were constructed in which each of the 19 serine or threonines that are glycosylated in FlaA was converted to an alanine. Eleven of the 19 mutants displayed no observable phenotype, but the remaining 8 mutants had two distinct phenotypes. Five mutants (mutations S417A, S436A, S440A, S457A, and T481A) were fully motile but defective in autoagglutination. Three other mutants (mutations S425A, S454A, and S460A) were reduced in motility and synthesized truncated flagellar filaments (Ewing et al., 2009).

4.1.2. Pseudomonas syringae pv. tabaci

Flagellin glycosylation of *Pseudomonas syringae* pv. *tabaci* 6605 has been reported at six serine residues, positioned at amino acids 143, 164, 176, 183, 193 and 201 (Taguchi et al., 2006). Mutants where 6 serine residues were converted to alanine individually were compared with the mutant containing the flagellin specific glycosyltransferase, *fgt1*. All mutants displayed reduced swarming ability, swimming speed, filament stability, and virulence (Taguchi et al., 2006; 2008; Takeuchi et al., 2008). In addition, reduction of the molecular weight of each mutant flagellin protein corresponded to the loss of a single carbohydrate chain moiety, and the degree of reduced biological functions were smaller than that of an all glycosylation-serine-replacement mutant (6 S/A).

4.2. Restoration of filament formation without glycosylation

Flagellin glycosylation of a thermophilic bacillus species was recently reported for *Bacillus* sp. PS3. Although there was low coverage of the flagellin sequence, at least four serine and threonine residues were identified as glycosylation sites (Hayakawa et al., 2009a). This potentially glycosylated flagellin protein was expressed in *B. subtilis Δhag* (flagellin deficient mutant strain) for complementation. The resulting transformant was non-motile, and the produced flagellin protein derived from *Bacillus* sp. PS3 was not glycosylated and accumulated intercellularly. However, spontaneously isolated flagellin mutants partially restored the motility and produced a truncated flagella filament without glycosylation (Hayakawa et al., 2009b and J. Hayakawa and M. Ishizuka, unpublished data). All characterized suppressing mutations contained single or double point mutations and about 30 residue intragenic duplications in the flagellin in the highly variable region (D2 and D3 domain) and the end of the α-helical structure (D1 domain). The positions of these mutations were in good accordance with the previously reported flagellin glycosylations sites from many other bacterial species. To our knowledge, this is the first report of a gain-of-function mutant of flagellin glycosylation.

5. Conclusions

Glycosylation is no longer a rare event regardless of whether bacteria or eukaryote are considered. Complete genomic information for several bacteria is now available and bioinformatic analyses demonstrated that bacterial flagellin glycosylation is widely spread over several genera. Many speculative functions of flagella glycosylation have been demonstrated, for example filament assembly (including flagellin export), filament stability, motility, virulence, gene regulation and mimicry with host-cell surface glycan structure. These glycosylation functions are similar regardless of the variety of eukaryote. In addition, the bacterial glycosylation pathway is becoming better defined; many genes which participate in flagellin glycosylation have been identified, but their number and loci are diverse in each bacterial species. Rapid increases in the knowledge of glycosyltransferases and glycan biosynthesis gene clusters will undoubtedly be achieved through glycoengineering with an aim to design a bacterial flagella motor for the development of a novel vaccine or drug-delivery-system.

Author details

Jumpei Hayakawa and Morio Ishizuka*
Department of Applied Chemistry, Faculty of Science and Engineering, Chuo University, Tokyo, Japan

Acknowledgement

The authors acknowledge In-Tech-Open Access Publisher for the kind invitation to contribute this chapter. This work was supported in part by Chuo University Grant for Special Research to M. I.

* Corresponding Author

6. References

Abu-Qarn M., Yurist-Doutsch, S., Giordano, A., Trauner, A., Morris, H. R., Hitchen, P., Medalia, O., Dell, A., & Eichler, J. J. (2007). *Haloferax volcanii* AglB and AglD are involved in N-glycosylation of the S-layer glycoprotein and proper assembly of the surface layer. *Journal of Molecular Biology*, Vol. 374, No. 5, (December 2007), pp. 1224-1236, ISSN 0022-2836

Aizawa S-I. (1996). Flagellar assembly in *Salmonella typhimurium*. *Molecular Microbiology*, Vol. 19, No. 1, (January 1996), pp. 1-5, ISSN 0950-382X

Allison, J. S., Dawson, M., Drake, D., & Montie, T. C. (1985). Electrophoretic separation and molecular weight characterization of *Pseudomonas aeruginosa* H-antigen flagellins. *Infection and immunity*, Vol. 49, No. 3, (September 1985), pp. 770-774, ISSN 0019-9567

Armitage, J. P. & Macnab, R. M. (1987). Unidirectional, intermittent rotation of the flagellum of *Rhodobacter sphaeroides*. *Journal of Bacteriology*, Vol. 169, No. 2, (February 1987), pp. 514-518, ISSN 0021-9193

Arnold, F., Bédouet, L., Batina, P., Robreau, G., Talbot, F., Lécher, P., & Malcoste, R. (1998). Biochemical and immunological analyses of the flagellin of *Clostridium tyrobutyricum* ATCC 25755. *Microbiology and immunology*, Vol. 42, No. 1, pp. 23-31, ISSN 0385-5600

Arora, S. K., Bangera, M., Lory, S., & Ramphal, R. (2001). A genomic island in *Pseudomonas aeruginosa* carries the determinants of flagellin glycosylation. *Proceedings of the National Academy of Sciences of the United States of America*, Vol. 98, No. 16, (July 2001), pp. 9342-9347, ISSN 0027-8424

Arora, S. K., Neely, A. N., Blair, B., Lory, S., & Ramphal, R. (2005). Role of motility and flagellin glycosylation in the pathogenesis of *Pseudomonas aeruginosa* burn wound infections. *Infection and immunity*, Vol. 73, No. 7, (July 2005), pp. 4395-4398, ISSN 0019-9567

Asakura, H., Churin, Y., Bauer, B., Boettcher, J. P., Bartfeld, S., Hashii, N., Kawasaki, N., Mollenkopf, H. J., Jungblut, P. R., Brinkmann, V., & Meyer, T. F. (2010). *Helicobacter pylori* HP0518 affects flagellin glycosylation to alter bacterial motility. *Molecular microbiology*, Vol. 78, No. 5, (December 2010), pp. 1130-1144, ISSN 0950-382X

Bardy, S. L., Ng, S. Y., & Jarrell, K. F. (2003). Prokaryotic motility structures. *Microbiology*, Vol. 149, No. 2, (February 2003), pp. 295-304, ISSN 1350-0872

Bardy, S. L., Ng, S. Y., & Jarrell, K. F. (2004). Recent advances in the structure and assembly of the archaeal flagellum. *Journal of Molecular Microbiology and biotechnology*, Vol. 7, No. 1-2, pp. 41-51, ISSN 1464-1801

Beatson, S. A., Minamino, T., & Pallen, M. J. (2006). Variation in bacterial flagellins: from sequence to structure. *Trends in Microbiology*, Vol. 14, No. 4, (April 2006), pp. 151-155, ISSN 0966-842X

Bédouet, L., Arnold, F., Robreau, G., Batina, P., Talbot, F., & Binet, A. (1998). Evidence for an heterogeneous glycosylation of the *Clostridium tyrobutyricum* ATCC 25755 flagellin. *Microbios*, Vol. 94, No. 379, pp. 183-192, ISSN 0026-2633

Brahamsha, B. & Greenberg, E. P. (1988). Biochemical and cytological analysis of the complex periplasmic flagella from *Spirochaeta aurantia*. *Journal of Bacteriology*, Vol. 170, No. 9, (September 1988), pp. 4023–4032, ISSN 0021-9193

Burch, A. Y., Shimada, B. K., Mullin, S. W., Dunlap, C. A., Bowman, M. J., & Lindow, S. E. (2012). *Pseudomonas syringae* coordinates production of a motility-enabling surfactant with flagellar assembly. *Journal of Bacteriology*, Vol. 194, No. 6, (March 2012), pp. 1287-1298, ISSN 0021-9193

Calo, D., Kaminski, L. & Eichler, J. (2010). Protein glycosylation in Archaea: sweet and extreme. *Glycobiology*, Vol. 20, No, 9, (September 2010), pp. 1065-1076, ISSN 0959-6658

Chaban, B., Ng, S. Y., Kanbe, M., Saltzman, I., Nimmo, G., Aizawa, S., & Jarrell, K. F. (2007). Systematic deletion analyses of the fla genes in the flagella operon identify several genes essential for proper assembly and function of flagella in the archaeon, *Methanococcus maripaludis*. *Molecular Microbiology*, Vol. 66, No. 3, (November 2007), pp. 596-609, ISSN 0950-382X

Charon, N. W. & Goldstein, S. F. (2002). Genetics of motility and chemotaxis of a fascinating group of bacteria: the *spirochetes*. *Annual Review of Genetics*, Vol. 36, (December 2002), pp. 47–73, ISSN 0066-4197

Che, F. S., Nakajima, Y., Tanaka, N., Iwano, M., Yoshida, T., Takayama, S., Kadota, I., & Isogai, A. (2000). Flagellin from an incompatible strain of *Pseudomonas avenae* induces a resistance response in cultured rice cells. *The Journal of biological chemistry*, Vol. 275, No. 41, (October 2000), pp. 32347-32356, ISSN 0021-9258

Chiku, K., Ishii, T., Ono, H., Yoshida, M., & Ichinose, Y. (2011). Identification of genes involved in the glycosylation of modified viosamine of flagellins in *Pseudomonas syringae* by mass spectrometry. Genes, Vol. 2, No. 4, pp. 788-803, ISSN 2073-4425

Deakin, W. J., Parker, V. E., Wright, E. L., Ashcroft, K. J., Loake, G. J., & Shaw, C. H. (1999). *Agrobacterium tumefaciens* possesses a fourth flagellin gene located in a large gene cluster concerned with flagellar structure, assembly and motility. *Microbiology*, Vol. 145, No. Pt 6, (June 1999), pp. 1397–1407, ISSN 1350-0872

DePamphilis, M. L. & Adler, J. (1971). Fine Structure and Isolation of the Hook-Basal Body Complex of Flagella from *Escherichia coli* and *Bacillus* subtilis. *Journal of Bacteriology*, Vol. 105, No. 1, (January 1971), pp. 384-395, ISSN 0021-9193

Ewing, C. P., Andreishcheva, E., & Guerry, P. (2009). Functional characterization of flagellin glycosylation in *Campylobacter jejuni* 81-176. *Journal of Bacteriology*, Vol. 191, No. 22, (November 2009), pp. 7086-7093, ISSN 0021-9193

Faulds-Pain, A., Birchall, C., Aldridge, C., Smith, W. D., Grimaldi, G., Nakamura, S., Miyata, T., Gray, J., Li, G., Tang, J. X., Namba, K., Minamino, T., & Aldridge, P. D. (2011). Flagellin redundancy in *Caulobacter crescentus* and its implications for flagellar filament assembly. *Journal of Bacteriology*, Vol. 193, No. 11, (June 2011), pp. 2695-2707, ISSN 0021-9193

Francis, N. R., Sosinsky, G. E., Thomas, D., & DeRosier, D. J. (1994). Isolation, characterization and structure of bacterial flagellar motors containing the switch complex. *Journal of Molecular Biology*, Vol. 235, No. 2, (January 1994), pp. 1261-1270, ISSN 0022-2836

Furmanek, A. & Hofsteenge, J. (2000). Protein C-mannosylation: facts and questions. *Acta biochimica Polonica*, Vol. 47, No. 3, pp. 781–789, ISSN 0001-527X

Ge, Y., Li, C., Corum, L., Slaughter, C. A., & Charon, N. W. (1998). Structure and expression of the FlaA periplasmic flagellar protein of *Borrelia burgdorferi*. *Journal of Bacteriology*, Vol. 180, No. 9, (May 1998), pp. 2418–2425, ISSN 0021-9193

Goon, S., Kelly, J. F., Logan, S. M., Ewing, C. P., & Guerry, P. (2003). Pseudaminic acid, the major modification on *Campylobacter* flagellin, is synthesized via the Cj1293 gene. *Molecular Microbiology*, Vol. 50, No. 2, (October 2003), pp. 659-671, ISSN 0950-382X

Gross, J., Grass, S., Davis, A. E., Gilmore-Erdmann, P., Townsend, R. R., & St. Geme, J. W. 3rd. (2008). The *Haemophilus influenzae* HMW1 adhesin is a glycoprotein with an unusual N-linked carbohydrate modification. *The Journal of biological chemistry*, Vol. 283, No. 38, (September 2008), pp. 26010-26015, ISSN 0021-9258

Guerry, P. (2007). *Campylobacter* flagella: not just for motility. *Trends in Microbiology*, Vol. 15, No. 10, (October 2007), pp. 456-461, ISSN 0966-842X

Guerry, P., Ewing, C. P., Schirm, M., Lorenzo, M., Kelly, J., Pattarini, D., Majam, G., Thibault, P., & Logan, S. (2006). Changes in flagellin glycosylation affect *Campylobacter* autoagglutination and virulence. *Molecular Microbiology*, Vol. 60, No. 2, (April 2006), pp. 299-311, ISSN 0950-382X

Gugolya, Z., Muskotál, A., Sebestyén, A., Diószeghy, Z. & Vonderviszt, F. (2003). Interaction of FliS flagellar chaperone with flagellin. *FEBS letters*, Vol. 535, No 1-3, (January 2003), pp. 66-70, ISSN 0014-5793

Haeuptle, M. A. & Hennet, T. (2009). Congenital disorders of glycosylation: an update on defects affecting the biosynthesis of dolichol-linked oligosaccharides. *Human mutation*, Vol. 30, No. 12, (December 2009), pp. 1628-1641, ISSN 1059-7794

Hayakawa, J., Kambe, T., & Ishizuka, M. (2009b). Amino acid substitutions and intragenic duplications of *Bacillus* sp. PS3 flagellin cause complementation of the *Bacillus* subtilis flagellin deletion mutant. *Bioscience, biotechnology, and biochemistry*, Vol. 73, No. 10, (October 2009), pp. 2348-5231, ISSN 0916-8451

Hayakawa, J., Kondoh, Y., & Ishizuka, M. (2009a). Cloning and characterization of flagellin genes and identification of flagellin glycosylation from thermophilic *Bacillus* species. *Bioscience, biotechnology, and biochemistry*, Vol. 73, No. 6, (June 2009), pp. 1450-1452, ISSN 0916-8451

Hirai, H., Takai, R., Iwano, M., Nakai, M., Kondo, M., Takayama, S., Isogai, A., & Che, F. S. (2011). Glycosylation regulates specific induction of rice immune responses by *Acidovorax avenae* flagellin. *The Journal of biological chemistry*, Vol. 286, No. 29, (July 2011), pp. 25519-25530, ISSN 0021-9258

Ielmini, M. V. & Feldman, M. F. (2011). *Desulfovibrio desulfuricans* PglB homolog possesses oligosaccharyltransferase activity with relaxed glycan specificity and distinct protein acceptor sequence requirements. *Glycobiology*, Vol. 21, No. 6, (June 2011), pp. 734-742, ISSN 0959-6658

Irikura, V. M., Kihara, M., Yamaguchi, S., Sockett, H., & Macnab, R. M. (1993). *Salmonella typhimurium fliG* and *fliN* mutations causing defects in assembly, rotation, and switching of the flagellar motor. *Journal of Bacteriology*, Vol. 175, No. 3, (February 1993), pp. 802-810, ISSN 0021-9193

Jarrell, K. F., Jones, G. M., Kandiba, L., Nair, D. B., & Eichler, J. (2010). S-layer glycoproteins and flagellins: reporters of archaeal posttranslational modifications. *Archaea*, Vol. 2010, (July 2010), ISSN 1472-3646

Johnson, R. C., Ferber, D. M., & Ely, B. (1983). Synthesis and assembly of flagellar components by *Caulobacter crescentus* motility mutants. *Journal of Bacteriology*, Vol. 154, No. 3, (June 1983), pp. 1137–1144, ISSN 0021-9193

Jones, G. M., Wu, J., Ding, Y., Uchida, K., Aizawa, SI., Robotham, A., Logan, S. M., Kelly, J., & Jarrell, K. F. (2012). Identification of Genes Involved in the Acetamidino Group Modification of the Flagellin N-linked Glycan of *Methanococcus maripaludis*. *Journal of Bacteriology*, published ahead of print 9 March 2012, ISSN 0021-9193

Josenhans, C., Ferrero, R. L., Labigne, A., & Suerbaum, S. (1999). Cloning and allelic exchange mutagenesis of two flagellin genes of *Helicobacter felis*. *Molecular microbiology*, Vol. 33, No. 2, (July 1999), pp. 350–362, ISSN 0950-382X

Josenhans, C., Vossebein, L., Friedrich, S., & Suerbaum, S. (2002). The *neuA/flmD* gene cluster of *Helicobacter pylori* is involved in flagellar biosynthesis and flagellin glycosylation. *FEMS Microbiology Letters*, Vol. 210, No. 2, (May 2002), pp. 165-172, ISSN 0378-1097

Kalmokoff, M. L., Allard, S., Austin, J. W., Whitford, M. F., Hefford, M. A., & Teather, R. M. (2000). Biochemical and genetic characterization of the flagellar filaments from the rumen anaerobe *Butyrivibrio fibrisolvens* OR77. *Anaerobe*, Vol. 6, pp. 93–109, ISSN 1075-9964

Kelly, J., Logan, S. M., Jarrell, K. F., VanDyke, D. J., & Vinogradov, E. (2009). A novel N-linked flagellar glycan from *Methanococcus maripaludis*. *Carbohydrate research*, Vol. 344, No. 5, (March 2009), pp. 648-653, ISSN 0008-6215

Kihara, M., Francis, N. R., DeRosier, D. J., & Macnab, R. M. Analysis of a FliM-FliN flagellar switch fusion mutant of *Salmonella typhimurium*. *Journal of Bacteriology*, Vol. 178, No. 15, (August 1996), pp. 4582-4589, ISSN 0021-9193

Kowarik, M., Young, N. M., Numao, S., Schulz, B. L., Hug, I., Callewaert, N., Mills, D. C., Watson, D. C., Hernandez, M., Kelly, J. F., Wacker, M., & Aebi, M. (2006). Definition of the bacterial N-glycosylation site consensus sequence. *The EMBO journal*, Vol. 25, No. 9, (May 2006), pp. 1957-1966, ISSN 0261-4189

Lányi, B. (1970). Serological properties of *Pseudomonas aeruginosa*. II. Type-specific thermolabile (flagellar) antigens. *Acta microbiologica Academiae Scientiarum Hungaricae*, Vol. 17, No. 1, pp. 35-48, ISSN 0001-6187

Lefèvre, C. T., Santini, C. L., Bernadac, A., Zhang, W. J., Li, Y., & Wu, L. F. (2010). Calcium ion-mediated assembly and function of glycosylated flagellar sheath of marine magnetotactic bacterium. *Molecular microbiology*, Vol. 78, No. 5, (December 2010), pp. 1304-1312, ISSN 0950-382X

Li, Z., Dumas, F., Dubreuil, D., & Jacques, M. (1993). A species-specific periplasmic flagellar protein of *Serpulina (Treponema) hyodysenteriae*. *Journal of Bacteriology*, Vol. 175, No. (December 1993), pp. 8000–8007, ISSN 0021-9193

Logan, S. M. (2006). Flagellar glycosylation - a new component of the motility repertoire? *Microbiology*, Vol. 152, No. 5, (May 2006), pp. 1249-1262, ISSN 1350-0872

Logan, S. M., Kelly, J. F., Thibault, P., Ewing, C. P., & Guerry, P. (2002). Structural heterogeneity of carbohydrate modifications affects serospecificity of *Campylobacter* flagellins. *Molecular Microbiology*, Vol. 46, No. 2, (October 2002), pp. 587-597, ISSN 0950-382X

Lyristis, M., Boynton, Z. L., Petersen, D., Kan, Z., Bennett, G. N., & Rudolph, F. B. (2000). Cloning, sequencing and characterization of the gene encoding flagellin, *flaC* and the posttranslational modification of flagellin, FlaC from *Clostridium acetobutylicum* ATCC824. *Anaerobe*, Vol. 6, pp. 69–79, ISSN 1075-9964

Macnab, R. M. (1977). Bacterial flagella rotating in bundles: a study in helical geometry. *Proceedings of the National Academy of Sciences of the United States of America*, Vol. 74, No. 1, (January 1977), pp. 221-225, ISSN 0027-8424

Malapaka, R. R., Adebayo, L. O., & Tripp, B. C. (2007). A deletion variant study of the functional role of the *Salmonella* flagellin hypervariable domain region in motility. *Journal of Molecular Biology*, Vol. 365, No. 4, (January 2007), pp. 1102-1116, ISSN 0022-2836

Mathews, M. A., Tang, H. L., & Blair, D. F. (1998). Domain Analysis of the FliM Protein of *Escherichia coli*. *Journal of Bacteriology*, Vol. 180, No. 21, (November 1998), pp. 5580-5590, ISSN 0021-9193

McCarter, L. L. (2001). Polar flagellar motility of the *Vibrionaceae*. *Microbiology and Molecular Biology Reviews*, Vol. 65, No. 3, (September 2001), pp. 445–462, ISSN 1092-2172

McCarter, L. L. (2004). Dual flagellar systems enable motility under different circumstances. *Journal of Molecular Microbiology and Biotechnology*, Vol. 7, No. 1-2, (May 2004), pp. 18–29, ISSN 1464-1801

McNally, D. J., Aubry, A. J., Hui, J. P., Khieu, N. H., Whitfield, D., Ewing, C. P., Guerry, P., Brisson, J. R., Logan, S. M., & Soo, E. C. (2007). Targeted metabolomics analysis of *Campylobacter coli* VC167 reveals legionaminic acid derivatives as novel flagellar glycans. *The Journal of biological chemistry*, Vol. 282, No. 19, (May 2007), pp. 14463-14475, ISSN 0021-9258

Mescher, M. F. & Strominger, J. L. (1976). Purification and characterization of a prokaryotic glucoprotein from the cell envelope of *Halobacterium salinarium*. *The Journal of biological chemistry*, Vol. 251, No. 7, (April 1976). pp. 2005–2014, ISSN 0021-9258

Moens, S., Michiels, K., Keijers, V., Van-Leuven, F., & Vanderleyden, J. (1995). Cloning, sequencing, and phenotypic analysis of *laf1*, encoding the flagellin of the lateral flagella of *Azospirillum brasilense* Sp7. *Journal of Bacteriology*, Vol. 177, No. 19, (October 1995), pp. 5419-5426, ISSN 0021-9193

Montie, T. C., Doyle-Huntzinger, D., Craven, R. C., & Holder, I. A. (1982). Loss of virulence associated with absence of flagellum in an isogenic mutant of *Pseudomonas aeruginosa* in the burned-mouse model. *Infection and immunity*, Vol. 38, No. 3, (December 1982), pp. 1296-1298, ISSN 0019-9567

Namba, K. & Vonderviszt, F. (1997). Molecular architecture of bacterial flagellum. *Quarterly Reviews of Biophysics*, Vol. 30, No. 1, (February 1997) pp. 1-65, ISSN 0033-5835

Nguyen, L. C., Yamamoto, M., Ohnishi-Kameyama, M., Andi, S., Taguchi, F., Iwaki, M., Yoshida, M., Ishii, T., Konishi, T., Tsunemi, K., & Ichinose, Y. (2009). Genetic analysis of genes involved in synthesis of modified 4-amino-4,6-dideoxyglucose in flagellin of *Pseudomonas syringae* pv. tabaci. *Molecular genetics and genomics*, Vol. 282, No. 6, (December 2009), pp. 595-605, ISSN 1617-4615

Nothaft, H., & Szymanski, C. M. (2010). Protein glycosylation in bacteria: sweeter than ever. *Nature reviews. Microbiology*, Vol. 8, No. 11, (November 2010), pp. 765-778, ISSN 1740-1526

Novotny, R., Schäffer, C., Strauss, J., & Messner, P. (2004). S-layer glycan-specific loci on the chromosome of *Geobacillus stearothermophilus* NRS 2004/3a and dTDP-L-rhamnose biosynthesis potential of *G. stearothermophilus* strains. *Microbiology*, Vol. 150, No. Pt 4, (April 2004), pp. 953-965, ISSN 1350-0872

Papaneophytou, C. P., Papi, R. M., Pantazaki, A. A., & Kyriakidis, D. A. (2012). Flagellin gene (*fliC*) of *Thermus thermophilus* HB8: characterization of its product and involvement to flagella assembly and microbial motility. *Applied Microbiology and biotechnology*, (February 2012), ISSN 0175-7598

Rabaan, A. A., Gryllos, I., Tomás, J. M., & Shaw, J. G. (2001). Motility and the polar flagellum are required for *Aeromonas caviae* adherence toHEp-2 cells. *Infection and immunity*, Vol. 69, No. 7, (July 2001), pp. 4257-4267, ISSN 0019-9567

Reid, S. D., Selander, R. K., & Whittam, T. S. (1999). Sequence diversity of flagellin (*fliC*) alleles in pathogenic *Escherichia coli*. *Journal of Bacteriology*, Vol. 181, No. 1, (January 1999), pp. 153-160, ISSN 0021-9193

Samatey, F. A., Imada, K., Nagashima, S., Vonderviszt, F., Kumasaka, T., Yamamoto, M., & Namba, K. (2001). Structure of the bacterial flagellar protofilament and implications for a switch for supercoiling. *Nature*, Vol. 410, No. 6826, (March 2001), pp. 331-337, ISSN 0028-0836

Schirm, M., Arora, S. K., Verma, A., Vinogradov, E., Thibault, P., Ramphal, R., & Logan, S. M. (2004a). Structural and genetic characterization of glycosylation of type a flagellin in *Pseudomonas aeruginosa*. *Journal of Bacteriology*, Vol. 186, No. 9, (May 2004), pp. 2523-2531, ISSN 0021-9193

Schirm, M., Kalmokoff, M., Aubry, A., Thibault, P., Sandoz, M., & Logan, S. M. (2004b). Flagellin from *Listeria monocytogenes* is glycosylated with beta-O-linked N-acetylglucosamine. *Journal of Bacteriology,* Vol. 186, No. 20, (October 2004), pp. 6721-6727, ISSN 0021-9193

Schirm, M., Schoenhofen, I. C., Logan, S. M., Waldron, K. C., & Thibault, P. (2005). Identification of unusual bacterial glycosylation by tandem mass spectrometry analyses of intact proteins. *Analytical chemistry,* Vol. 77, No. 23, (December 2005), pp. 7774-7782, ISSN 0003-2700

Schirm, M., Soo, E. C., Aubry, A. J., Austin, J., Thibault, P., & Logan, S. M. (2003). Structural, genetic and functional characterization of the flagellin glycosylation process in *Helicobacter pylori. Molecular Microbiology,* Vol. 48, No. 6, (June 2003), pp. 1579-1592, ISSN 0950-382X

Schoenhofen, I. C., Vinogradov, E., Whitfield, D. M., Brisson, J. R., & Logan, S. M. (2009). The CMP-legionaminic acid pathway in *Campylobacter*: biosynthesis involving novel GDP-linked precursors. *Glycobiology,* Vol. 19, No. 7, (July 2009), pp. 715-725, ISSN 0959-6658

Scott, A. E., Twine, S. M., Fulton, K. M., Titball, R. W., Essex-Lopresti, A. E., Atkins, T. P., & Prior, J. L. (2011). Flagellar glycosylation in *Burkholderia pseudomallei* and *Burkholderia thailandensis. Journal of Bacteriology,* Vol. 193, No. 14, (July 2011), pp. 3577-3587, ISSN 0021-9193

Sebaihia, M., Peck, M. W., Minton, N. P., Thomson, N. R., Holden, M. T., Mitchell, W. J., Carter, A. T., Bentley, S. D., Mason, D. R., Crossman, L., Paul, C. J., Ivens, A., Wells-Bennik, M. H., Davis, I. J., Cerdeño-Tárraga, A. M., Churcher, C., Quail, M. A., Chillingworth, T., Feltwell, T., Fraser, A., Goodhead, I., Hance, Z., Jagels, K., Larke, N., Maddison, M., Moule, S., Mungall, K., Norbertczak, H., Rabbinowitsch, E., Sanders, M., Simmonds, M., White, B., Whithead, S., & Parkhill, J. (2007). Genome sequence of a proteolytic (Group I) *Clostridium botulinum* strain Hall A and comparative analysis of the clostridial genomes. *Genome research,* Vol. 17, No. 7, (July 2007), pp. 1082-1092, ISSN 088-9051

Shen, A., Kamp, H. D., Gründling, A., & Higgins, D. E. (2006). A bifunctional O-GlcNAc transferase governs flagellar motility through anti-repression. Genes and development, Vol. 20, No. 23, (December 2006), pp. 3283-3295, ISSN 0890-9369

Shigematsu, M., Meno, Y., Misumi, H., & Amako, K. (1995). The measurement of swimming velocity of *Vibrio cholerae* and *Pseudomonas aeruginosa* using the video tracking methods. *Microbiology and immunology,* Vol. 39, No. 10, pp. 741-744, ISSN 0385-5600

Sleytr, U. B. & Thorne, K. J. (1976). Chemical characterization of the regularly arranged surface layers of *Clostridium thermosaccharolyticum* and *Clostridium thermohydrosulfuricum. Journal of Bacteriology,* Vol. 126, No. 1, (April 1976), pp. 377–383, ISSN 0021-9193

Sleytr, U. B. (1975). Heterologous reattachment of regular arrays of glycoproteins on bacterial surfaces. *Nature*, Vol. 257, No. 5525, (October 1975), pp. 400–402, ISSN 0028-0836

Sockett, H., Yamaguchi, S., Kihara, M., Irikura, V. M. & Macnab, R. M. (1992). Molecular analysis of the flagellar switch protein FliM of *Salmonella typhimurium*. *Journal of Bacteriology*, Vol. 174, No. 3, (February 1992), pp. 793-806, ISSN 0021-9193

Steine, K., Novotny, R., Patel, K., Vinogradov, E., Whitfield, C., Valvano, M. A., Messner, P., & Schäffer, C. (2007). Functional characterization of the initiation enzyme of S-layer glycoprotein glycan biosynthesis in *Geobacillus stearothermophilus* NRS 2004/3a. *Journal of Bacteriology*, Vol. 189, No. 7, (April 2007), pp. 2590-2598, ISSN 0021-9193

Sun, Y., Wen, J., Wu, M., Chen, H., & He, C. (2009). Deletion mutation of *rbfCxoo*, encoding a putative glycosyltransferase in *Xanthomonas oryzae* pv. *oryzae* leads to enhanced virulence expression. *Acta microbiologica Sinica*, Vol. 49, No. 6, (June 2009), pp. 740-745, ISSN 0001-6209

Szymanski, C. M. & Wren, B. W. (2005). Protein glycosylation in bacterial mucosal pathogens. *Nature reviews. Microbiology*, Vol. 3, No. 3, (March 2005), pp. 225-237, ISSN 1740-1526

Szymanski, C. M., Burr, D. H., & Guerry, P. (2002). *Campylobacter* protein glycosylation affects host cell interactions. *Infection and immunity*, Vol. 70, No. 4, (April 2002), pp. 2242-2244, ISSN 0019-9567

Tabei, S. M., Hitchen, P. G., Day-Williams, M. J., Merino, S., Vart, R., Pang, P. C., Horsburgh, G. J., Viches, S., Wilhelms, M., Tomás, J. M., Dell, A., & Shaw, J. G. (2009). An *Aeromonas caviae* genomic island is required for both O-antigen lipopolysaccharide biosynthesis and flagellin glycosylation. *Journal of Bacteriology*, Vol. 191, No. 8, (April 2009), pp. 2851-2863, ISSN 0021-9193

Taguchi, F., Shibata, S., Suzuki, T., Ogawa, Y., Aizawa, S., Takeuchi, K., & Ichinose, Y. (2008). Effects of glycosylation on swimming ability and flagellar polymorphic transformation in *Pseudomonas syringae* pv. *tabaci* 6605. *Journal of Bacteriology*, Vol. 190, No. 2, (January 2008), pp. 764-768, ISSN 0021-9193

Taguchi, F., Shimizu, R., Inagaki, Y., Toyoda, K., Shiraishi, T., & Ichinose, Y. (2003). Post-translational modification of flagellin determines the specificity of HR induction. *Plant and cell physiology*, Vol. 44, No. 3, (March 2003), pp. 342-349, ISSN 0032-0781

Taguchi, F., Takeuchi, K., Katoh, E., Murata, K., Suzuki, T., Marutani, M., Kawasaki, T., Eguchi, M., Katoh, S., Kaku, H., Yasuda, C., Inagaki, Y., Toyoda, K., Shiraishi, T., & Ichinose, Y. (2006). Identification of glycosylation genes and glycosylated amino acids of flagellin in *Pseudomonas syringae* pv. *tabaci*. *Cellular Microbiology*, Vol. 8, No. 6, (June 2006), pp. 923-938, ISSN 1462-5814

Taguchi, F., Yamamoto, M., Ohnishi-Kameyama, M., Iwaki, M., Yoshida, M., Ishii, T., Konishi, T., & Ichinose, Y. (2010). Defects in flagellin glycosylation affect the virulence of *Pseudomonas syringae* pv. *tabaci* 6605. *Microbiology*, Vol. 156, No. Pt 1, (January 2010), pp. 72-80, ISSN 1350-0872

Takeuchi, K., Ono, H., Yoshida, M., Ishii, T., Katoh, E., Taguchi, F., Miki, R., Murata, K., Kaku, H., & Ichinose, Y. (2007). Flagellin glycans from two pathovars of *Pseudomonas syringae* contain rhamnose in D and L configurations in different ratios and modified 4-amino-4,6-dideoxyglucose. *Journal of Bacteriology,* Vol. 189, No. 19, (October 2007), pp. 6945-6956, ISSN 0021-9193

Takeuchi, K., Taguchi, F., Inagaki, Y., Toyoda, K., Shiraishi, T., & Ichinose, Y. (2003). Flagellin glycosylation island in *Pseudomonas syringae* pv. *glycinea* and its role in host specificity. *Journal of Bacteriology,* Vol. 185, No. 22, (November 2003), pp. 6658-6665, ISSN 0021-9193

Thibault, P., Logan, S. M., Kelly, J. F., Brisson, J. R., Ewing, C. P., Trust, T. J., & Guerry, P. (2001). Identification of the carbohydrate moieties and glycosylation motifs in *Campylobacter jejuni* flagellin. *The Journal of biological chemistry,* Vol. 276, No. 37, (September 2001), pp. 34862-34870, ISSN 0021-9258

Twine, S. M., Reid, C. W., Aubry, A., McMullin, D. R., Fulton, K. M., Austin, J., & Logan, S. M. (2009). Motility and flagellar glycosylation in *Clostridium difficile. Journal of Bacteriology,* Vol. 191, No. 22, (November 2009), pp. 7050-7062, ISSN 0021-9193

Twine, S. M., Paul, C. J., Vinogradov, E., McNally, D. J., Brisson, J. R., Mullen, J. A., McMullin, D. R., Jarrell, H. C., Austin, J. W., Kelly, J. F., & Logan, S. M. (2008). Flagellar glycosylation in *Clostridium botulinum. The FEBS journal,* Vol. 275, No. 17, (September 2008), pp. 4428-4444, ISSN 742-464X

VanDyke, D. J., Wu, J., Logan, S. M., Kelly, J. F., Mizuno, S., Aizawa, S., & Jarrell, K. F. (2009). Identification of genes involved in the assembly and attachment of a novel flagellin N-linked tetrasaccharide important for motility in the archaeon *Methanococcus maripaludis. Molecular Microbiology,* Vol. 72, No. 3, (May 2009), pp. 633-644, ISSN 0950-382X

Varki, A. (1993). Biological roles of oligosaccharides: all of the theories are correct. *Glycobiology,* Vol. 3, No. 2, (April 1993), pp. 97-130, ISSN 0959-6658

Verma, A., Schirm, M., Arora, S. K., Thibault, P., Logan, S. M., & Ramphal, R. (2006). Glycosylation of b-Type flagellin of *Pseudomonas aeruginosa*: structural and genetic basis. *Journal of Bacteriology,* Vol. 188, No. 12, (June 2006), pp. 4395-4403, ISSN 0021-9193

Voisin, S., Houliston, R. S., Kelly, J., Brisson, J. R., Watson, D., Bardy, S. L., Jarrell, K. F., & Logan, S. M. (2005). Identification and characterization of the unique N-linked glycan common to the flagellins and S-layer glycoprotein of *Methanococcus voltae. The Journal of biological chemistry,* Vol. 280, No. 17, (April 2005), pp. 16586–16593, ISSN 0021-9258

Wacker, M., Linton, D., Hitchen, P. G., Nita-Lazar, M., Haslam, S. M., North, S. J., Panico, M., Morris, H. R., Dell, A., Wren, B. W., & Aebi, M. (2002). N-linked glycosylation in *Campylobacter jejuni* and its functional transfer into *E. coli. Science,* Vol. 298, No. 5599, (November 2002), pp. 1790-1793, ISSN 0036-8075

Weerapana, E. & Imperiali, B. (2006). Asparagine-linked protein glycosylation: from eukaryotic to prokaryotic systems. *Glycobiology*, Vol. 16, No. 9, (June 2006), pp. 91R-101R, ISSN 0959-6658

Westerlund-Wikström, B. (2000). Peptide display on bacterial flagella: principles and applications. *International journal of medical Microbiology*, Vol. 290, No. 3, (July 2000), pp. 223-230, ISSN 1438-4221

Wieland, F., Paul, G., & Sumper, M. (1985). *Halobacterial* flagellins are sulfated glycoproteins. *The Journal of biological chemistry*, Vol. 260, No. 28, (December 1985), pp. 15180–15185, ISSN 0021-9258

Woods, R. D., Takahashi, N., Aslam, A., Pleass, R. J., Aizawa, S.-I., & Sockett, R. E. (2007). Bifunctional nanotube scaffolds for diverse ligands are purified simply from *Escherichia coli* strains coexpressing two functionalized flagellar genes. *Nano letters*, Vol. 7, No. 6, (June 2007), pp. 1809-1816, ISSN 1530-6984

Wu, L., Wang, J., Tang, P., Chen, H., & Gao, H. (2011). Genetic and molecular characterization of flagellar assembly in *Shewanella oneidensis*. *PloS one*, Vol. 6, No. 6, pp. e21479, ISSN 1932-6203

Wyss, C. (1998). Flagellins, but not endoflagellar sheath proteins, of *Treponema pallidum* and of pathogen-related oral spirochetes are glycosylated. *Infection and immunity*, Vol. 66, No. 12, (December 1998), pp. 5751–5754, ISSN 0019-9567

Xu, J., Platt, T. G., & Fuqua, C. (2012). Regulatory linkages between flagella and surfactant during swarming behavior: lubricating the flagellar propeller? *Journal of Bacteriology*, Vol. 194, No. 6, (March 2012), pp. 1283-1286, ISSN 0021-9193

Yonekura, K., Maki-Yonekura, S., & Namba, K. (2005). Building the atomic model for the bacterial flagellar filament by electron cryomicroscopy and image analysis. *Structure*, Vol. 13, No. 3, (March 2005), pp. 407-412, ISSN 0969-2126

Yonekura, K., Maki-Yonekura, S., & Namba, K. (2003). Complete atomic model of the bacterial flagellar filament by electron cryomicroscopy. *Nature*, Vol. 424, No. 6949, (August 2003), pp. 643-650, ISSN 0028-0836

Yoshioka, K., Aizawa, S-I., & Yamaguchi, S. (1995). Flagellar filament structure and cell motility of *Salmonella typhimurium* mutants lacking part of the outer domain of flagellin. *Journal of Bacteriology*, Vol. 177, No. 4, (February 1995), pp. 1090-1093, ISSN 0021-9193

Young, N. M., Brisson, J. R., Kelly, J., Watson, D. C., Tessier, L., Lanthier, P. H., Jarrell, H. C., Cadotte, N., St Michael, F., Aberg, E., & Szymanski, C. M. (2002). Structure of the N-linked glycan present on multiple glycoproteins in the Gram-negative bacterium, *Campylobacter jejuni*. *The Journal of biological chemistry*. Vol. 277, No. 45, (November 2002), pp. 42530-43549, ISSN 0021-9258

Zampronio, C. G., Blackwell, G., Penn, C. W., & Cooper, H. J. (2011). Novel glycosylation sites localized in *Campylobacter jejuni* flagellin FlaA by liquid chromatography electron capture dissociation tandem mass spectrometry. *Journal of proteome research*, Vol. 10, No. 3, (March 2011), pp. 1238-1245, ISSN 1535-3893

Zhang, W. J., Santini, C. L., Bernadac, A., Ruan, J., Zhang, S. D., Kato, T., Li, Y., Namba, K., & Wu, L. F. (2012). Complex spatial organization and flagellin composition of flagellar propeller from marine magnetotactic ovoid strain MO-1. *Journal of molecular biology,* Vol. 416, No. 4, (March 2012), pp. 558-570, ISSN 0022-2836

Chapter 7

Glycosylating Toxin of *Clostridium perfringens*

Masahiro Nagahama, Masataka Oda and Keiko Kobayashi

Additional information is available at the end of the chapter

1. Introduction

Clostridium perfringens type C strains that produce various toxins cause hemorrhagic noxious ulceration or mucousal necrosis of the small intestine in humans, pigs, cattle and chickens (Sakurai et al. 1997, Sakurai and Nagahama 2006). In humans, the bacteria cause necrotic enteritis, which is termed "pig-bel" (Sakurai and Nagahama 2006). *C. perfringens* has been classified into five types, A to E, according to the toxinogenicity of major extracellular toxins designated alpha-, beta-, epsilon- and iota-toxins. The *C. perfringens* strains defined as type C show alpha- and beta-, but not epsilon- and iota-toxigenicities (Sakurai and Nagahama 2006). Type C strains produce alpha-toxin, beta-toxin, beta2-toxin, and perfringolysin O. Beta-toxin is known to be the primary pathogenic factor of necrotic enteritis in the type C strains (Tweten 2005, Sakurai and Nagahama 2006). Beta2-toxin is a toxin discovered in *C. perfringens* type C isolated from piglets with necrotic enteritis, and speculated to be important because its gene has been detected in most *C. perfringens* type C strains recovered from animals with clinical disease (Manteca et al. 2002, Waters et al. 2003).

Most of the toxins produced by *C. perfringens* type C are toxic to particular cells or cell lines. Beta-toxin possesses lethal, dermonecrotic, and pressor activities. Direct evidence concerning the biological activities of beta-toxin had been lacking due to the absence of a susceptible cell line for experiments *in vitro*. However, a cell line that is susceptible to the toxin was found (Nagahama et al. 2003). The toxin induced swelling and lysis on HL60 cells by binding to lipid rafts and forming a functional oligomer. Beta2-toxins is lethal for mice and cytotoxic for the cell line I407 cells inducing cell rounding and lysis without affecting the actin cytoskeleton (Gibert et al. 1997). On the other hand, delta-toxin is toxic to various rabbit immune cells, i.e. alveolar macrophages, peritoneal appendix cells, bone marrow cells, splenocytes and thymocytes (Jolivet-Reynaud et al. 1982). Theta-toxin is the prototypic member of the cholesterol-dependent cytolysin (CDC) family that includes listeriolysin O (LLO), streptolysin O (SLO), pneumolysin, and others (Tweten 2005). Theta-toxin exhibits cytotoxic effects on macrophages to escape from phagosomes in consort with alpha-toxin. Alpha-toxin is essential for growth and spread of infection in the host (Sakurai et al. 2004), and it helps *C. perfringens* avoid the host defense mechanism by altering the normal traffic of the host phagocytes (Ochi et al. 2002, Oda et al. 2006). The role played by alpha-toxin in pathogenesis is dictated by its ability to interact with membranes, whether from outside the cell or while inside the phagosomes (Naylor et al. 1998). Alpha-toxin possesses phospholipase C (PLC), sphingomyelinase (SMase) and biological activities causing hemolysis, lethality and dermonecrosis (Sakurai et al. 2004). On the other hand, enterotoxin, which has molecular masses of 35 kDa as a monomer (Duffy et al. 1982) and 90–200 kDa as aggregate forms with eukaryotic proteins (Singh et al. 2001), is known to be highly cytotoxic to Vero and Caco-2 cells (Gao and McClane 2012). Among the *C. perfringens* type C toxins reported previously, only enterotoxin is toxic to Vero cells, and no toxin is known to have a molecular mass of around 200 kDa without eukaryotic proteins.

TpeL, a recently-described novel member of the family of large clostridial cytotoxins, was found in *C. perfringens* type C. TpeL (named TpeL: *C. perfringens* large cytotoxin) is a truncated homologue of *Clostridium difficile* TcdA and TcdB (Amimoto et al. 2007). TpeL was identified in the culture supernatant of *C. perfringens* strain CP4 and is thought to be associated with necrotic enteritis (Amimoto et al. 2007). Coursodon et al. (2012) reported that TpeL may potentiate the effect of other virulence attributes of necrotic enteritis caused by *C. perfringens*.

2. The family of clostridial glucosylating toxins

Cytosolic mono-O-glucosylation is an important molecular mechanism by which various bacterial protein toxins and effectors target eukaryotic cells. *C. difficile* TcdA and TcdB , *Clostridium sordellii* lethal toxin (TcsL) and *Clostridium novyi* alpha-toxin (TcnA) are important pathogenic factors of the family of large clostridial toxins (LCTs). The pathogenicity of *C. difficile* is based upon the action of at least one of the two major exotoxins (TcdA and TcdB) (Voth and Ballard 2005). TcdA and TcdB are the main cause of antibiotic-associated diarrhea and pseudomembraneous colitis (Voth and Ballard 2005), as a consequence of treatment with antibiotics, which destroy the normal microflora of the gut and allow colonization and proliferation of *C. difficile* bacteria (Bartlett et al. 1977, Kelly and LaMont 2008). Although the precise pathogenic mechanisms of induction of diarrhea and colitis are not known, it is generally accepted that the toxin-induced glucosylation of Rho GTPases is central to the action of the *C. difficile* toxins (Kelly and LaMont 2008). TcsL is implicated in toxic shock syndrome after medical-induced abortion (Ho et al. 2009) and TcnA causes gas gangrene syndrome (Tsokos et al. 2008). All these toxins are 50 to 90% identical in their amino acid sequences. They are large proteins of 250 to 308 kDa. This family has now more than 30 members, including putative glycosyltransferases from *C. perfringens, Escherichia coli, Citrobacter rodentium, Photobacterium profundum, Pseudomonas fluorescens* and various species of *Chlamydia* and *Chlamidophila*.

2.1. Structure of LCTs

LCTs are single protein chains containing four functional domains and share 26 to 76 % sequence identity and are structurally and functionally organized (Busch and Aktories

2000). LCTs are composed of four domains, the glucosylating enzymatic A-domain, the autocatalytic processing C-domain, the translocating D-domain and the binding B-domain (Fig. 1). The one third C-terminal part exhibits multiple repeated sequences (31 short repeats and 7 long repeats in TcdA), which are involved in the recognition of a cell-surface receptor. A trisaccharide (Gal-α1-3Gal-β1-4GlcNac) has been found to be the motif recognized by TcdA. Related carbohydrates could be involved as TcdA receptor. The gp96, a member heat shock protein family, has been proposed to bind TcdA to the plasma membrane of enterocytes (Na et al. 2008). Because this B-domain exhibits sequence similarity to the carbohydrate binding region of the glucosyltransferase from *Streptococcus mutans*, it was suggested early on that this part of the toxin is involved in binding to a carbohydrate-containing receptor. The crystal structure of the C-terminal binding domain of

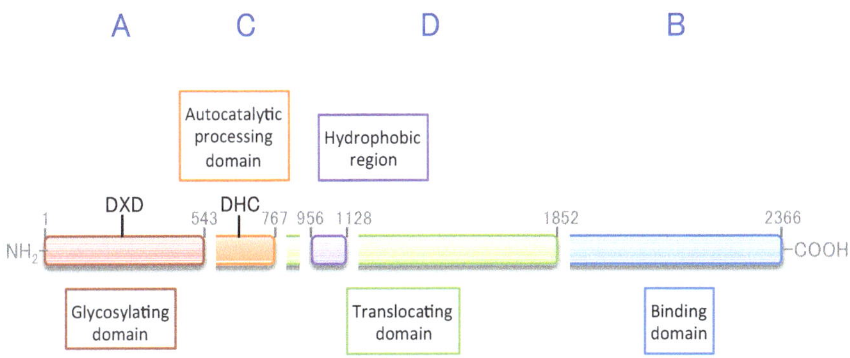

Figure 1. Domain organization of clostridial glucosylating toxins.

Clostridial glucosylating toxins are composed of four domains, the glucosylating enzymatic domain (A-domain) , the autocatalytic processing domain (C-domain), the translocating domain (D-domain) and the binding domain (B-domain). The A-domain has the glucosyltransferase activity. The B-domain containing polypeptide repeats is responsible for specific receptor-binding. The C-domain participates in the auto-catalytic cleavage of LCTs and is a cysteine-protease with the catalytic residues DHC. $InsP_6$ is required for activation of the cysteine protease. The D-domain is likely involved in the delivery of the A-domain into the cytosol. This domain includes a hydrophobic region suggested to be important for insertion of the toxin into endosome membranes.

TcdA has been determined (Ho et al. 2005, Greco et al. 2006), showing a solenoid-like structure with 32 repeats consisting of 15–21 amino acid residues and seven repeats consisting of 30 residues. The repeats form β-hairpins, arranged in pairs with each adjacent pair of hairpins rotated by 120° to the next pair, resulting in a screw-like structure of a left-handed β-solenoid helix (Greco et al. 2006). Co-crystallization with a derivate of the trisaccharide α-Gal(1,3)β-Gal(1,4)βGlcNAc confirmed the carbohydrate binding capacity of the domain. In this complex there are two carbohydrate-binding regions. However, in the full-length C-terminal fragment there are seven of these potential binding domains that are highly conserved, giving it a high binding capacity. Although there is little information about the binding domain of TcdB, it is believed that TcdB uses different receptors to bind to target cell surfaces than TcdA (Jank et al. 2007a).

The cysteine protease C-domain is located between residues 543 and 769 in TcdA and between residues 543 and 767 in TcdB (Egerer et al. 2007, Giesemann et al. 2008). It was shown that cleavage of the toxin occurs auto-catalytically by a cysteine protease activity, which is harbored in the C-domain, covering residues 544–955, directly down-stream of the glucosyltransferase domain (Egerer et al. 2007, Giesemann et al. 2008). Cys-698 and His-653 have been shown to be part of the catalytic dyad, which in addition to Asp-587 might participate in the auto-cleavage reaction (Egerer et al. 2007). The crystal structure of C-domain (543–809) from TcdA was resolved in the presence of inositol hexakisphosphate ($InsP_6$) (Pruitt et al. 2009). $InsP_6$ binds to the C-Domain, causing a conformational change that activates the auto-catalytic activity (Egerer et al. 2009). This locates between the enzymatic A-domain and the delivery D-domain, playing a role in proteolytic cleavage of the toxin. The toxins undergo autoproteolysis allowing only the enzymatic A-domain to be released into the cytosol in the presence of $InsP_6$. Once the target cell has taken up the LCT via receptor-mediated endocytosis at the B-domain, the toxin undergoes autoproteolysis in order to allow the A-domain to pass across the endosomal membrane into the cytosol (Kreimeyer et al. 2011).

D-domain, which is located between residues 955 and 1852, is a large hydrophobic region that makes up almost 50% of the total size of the toxin (Barth et al. 2001, Qa'Dan et al. 2000). However, the exact function of D-domain is unknown. It is characterized by a hydrophobic stretch which is most probably responsible for membrane penetration (transmembrane prediction) (von Eichel-Streiber et al. 1992). Therefore, this region is speculated to as the translocation domain. Deletion studies proved the importance of the hydrophobic region for toxin activity (Barroso et al. 1994). A small region in the primary sequence between residues 965 and 1128 is characterized by hydrophobic amino acids and is suggested to play a role in formation of transmembrane structure during pore formation and translocation of the toxin into the cytosol (Voth and Ballard 2005). Pore formation induced by the toxin has been shown in artificial lipid membranes (Barth et al. 2001, Giesemann et al. 2006). However, so far it is not clear how pore formation relates to the delivery of the toxin into the cytosol.

The biologically active domain, A-domain, harboring the glucosyltransferase activity, is translocated into the cytosol, comprises the first 543 aa (Pfeifer et al. 2003, Rupnik et al. 2005). Therefore, cleavage of the toxin is required. The 3D-structure of A-domain showed that it was closely related to other bacterial glucosyltransferases belonging to the glucosyltransferase A family (Reinert et al. 2005). The catalytic core consists of 234 aa and is formed by a mixed α/β-fold with mostly parallel β-strands as the central part. The more than 300 additional residues are mainly helices, of which the first four N-terminal helices are most probably involved in membrane association, therefore assuring close proximity of the enzyme with its substrates. The structure of the central core is similar to that of glucosyltransferase A family (Reinert et al. 2005, Ziegler et al. 2008). Characteristic for glucosyltransferase A family members is the DXD motif involved in complexation of manganese ions, UDP and glucose. Mutation of these essential aspartate residues leads to inactivation of the toxin (Giesemann et al. 2008). The first aspartic residue of the DXD motif binds to ribosyl and glucosyl moieties of UDP-glucose and the second aspartic residues binds to divalent cation (mainly manganese ions) which increases the hydrolase activity and/or the binding of UDP-glucose (Just et al. 2000). Other amino acids in TcdB having an essential role in the enzymatic activity have been identified such as Trp-102 which is involved in the binding of UDP-glucose, Asp-270, Arg-273, Tyr-284, Asn-384, Trp-520, as well as Ile-383 and Glu-385 being important for the specific recognition of UDP-glucose (Busch et al 2000a and 2000b, Jank et al. 2005, Jank et al. 2007a,). Differences in α-helices probably account for the substrate specificity of each toxins (Reinert et al. 2005, Ziegler et al. 2008). Chimeric molecules between TcdB and TcsL have been used to identify the sites of Rho-GTPase recognition. Amino acids 408 to 468 of TcdB ensure the specificity for Rho, Rac and Cdc42, whereas in TcsL, the recognition of Rac and Cdc42 is mediated by residues 364 to 408, and that of Ras proteins by residues 408 to 516 (Voth and Ballard 2005, Jank et al. 2007a). The four N-terminal helices which mediates the binding of TcsL to phosphatidylserine, are possibly involved in membrane interaction (Mesmin et al. 2004). Amino acids 22-27 of Rho and Ras GTPase which are part of the transition of the α1-helix to the swich 1 region, are the main domain recognized by the glucosylating toxins (Müller et al. 1999). The cosubstrate for the bacterial glucosyltransferases is UDP-glucose; only TcnA utilizes UDP-N-acetylglucosamine (UDP-GlcNAc) (Selzer et al. 1996). This difference in cosubstrate specificity is based on sterical hindrance by bulky amino acids (e.g. Ile383/Gln385 in TcdB) blocking the catalytic pocket for the larger UDP-GlcNAc. In TcnA, small serine and alanine residues at the corresponding positions allow UDP-GlcNAc to enter the catalytic cleft (Jank et al. 2005). Little is known so far about the molecular/structural determinants underlying the differences in substrate recognition by different glucosylating toxins.

2.2. Internalization of LCTs

LCTs enter eukaryotic target cells through receptor-mediated endocytosis according to the 'short trip model' of bacterial exotoxin uptake (Sandvig et al. 2004). The cytotoxicity of the toxins are blocked by endosomal and lysosomal acidification inhibitors (monensin, bafilomycin A1, ammonium chloride) and the inhibiting effects can be by-passed by an extracellular acidic pulse (Popoff et al. 1996, Qa'Dan et al. 2000, Barth et al. 2001, Popoff and Bouvet 2009). As shown in Fig. 2, on binding to host cell receptors (Karlsson 1995, Giesemann et al. 2008), the toxins are endocytosed (Voth and Ballard 2005). After endocytosis, the acidification of early endosomes by the vesicular H^+-ATPase induces a conformational change characterized by an increase in hydrophobicity, leading to membrane insertion (Qa'Dan et al. 2000, Barth et al. 2001, Voth and Ballard 2005). It reported that at low pH, LCTs induce channel formation in cell membranes and artificial lipid bilyers (Qa'Dan et al. 2000 and 2001, Giesemann et al. 2006). Membrane cholesterol seems critical for TcdA pore formation (Giesemann et al. 2006). Then, the hydrophobic region enables to form a pore through which the catalytic domain can translocate into the cytosol. Pore formation under acidic conditions has been demonstrated for TcdA and B (Barth et al. 2001,

Giesemann et al. 2006). The exact mode of translocation remains to determine. The translocation-ligand domain remains associated with endosomal membranes and only the catalytic-DXD domain penetrates into the cytosol (Pfeifer et al. 2003, Rupnik et al. 2005). The N-terminal catalytic domain (A-domain) is then delivered from the early endosomes into the cytosol by auto-proteolytic activity stimulated by $InsP_6$ (Reineke et al. 2007, Egerer et al. 2007). This autoproteolytic activity is induced by $InsP_6$ and/or dithiothreitol and is responsible for the separation of the catalytic domain from the holotoxin (Egerer et al. 2007, Reineke et al. 2007). A cysteine protease domain (C-domain containing putative catalytic residues, DHC) has been identified close to the cutting site in TcdB (amino acid 544-955), which is conserved in all LCTs (Egerer et al. 2007, Reineke et al. 2007, Egerer et al. 2009). Cys-700, His-655 and Asp-589 have been identified as the catalytic triad. It reported that a cysteine protease catalytic triad is involved in processing of the toxin and auto-cleavage is essential for toxin activity (Egerer et al. 2007). Following its translocation and release, the catalytic-DXD fragment (A-domain) acts on its cytosolic targets, the GTPases of the Rho/Rac family, leading to the observed blockade of signal transduction processes and, consequently, the disaggregation of the cytoskeleton and cell death (Just and Gerhard 2004, Belyi and Aktories 2010). TcdA and B target Rho GTPases (Rho, Ras and Cdc42), which are molecular switches involved in numerous signal processes, in particular, the regulation of the actin cytoskeleton. Once the toxins enter the cytosol, they catalyse the addition of UDP-Glc (UDP-glucose) to Thr-37 (monoglucosylation) in Rho GTPase leading to depolymerization of actin filaments, disruption of the cytoskeleton and eventually cell rounding and cell death (Jank et al. 2007b, Belyi and Aktories 2010) (Fig. 2).

2.3. Glucosylation of Rho GTPase by LCTs

In the cytosol, LCTs glucosylate small GTPases of the Rho and Ras superfamily (Popoff et al. 1996, Selzer et al. 1996, Belyi and Aktories 2010). Small GTP-binding proteins involve organization of the cytoskeleton and control cellular activity of numerous other cellular enzymes. Rho proteins are molecular switches involved in various signal processes, including actin cytoskeleton regulation, cell cycle progression, gene transcription, and control of the activity of many enzymes like protein and lipid kinases, phospholipases, and nicotanimide adenine dinucleotide-oxidase (Etienne-Manneville and Hall 2002, Burridge and Wennerberg, 2004). In respect to their role in host–pathogen interactions, Rho proteins essentially participate in epithelial barrier functions and cell–cell contact, in immune cell

LCTs bind with their B-domain to the receptor of target cells. After endocytosis, the toxin inserts into the endosome membrane most likely involving the hydrophobic part of the D-domain. The acidic pH of the endosome triggers the first conformational change and results in pore formation of the ligand-translocation domain. Cytosolic $InsP_6$ interacts with the cysteine protease C-domain and induces a second conformational change, activating the protease function. This results in cleavage of the toxin and release of the glucosyltransferase A-domain into the cytosol. In the cytosol, Rho GTPases are glucosylated and thereby inactivated. Inactive Rho can not interact with a numerous variety of effectors and induce multiple signaling events.

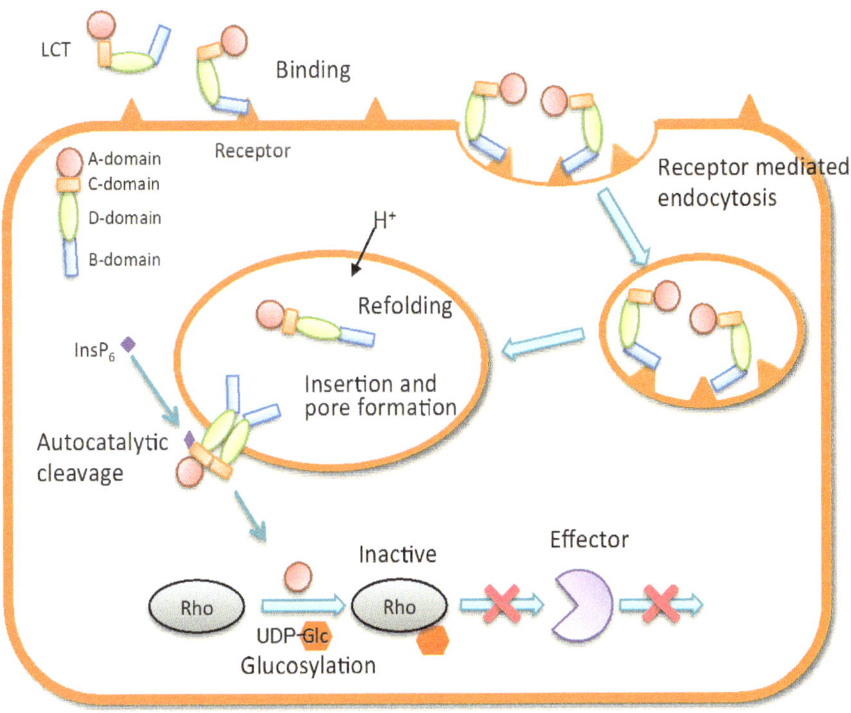

Figure 2. Model of entry and intracellular modification of LCTs

Migration, phagocytosis, cytokine production, wound repair, immune cell signaling, and superoxide anion production. Modification of small GTP-binding proteins by LCTs arises at a Thr-35/37, depending on the Rho GTPase isoforms (Fig. 3) (Belyi and Aktories 2010). Marked differences in substrate specificity have been recognized among the various LCTs. TcdA and B glucosylate Rho, Rac and Cdc42 at Thr-37, whereas TcsL glucosylates Ras at Thr-35, Rap, Ral and Rac at Thr-37 and TcsH glucosylates Rho, Rac, Cdc42 (Fig. 4). LCTs catalyze the glucosylation of 21 kDa small GTP-binding proteins from UDP-glucose, except TcnA which uses UDP-N-acetylglucosamine as cosubstrate. TcnA glucosylates Rho, Rac, Cdc42 (Fig. 4). LCTs cleave the cosubstrate and transfer the glucose moiety to the acceptor amino acid of the Rho proteins (Popoff et al. 1996, Popoff and Bouvet 2009, Belyi and Aktories 2010) (Fig. 3). The conserved Thr, which is glucosylated, is located in switch 1. Thr-35/37 is involved in the coordination of Mg^{2+} and subsequently to the binding of the two phosphates of GTP. The hydroxyl group of Thr-35/37 is exposed to the surface of molecule in its GDP-bound

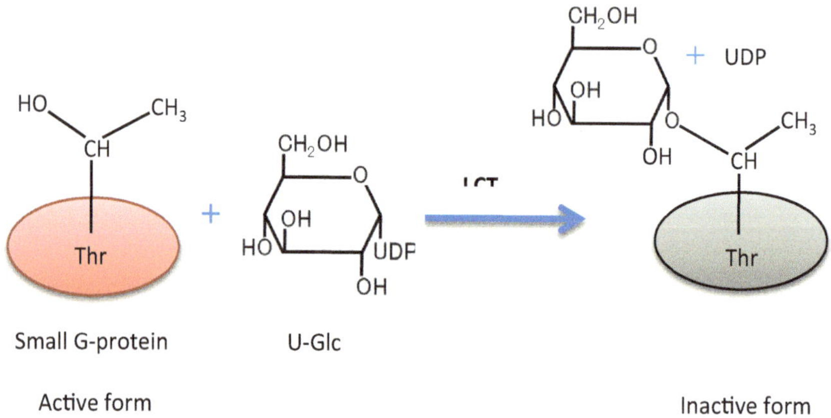

Figure 3. Model of glucosyltransferase activity of LCTs.

LCT	MW(kDa)	Sugar donor	Protein substrate
TcdA	308	U-Glc	Rho, Cdc42, Rac1
TcdB	270	U-Glc	Rho, Cdc42, Rac1
TcsL	271	U-Glc	Rac1, H-Ras, Rap, Ral
TcsH	300	U-Glc	Rho, Rac,Cdc42
TcnA	250	U-Glc, U-NAG	Rho, Rac, Cdc42
TpeL	191	U-Glc, U-NAG	Rac1, H-Ras, Rap, Ral

Figure 4. Protein substrates and cosubstrates of the LCTs.

TcdA: *Clostridium difficile* toxin A, TcdB: *Clostridium difficile* toxin B, TcsL: *Clostridium sordellii* lethal toxin, TcsH: *Clostridium sordellii* hemorrhagic toxin, TcnA: *Clostridium novyi* alpha-toxin, TpeL: *Clostridium perfringnes* large cytotoxin. U-Glc: UDP-glucose, U-NAG: UDP-*N*-acetylglucosamine form, which is the only accessible substrate of LCTs. Glucosylation of Rho or Ras GTPases inhibits activation of the GTPases by GEFs and blocks interaction with their effectors (Sehr et al. 1998, Vetter et al. 2000) as well as the cycling of Rho GTPases between the membrane localization and cytosolic localization (Belyi and Aktories 2010). Glucosylated Rho proteins are located at the membrane. Most importantly, the toxin-induced glucosylation inhibits the active conformation of Rho/Ras GTPases (Vetter et al. 2000, Geyer et al. 2003). Glucosylation of Thr-35 completely prevents the recognition of the downstream effector, blocking the G-protein in the inactive form (Popoff and Bouvet 2009).

2.4. Cellular effects of LCTs.

The inactivation of Rho proteins by LCTs-induced glucosylation causes extensive morphological changes, with loss of actin stress fibers, reorganization of the cortical actin, disrupution of the intercellular junctions and thus increase in cell barrier permeability. The actin cytoskeleton of intoxicated cells is redistributed, causing shrinking and rounding up of most cell types, which is initially accompanied by the formation of neurite-like retraction fibers. Finally, the retraction fibers disappear and the cells detach from the dishes (Ottlinger and Lin 1988, Popoff and Bouvet 2009). Inactivation of Rac assumes to be critical player in disorganization of actin cytoskeleton (Halabi-Cabezon et al. 2008). Numerous other cellular responses to the inactivation of Rho and Ras proteins by LCTs have been described, which are all caused by inhibition of the various functions of the small GTPase. They include inhibition of secretion (Prepens et al. 1996), phospholipase D activity (Schmidt et al. 1996), apoptosis (Brito et al. 2002, Voth and Ballard 2005), chemoattractant receptor signaling (Servant et al. 2000), phagocytosis (Caron and Hall 1998) and alteration of endothelial barrier function (Hippenstiel et al. 1997). However, the role of LCTs on pathogenesis including diarrhea and pseudomembranous colitis are still unknown.

Tissue damage and inhibition of the barrier function of enterocytes might explain the fluid response in toxin-induced diarrhea. TcdA and B affect the morphology and function of tight junctions and associated proteins (ZO-1, ZO-2, occuludin, claudin) to decrease transepithelial resistance, whereas E-cadherin junctions show little alteration (Nusrat et al. 2001, Chen et al. 2002, Aktories and Barbieri 2005). F-actin restruction is accompanied by the dissociation of occluding, ZO-1 and ZO-2 from lateral tight junctions without affecting adherens junctions. These data indicate that Rho proteins play an important role in tight junction regulation. On the other hand, TscH, which mainly modifies Rac, alter the permeability of intestinal cell monolayers causing a redistribution of E-cadherin whereas tight junctions are not significantly affected (Boehm et al. 2006).

TcdA and B induce apoptosis as a consequence of Rho glucosylation and caspase activation (Nottrott et al. 2007, Gerhard et al. 2008) or possibly cell necrosis (Genth et al. 2008). TcdB and TscH cause apoptosis by targeting mitochondria (Petit et al. 2003, Matarrese et al. 2007). In addition, the inactivation of Rho blocks various cellular functions including exocytosis, endocytosis, activation of lymphocyte, phagocytosis in macrophages, control of NADPH oxidase, activation of phospholipase D, contraction of smooth muscle, activation of the pro-apoptotic RhoB, and transcriptional activation via JNK/or p38 (Just and Gerhard 2004, Gerhard et al. 2005, Huelsenbeck et al. 2007, Popoff and Bouvet 2009). TcdA and B induce a large nflammatory response in the gut. There is massive infiltration of neutrophils, and release of many cytokines (Jefferson et al. 1999).

TcdA-caused p38 activation induces the production of IL-8 and IL-1beta, necrosis of monocytes, and inflammation of intestinal mucosa (Popoff and Bouvet 2009). TcsL glycocylates Ras and inhibits the MAP-kinase cascade and PLD regulation. On the other hand, the role of those inhibition on cytotoxicity is still unclear (Schmidt et al. 1998, El Hadj et al. 1999).

3. Characterization of TpeL

C. perfringens type C has been identified as a causative agent of necrotizing enterocolitis associated with diarrhea and dysentery in infant animals (Tweten 2005, Sakurai and Nagahama 2006). Type C strain produces various toxins responsible for the pathogenesis (Sakurai and Nagahama 2006). Amimoto et al. (2007) fractionated the culture filtrate of *C. perfringens* type C strain MC18, and they discovered an unknown toxin (TpeL) that was lethal to mice. The toxin was cytotoxic to Vero cells, and its molecular mass was estimated to be about 180 kDa by SDS-PAGE analysis (Amimoto et al. 2007). These characteristics completely differed from the previously reported toxins. TpeL was purified with HPLC and affinity column coupled with the mAb. Coursodon et al. (2012) reported that TpeL-producing strains of *C. perfringens* type A are highly virulent for broiler chicks. Paredes-Sabja et al. (2011) reported that TpeL is also expressed during sporulation and is a sporulation-regulated *C. perfringens* toxin.

3.1. Toxicity of TpeL

TpeL was lethal to mice and toxic to Vero cells (Amimoto et al. 2007). The lethal activity of the purified TpeL in mice was determined as 62 MLD/mg (one was 16 μg) and 91 LD_{50}/mg (one LD_{50} was 11 μg) by intravenous injection (Amimoto et al. 2007). TpeL showed obvious cytotoxicity in Vero cells, and the specific activity was 6.2 x 10^5 CU/mg (one CU was 1.6 ng) (Amimoto et al. 2007). Morphological changes induced by TpeL in Vero cells. The cytopathic effect induced by a low dose of TpeL was characterized by the enlargement of cells and appearance of rounded cells. Vero cells treated with a high dose of TpeL initially manifested similar changes to those treated with the low dose, then formed aggregates, and eventually detached from the well surface (Amimoto et al. 2007).

Vero cells were incubated with SLO (100 ng/ml) alone (A) or a combination of TpeL1-525 (10 μg/ml) with SLO (100 ng/ml) (B) at 37 °C for 15 min. Pictures were taken after 120 min of resealing. (C) Vero cells were incubated with various amounts of TpeL1-525 with SLO (100 ng/ml) at 37 °C for 15 min. After 120 min of resealing, pictures of cells were taken and the percentage of rounded cells was determined. Value of three experiments were given a mean ± the standard deviation (SD).

To clarify the biological acitivty of TpeL, we prepared a recombinant glycosyltransferase domain, TpeL1-525 (covering amino acids 1 to 525) because native TpeL is labile and is difficult to purify from the culture supernatant of *C. perfringens* type C, and the recombinant full-length TpeL was poorly expressed in *Escherichia coli* (Nagahama et al. 2011). As TpeL1-525 did not possess binding domain, we used the streptolysin O (SLO) delivery systems (Nagahama et al. 2011). As shown in Fig. 5(B), in the presence of SLO, TpeL1-525 caused the cell rounding like the native toxin. TpeL1-525 at 1 – 10 μg/ml in the presence of SLO induced cell rounding in a dose-dependent manner (Fig. 5(C)). The cells finally detached from the wells. On the other hands, the cells eventually detached from the well. Furthermore, when TpeL1-525 at a concentration of 1 to 10 μg/ml was delivered to the cells by SLO, cell viability decreased in a dose-dependent manner. The cytotoxicity was inhibited by the anti-TpeL antibody, and heat-inactivated TpeL1-525 was not cytotoxic. On the other hand, TpeL1-525 by itself did not have the cytotoxic effects. The results indicated that the N-terminal region of TpeL plays a role in the cytotoxicity and the C-terminal region is responsible for the binding of cells. The morphological alteration of cultured cells induced by TpeL is similar to that caused by TcdB and TcsL.

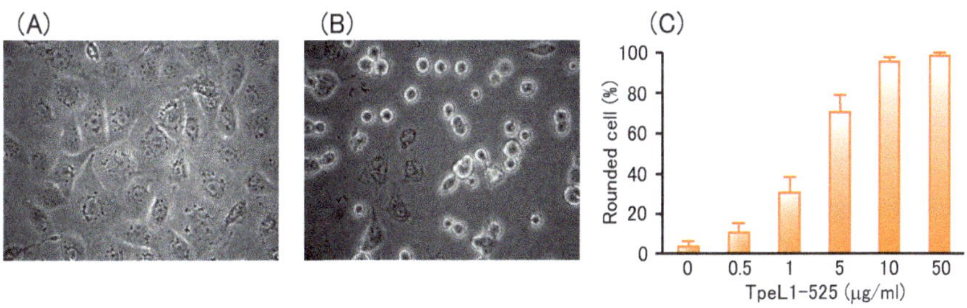

Figure 5. Morphological changes of Vero cells upon treatment with SLO plus TpeL1-525.

3.2. *tpeL* gene

The gene encoding TpeL revealed the presence an ORF of 4953 bases (Amimoto et al. 2007). The *tpeL* gene encoded 1651 amino acid residues and the molecular mass of TpeL calculated from the deduced amino acid sequence was 191 kDa. A signal peptide region was not found within the ORF. The deduced amino acid sequence shared homology with TcdA, TcdB, TcsL and TcnA, called LCTs. The homology scores were 39 % to TcdA, 38 % to TcdB, 39% to TcsL and 30% to TcnA. The amino acid sequence of TpeL is shorter than that of any of these toxins, and the homologous region was located at the N-terminal site of the LCTs. A DXD motif in N-terminal region of LCTs is essential for glycosyltransferase activity, and W-102 in TcsL is an essential amino acid residue for the enzyme activity (Busch et al. 2000a). TpeL conserved the DXD motif and W-102 of TcsL. However, the C-terminal carbohydrate-binding sites of LCTs (von Eichel-Streiber et al. 1992) were not conserved.

The *tpeL* gene was detected not only in type C strains isolated in recent years but also ATCC 3626, a type B strain preserved for many years. It therefore suggest that the *tpeL* gene has been conserved in *C. perfringens* DNA for a long time. Interestingly, beta-toxin-gene-positive strains completely coincided with *tpeL*-positive strains among the 18 strains examined

(Amimoto et al. 2007). Complete chromosomal and plasmid sequences of *C. perfringens* type A strain 13 are available (Shimizu et al. 2002). Also, there is no *tpeL* gene sequence within the data. It has been pointed out that beta- and epsilon-toxin genes carried by plasmids are sometimes lost during the passage of the strains (Katayama et al. 1996, Gibert et al. 1997). So, when the strain loses the plasmids, it changes to type A. In genetic study, total DNAs were used in the cloning and detection of the *tpeL* gene (Amimoto et al. 2007). Sayeed et al. (2010) reported that *tpeL* gene is located approximately 3 kb downstream of the plasmid-borne *cpb* gene. Gurjar et al. (2010) also reported that *tpeL* gene is localized to the plasmids containing *cpb* gene of *cpe*-negative type C isolates.

3.3. Glucosylation of small G proteins by TpeL

Rac1 is the only substrate GTPase inactivated by all LCTs. When the cells were treated with TpeL in the presence of SLO, glycosylation of cellular Rac1 was confirmed by Western blotting with the glycosylation-sensitive anti-Rac1 (Mab102) (Genth et al. 2006, Nagahama et al. 2011). TpeL and TcsL (Voth and Ballard 2005) act on Rac1 and the Ras subfamily but not RhoA. Furthermore, the isomeric TcdB from the variant *C. difficile* serotype F strain 1470 (TcdBF) that glucosylates Rac1 but not RhoA, has a cytopathic effect (Jank et al. 2007b). Halabi-Cabezon et al. (2008) reported that the glucosylation of Rac1 rather than RhoA correlates with the cytopathic effect of TcdB. It has been reported that Rac1 plays a critical role in the organization of the actin cytoskeleton (Bosco et al. 2009). These results strongly suggest that glycosylation of Rac1 is critical for the cytopathic effect of TpeL.

TpeL uses UDP-Glc and UDP-GlcNAc as sugar donors (Nagahama et al. 2011). All other LCTs use a single UDP-hexose (Belyi and Aktories 2010). The crystal structure provides evidence that two amino acids in the vicinity of the catalytic cleft are responsible for the specificity (Jank et al. 2005). TcdA and B, which both use UDP-Glc, have isoleucine and glutamine in the equivalent positions (Ile-383 and Gln-385 in TcdB), whereas TcnL, which uses UDP-GlcNAc, has serine and alanine residues at the respective positions (Jank et al. 2005). It has been reported that the bulkier side chains of Ile-383 and Gln-385 in TcdB limit the space of the catalytic cleft for the binding of UDP-GlcNAc and the exchange of these side chains with smaller groups changes the cosubstrate specificity from UDP-Glc to UDP-GlcNAc (Jank et al. 2005). TpeL has the smaller side chain Ala-383 and the bulkier side chain Gln-385 at the respective positions (Amimoto et al. 2007). We speculate that Ala-383 and Gln-385 in TpeL may stabilize the binding of UDP-Glc and UDP-GlcNAc and favor the acceptance of UDP-Glc and UDP-GlcNAc as the cosubstrates (Nagahama et al. 2011).

The sequential glycosylation of Rac1 by TpeL followed by TcdB, and vice versa indicates that both toxins share the same acceptor amino acid in Rac1. The acceptor amino acid of TcdB-glycosylated Rac1 has been determined as Thr-35 (Belyi and Aktories 2010). TpeL inactivates Rac1 through the glycosylation of Thr-35 (Nagahama et al. 2011).

TpeL glycosylated Rac1, as well as the Ras subfamily consisting of Ha-Ras, RalA, and Rap1B, but not RhoA and Cdc42 (Fig. 4). Important differences in substrate specificity have been detected among the various LCTs. Whereas TcdA, TcdB, and TcnA modify most RhoA, Rac1, and Cdc42 isoforms, TcsL glucosylates Rac1 but not RhoA or Cdc42 (Voth and Ballard 2005). On the other hand, TcsL also modifies the Ras subfamily, including Ras, Rap, and Ral isoforms (Voth and Ballard 2005). Thus, TpeL modifies similar substrates to TcsL. It was reported that Arg-455, Asp-461, Lys-463, and Glu-472 and residues of helix $\alpha17$ (e.g., Glu-449) of TcdB are essential for enzyme-RhoA recognition (Jank et al. 2007b). Changing the respective amino acid residues in TcsL to those of TcdB reduced glycosylation of Ras by TcsL (Jank et al. 2007a). Furthermore, the introduction of helix $\alpha17$ of TcdB into TcsL caused a reduction in the glycosylation of Ras subfamily proteins but permitted the glycosylation of RhoA, indicating that helix $\alpha17$ is involved in RhoA's recognition by TcdB (Jank et al. 2007b). Glu-449, Lys-463, and Glu-472 in TcdB correspond to Lys, Arg, and Gly residues in TcsL and TpeL. Arg-455 in TcdB corresponds to Lys in TcsL and Gly in TpeL (Amimoto et al. 2007). The difference in those amino acid residues may be involved in recognizing small GTPases by TpeL. Additional residues in LCTs are needed for the recognition of small GTPases.

4. Conclusion

Infection with TpeL-positive *C. perfringens* strains may yield disease with a more rapid course and higher case fatality rate in broiler chicks. Thus, TpeL may potentiate the effect of other virulence attributes of necrotic enteritis strains of *C. perfringens.* TpeL from *C. perfringens* has been identified as a glycosyltransferase using UDP-GlcNAc and UDP-Glc as cosubstrates. The substrates of TpeL are confined to Rac1 and Ras subfamily proteins. The modification of Thr-35 on Rac1 induces cytopathic effects.

Author details

Masahiro Nagahama, Masataka Oda and Keiko Kobayashi
Tokushima Bunri University, Japan

5. References

Aktories, K. and J. T. Barbieri. (2005) Bacterial cytotoxins: targeting eukaryotic switches. *Nat Rev Microbiol* 3: 397-410.

Amimoto, K., T. Noro, E. Oishi and M. Shimizu. (2007) A novel toxin homologous to large clostridial cytotoxins found in culture supernatant of *Clostridium perfringens* type C. *Microbiology* 153: 1198-1206.

Barroso, L. A., J. S. Moncrief, D. M. Lyerly and T. D. Wilkins. (1994) Mutagenesis of the *Clostridium difficile* toxin B gene and effect on cytotoxic activity. *Microb Pathog* 16: 297–303.

Barth, H., G. Pfeifer, F. Hofmann, E. Maier, R. Benz and K. Aktories. (2001) Low pH-induced formation of ion channels by *Clostridium difficile* Toxin B in target cells. *J Biol Chem* 276: 10670-10676.

Bartlett, J. G., A. B. Onderdonk, R. L. Cisneros and D. L. Kasper. (1977) Clindamycin-associated colitis due to a toxin-producing species of *Clostridium* in hamsters. *J Infect Dis* 136: 701–705.

Belyi, Y. and K. Aktories. (2010) Bacterial toxin and effector glycosyltransferases. *Biochim Biophys Acta* 1800: 134-143.

Brito, G. A, J. Fujji, B. A. Carneiro-Filho, A. A. Lima, T. Obrig and R. L. Guerrant. (2002) Mechanism of *Clostridium difficile* toxin A-induced apoptosis in T84 cells. *J Infect Dis* 186: 1438-1447.

Boehm, C., M. Gibert, B. Geny, M. R. Popoff and P. Rodriguez. (2006) Modification of epithelial cell barrier permeability and intercellular junctions by *Clostridium sordellii* lethal toxins. *Cell Microbiol* 8: 1070-1085.

Bosco, E. E., J. C. Mulloy and Y. Zheng. (2009) Rac1 GTPase: a "Rac" of all trades. *Cell Mol Life* Sci 66: 370-374.

Burridge, K. and K. Wennerberg. (2004) Rho and Rac take center stage. *Cell* 116: 167-179.

Busch, C. and K. Aktories. (2000) Microbial toxins and the glucosylation of Rho family GTPases. *Curr Opin Struct Biol* 10: 528–535.

Busch, C., F. Hofmann, R. Gerhard and K. Aktories. (2000a) Involvement of a conserved tryptophan residue in the UDP-glucose binding of large clostridial cytotoxin glycosyltransferases. *J Biol Chem* 275: 13228-13234.

Busch, C., K. Schömig, F. Hofmann and K. Aktories. (2000b) Characterization of the catalytic domain of *Clostridium novyi* alpha-toxin. *Infect Immun* 68: 6378-6383.

Caron, E. and A. Hall. (1998) Identification of two distinct mechanisms of phagocytosis controlled by different Rho GTPases. *Science* 282: 1717-1721.

Chen, M. L., C. Pothoulakis and J. T. LaMont. (2002) Protein kinase C signaling regulates ZO-1 translocation and increased paracellular flux of T84 colonocytes exposed to *Clostridium difficile* toxin A. *J Biol Chem* 277: 4247-4254.

Coursodon, C. F., R. D. Glock, K. L. Moore, K. K. Cooper and J. G. Songer. (2012) TpeL-producing strains of *Clostridium perfringens* type A are highly virulent for broiler chicks. *Anaerobe* 18: 117-121.

Duffy. L. K., J. L. McDonel, B. A. McClane and A. Kurosky. (1982) *Clostridium perfringens* type A enterotoxin: characterization of the amino-terminal region. *Infect Immun* 38: 386-388.

Egerer, M., T. Giesemann, T. Jank, K. J. Satchell and K. Aktories. (2007) Auto-catalytic cleavage of *Clostridium difficile* toxins A and B depends on cysteine protease activity. *J Biol Chem* 282: 25314-25321.

Egerer, M., T. Giesemann, C. Herrmann and K. Aktories. (2009) Autocatalytic processing of *Clostridium difficile* toxin B. Binding of inositol hexakisphosphate. *J Biol Chem* 284: 3389–3395.

El Hadj, N. B., M. R. Popoff, J. C. Marvaud, B. Payrastre, P. Boquet and B. Geny. (1999) G-protein-stimulated phospholipase D activity is inhibited by lethal toxin from *Clostridium sordellii* in HL-60 cells. *J Biol Chem* 274: 14021-14031.

Etienne-Manneville, S. and A. Hall. (2002) Rho GTPases in cell biology. *Nature* 420: 629-635.

Gao, Z. and B. A. McClane (2012) Use of *Clostridium perfringens* Enterotoxin and the Enterotoxin Receptor-Binding Domain (C-CPE) for Cancer Treatment: Opportunities and Challenges. *J Toxicol* 2012: 981626.

Genth, H., J. Huelsenbeck, B. Hartmann, F. Hofmann, I. Just and R. Gerhard. (2006) Cellular stability of Rho-GTPases glucosylated by *Clostridium difficile* toxin B. *FEBS Lett* 580: 3565-3569.

Genth, H., S. C. Dreger, J. Huelsenbeck and I. Just. (2008) *Clostridium difficile* toxins: more than mere inhibitors of Rho proteins. *Int J Biochem Cell Biol* 40: 592-597.

Gerhard, R., H. Tatge, H. Genth, T. Thum, J. Borlak, G. Fritz and I. Just. (2005) *Clostridium difficile* toxin A induces expression of the stress-induced early gene product RhoB. *J Biol Chem* 280: 1499-1505.

Gerhard, R., S. Nottrott, J. Schoentaube, H. Tatge, A. Olling and I. Just. (2008) Glucosylation of Rho GTPases by *Clostridium difficile* toxin A triggers apoptosis in intestinal epithelial cells. *J Med Microbiol* 57: 765-770.

Geyer, M., C. Wilde, J. Selzer, K. Aktories and H. R. Kalbitzer. (2003) Glucosylation of Ras by *Clostridium sordellii* lethal toxin: consequences for the effector loop conformations observed by NMR spectroscopy. *Biochemistry* 42: 11951–1195.

Gibert, M., C. Jolivet-Reynaud and M. R. Popoff. (1997) Beta2-toxin, a novel toxin produced by *Clostridium perfringens*. *Gene* 203: 65–73

Giesemann, T., T. Jank, R. Gerhard, E. Maier, I. Just, R. Benz and K. Aktories. (2006) Cholesterol-dependent pore formation of *Clostridium difficile* toxin A. *J Biol Chem* 281: 10808–10815.

Giesemann, T., M. Egerer, T. Jank and K. Aktories. (2008) Processing of *Clostridium difficile* toxins. *J Med Microbiol* 57: 690-696.

Greco, A., J. G. Ho, S. J. Lin, M. M. Palcic, M. Rupnik and K. K. Ng. (2006) Carbohydrate recognition by *Clostridium difficile* toxin A. *Nat Struct Mol Biol* 13: 460–461.

Gurjar, A., J. Li and B. A. McClane. (2010) Characterization of toxin plasmids in *Clostridium perfringens* type C isolates. *Infect Immun* 78: 4860-4869.

Halabi-Cabezon, I., J. Huelsenbeck, M. May, M. Ladwein, K. Rottner, I. Just and H. Genth. (2008) Prevention of the cytopathic effect induced by *Clostridium difficile* Toxin B by active Rac1. *FEBS Lett* 582: 3751-3756.

Hippenstiel, S., S. Tannert-Otto, N. Vollrath, M. Krüll, I. Just, K. Aktories, C. von Eichel-Streiber and N. Suttorp. (1997) Glucosylation of small GTP-binding Rho proteins disrupts endothelial barrier function. *Am J Physiol Lung Cell Mol Physiol* 272: L38-L43.

Ho, C. S., J. Bhatnagar, A. L. Cohen, J .K. Hacker, S. B. Zane, S. Reagan, M. Fischer, W. J. Shieh, J. Guarner, S. Ahmad, S. R. Zaki and L. C. McDonald (2009) Undiagnosed cases of fatal Clostridium-associated toxic shock in Californian women of childbearing age. *Am J Obstet Gynecol* 201: 459.e1-7.

Ho, J. G., A. Greco, M. Rupnik and K. K. Ng. (2005) Crystal structure of receptor-binding C-terminal repeats from *Clostridium difficile* toxin A. *Proc Natl Acad Sci USA* 102: 18373–18378.

Huelsenbeck, J., S. C. Dreger, R. Gerhard, G. Fritz, I. Just and H. Genth. (2007) Upregulation of the immediate early gene product RhoB by exoenzyme C3 from *Clostridium limosum* and toxin B from *Clostridium difficile. Biochemistry* 46: 4923-4931.

Jank, T., D. J. Reinert, T. Giesemann, G. E. Schulz and K. Aktories. (2005) Change of the donor substrate specificity of *Clostridium difficile* toxin B by site-directed mutagenesis. *J Biol Chem* 280: 37833-37838.

Jank, T., T. Giesemann and K. Aktories. (2007a) *Clostridium difficile* glucosyltransferase Toxin B-essential amino acids for substrate binding. *J Biol. Chem* 282: 35222-35231.

Jank, T., T. Giesemann and K. Aktories. (2007b) Rho-glucosylating *Clostridium difficile* Toxins A and B: new insights into structure and function. *Glycobiology* 17: 15R-22R.

Jefferson, K. K., M. F. Smith Jr. and D. A. Bobak. (1999) Roles of intracellular calcium and NF-kappa B in the *Clostridium difficile* toxin A-induced up-regulation and secretion of IL-8 from human monocytes. *J Immunol* 163: 5183-5191.

Jolivet-Reynaud, C., J. M. Cavaillon and J. E. Alouf. (1982) Selective cytotoxicity of *Clostridium perfringens* delta toxin on rabbit leukocytes. *Infect Immun* 38: 860-864.

Just, I., F. Hofmann and K. Aktories. (2000) Molecular mechanism of action of the large clostridial cytotoxins, in Bacterial Protein Toxins. K. Aktories and I. Just, Editors. Springer, Berlin, p. 307-331.

Just, I. and R. Gerhard. (2004) Large clostridial cytotoxins. *Rev Physiol Biochem Pharmacol* 152: 23–47.

Karlsson, K. A. (1995) Microbial recognition of target-cell glycoconjugates. *Curr Opin Struc Bio* 5: 622–635.

Katayama, S., B. Dupuy, G. Daube, B. China and S. T. Cole. (1996) Genome mapping of *Clostridium perfringens* strains with I-CeuI shows many virulence genes to be plasmid-borne. *Mol Gen Genet* 251: 720–726.

Kelly, C. P. and J. T. LaMont. (2008) *Clostridium difficile* - more difficult than ever. *N Engl J Med* 359: 1932–1940.

Kreimeyer, I., F. Euler, A. Marckscheffel, H. Tatge, A. Pich, A. Olling, J. Schwarz, I. Just and R. Gerhard. (2011) Autoproteolytic cleavage mediates cytotoxicity of *Clostridium difficile* Toxin A. *Naunyn Schmiedebergs Arch Pharmacol* 383: 253-262.

Manteca, C., G. Daube, T. Jauniaux, A. Linden, V. Pirson, J. Detilleux, A. Ginter, P. Coppe, A. Kaeckenbeeck and J. G. Mainil. (2002) A role for the *Clostridium perfringens* beta2 toxin in bovine enterotoxaemia? *Vet Microbiol* 86: 191-202.

Matarrese, P., L. Falzano, A. Fabbri, L. Gambardella, C. Frank, B. Geny, M. R. Popoff, W. Malorni and C. Fiorentini. (2007) *Clostridium difficile* toxin B causes apoptosis in epithelial cells by thrilling mitochondria. Involvement of ATP-sensitive mitochondrial potassium channels. *J Biol Chem* 282: 9029-9041.

Mesmin, B., K. Robbe, B. Geny, F. Luton, G. Brandolin, M. R. Popoff and B. Antonny. (2004) A phosphatidylserine-binding site in the cytosolic fragment of *Clostridium sordellii* lethal toxin facilitates glucosylation of membrane-bound Rac and is required for cytotoxicity. *J Biol Chem* 279: 49876-49882.

Müller, S., C. von Eichel-Streiber and M. Moos. (1999) Impact of amino acids 22-27 of Rho-subfamily GTPases on glucosylation by the large clostridial cytotoxins TcsL-1522, TcdB-1470 and TcdB-8864. *Eur J Biochem* 266: 1073-1080.

Na, X, H. Kim, M. P. Moyer, C. Pothoulakis and J. T. LaMont. (2008) gp96 is a human colonocyte plasma membrane binding protein for *Clostridium difficile* toxin A. *Infect Immun* 76: 2862-2871.

Nagahama, M., S. Hayashi, S. Morimitsu and J. Sakurai. (2003) Biological activities and pore formation of *Clostridium perfringens* beta toxin in HL 60 cells. *J Biol Chem* 278: 36934-36941.

Nagahama, M., A. Ohkubo, M. Oda, K. Kobayashi, K. Amimoto, K. Miyamoto and J. Sakurai. (2011) *Clostridium perfringens* TpeL glycosylates the Rac and Ras subfamily proteins. *Infect Immun* 79: 905-910.

Naylor, C. E., J. T. Eaton, A. Howells, N. Justin, D. S. Moss, R. W. Titball and A. K. Basak. (1998) Structure of the key toxin in gas gangrene. *Nat Struc. Biol* 5: 738–746.

Nottrott. S., J. Schoentaube, H. Genth, J. Just and R. Gerhard. (2007) *Clostridium difficile* toxin A-induced apoptosis is p53-independent but depends on glucosylation of Rho GTPases. *Apoptosis* 12: 1443-1453.

Nusrat, A., C. von Eichel-Streiber C, J. R. Turner, P. Verkade, J. L. Madara and C. A. Parkos. (2001) *Clostridium difficile* toxins disrupt epithelial barrier function by altering membrane microdomain localization of tight junction proteins. *Infect Immun* 69: 1329-1336.

Ochi, S., T. Miyawaki, H. Matsuda, M. Oda, M. Nagahama and J. Sakurai. (2002) *Clostridium perfringens* alpha-toxin induces rabbit neutrophil adhesion. *Microbiology* 148: 237–245.

Oda, M., S. Ikari, T. Matsuno, Y. Morimune, M. Nagahama and J. Sakurai. (2006) Signal transduction mechanism involved in *Clostridium perfringens* alpha-toxin-induced superoxide anion generation in rabbit neutrophils. *Infect Immun* 74: 2876-2886.

Ottlinger, M. E. and S. Lin. (1988) *Clostridium difficile* toxin B induces reorganization of actin, vinculin, and talin in cultured cells. *Exp Cell Res* 174: 215-229.

Paredes-Sabja, D., N. Sarker and M. R. Sarker. (2011) *Clostridium perfringens* tpeL is expressed during sporulation. *Microb Pathog* 51: 384-388.

Petit, P., J. Bréard, V. Montalescot, N. B. El Hadj, T. Levade, M. Popoff and B. Geny. (2003) Lethal toxin from *Clostridium sordellii* induces apoptotic cell death by disruption of mitochondrial homeostasis in HL-60 cells. *Cell Microbiol* 5: 761-771.

Pfeifer, G., J. Schirmer, J. Leemhuis, C. Busch, D. K. Meyer, K. Aktories and H. Barth. (2003) Cellular uptake of *Clostridium difficile* toxin B: translocation of the N-terminal catalytic domain into the cytosol of eukaryotic cells. *J Biol Chem* 278: 44535–44541.

Popoff, M. R., F. Chaves-Olarte, E. Femichez, C. von Eichel-Streiber, M. Thelestam, P. Chardin, D. Cussac, B. Antonny, P. Chavrier, G. Flatau, M. Giry, J. de Gunzburg and P. Boquet. (1996) Ras, Rap, and Rac small GTP-binding proteins are targets for *Clostridium sordellii* lethal toxin glucosylation. *J Biol Chem* 271: 10217-10224.

Popoff, M. R. and P. Bouvet. (2009) Clostridial toxins. *Future Microbiol* 4: 1021-1064.

Prepens, U., I. Just, C. von Eichel-Streiber and K. Aktories. (1996) Inhibition of Fc epsilon-RI-mediated activation of rat basophilic leukemia cells by *Clostridium difficile* toxin B (monoglucosyltransferase). *J Biol Chem* 271: 7324-7329.

Pruitt, R. N., B. Chagot, M. Cover, W. J. Chazin, B. Spiller and D. B. Lacy (2009) Structure-function analysis of inositol hexakisphosphate-induced autoprocessing in *Clostridium difficile* Toxin A. *J Biol Chem* 284: 21934-21940.

Qa'Dan, M., L. M. Spyres and J. D. Ballard. (2000) pH-induced conformational changes in *Clostridium difficile* Toxin B. *Infect Immun* 68: 2470-2474.

Qa'Dan, M., L. M. Spyres and J. D. Ballard. (2001) pH-enhanced cytopathic effects of *Clostridium sordellii* lethal toxin. *Infect Immun* 69: 5487-5493.

Reineke, J., S. Tenzer, M. Rupnik, A. Koschinski, O. Hasselmayer, A. Schrattenholz, H. Schild and C. von Eichel-Streiber. (2007) Autocatalytic cleavage of *Clostridium difficile* toxin B. *Nature* 446: 415-419.

Reinert, D. J., T. Jank, K. Aktories and G. E. Schulz. (2005) Structural basis for the function of *Clostridium difficile* toxin B. *J Mol Biol* 351: 973–981.

Rupnik, M., S. Pabst, M. Rupnik, C. von Eichel-Streiber, H. Urlaub and H. D. Soling. (2005) Characterization of the cleavage site and function of resulting cleavage fragments after limited proteolysis of *Clostridium difficile* toxin B (TcdB) by host cells. *Microbiology* 151: 199-208.

Sakurai, J., M. Nagahama and S. Ochi. (1997) Major toxins of *Clostridium perfringens. J Toxicol Toxin Rev* 16, 195–214.

Sakurai, J., M. Nagahama and M. Oda. (2004) *Clostridium perfringens* alpha-toxin: characterization and mode of action. *J Biochem* 136: 569-574.

Sakurai, J. and M. Nagahama. (2006) *Clostridium perfringens* beta-toxin: characterization and action. *Toxin Rev* 25: 89-108.

Sandvig, K., B. Spilsberg, S. U. Lauvrak, M. L. Torgersen, T. G. Iversen and B. van Deurs. (2004). Pathways followed by protein toxins into cells. *Int J Med Microbiol* 293: 483–490.

Sayeed, S., J. Li and B. A. McClane. (2010) Characterization of virulence plasmid diversity among *Clostridium perfringens* type B isolates. *Infect Immun* 78: 495-504.

Schmidt, M., U. Rümenapp, C. Bienek, J. Keller, C. von Eichel-Streiber and K. H. Jakobs. (1996) Inhibition of receptor signaling to phospholipase D by *Clostridium difficile* toxin B. Role of Rho proteins. *J Biol Chem* 271: 2422-2426.

Schmidt, M., M. Voss, M. Thiel, B. Bauer, A. Grannass, E. Tapp, R. H. Cool, J. de Gunzburg, C. von Eichel-Streiber, and K. H. Jakobs. (1998) Specific inhibition of phorbol ester-stimulated phospholipase D by *Clostridium sordellii* lethal toxin and *Clostridium difficile* toxin B-1470 in HEK-293 cells. *J Biol Chem* 273: 7413-7422.

Sehr, P., G. Joseph, H. Genth, I. Just, E. Pick and K. Aktories. (1998) Glucosylation and ADP-ribosylation of Rho proteins - Effects on nucleotide binding, GTPase activity, and effector-coupling. *Biochemistry* 37: 5296–5304.

Selzer, J., F. Hofmann, G. Rex, M. Wilm, M. Mann, I. Just and K. Aktories. (1996). *Clostridium novyi* α-toxin-catalyzed incorporation of GlcNAc into Rho subfamily proteins. *J Biol Chem* 271: 25173–25177.

Servant, G., O. D. Weiner, P. Herzmark, T. Balla, J. W. Sedat and H. R. Bourne. (2000) Polarization of chemoattractant receptor signaling during neutrophil chemotaxis. *Science* 287: 1037-1040.

Shimizu, T., K. Ohtani, H. Hirakawa, K. Ohshima, A. Yamashita, T. Shiba, N. Ogasawara, M. Hattori, S. Kuhara and H. Hayashi. (2002) Complete genome sequence of *Clostridium perfringens*, an anaerobic flesh-eater. *Proc Natl Acad Sci USA* 99: 996–1001.

Singh, U., L. L. Mitic, E. U. Wieckowski, J. M. Anderson and B. A. McClane. (2001) Comparative biochemical and immunocytochemical studies reveal differences in the effects of *Clostridium perfringens* enterotoxin on polarized CaCo-2 cells versus Vero cells. *J Biol Chem* 276: 33402-33412.

Tsokos M, S. Schalinski, F. Paulsen, J. P. Sperhake , K. Puschel and I. Sobottka. (2008) Pathology of fatal traumatic and nontraumatic clostridial gas gangrene: a histopathological, immunohistochemical, and ultrastructural study of six autopsy cases. *Int J Legal Med* 122: 35–41.

Tweten, R. K. (2005) Cholesterol-dependent cytolysins, a family of versatile pore-forming toxins. *Infect Immun* 73: 6199– 6209.

Vetter, I. R., F. Hofmann, S. Wohlgemuth, C. Herrmann and I. Just. (2000) Structural consequences of mono-glucosylation of Ha-Ras by *Clostridium sordellii* lethal toxin. *J Mol Biol* 301: 1091–1095.

von Eichel-Streiber, C., M. Sauerborn and H. K. Kuramitsu. (1992) Evidence for a modular structure of the homologous repetitive C-terminal carbohydrate-binding sites of *Clostridium difficile* toxins and *Streptococcus mutans* glucosyltransferases. *J Bacteriol* 174: 6707-6710.

Voth, D. E. and J. D. Ballard (2005). *Clostridium difficile* toxins: mechanism of action and role in disease. *Clin Microbiol Rev* 18: 247-263.

Waters, M., A. Savoie, H. S. Garmory, D. Bueschel, M. R. Popoff, J. G. Songer, R. W. Titball, B. A. McClane and M. R. Sarker. (2003) Genotyping and phenotyping of beta2-toxigenic *Clostridium perfringens* fecal isolates associated with gastrointestinal diseases in piglets. *J Clin Microbiol* 41: 3584-3591.

Ziegler, M. O., T. Jank, K. Aktories and G. E. Schulz. (2008) Conformational changes and reaction of clostridial glycosylating toxins. J Mol Biol 377: 1346-1356.

Chapter 8

N-Glycosylation of the 66-kDa Cell-Wall Glycoprotein of a Red Microalga

Oshrat Levy-Ontman

Additional information is available at the end of the chapter

1. Introduction

1.1. Cell-wall polysaccharide of *Porphyridium* sp.

1.1.1. Chemical studies

The cells of the red microalga *Porphyridium* sp. are encapsulated within a polysaccharide, one of the main products of this alga and that is vital for its survival [1-4]. Due to the unique properties of this polysaccharide, which has tremendous value in the biotechnology field, strenuous, interdisciplinary efforts have been devoted in the past few decades to the study of its chemical structure, rheological properties, and bioactivities [3].

The red microalga *Porphyridium* sp. cell-wall polysaccharide comprises negatively charged heteropolymers with a relatively high molecular mass, apparently in the $2*10^6$ Da range [5-8]. The precise pk_a value of dissociation is very low and is somewhat difficult to determine due to the heterogeneous nature of the molecules. The polysaccharide is composed of 10 different monosugars, proteins, and sulfate groups (~7.6 %w/w) [5-6, 8-10]. The prominent sugars are glucose:galactose:xylose in a molar ratio of 1:1.9:3.2 , respectively [10]. The minor sugars include rhamnose, arabinose, mannose, and methylated monosugars [5-6, 8, 10]. Moreover, it is anionic due to the presence of uronic acid groups and half-ester sulfate groups [1, 5-6, 11], the latter of which are attached to the 3 or 6 position of glucose and galactose [11].

During growth, the external part of the polysaccharide (known as the "soluble fraction") is released to the surrounding aqueous medium and accumulates in the medium, while the remainder, i.e., most of the polysaccharide (~ 50-70%, known as the "bound fraction"), remains attached to the cell [1-4, 12]. When red microalgae are grown in a liquid medium, the viscosity of the medium increases continuously as the polysaccharides are released from the cell surface [1, 13]. As a result, during the logarithmic phase of growth polysaccharide capsules are thinnest, while during the stationary phase they are thickest. Cell-wall polysaccharide production (quantity and quality) has been found to be affected by environmental conditions and genetic modifications [4]. For example, growing *Porphyridium* sp. in a medium deprived of nitrate and sulfate enhances production and solubilization of the polysaccharide [14]. Moreover, these conditions have also been shown to change polysaccharide compositions [14].

The precise structure of the *Porphyridium* sp. cell-wall polysaccharide is not fully understood due to its complexity and the lack of known specific enzymes that degrade it [3]. Although several studies have been conducted on *Porphyridium* sp. polysaccharide structure, most have focused mainly on the soluble fraction of the polysaccharide. The soluble polysaccharide of *Porphyridium* sp. was found to have a basic building block comprising aldobiouronic acid 3-*O*-(α -D-glucopyranosyluronic acid)-L-galactopyranose disaccharide [15]. This building block has also found in other red microalgal cell-wall polysaccharides, i.e., *Porphyridium aerugineum* and *Dixoniella grisea* [15]. Moreover, it was also discovered to be part of a bigger structure, composed of a larger, linear building block that contains (1→2 or 1→4)-linked xylopyranosyl, (1→3)-linked galactopyranosyl, and (1→3)-linked glucopyranosyl or glucopyranosyluronic acid residues [16]. Two oligosaccharides isolated from the bound fraction were also investigated. They were shown to comprise three major neutral monosaccharides – Xylose, Glucose Galactose, and GlcA – the last of which did not contain the disaccharide building block that was found in the soluble polysaccharide fraction [17].

1.1.2. *Rheological studies*

Their physicochemical natures make the red microalgal polysaccharides potentially valuable candidates for various industrial applications. One of the most important properties of the polysaccharides is their capacity to yield highly viscous solutions, comparable with those of industrial polysaccharides such as xanthan and carageenan, under relatively low polymer concentrations [4]. The *Porphyridium* sp. polysaccharide is composed of an oriented single, two-fold helical structure with a pitch of 1.6 nm (i.e., a single chain helix with two chemical repeats, which are probably the aldobiouronic acid) [18-19]. Chain stiffness is in a range comparable to that of rigid helicels such as xanthan gum and DNA [18].

A heteropolyelectrolyte, the polysaccharide of *Porphyridium* sp. shows marked shear thinning (with no evidence of a Newtonian plateau, typical of a structured medium), thixotropic, and elastic behaviors [18-19]. It was suggested that the shear thinning behavior and the capability to yield highly viscous solutions under relatively low polymer concentrations can be attributed to the ability of the polysaccharide to form a weakly cross-linked, elastic, gel-like network structure (that breaks down under shear).

The similarity of the physicochemical properties of red microalgal polysaccharides to those of other polysaccharides currently used in industry as gelling agents, thickeners, stabilizers, and emulsifiers (such as xanthan) make the red microalgal polysaccharides a valuable alternative source to these existing industrial polysaccharides. One of their unique advantages over those of other phycocolloids, mainly for commercial applications, is their stability over wide temperature, pH value, light, and salinity ranges [4, 6, 18-22].

There is evidence that upon heating the *Porphyridium* sp. cell-wall polysaccharide in solution, the weak elastic gel, transforms into a stronger elastic network that is reversible by cooling [19]. Aqueous preparations that gel with heating and melt upon cooling are rare and may have unique industrial applications. This phenomenon demonstrates again the notable stability attributed to the polysaccharides that differentiate them from other known phycocolloids, e.g., agar-agar. Another significant property of the red microalgal polysaccharides that confers on them great industrial value is their effectiveness in drag reduction [4, 23]. Indeed, they were proposed as ideal for cargo ships to lessen the drag felt by the marine transport vessels, thereby reducing the needed propulsion power and fuel costs or, alternatively, increasing ship speeds [24].

The *Porphyridium* sp. cell-wall polysaccharide was found to adsorb onto mica surfaces (negatively charged), forming ultrathin coating layers in the nanometer range [21-22]. The polysaccharide layer appeared to remain highly mobile at the surface, as flexible microfibrils (~ 10 nm in width and 1 to 2 nm in height), [21]. However, hyaluronic acid under the same conditions did not show any sign of adsorption onto the mica surfaces [22].

One of the most outstanding properties of *Porphyridium* sp. polysaccharide lubricating films is that only a subnanometric (0.5-1 nm) monolayer is needed to provide a stable, low-friction coefficient, robustness (high load carrying capacity), and good wear protection, and the friction force exhibits a weak dependence on sliding velocity [22]. Moreover, pressing/shearing was shown to affect the adhesion such that the application of pressure plays an important role in reordering the polysaccharide molecules between two surfaces, binding them together to protect them from damage and control their friction.

In comparison to the other biopolymers investigated so far, the *Porphyridium* sp. polysaccharide at once possesses most of the tribological requirements for efficient biolubrication, e.g., steady low friction, stability at high pressures, stability at high and low velocities, and wear protection and stability over large shearing distances. In addition, the polysaccharide was shown to be superior to hyaluronic acid as biolubricants in terms of stability, friction reduction, and adsorption [21-22]. Also, the polysaccharide was affected neither by hyaluronidase activity, in contrast to hyaluronic acid [22], nor by carbohydrolases [25]. In summary, these properties combine to make the red microalgal polysaccharide of *Porphyridium* sp. ideal for applications in various industries, i.e., marine transport, biomedical, cosmetics, and nutrition.

1.1.3. Bioactivities

Among its bioactivities whose potential has been at least partially realized are the *Porphyridium* sp. polysaccharide's anti-inflammatory and anti-irritating activities [26]. Found to be generally well-suited for a variety of skin applications, the polysaccharide was shown to have a marked soothing affect on the irritation associated with common skin inflammations. Moreover, using TBA and FOX methods, the polysaccharide was also shown to have potent antioxidant properties [27]. Indeed, these proven polysaccharide bioactivities led to its current application as an anti aging agent by a leading global cosmetics company and have promoted further research aimed at discovering additional uses.

In addition to bioactivities with dermal applications, the polysaccharide was also found to possess anti-viral activity against animal viruses [28-33]. Moreover, it was shown to significantly inhibit infection with retroviruses (murine leukemia virus, HIV-1, and HIV-2) and cellular transformation by murine sarcoma virus [29].

The *Porphyridium* sp. polysaccharide can also be used as a nutritional agent, as the fibers it contains constitute a viable dietary supplement. Animal feeding experiments have shown that rats whose diets were supplemented with low concentrations of polysaccharide had considerably lower levels of serum cholesterol, triglycerides, and very low-density lipoprotein (VLDL) [34-36]. This diet also resulted in an increase in feces mass (by 130%) and in bile acid excretion (5.1 fold or more). Moreover, rats fed the polysaccharide exhibited longer small intestines (by 17%) and colons (by 8.5%) [36]. An important finding is the complete absence of toxic effects following the *Porphyridium* sp. polysaccharide diet in comparison to diets based on other known sulphated polysaccharides that were found to be toxic [37]. It was thus suggested that the *Porphyridium* sp. polysaccharide could be produced and marketed commercially as a dietary fiber supplement [35-36].

It is noteworthy to add that the beneficial bioactivities and fluid dynamic behavior observed in *Porphyridium* sp. polysaccharide are probably the direct results of the their general role in their natural surroundings, isolated from the sea sand, where the environmental conditions are subject to harsh, widely fluctuating conditions, i.e., extreme light and drought during ebb tides. Most likely the unique polysaccharide structure is responsible for these special properties. Indeed, the polysaccharide form the boundary between the cell and its surroundings, functioning as a capsular defense barrier.

2. The 66-kDa glycoprotein

Almost no work has been reported on the cell-wall proteins of red microalgae. However, a number of non-covalently-bound cell-wall proteins were detected in SDS polyacrylamide gel electrophoresis (SDS-PAGE) when the cell-wall polysaccharide complex of *Porphyridium* sp. was loaded on the gel after it had been boiled in sample buffer containing SDS and β-mercaptoethanol (Figure 1). The most prominent of those proteins is named after its molecular mass: the 66-kDA cell-wall glycoprotein. The total mass of all N-glycans attached to the protein was estimated at 8 kDa [38-39].

The 66-kDa glycoprotein was found to be non-covalently, tightly bound to the polysaccharide [38-39]. Although it could not be co-eluted with the polysaccharide in size-exclusion chromatography (SEC) by increasing NaCl concentrations (0.25-1.5M), it could be partially dissociated from the polysaccharide by SEC in the presence of 2M guanidine hydrochloride. Furthermore, the glycoprotein could not be completely separated from the polysaccharide that had been denaturated by boiling, in buffer containing SDS and β-mercaptoethanol before loading it into an SEC column. Western blot analysis (using polyclonal antiserum raised against the 66-kDa glycoprotein) revealed that the glycoprotein is specific to *Porphyridium* sp. and its closely related isolates, but it is not detected in other red microalgae, blue-green algae, or plants [38-39]. Indirect immunofluorescent assay and immune-gold labeling with the antiserum showed that the 66-kDa glycoprotein is located in the Golgi and on the cell surface of *Porphyridium* sp.

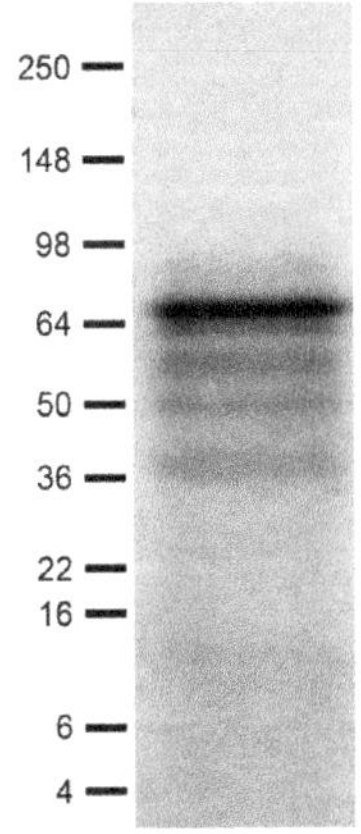

Figure 1. Cell-wall proteins of the *Porphyridium* sp. polysaccharide. The polysaccharide (36 μg) was subjected to SDS-PAGE and stained with Coomassie blue.

The 66-kDa glycoprotein was also detected in genetically spontaneous mutants that are resistant to the cellulose biosynthesis inhibitor 2,6-dichlorobenzonitrile (DCB) or in physiologically modified cell-wall complexes of *Porphyridium* sp. (from sulfate, nitrate, calcium starved cultures) [38-39].

By means of an *in vitro* assay, it was demonstrated that the 66-kDa glycoprotein binds to the cell-wall polysaccharide of *Porphyridium* sp. Furthermore, it also binds to the cell-wall polysaccharides of two other species of red microalgae, *Dixoniella grisea* and *Porphyridium aerugineum,* and to λ-carrageenan from a red seaweed. But it does not bind to the other polysaccharides examined, i.e., dextran, dextran sulfate, xylan, and xanthan gum [38-39].

Sequencing of a cDNA clone encoding the 66-kDa glycoprotein revealed that this is a novel protein, with four potential N-glycan sites, which does not show similarity to any protein in the public domain databases.

Although the sequencing clone revealed this glycoprotein to be a novel protein, it does show structural similarities, within the carbohydrate-binding domain (CBD), to some protein superfamilies, namely, glycosyltransferases, pectin lyase-like, sialidases, and conA-like lectins/glucanases in the SCOP and PROSITE databases, indicating a possible role of the 66-

kDa glycoprotein in cell-wall polysaccharide synthesis/modification [38-39]. In addition, two amino acid sequences of the N-terminus and several internal peptides showed some homology to endo β 1-4 xylanase [38-39]. Moreover, this protein was found in the early stages of the cell-wall cycle as an intermediate product [40-41] and in all mutants characterized by modified cell walls [39], which indicates that it may be involved in polysaccharide biosynthesis. In addition, the glycoprotein was shown to play a role in biorecognition [42]: *Porphyridium* sp. cells that were treated with antibodies to the 66-kDa glycoprotein were not recognized by the microalga's predator, the dinoflagellate *Crypthecodinium cohnii* [42].

3. N-Glycan structures of the 66-kDa glycoprotein

The primary structures of the 66-kDa N-glycan have been investigated by various methodologies. Preliminary characterization of glycan moieties attached to the 66-kDa protein was done by lectin array analysis. The SDS-PAGE–resolved polysaccharide proteins (containing the 66-kDa glycoprotein) were blotted onto nitrocellulose membranes and probed with lectin-conjugated-biotin and streptavidin-conjugated-HRP according to Gravel [43]. The glycoprotein was detected by the lectins ConA (Concanavalin A), GNA (Galanthus nivalis lectin), and GSL I (*Griffonia* (Bandeiraea) simplicifolia lectin I (Figure 2). ConA has high affinity to α-D-mannose and lower affinity to α-D glucose [44]. GNA recognized a terminal mannose via α(1-3), α(1-6) or α(1-2) to another mannose residues [45]. The positive reactions with the lectins ConA and GNA suggest the presence of N-glycosidically-linked "high mannose" or hybrid"-type glycan chains while that with GSL I indicates the possible presence of O-linked chains comprising α-Gal/α-GalNac monosaccharides. In contrast, the glycoprotein could not be detected by the lectins DSA (Datura stramonium), AAA (*Aleuria aurantia* agglutinin), RCA I (*Ricinus communis* agglutinin I), PNA (Peanut Agglutinin), WGA (Wheat Germ Agglutinin), SNA (*Sambucus nigra*), or MAA (*Maackia amurensis* lectin, suggesting that it lacks the Galβ(1-4)GlcNAc, GlcNAc-Ser/Thr, α(1-6)-linked fucose, terminal β-D-galactose, Gal β(1-3) GalNAc, GlcNAcβ(1-4)GlcNAc, sialic acid terminally linked to α(2-6)Gal, or GlcNAc and sialic acid terminally linked to α(2-3) Gal groups, respectively.

Other direct, well-known methods for N-glycan analysis have also been conducted as follows. The glycoprotein was separated using a funnel-shaped polyacrylamide gel under conditions described previously [38-39]. The 66-kDa glycoprotein was detected by Coomassie blue staining, and its N-glycans were separated using in-gel digestion with PNGase F according to Küster et al. [46]. Following several cleaning steps [47], part of the separated N-glycans were labeled with the fluorescence agent 2AB according to the method described by Bigge et al. [48], and the rest were kept for mass spectrometry analysis. An NP-HPLC analysis of the 2AB-labeled N-glycans revealed four main peaks, indicating a minimum of four different N-glycans in the sugar moieties of the 66-kDa glycoprotein (ranging in size from 7 to 8.5 GU values in terms of the glucose ladder standard) [47]. To test whether the N-glycan moieties contain other types of sugars, i.e., those containing a 3-linked fucose attached to the reducing terminal GlcNAc residue, the 66-kDa glycoprotein was digested with PNGase A [47]. Following labeling of the PNGAse-A–released N-glycans, they were run on an NP-HPLC. The resulting NP-chromatogram was identical to that of the PNGase-F–released glycans, indicating the absence of a core α1,3-linked Fuc.

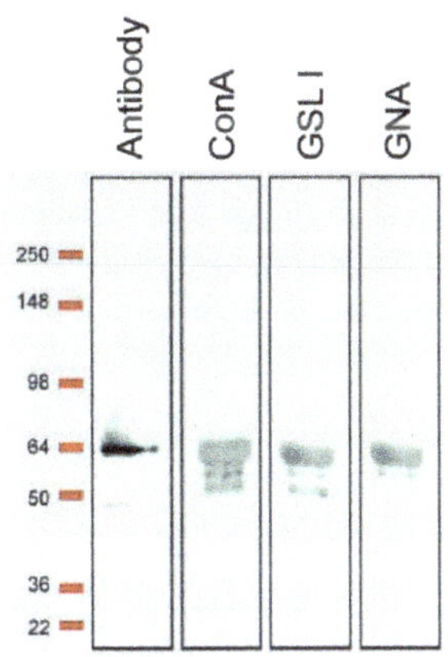

Figure 2. Lectin analysis of the 66-kDa cell-wall glycoprotein. The cell-wall polysaccharide (36 μg) was subjected to SDS-PAGE. Following electrophoresis, the proteins were blotted onto a nitrocellulose membrane and probed with ConA, GNA, and GSL I.

To elucidate the N-glycan structures, the next step was to use an exoglycosidase array of enzymes that normally cleave the non-reducing end of typical N-glycans. Following the digestion of the 2AB- labeled PNGase-F–released N-glycans with an exoglycosidase array (ABS, BTG, SPH, BKF, XYL, JBM), the glycan NP-HPLC chromatogram did not change in comparison to that prior to digestion, indicating that the mixture of N-linked glycans obtained from the 66-kDa glycoprotein of *Porphyridium* sp. differs from glycans known to date [47]. To obtain more information about N-glycan structures, the labeled and unlabeled N-glycans were analyzed by mass spectrometry (positive-ion MALDI-TOF MS and negative-ion ESI-MS). As expected, the results did not match the typical mass values of other known, investigated N-glycans. Moreover, the 2AB-labeled N-glycan fraction released after PNGase F/A was also run on WAX-HPLC, and all the glycans were found to be neutral [47].

Traditionally, the gold standard for such studies would have been to include GC/MS and/or NMR data, which is not the case here. Since the glycoprotein is associated with the soluble polysaccharide, first it has to be separated from the polysaccharide (loading volume was 1.7 ml). Working with the polysaccharide is tedious and time consuming due to its high viscosity with excessive shear thinning. Moreover, the polysaccharide contains numerous other compounds, which dictated that we first dialyze it against double distilled water and dilute its concentration to 0.3 w/v. At the gel loading point, the polysaccharide concentration was lower since it was diluted again with Laemmli sample buffer (lowering the final concentration to about 0.2 w/v). Each gel run yielded one band (1 cm × 0.5 mm) that contained about 30 μg of the glycoprotein and a relatively small amount of total N-glycans (about 500 pmol, calculated by 2AB calibration standards). To analyze the glycan structures using GC-MS, it was vital to collect enough material. Glycans from 40 gel pieces were collected, separated from the gel pieces, and cleaned. We hydrolyzed the glycans by rigorous acid conditions, methylated them, and tried to compare their GC-MS spectra to that of known methylated monosaccharide standards. The methylated monosaccharide GC-MS spectra derived from the unknown glycans contained a lot of background noise that presumably hid the monosaccharide peaks. The noise probably derived from the preparation of the polyacrylamide gel pieces together with the steps leading to the GC-MS analysis. We now understand that conventional strategies (collection of small glycan amounts from relatively large gel pieces) are insufficient in this research. As a result, it was necessary to use indirect methodologies to obtain structural information. In future research, we hope to develop a method for producing uncontaminated proteins from the polysaccharide.

To understand the N-glycan compositions suggested by the mass values, the identities of the constituent monosaccharides of the N-glycans were determined. The N-glycans were hydrolyzed and labeled with 2AB and analyzed using a combination of MS spectrometry and a comparison of monosaccharide standards to the hydrolyzed monosaccharide chromatograms of the N-glycans obtained by NP/RP-HPLC [47]. The analysis indicated that each of the N-glycans derived from the 66-kDa glycoprotein comprised the same four monosaccharides: GlcNAc, mannose, probably 6-*O*-MeMan and xylose. Integrating monosaccharide identity data with the MS analysis, a sugar composition can be determined for each N-glycan feature (Table 1):

N-Glycan calculated mass (Da)	Composition			
	Mannose	MeMan	GlcNAc	Xylose
1894.70	5	3	2	1
2026.75	5	3	2	2
2056.76	6	3	2	1
2188.80	6	3	2	2

Table 1. Compositions of the N-glycans from the 66-kda glycoprotein [47]

The N-glycans were also released by Endo-H and then analyzed by NP-HPLC. To estimate the difference in glucose unit values between the PNGase-F– and Endo-H–released N-glycans from the 66-kDa glycoprotein, the NP-HPLC chromatograms of the two preparations were compared to the NP-HPLC chromatogram of a known standard of PNGase-F– and Endo-H–released N-glycans derived from RNase B. The differences in the elution times (in the NP-chromatogram) of the 66-kDa glycoprotein fractions compared to those of the RNase B fractions showed the same pattern. The NP-HPLC chromatogram of the N-glycans released from the RNAse B glycoprotein (data not shown) indicated that the size difference between 2AB-labeled glycans released by the action of PNGase F and Endo H is very small (0.16 GU or less). For example, the size of the oligomannose structure with five mannose residues, which was separated from RNase B by PNGase F action, exceeded that obtained by Endo H digestion by a mere 0.16 GU. It was also shown that the size difference between the PNGase-F– and Endo-H–released glycans decreased as glycan size increased: Oligomannose structures with eight or nine mannose residues exhibited no size difference between PNGase-F– and Endo-H–released materials. The small difference in glycan size (in GU values) between the smallest glycan, released by PNGase F and yielding the smallest GU value (*m/z* 1895), and the major peak obtained in the NP-HPLC chromatogram after Endo H digestion, indicates that the glycan features are the same as those associated with N-glycans from RNase B. A comparison of the sizes of the minor peaks for the derived glycans released by Endo H compared to those released by PNGase F digestion (corresponding to 2027, 2057, 2189 Da) was in good agreement with the differences found in the measurements of the RNase B fraction. The yield of these glycans released by Endo H was found to be much lower than expected compared to the major glycan released by Endo H. This observation indicates the possible existence of structures with positions that interfere with Endo H activity. In addition, to verify that the glycans released after Endo H were derived from the same N-glycans released after PNGase F digestion, a comparison of analyses of their masses by MS indicated that they possess the same glycan features.

To obtain a more detailed analysis, unlabeled oligosaccharides released by PNGase F were subjected to negative mode MS/MS [47]. The negative ion MS/MS spectra were typical of neutral glycans run as phosphate adducts (phosphate was the anion used to ionize the compounds) [47]. Spectra were interpreted according to published data [49-52]. All spectra contained a major ion 259 mass units below that of the molecular ion and consistent with a ^{2,4}A fragmentation (Domon and Costello [53] nomenclature) of the core HexNAc (Scheme 1, loss of 161 mass units and the phosphate adduct) following abstraction of the 3-proton by the phosphate. This mass loss showed no substitution of the core GlcNAc.

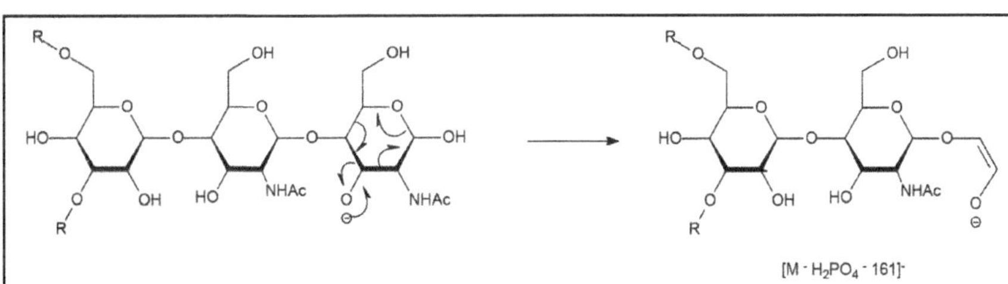

Scheme 1. Fragmentation mechanism in the GlcNAc ring of the chitobiose core

A second ion, 60 mass units below this ion, was also present in all compounds and corresponds to a B_R cleavage (the subscript is used here to refer to the "reducing terminus") (Scheme 2) consistent with a β(1→4)-linkage.

The spectra of the compounds weighing 1991 and 2153 Da contained an additional ion, 203 mass units below that of the $^{2,4}A_R$ ion, corresponding to a similar cleavage of the penultimate GlcNAc (Scheme 3).

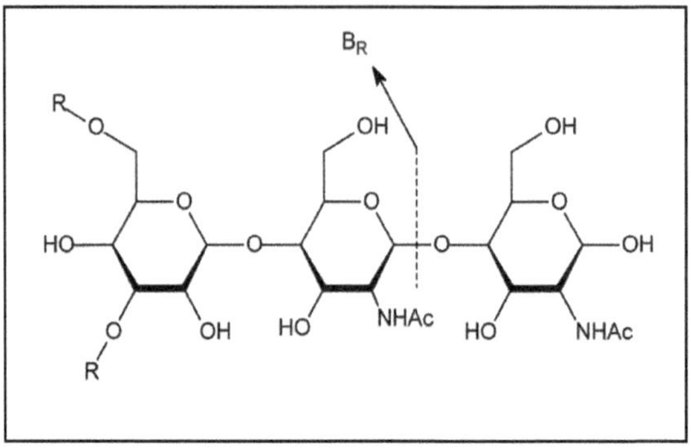

Scheme 2. Fragmentation mechanism between the two GlcNAc residues of the chitobiose core

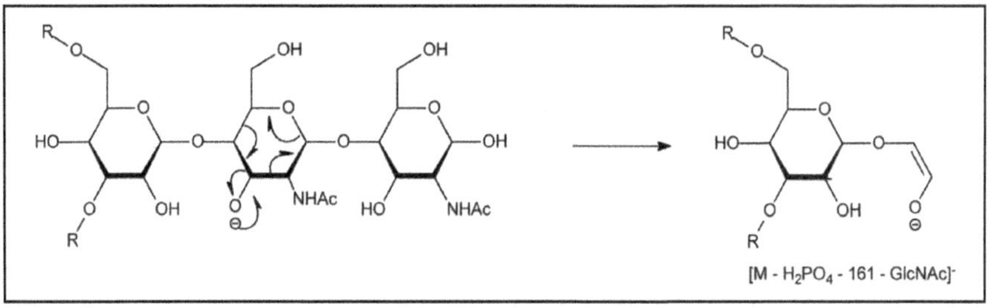

Scheme 3. Fragmentation mechanism in the penultimate GlcNAc of the chitobiose core

The spectra of compounds weighing 2123 and 2285 Da, that had an extra xylose residue did not contain this ion, suggesting that the xylose was attached to the 3-oxygen of the penultimate GlcNAc, blocking the abstraction of a proton at this site and accounting for the absence of the $^{2,4}A_{R-1}$ ion [47].

Normally, xylose is found attached to the 2-position of the branching mannose. However, the negative ion MS/MS spectrum of $[Man]_2[GlcNAc]_2[Xyl]_1[Fuc]_1$ from horseradish peroxidase, which contains such a 2-linked xylose, contained an abundant ion corresponding to the $^{2,4}A_{R-1}$ fragment (*m/z* 677) consistent with the 3-proton being available for abstraction [47]. Thus, it appears that the compounds with two xylose residues have one xylose attached to the 3-position of the penultimate GlcNAc residue [47].

The negative ion MS/MS spectra of all four compounds were virtually identical. The group of ions at *m/z* 1131, 1113, 1059, and 1029 (weak) are similar to those from high-mannose glycans and correspond to D, $[D\text{-}18]^-$, $^{O,3}A_{R-2}$, and $^{O,4}A_{R-2}$, respectively (Scheme 4)[47].

The similarity of these ions to those in the high-mannose glycans again suggests no xylose substitution on the core mannose. The mass of the D ion, which contains the 6-antenna, indicated a composition of $[Hex]_4[MeHex]_2[Xyl]_1$ leaving after subtraction of the core GlcNAc residues. The similarity of the spectra to those of the high-mannose glycans suggests a similar topology, and therefore, the two branches of the 6-antenna contain Hex-

MeHex and Xyl-Hex-MeHex compositions. The composition of the ion at *m/z* 631 appears to be $[Hex]_2[MeHex]_1[Xyl]_1$, which is consistent with that of a D′ ion [the linkage (3 and 6) around the mannose attached to the 6-position of the core mannose is the same as that of the core mannose itself, Scheme 5].

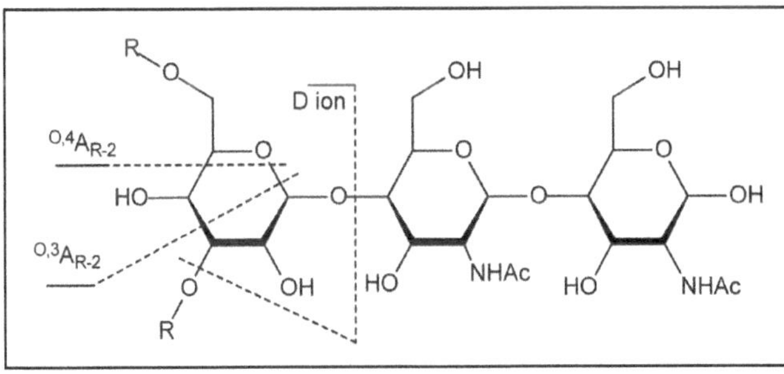

Scheme 4. Fragmentation mechanism between the branching Mannose and the penultimate GlcNAc of the core core (D, $^{O,3}A_{R-2}$ and $^{O,4}A_{R-2}$ ions)

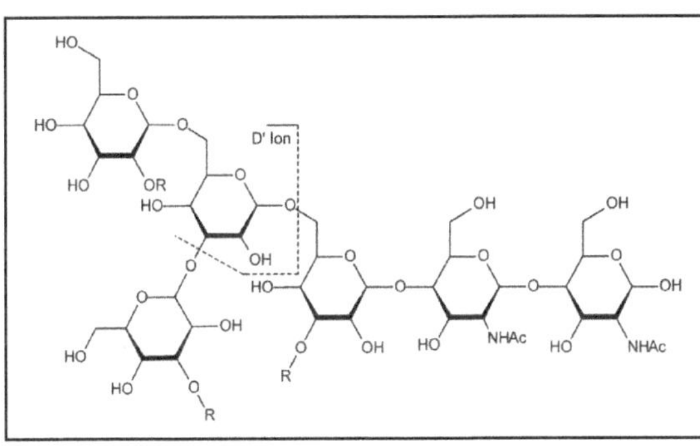

Scheme 5. Fragmentation mechanism in the 6-branch mannose, creating the D′ ion.

To further elucidate the glycan structures, each of the 2AB-labeled glycans were also analyzed by positive MS/MS. The results of the positive MS/MS spectra were in good agreement with those of the negative spectra, indicating that each of the N-glycans possesses the same core structure with a composition comprising $[MeMan]_2[Man]_4$ $[Xyl]_1[GlcNAc]_2$ [47]. It can also be suggested that the glycans with the additional xylose residues (2026, 2188 Da) are attached to the penultimate GlcNAc. The major 2AB Endo-H–released glycan was also analyzed by positive MS/MS, indicating the existence of two isomers in the fractions. The positive MS/MS spectra also indicated that different isomers exist in two of the glycan features [47].

Based on a combination of the two MS/MS spectra, the following structures were suggested [47] (Table 2):

N-Glycan calculated mass (Da)	Suggested structure	
1894.70	Isomer 1 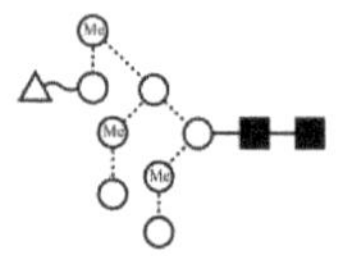	Isomer 2 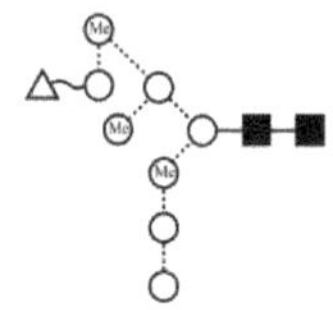
2026.75	Isomer 1 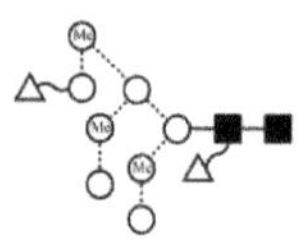	Isomer 2 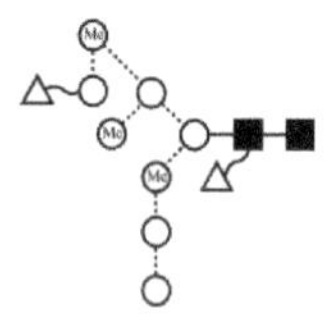
2056.76	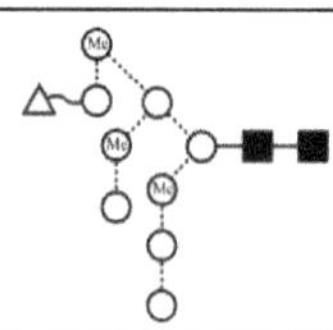	
2188.80	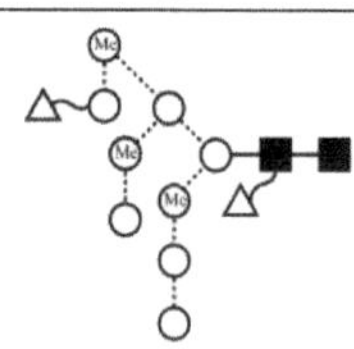	

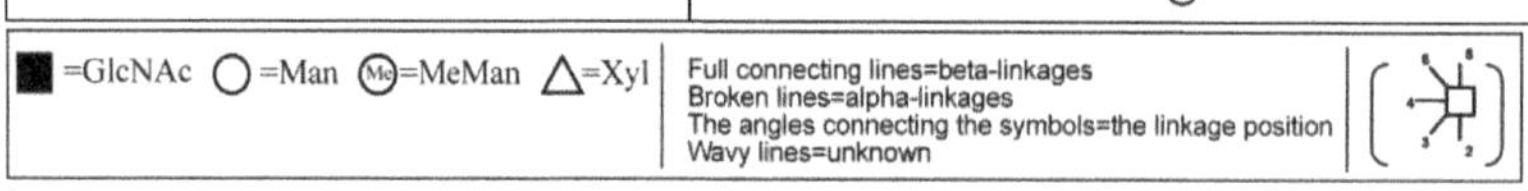

Table 2. Suggested structures of N-glycans separated from the 66-kDa glycoprotein within the-cell wall polysaccharide of *Porphyridium* sp.

All these diverse glycan structures were found to have oligomannose topologies, containing unique motifs that differentiate them from other, known N-linked glycan structures found to date in other organisms, including the 6-methylation of mannose residues inside the glycan chain and the xylose attached in different positions, both of which have never before been reported [47].

4. Effect of growth conditions on the cell-wall glycoproteins and on N-glycans within the 66-kDa glycoprotein

Since different physiological conditions were found to influence polysaccharide production [4] and since the 66-kDa glycoprotein is part of the polysaccharide structure, the study of cell-wall glycoprotein production and its N-glycosylation may help us understand the biosynthesis process and function of the polysaccharide. As a result, in addition to N-glycan structure determination of the 66-kDa cell-wall glycoprotein, the effect of growth conditions along with the starvation of sulfate, nitrogen and calcium or the enrichment of sulfate were also tested on the composition and structure of the N-glycan moieties. Prior to the experiments, *Porphyridium* sp. cells were cultivated in accordance with the treatment conditions as follows: The sulfate-enriched growth medium contained four-fold sulfate concentration compared to the original medium (ASW). Sulfate starvation cultures were cultivated for five cycles (five days each) in a medium containing 1/100 of the sulfate concentration of the regular medium. Cells subjected to starvation of either nitrogen or calcium were cultivated in a deficient medium free of nitrate or calcium (without KNO_3 or $CaCl_2$, respectively). The *Porphyridium* sp. control cells were cultivated in ASW medium.

After two weeks of growth, all of the cultures (sulfate enrichment or sulfate and nitrogen starvation, or regular medium), which were in the stationary phase, were centrifuged and the supernatant, which contained the polysaccharide, was isolated and dialyzed and concentrated to a final concentration of 0.3 w/v. The amount of cell-wall proteins within the concentrated polysaccharide (1.7 ml) was determined for each of the treatments by Lowry analysis [54]. To isolate the 66-kDa glycoprotein, the concentrated polysaccharide in the different treatments (1.7 ml) was run through SDS-PAGE electrophoresis, and the N-glycans were released by PNGase F following 2AB labeling. The NP-HPLC results for the N-glycans released from the 66-kDa protein were compared between the different treatments. In each NP-HPLC chromatogram, the molar ratios between the sugar features were determined with Empower HPLC software, which calculates the area under each peak, an indication of the sugar molar rate. The fluorescence rate was calibrated to mole amounts using 2-AB calibration standards. Each experiment was repeated twice.

Polysaccharide protein amounts produced under sulfate or nitrogen starvation treatments were 90% less than under the control or sulfate enrichment condition. Accordingly, the N-glycan amounts measured within the 66-kDa glycoproteins produced under these starvation conditions were also low (50 pmol compared to 500 pmol in the control and 450 pmol under the sulfate enrichment conditions). In addition, there was no difference compared to the control in either the cell-wall protein or N-glycan amounts measured within the 66-kDa protein produced under calcium starvation conditions. The NP-HPLC chromatogram of the 2AB- labeled N-glycans released from the 66-kDa glycoprotein, which were separated from algal cultures grown in the different treatments, are shown in Figure 3. The molar rate percentage of the various peaks detected in the different treatments are described below (Table 3). There is no significant change between the NP-HPLC chromatogram and the molar ratio of the different glycans produced under the sulfate enrichment, calcium starvation or control condition. However, an additional N-glycan feature was detected in the algae grown under sulfate enrichment conditions (designated in *, Figure 3A). It is interesting to note that under sulfate starvation conditions, the largest N-glycans were not found in the NP-HPLC (peaks 6 and 7, Figure 3). Similar to the findings under sulfate starvation, the largest N-glycan was not detected in the NP-HPLC chromatogram of the nitrogen starvation chromatogram (peak 7, Figure 3B). Therefore, the effects of nitrogen and sulfate starvation on N-glycosylation of the 66-kDa protein were found to be similar. These observations are in agreement with former studies [55-57], where it was reported that in both starvation conditions, the cells directed most of their energy toward the synthesis of cell- wall polysaccharide, an activity that is probably important for its survival. The decrease in 66-kDa protein production and in its N-glycan composition in both starvation conditions was expected, because under these conditions, the cells inhibit protein synthesis to the benefit of polysaccharide production. Although amounts of the glycoprotein under these starvation regimes are much lower than for the control, it is still being produced, just not at levels observed in the control cells, a finding that hints at the protein's vitality to cell survival. Since polysaccharide compositions in the sulfate/nitrogen deficient conditions (particularly the increased methyl hexose amounts) were found to differ from that of the control [57], the 66-kDa protein's role in polysaccharide production cannot be ruled out (i.e., it could be part of a specific polysaccharide process that does not occur under these starvation conditions).

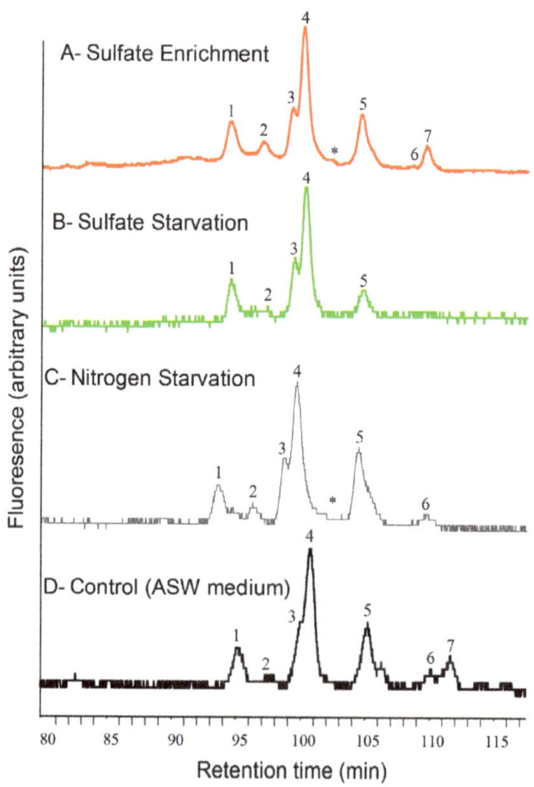

Figure 3. NP-HPLC chromatograms of N-glycans released from the 66-kDa glycoprotein produced in different treatments: A – sulfate enrichment, B – Sulfate starvation, C – Nitrogen starvation, D – Control/ASW medium

Since polysaccharide quantity in the medium was also found to be affected by growth phase (thinnest in logarithmic phase vs. thicker in stationary phase), cell-wall protein production and the 66-kDa glycoprotein N-glycans were studied as described above. In contrast to the nitrogen/sulfate starvation treatments, no difference was observed between the two phases of growth either in cell-wall protein production or in the 66-kDa N-glycan chromatograms. This observation lends credence to the hypothesis suggested by Ramus [58], i.e., in the stationary phase, polysaccharide production is not increased, but rather, its level of production exceeds its dissolution into the medium. If polysaccharide production were actually increasing throughout the stationary phase, then we would expect the corresponding increased energy consumption to be at the expense of protein production, as found in the algae grown under the sulfate/nitrogen starvation conditions. That was not the case here, where no difference in the cell-wall glycoprotein amount was observed between these two growth phases.

Treatment	Molar rate percentage of peak**							
	1-1894.70 Da	2- n.d	3- 2026.75 Da	4- 2056.76 Da	n.d - *	5- 2188.8 Da	6- n.d	7- n.d
Sulfate Enrichment	13.67	8.09	12.95	37.6	1.56	20.86	Little	5.28
Sulfate Starvation	16.59	1.64	18.21	49.59	n.f	13.97	n.f	n.f
Nitrogen Starvation	11.62	3.62	13.32	39.86	n.f	28.83	2.75	n.f
Calcium starvation	11.70	4.96	8.65	41.31	n.f	20.54	4.07	8.76
Logarithmic phase	13.81	4.08	12.23	35.42	n.f	24.74	2.55	7.17
Control	13.01	4.32	12.93	35.08	n.f	25.68	2.21	6.77

** Each peak matches its GU value in line with those shown in the NP-HPLC (in Figure 3). The molar rate was calculated by Water Empower software.
n.f- not found
n.d- not defined

Table 3. Molar rate percentage of the different N-glycans released from the 66-kDa glycoprotein that was isolated from the polysaccharide produced under different treatments.

5. Significance

Several years of intensive, multidisciplinary research have been directed at red microalgae, particularly *Porphyridium* sp. Among the various chemicals produced by *Porphyridium* sp., sulphated polysaccharides have perhaps garnered the most attention because of their potentially high value in biotechnological applications [2-4, 59-60]. However, little attention has been devoted to elucidating the glycosylation process in red microalgae. To date, our study is the first to report the structures of several N-glycans from a specific red microalga species, *Porphyridium* sp. [47]. This knowledge is important for both basic and applied research. An understanding of the way in which the sugar moieties of glycoproteins are bound to the microalgal proteins will elucidate glycosylation pathways, in the process revealing the enzymes involved, and it will contribute to an understanding of the role(s) of the sugar moieties in microalgal glycoproteins. The findings of this study will thus facilitate the identification of glycan biosynthetic components, thereby making an invaluable contribution to a comprehensive understanding of N-glycosylation in red microalgae. Since the N-glycan structures within the cell-wall glycoprotein were found to be novel, one particularly intriguing research direction will be to test whether these glycosylation structures are unique to the formation of the cell-wall polysaccharide. Alternatively, perhaps they are part of the general glycosylation process in these red microalgae cells exclusively or in a variety of red microalgae species.

Importantly, the technology for growing this species in controlled environments, both in small-scale laboratory facilities and in large-scale, semi-industrial systems, is already well-developed. A stable chloroplast transformation system [62], and recently, a nuclear transformation system, have been developed [63], the latter of which has paved the way for the expression of foreign genes in red algae and has far-reaching biotechnological implications. A growing number of scientists around the world are building a novel assortment of pharmaceutical products using algae as cell factories [64-66]. However, although they are well suited to the large-scale production of recombinant proteins, algae have not been extensively utilized for protein expression [66-67]. There are a number of advantages in cultivating algae as a platform for producing therapeutic proteins. Relatively simple and cheap to grow, algae are also amenable to cultivation under a variety of growth conditions. In addition, they are energy efficient, have a minimal negative impact on the environment, and they are easy to collect and purify. It is, therefore, of the utmost importance to evaluate the glycans attached to any recombinant protein expressed in any system. Since glycosylation may affect the biological role(s) of proteins or elicit an immunogenic response, knowledge of the structure of the microalgal N-glycans is essential for these applications. Moreover, knowledge of glycosylation patterns in algae will enable us to evaluate the potential of red microalgae species, particularly of *Porphyridium* sp., to be used as hosts and as potential alternatives to other plant-derived, transgenic therapeutic proteins. Furthermore, to fully exploit the inherent biotechnological potential of algae, it is important to initiate an overarching research program on the glycosylation pathways in algae that will include in-depth study of the enzymes involved. On the basis of the results toward elucidation of N-glycoslation pathways in red microalgae, we will be able to suggest glycosylation pathway manipulations to produce therapeutic proteins with ideal glycosylation patterns. In addition, the study can provide information about the evolutionary status of the red microalgae, since the N-glycans of the red microalgae combine not only the structural features of eukaryotes and prokaryotes, they also contain additional elements (e.g., the *O*-methylhexose and the pentose modifications) never before reported in other organisms.

6. Suggested biochemical processes of N-glycosylation

Our study to elucidate 66-kDa glycoprotein N-glycan structures found that those released from the 66-kDa protein possess oligomannose topology. The oligomannose topology may imply the existence of a conserved N-glycosylation pathway in red microalgae that takes place in the ER – which is common to eukaryotic organisms – and that includes the building of the N-glycan on the lipid substrate-dolichol-phosphate and its transfer to the protein. The results of other studies, such as that by Fishcer [68], also hint at the existence of this conserved pathway. Supporting evidence is based on homology searches for N-glycosylation protein sequences using the TBLASTN function on the algae DNA scaffold contigs database. Homologs were found for all N-glycosylation protein sequences in the ER pathways in the algae, thus suggesting that the pathway is conserved in *Porphyridium* sp. as it is in other organisms (animals, plants, yeast, etc.).

All the N-glycans investigated seem to go through the same intermediate glycan feature within their glycosylation biosynthesis, that probably have similar basic form based on the Man-9 topology structure constructed along the pathways typical of the ER. However, other enzymes, not typical to N-glycosylation pathways investigated so far, are involved in this pathway (e.g., the xylose, mannose, and methylated tranferase enzymes). However, we do not know at what stage the methyl and xylose groups were added to the mannoses during biosynthesis. In addition, we do not know if the glycan is assembled by incorporation of methyl-mannose rather than plain mannose or where the methyl groups are added to the intact high-mannose glycans.

If the methylated mannoses were incorporated into the assembled core oligosaccharide (parallel to $[Glc]_3[GlcNAc]_2[Man]_9$) via the same conserved pathway in the ER, the following mechanism can be suggested. The assembled core oligosaccharide (containing methylated groups) is transferred onto a nascent polypeptide imported into the ER (because of its signal sequence). This step is probably catalyzed by an enzyme complex (oligosaccharide transferase). Following Glucosidase I and II actions, the 3 glucose residues are cleaved from the end of the 6-branch, which initiates a process called glycan-mediated chaperoning. The last sugar that is removed in the ER is a mannose that is trimmed by an α-1,2-mannosidase through the action of ER mannosidase I (ManI, scheme 6), an enzyme that is also normally active in N-glycosylation processes, creating Man-8. However, this enzyme seems to be partially activated, as it does not cleave all the mannose residues. This phenomenon can be explained based on the methyl group structure of the oligosaccharide, which may interfere with enzyme cleavage. As a result, two different glycoprotein structures exit the ER (Figure 4):

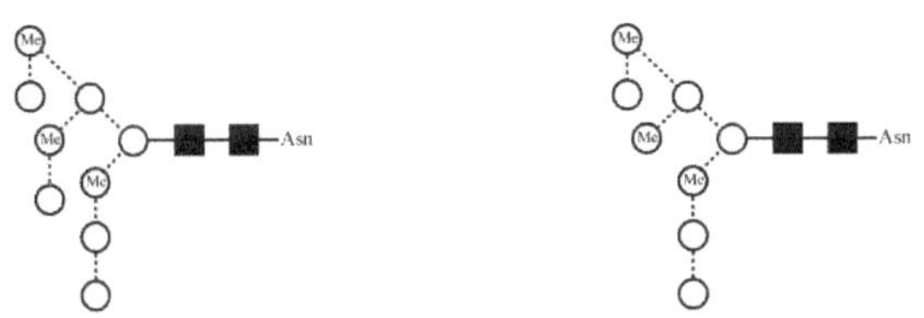

Figure 4. Two suggested intermediate features that are leaving the ER, Followed the ER they are getting their final structures within the Golgi apparatus (GA) by different enzymes.

Other changes are probably made in the GA by various xylose transferases (XylT) and a specific mannosidase (Man-2, scheme 6), the latter of which may only be able to remove the terminal mannose on the 3-antenna of the mannose-9 analog. Moreover, this enzyme may be a Golgi endomannosidase [69-70] that specifically cleaves the α1-2 linkage between the glucose-substituted mannose residue and the more internal portion of its polymannose branch, leading to the formation of the $[Man]_8[GlcNAc]_2$ (Man 8A) isomer [71].

Based on an assumption about N-glycosylation of the 66-kDa protein, namely, that processes occurring in the red microalgal ER are conserved as in eukaryote cells, a scheme for the mechanism of 66-kDa protein N-glycosylation is presented (Scheme 6).

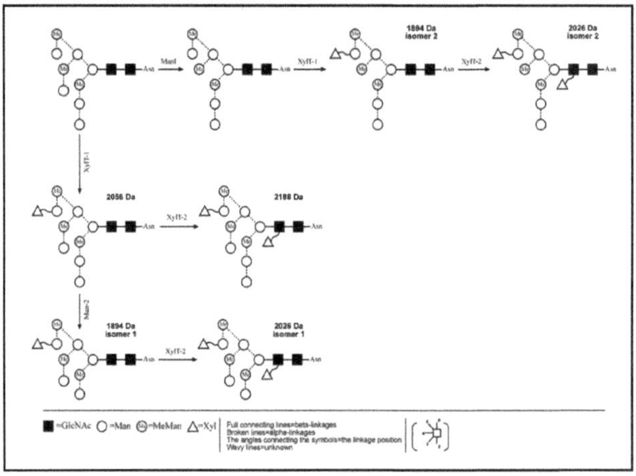

Scheme 6. Suggested mechanism of 66-kDa protein N-glycosylation, after formation of the basic core structure $[GlcNAc]_2[MeMan]_9$.

Another mechanism for N-glycosylation in red microalgae may be suggested based on the assumption that mannose methylation takes place in the GA, after mannose incorporation into the assembled ER core oligosaccharide. The following mechanism (Scheme 7) is based on the additional assumption that the conserved ER pathway of red microalgae functions much the same as in most eukaryotes, including synthesis of a lipid-linked oligosaccharide, transfer of glucose trimming in the ER, and subsequent cycles of glucose re-addition and removal involved in protein-folding quality control. After core oligosaccharide construction in the ER (following mannosidase I (ManI) cleavage), the ER oligosaccharide is further modified in the GA. The pathway present in the Golgi probably includes the cleavage of 3 mannose sugars by an α-1,2-mannosidase to produce $[Man]_5[GlcNAc]_2$, the known substrate for N-acetylglucosaminyltransferase I, which adds a single N-acetylglucosamine (GlcNAc) sugar onto the terminal of the 1,3-mannose in the mammalian glycosylation pathway. However, the structures found in this study indicate that the $[Man]_5[GlcNAc]_2$ is the substrate for the methyl-transferase (MeT) enzyme. Following the addition of methyl groups to the non-reducing end of the substrate, more changes occur, including the addition of mannose and xylose residues to the oligosaccharide mediated by specific transferases (xylose tranferases designated as XylT-1 and XylT-2, and mannose transferases as ManT, scheme 7).

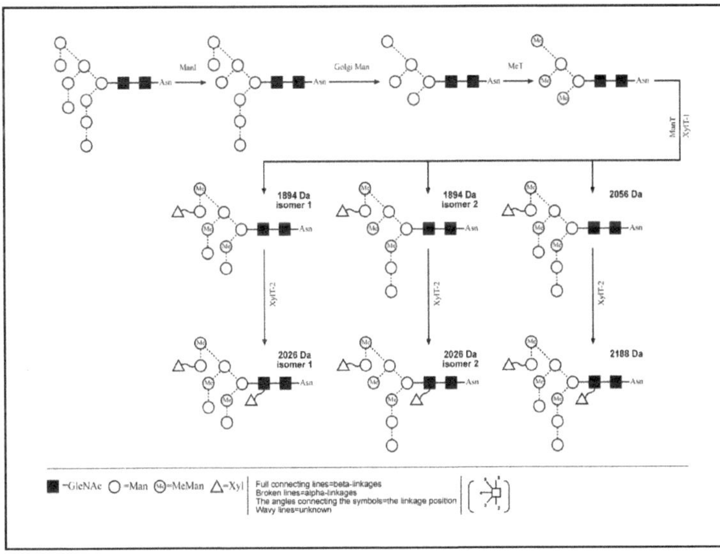

Scheme 7. Suggested mechanism of 66-kDa protein N-glycosylation, after formation of the typical eukaryote core structure comprising $[GlcNAc]_2[MeMan]_9$ and based on the assumption that mannose methylation takes place in the GA.

In both the suggested glycosylation pathways, xylose transferases, novel in N-glycosylation pathways, play prominent roles in the glycosylation. Since the ER pathway is probably conserved, it may be assumed that mannose methylation and xylose addition take place in the GA. Moreover, these novel enzymes are probably identical to those involved in the cell-wall polysaccharide biosynthesis that occurs in the GA.

A previous study of the evolutionary conservation of genes that participate in the N-glycosylation pathway in *Porphyridium* sp. showed that the protein sequences had relatively high similarity (40%) to orthologous sequences from red and green algae, diatoms, mammals and yeast [68]. These data are indicative of the extent of conservation of the N-glycosylation pathway and of its general importance in eukaryotes, particularly in photosynthetic organisms. The phylogenetic status of the algae can also be discussed based on the structure of the bodies involved in the N-glycosylation, e.g., ER and Golgi. The Golgi bodies and ER of red microalgae have not been extensively studied. Ultrastructural studies of these cells have produced little information. Both the smooth and rough ER appear to be present, although typically not in large quantities, in red algal unicells that have been studied. One characteristic shared by all unicells studied is a smooth ER system lining the entire interior of the plasma membrane [72-75]. The wide region between this peripheral ER system and the plasma membrane is 100-150 nm and free of major organelles (including ribosomes), but it does appear to contain a fibrous substance [76]. At irregular intervals, tubules arise at right angles from the ER toward the plasma membrane. Some appear to fuse with the plasma membrane, suggesting direct communication between the ER and the cell exterior [72]. Evans et al. [77] suggested that this system may play a role in mucilage production.

The presence of structures indicative of a eukaryotic organism may also imply that the ER-based glycosylation occurs similar to how it does in other eukaryote organisms. In addition, all red algae also contain a typical eukaryotic GA, comprising 4 to 15 cisternae [72] that are especially prominent during sporogenesis. GA numbers, size and morphology may vary with the cell cycle or culture conditions [75], i.e., in the logarithmic phase of growth of *Porphyridium*, the Golgi bodies are larger and more numerous (due to its higher number of dictyosomes, which are larger and have distended cisternae) than in stationary phase cells [75, 78]. Whereas in most eukaryotes, the cis-Golgi is associated with the ER, in red microalgae it may be involved with other bodies. Some red microalgal ultrastructure micrographs show a thin line of an apparently fibrillar substance between the forming face of the Golgi body and its associated organelles [79], representing a possible cytoskeleton element, such as actin. Moreover, it may be responsible for maintaining the associations between dictyosomes and other organelles.

Golgi involvement in the N-glycosylation pathway has yet to be elucidated. However, some reports have been published about the relationship between cell-wall polysaccharide biosynthesis and the Golgi. The GA of *Porphyridium* sp. [12, 58, 80] and of other red microalgae species [77, 81] were found to be involved in the synthesis of the cell-wall polysaccharide. Polysaccharide synthesis in *P. aerugineum* and *Rhodella reticulata* and their subsequent packaging into vesicles takes place in the Golgi [77, 82]. The vesicles are transported, fuse with the plasma membrane, and then secrete their contents on the cell surface [12, 58, 77]. Involvement of the GA in polysaccharide biosynthesis may indicate the existence of an unusual algal glycosylation process, i.e., enzymes responsible for polysaccharide biosynthesis also act on other glycol-substrates, in our case N-glycans.

The immunological natures of the additions unique to the red microalgal polysaccharide, including methylated and xylose residues, need to be determined. Xylose residues are found in N-glycans from plants [83], insects [84], molluscs [85], and rarely in parasitic helminths [86], but not normally in mammals [47]. In addition, the position and linkage of xylose (attached to the 2-position of the core branching mannose) is the same in all the organisms mentioned above. In this study, we found, for the first time, a xylose residue attached to the mannose of the 6-antenna and 1→3-linked to the penultimate GlcNAc of the core. These xylose residues are attached to a different monosaccharide (and in a different linkage position) than known glycans. Therefore, it is not known whether the xylose residues reported here have allergenic natures similar to those of the xylose residues found in other known glycans [87-88]. In addition, we also do not know how the additional methyl groups affect the protein and its immunogenic response.

The many remaining questions about N-glycosylation in the cell wall of red microalgae prevent the full potential of *Porphyridium* sp. to serve as a host for therapeutic protein production from being realized. For example, is not known whether the unusual N-glycan structures are typical specifically to the 66-kDa glycoprotein (that is part of the polysaccharide) or whether they represent glycosylation structures characteristic of all algal N-glycosylation processes. Therefore, microalgal potential as a protein production machine cannot be evaluated without additional and extensive research, preferably with a multidisciplinary approach.

Author details

Oshrat Levy-Ontman
Department of Chemical Engineering, Sami Shamoon College of Engineering, Beer Sheva, Israel

Acknowledgement

I would like to thank Prof S. Arad for her support of this research.

7. References

[1] Arad (Malis) S (1988) Production of sulfated polysaccharides from red unicellular algae. In: Stadler T, Mollion J, Verduset MC, editors. Algal Biotechnology–An Interdisciplinary Perspective. London: Elsevier Applied Science. pp. 65-87.
[2] Arad (Malis) S (1999) Polysaccharides of red microalgae. In: Cohen Z, editor. Chemicals from Microalgae. London: Taylor and Francis. pp. 282-291.
[3] Arad (Malis) S, Levy-Ontman O (2010) Red microalgal cell-wall polysaccharides: biotechnological aspects. Curr. Opin. Biotech. 21: 358-364.
[4] Arad (Malis) S, Levy-Ontman O (2012) Sulfated Polysaccharides in the Cell Wall of Red Microalgae. In: Sabu T, Dominique D, Christophe C, Jyotishkumar P, editors. Structure, Properties and Applications (Chapter 2) Biopolymer-based materials: from blends and composites to gels and complex networks. Verlag: Wiley-VCH.
[5] Percival EP, Foyle RAJ (1979) The extracellular polysaccharides of *Porphyridium cruentum* and *Porphyridium aerugineum*. Carbohydr. Res. 72: 165-176.
[6] Geresh S, Arad (Malis) S (1991) The extracellular polysaccharides of the red microalgae: chemistry and rheology. Bioresourc. Technol. 38: 195-201.
[7] Geresh S, Adin I, Yarmolinsky E, Karpasas M (2002) Characterization of the extracellular polysaccharide of *Porphyridium* sp.: molecular weight determination and rheological properties. Carbohydr. Polym. 50: 183–189.
[8] Geresh S, Lupescu N, Arad (Malis) S (1992) Fractionation and partial characterization of the sulfated polysaccharide of *Porphyridium*. Phytochemistry 31: 4181-4186.
[9] Heaney-Kieras J, Chapman, DJ (1976) Structural studies on the extracellular polysaccharide of the red alga *Porphyridium cruentum*. Carbohydr. Res. 52: 169-177.
[10] Arad (Malis) S, Kolani R, Simon-Berkovitch B, Sivan A (1994) Inhibition by DCB of cell-wall polysaccharide formation in the red microalga *Porphyridium* sp. (Rhodophyta) Phycologia 33: 158–162.
[11] Lupescu N, Arad (Malis) S, Geresh S, Bernstein M, Glazer R (1991) Structure of some sulfated sugars isolated after acid hydrolysis of the extracellular polysaccharide of *Porphyridium* sp. a unicellular red alga. Carbohydr. Res. 210: 349–352.

[12] Ramus J (1986) Rhodophytes unicells: biopolymer physiology and production. In: Barclay WR, McIntosh RP, editors. Algal Biomass Technology. Berlin: J. Cramer. pp. 51–55.

[13] Ramus J (1973) Cell surface polysaccharides of the red alga *Porphyridium*. In: Loewus F, editor. Biogenesis of Plant Cell Wall Polysaccharides. New York : Academic Press. pp. 333–359.

[14] Ucko M, Geresh S, Simon-Berkovitch B, Arad (Malis) S (1994) Predation by a dinoflagellate on a red microalga with a cell wall modified by sulfate and nitrate starvation. Mar. Ecol. Prog. Ser. 104: 293–298.

[15] Geresh S, Dubinsky O, Arad (Malis) S, Christiaen D, Glaser R (1990) Structure of 3-O-(alpha-D-glucopyranosyluronic acid)-L-galactopyranose, an aldobiouronic acid isolated from the polysaccharides of various unicellular red algae Carbohydr. Res. 208: 301–305.

[16] Geresh S, (Malis) Arad S, Levy-Ontman O, Zhang W, Tekoah Y, Glaser R (2009) Isolation and characterization of poly- and oligosaccharides from the red microalga *Porphyridium* sp. Carbohyd. Res. 344: 343–349.

[17] Gloaguen V, Ruiz G, Morvan H, Mouradi-Givernaud A, Maes E, Krausz P, Strecker G (2004) The extracellular polysaccharide of *Porphyridium* sp. an NMR study of lithium-resistant oligosaccharidic fragments. Carbohyd. Res. 339: 97–103.

[18] Eteshola E, Gottlieb M, Arad (Malis) S (1996) Dilute solution viscosity of red microalga exopolysaccharide Chem. Eng. Sci. 51: 1487–1494.

[19] Eteshola E, Karpasas M, Arad (Malis) S, Gottlieb M (1998) Red microalga exopolysaccharides: 2. Study of the rheology, morphology and thermal gelation of aqueous preparations. Acta. Polym. 49: 549–556.

[20] Ramus J, Kenney BE (1989) Shear degradation as a probe of microalgal exopolymer structure and rheological properties. Biotechnol. Bioeng. 34: 1203–1208.

[21] Gourdon D, Lin Q, Oroudjev E, Hansma H, Golan Y, Arad S, Israelachvili J (2008) Adhesion and stable low friction provided by a subnanometer-thick monolayer of a natural polysaccharide. Langmuir 24: 1534–1540.

[22] Arad (Malis) S, Rapoport L, Moshkovich A, van-Moppes D, Karpasas M, Golan R, Golan Y (2006) Superior biolubricant from a species of red microalga. Langmuir 22: 7313–7317.

[23] Gasljevic K, Hall K, Chapman D, Matthys EF (2008) Drag-reducing polysaccharides from marine microalgae: species productivity and drag reduction effectiveness. J. Appl. Phycol. 20: 299–310.

[24] Gasljevic K, Matthys EF (2007) Ship drag reduction by microalgal biopolymers: A fisibility analysis. J. Ship. Res. 51: 326–337.

[25] Arad (Malis) S, Keristovsky, G, Simon B, Barak Z, Geresh S (1993) Biodegradation of the sulphated polysaccharide of *Porphyridium* by soil bacteria. Phytochemistry 32: 287–290.

[26] Matsui MS, Muizzuddin N, Arad S, Marenus K (2003) Sulfated polysaccharides from red microalgae have anti-inflammatory properties *in vitro* and *in vivo*. Appl. Biochem. Biotechnol. 104: 13–22.

[27] Tannin-Spitz T, Bergman M, van-Moppes D, Grossman S, Arad (Malis) S (2005) Antioxidant activity of the polysaccharide of the red microalga *Porphyridium* sp. J. Appl. Phycol. 17: 215–222.
[28] Minkova K, Michailov Y, Toncheva-Panova T, Houbavenska N (1996) Antiviral activity of *Porphyridium cruentum* polysaccharide. Pharmazie 51: 3-7.
[29] Talyshinsky M, Souprun Y, Huleihel M (2002) Anti-viral activity of red microalgal polysaccharides against retroviruses. Cancer Cell Int. 2: 1-8.
[30] Huleihel M, Ishanu V, Tal J, Arad (Malis), S (2002) Activity of *Porphyridium* sp. polysaccharide against herpes simplex viruses *in vitro* and *in vivo*. J. Biochem. Biophys. Meth. 50: 189-200.
[31] Huleihel M, Vladimir Z, Tal J, Arad S (2001) Antiviral effect of red microalgal polysaccharides on *Herpes simplex* and *Varicella* viruses. J. Appl. Phycol. 13: 127-134.
[32] Arad (Malis) S, Ginzberg A, Huleihel M (2006) Antiviral activity of sulfated polysaccharides of marine red algae. In: Fingerman M, editor. Recent Advances in Marine Biotechnology: Biomaterials from Aquatic and Terrestrial Organisms. Enfield : Science Publishers Inc. pp. 37-62.
[33] Fabregas J, García D, Fabregas J, Garcia D, Fernandez-Alonso M, Rocha AI, Gómez-Puertas P, Escribano JM, Otero A, Coll JM (1999) *In vitro* inhibition of the replication of *haemorrhagic septicaemia* virus and African swine fever virus by extracts from marine microalgae. Antiviral Res. 44: 67-73.
[34] Dvir I, Geva O, Chayoth R, Sod-Moriah U, Shany S, Arad S, Madar Z (1996) Polysaccharides and algal biomass as new sources of dietary fibers and their physiological effects in rats. FASEB J. 10: 3012-3012.
[35] Dvir I, Chayoth R, Sod-Moriah U, Shany S, Nyska A, Stark AH, Madar Z, Arad (Malis) S (2000) Soluble polysaccharide and biomass of red microalga *Porphyridium* sp. alter intestinal morphology and reduce serum cholesterol in rats. Br. J. Nutri. 84: 469-476.
[36] Dvir I, Stark AH, Chayoth R, Madar Z, Arad (Malis) S (2009) Hypocholesterolemic effects of nutraceuticals produced from the red microalga *Porphyridium* sp. Nutrients 1: 156–167.
[37] Raghavendran HR, Srinivasan P (2008) Effect of crude sulphated polysaccharide from brown algae against acetaminophen-induced toxicity in rats. Can. J. Physiol. Pharmacol. 86: 660-666.
[38] Shrestha, RP, Weinstein Y, Bar-Zvi D, Arad SM (2004) A glycoprotein noncovalently associated with cell-wall polysaccharide of the red microalga *Porphyridium* sp. (Rhodophyta). J. Phycol. 40: 568-580.
[39] Shrestha RP (1999) A non covalenty bound cell-wall glycoprotein of the red microalga *Porphyridium* sp.: characterization and functions. Ph.D. Thesis, Submitted to Ben-Gurion University of the Negev, Israel.
[40] Simon-Bercovitch B, Bar-Zvi D, and Arad (Malis) S (1999) Cell-wall formation during the cell cycle of *Porphyridium* sp. (Rhodophyta). J. Phycol. 35: 78-83.
[41] Simon-Bercovitch B. (1997) Cell-wall formation in the red microalga *Porphyridium* sp. Ph.D. Thesis, Submitted to Ben-Gurion University of the Negev, Israel.
[42] Ucko, M, Shrestha RP, Mesika P, Bar-Zvi D, Arad (Malis) S (1999) Glycoprotein moiety in the cell wall of the red microalga *Porphyridium* sp. (Rhodophyta) as the

biorecognition site for the *Crypthecodinium cohnii*-like dinoflagellate. J. Phycol. 35: 1276-1281.

[43] Gravel P (1996) Identification of glycoproteins on nitrocellulose membranes using lectin blotting. In: Walker JM, editor. The protein protocols Handbook. Totowa: Humana Press. pp.779-794.

[44] Van Damme, EJM, Peumans WJ, Pusztai A, Bardocz S (1998) Handbook of plant lectins: properties and biomedical application. Chichester: John Wiley and Sons. pp. 452.

[45] Shibuya N, Goldstein IJ, Van Damme EJM, Peumans WJ (1988) Binding properties of mannose specific lectin from the snowdrop (*Galanthus nivalis* Bulb.) J. Biol. Chem. 263: 728-734.

[46] Küster B, Wheeler SF, Hunter AP, Dwek RA, Harvey DJ (1997) Sequencing of N-linked oligosaccharides directly from protein gels: In-gel deglycosylation followed by matrix-assisted laser desorption/ionisation mass spectrometry and normal-phase high performance liquid chromatography. Anal. Biochem. 250: 82-101.

[47] Levy-Ontman O, (Malis) Arad S, Harvey DSJ, Parsons TB, Fairbanks A, Tekoah Y. (2011) Unique N-glycan moieties of the 66-kDa cell wall glycoprotein from the red microalga *Porphyridium* sp. J. Biol. Chem. 286: 21340–21352.

[48] Bigge JC, Patel TP, Bruce JA, Goulding PN, Charles SM, Parekh RB (1995) Nonselective and efficient fluorescent labelling of glycans using 2-amino benzamide and anthranilic acid. Anal. Biochem. 230: 229-238.

[49] Harvey DJ (2005) Fragmentation of negative ions from carbohydrates: Part 1; Use of nitrate and other anionic adducts for the production of negative ion electrospray spectra from *N*-linked carbohydrates J. Am. Soc. Mass Spectrom. 16: 622-630.

[50] Harvey DJ (2005) Fragmentation of negative ions from carbohydrates: Part 2, Fragmentation of high-mannose *N*-linked glycans. J. Am. Soc. Mass Spectrom. 16: 631-646.

[51] Harvey DJ (2005) Fragmentation of negative ions from carbohydrates: Part 3, Fragmentation of hybrid and complex *N*-linked glycans. J. Am. Soc. Mass Spectrom. 16: 647-659.

[52] Harvey DJ, Royle L, Radcliffe CM, Rudd PM, Dwek RA (2008) Structural and quantitative analysis of *N*-linked glycans by MALDI and negative ion nanospray mass spectrometry. Anal. Biochem. 376: 44-60.

[53] Domon B, Costello CE (1988) A systematic nomenclature for carbohydrate fragmentations in FAB-MS/MS spectra of Glycoconjugates. Glycoconj. J. 5: 397-409.

[54] Lowry OH, Rosebrough NJ, Farr AL, Randall RJ (1951) Protein measurement with the folin phenol reagent. J. Biol. Chem. 193: 265-275.

[55] Arad (Malis) S, (Dahan) Friedman O, Rotem A (1988) Effect of nitrogen on polysaccharide production in *Porphyridium* sp. Appl. Environ. Microbiol. 54: 2411–2414.

[56] Adda M, Merchuk JC, (Malis) Arad S (1986) Effect of nitrate on growth and production of cell-wall polysaccharide by the unicellular red alga *Porphyridium*. Biomass 10: 131–140.

[57] Ucko M, Geresh S, Simon-Berkovitch B, Arad (Malis) S (1994) Predation by a dinoflagellate on a red microalga with a cell wall modified by sulfate and nitrate starvation. Mar. Ecol. Prog. Ser. 104: 293–298.

[58] Ramus J (1972) The production of extracellular polysaccharide by the unicellular red alga *Porphyridium aerugineum*. J. Phycol. 15: 97-111.
[59] Arad (Malis) S, Richmond A (2004) Industrial production of microalgal cell-mass and secondary products-species of high potential: *Porphyridium* sp. In: Richmond A, editor. Handbook of Microalgal Culture: Biotechnology and Applied Phycology. Carlton: Blackwell Science. pp. 289-297.
[60] Lapidot M, Shrestha RP, Weinstein Y, Arad (Malis) S (2010) Red microalgae: from basic known-how to Biotechnology. In: Seckbach J, Chapman D, editors. Red algae in the genomic age, vol. 13 of Cellular Origins, Life in Extreme Habitats and Astrobiology series. Dordrecht: Springer. pp. 11-19.
[61] Matsui MS, Muizzuddin N, Arad S, Marenus K (2003) Sulfated polysaccharides from red microalgae have anti-inflammatory properties *in vitro* and *in vivo*. Appl. Biochem. Biotechnol. 104: 13-22.
[62] Lapidot M, Raveh D, Sivan A, Arad (Malis) S, Shapira M (2002) Stable chloroplast transformation of the unicellular red alga *Porphyridium* sp. Plant Physiol. 129: 7-12.
[63] Plesser E (2009) Molecular characterization of the Sulfotransferase from the red microalga *Porphyridium* sp. PhD thesis, Ben-Gurion University of the Negev, Beer-Sheva, Israel.
[64] Giddings G, Allison G, Brooks D and Carter A (2000) Transgenic plant as factories for biopharmaceuticals. Nature Biotechnol. 18: 1151-1155.
[65] Walmsley AM, Arntzen CJ (2003) Plant cell factories and mucosal vaccines. Curr. Opin. Biotechnol. 14: 145-150.
[66] Rasala BA, Muto M, Lee PA, Jager M, Cardoso RM, Behnke CA, Kirk P, Hokanson CA, Crea R, Mendez M, Mayfield SP (2010) Production of therapeutic proteins in algae, analysis of expression of seven human proteins in the chloroplast of *Chlamydomonas reinhardtii*. Plant Biotechnol. J. 6: 719-33.
[67] Franklin SE, Mayfield SP (2005) Recent developments in the production of human therapeutic proteins in eukaryotic algae. Expert Opin. Biol. Ther. 5: 225-235.
[68] Fisher M (2011) Elucidation of the N-Glycosylation Pathway in Red Microalgae *Porphyridium* sp. Ph.D. Thesis, Submitted to Ben-Gurion University of the Negev, Israel.
[69] Lubas WA, Spiro RG (1987) Evaluation of the Role of Rat Liver Golgi Endo-α-D-mannosidase in Processing N-Linked Oligosaccharides. J. Biol. Chem. 262: 3775-3781.
[70] Lubas WA, Spiro RG (1988) Evaluation of the role of rat liver Golgi endo-α-D-mannosidase in processing N-linked oligosaccharides. J. Biol. Chem. 263: 3990-3998.
[71] Moore SEH, Spiro RG (1990) Demonstration that Golgi endo-α-D-mannosidase provides a glucosidase-independent pathway for the formation of complex N-linked oligosaccharides of glycoproteins. J. Biol. Chem. 265: 13104-13112.
[72] Pueschel CM (1990) Cell structure. In: Cole KM, Sheath RG, editors. Biology of the Red Algae. Cambridge: Cambridge University Press. pp. 7-41.
[73] Patrone LM, Broadwater ST, Scott JL (1991) Ultrastructure of vegetative and dividing cells of the uni-cellular red algae *Rhodella violacea* and *Rhodella maculata*. J. Phycol. 27: 742-753.

[74] Scott J, Broadwater S, Gabrielson P, Thomas J, Saunders B (1992). Ultrastructure of vegetative organization and cell division in the unicellular red alga *Dixoniella grisea* gen. nov. (Rhodophyta) and a consideration of the genus *Rhodella*. J. Phycol. 28: 649-660.

[75] Seckbach J (1994) Evolutionary pathways and enigmatic algae: *Cyanidium caldarium* (Rhodophyta) and related cells. In: Broadwater ST, Scott JL, editors. Ultrastructure of unicellular red algae. Dordrecht: Kluwer Academic Publishers. pp. 215-230.

[76] Broadwater ST, Scott JL, Garbary DJ (1992) Cytoskeleton and mitotic spindle in red algae In: Menzel D, editor. The cytoskeleton of the Algae. Boca Raton: CRC Press: pp. 93-112.

[77] Evans LV, Callow ME, Pervical E, Fareed V (1974) Studies on the synthesis and composition of extracellular mucilage in the unicellular red alga *Rhodella*. J. Cell Sci. 16: 1-21.

[78] Ramus J, Robins DM (1975) The correlation of Golgi activity and polysaccharide secretion in *Porphyridium*. J. Phycol. 11: 70-74.

[79] Alley CD, Scott JL (1977) Unusual dictyosome morphology and vesicle formation in tetrasporangia of the marine alga *Polysiphonia denudate*. J. Ultrastruct. Res. 58: 289-298.

[80] Keidan M, Friedlander M, Arad (Malis) S (2009) Effect of Brefeldin A on cell-wall polysaccharide production in the red microalga *Porphyridium* sp. (Rhodophyta) through its effect on the Golgi apparatus. Phycologia 21: 707-717.

[81] Gantt E, Conti SF (1965) The ultrastructure of *Porphyridium cruentum*. J. Cell Biol. 26: 365-381.

[82] Ramus J (1976) Cell surface polysaccharides of the red alga *Porphyridium*. Plant and cell wall polysaccharides pp.333-357

[83] Lerouge P, Cabanes-Macheteau M, Rayon C, Fitchette-Lainé AC, Gomord V, Faye L (1998) N-glycoprotein biosynthesis in plants: recent developments and future trends. Plant Mol. Biol. 38: 31-48.

[84] Altmann F, Staudacher E, Wilson IB, Marz L (1999) Insect cells as hosts for the expression of recombinant glycoproteins. Glycoconj. J. 16: 109-123.

[85] Kamerling JP, Vliegenthart JFG. (1997) Hemocyanins. In: Montreuil J, Vliegenthart JFG, Schachter H, editors. Glycoproteins II. Amsterdam: Elsevier Science B.V. pp. 123-140.

[86] Khoo KH, Chatterjee D, Caulfield JP, Morris HR, Dell A (1997) Structural mapping of the glycans from the egg glycoproteins of *Schistosoma mansoni* and *Schistosoma japonicum*: identification of novel core structures and terminal sequences. Glycobiology 7: 663-677.

[87] Garcia-Casado G, Sanchezmonge R, Chrispeels MJ, Armentia A, Salcedo G, Gomez L (1996) Role of complex asparagine-linked glycans in the allergenicity of plant glycoproteins. Glycobiology 6, 471–477.

[88] Van Ree R, Cabanes-Macheteau M, Akkerdaas J, Milazzo JP, Loutelier-Bourhis C, Rayon C, Villalba M, Koppelman S, Aalberse R, Rodriguez R, Faye L, Lerouge P (2000) α(1,2)-xylose and α(1,3)-fucose residues have a strong contribution in IgE binding to plant glycoallergens. J. Biol. Chem. 275, 11451-11458.

Section 3

Glycosylation and Disease

Chapter 9

Plant-Derived Agents with Anti-Glycation Activity

Mariela Odjakova, Eva Popova,
Merilin Al Sharif and Roumyana Mironova

Additional information is available at the end of the chapter

1. Introduction

1.1. Glycation and consequences

Glycation or the Maillard reaction is the non-enzymatic adduct formation between amino groups (predominantly the ε-amino group of lysine and the guanidine group of arginine) [1, 2] and carbonyl groups of reducing sugars or other carbonyl compounds. This reaction is subdivided into three main stages: early, intermediate, and late. In the early stage, glucose (or other reducing sugars such as fructose, pentoses, galactose, mannose, xylulose) react with a free amino group of biological amines, to form an unstable aldimine compound, the Shiff base. Then through an acid-base catalysis, this labile compound undergoes a rearrangement to a more stable early glycation product known as Amadori product [3]. Because the Maillard reaction is non-enzymatic, the variables which regulate its velocity *in vivo* are the glucose and protein concentrations, the half-life of the protein, its reactivity in terms of free amino groups, and the cellular permeability to glucose.

In the intermediate stage, *via* dehydratation, oxidation and other chemical reactions, the Amadori product degrades to a variety of reactive dicarbonyl compounds such as glyoxal, methylglyoxal, and deoxyglucosones which, being much more reactive than the initial sugars, act as propagators of the reaction, again reacting with free amino groups of biomolecules. In the late stage of the glycation process through oxidation, dehydratation and cyclization reactions, irreversible compounds, called Advanced glycation end products (AGEs) are formed. The AGEs are yellow-brown, often fluorescent and insoluble adducts that accumulate on long-lived proteins thus compromising their physiological functions [4]. Glycation of proteins can interfere with their normal functions by disrupting molecular conformation, altering enzymatic activity, reducing degradation capacity, and interfering with receptor recognition [5]. AGE-modified proteins lose their specific functions and undergo accelerated degradation to free AGEs such as 2-(2-furoyl)-4(5)-furanyl-1H-imidazole (FFI), imidazolone, N-ε-carboxy-methyl-lysine (CML), N-ε-carboxy-ethyl-lysine (CEL), glyoxal-lysine dimmer (GOLD), methyl-glyoxal-lysine dimer (MOLD), and others. Moreover, AGEs can also act as cross-linkers between proteins, resulting in the production of proteinase-resistant aggregates [6].

The formation of AGEs progressively increases with normal aging and age-dependent AGEs have been shown to accumulate in human cartilage, skin collagen and pericardial fluid [7]. Long-lived proteins such as lens crystallins and especially collagens contain numerous lysine, hydroxylysine and arginine residues, have a slow turn over, and are prone to age-related accumulation of glycation damage [8]. Besides accumulation during healthy aging, AGEs are formed at accelerated rates in diabetes [9]. They are markers and also important causative factors for the pathogenesis of diabetes [10], cataracts [11], atherosclerosis [12], diabetic nephropathy [13], and neurodegenerative diseases, including Alzheimer's disease [14]. Three routes have been proposed for AGEs formation: 1) autoxidative pathway in which sugars give rise to reactive products by autoxidation, 2) Amadori rearrangement, and 3) from the Shiff base. Reactive oxygen species (ROS) in the presence of trace levels of catalytic redox-active transition metal ions also contribute to AGEs formation. The process includes oxidative steps and is therefore called glycoxidation [15].

Except these endogenous AGEs, humans are also exposed to exogenous AGEs which are ingested with food. Some approaches for food processing promote the Maillard reaction and the development of browning products [16]. The formation of Maillard reaction products (MRPs) depends on the processing temperature and duration, and is greatly accelerated by long exposure to heat [17]. Food treatments such as frying or baking have a greater impact on the formation of MRPs than boiling [18]. MRPs are inherent to Western diet [19] and fructose intake was largely elevated in recent years because of an increased consumption of soft drinks and processed foods. Fructose and its metabolites, can initiate the non-enzymatic fructosylation of proteins. Moreover, among the various physiological sugars, fructose undergoes a more rapid oxidative degradation and is a more potent protein glycating agent than glucose [20]. During chronic hyperglycemia, excessive glucose uptake in tissues also affects the key enzyme aldose reductase (AR) in the polyol pathway. This leads to the reduction of various sugars to sugar alcohols, such as glucose to sorbitol, followed by nicotinamide adenine dinucleotide (NADH)-dependent sorbitol dehydrogenase-catalyzed fructose production. Increased fructose formation in turns leads to the production of reactive carbonyl species which are key factors in AGEs formation [21]. Furthermore, sorbitol and its metabolites accumulate in the nerves, retina, kidney, and lens due to poor penetration across membranes and inefficient metabolism, resulting in the development of diabetic complications [22]. Advanced glycation end products formed inside the body or ingested with food can also interact with specific receptors and/or binding proteins thus activating a series of intracellular signalling pathways, which are implicated in diabetic complications. Interaction of AGEs with receptors for advanced glycation end products (RAGE) can trigger signaling events through p38 MAP Kinase, nuclear factor kappa-B

(NF-κB), P21 Ras and Jak/STAT pathways. The cellular response can involve the overexpression of cell adhesion molecules such as vascular cell adhesion molecule-1(VCAM-1), and the production of cytokines (interleukin-2, interleukin-6 and tumor necrosis factor-α) and vascular endothelial growth factor (VEGF) [23-25]. Various studies have shown that diabetes mellitus is associated with an increased production of free radicals and also with a decrease in the antioxidant potential leading to oxidative stress. Thus, the disturbed balance between radical formation and radical neutralization leads to oxidative damage of cell components such as proteins, lipids, and nucleic acids [26]. During glycoxidative stress, NF-κB activates the production of TNF-α which in turn leads to enhanced ROS production. In such a way AGEs formation keeps the oxidative stress ongoing [27, 28].

ROS exert a multitude of effects. They are both harmful by-products and cellular messengers. As messengers, ROS are involved in a network of intracellular and intercellular communication pathways and that is why mitochondrial production of ROS plays a crucial role in the pathogenesis of type II diabetes, neurodegenerative and cardiovascular diseases [29]. It has been reported that direct exposure of endothelial cells to hyperglycaemic concentrations of glucose increases the formation of ROS. Activation of the enzyme NADPH oxidase is strongly implicated in this process [30]. NADPH oxidase may be activated through an increased diacylglycerol mediated activation of protein kinase C (PKC) [31]. High concentrations of glucose [32] and ROS [33] also have been reported to activate PKC. Since hyperglycaemia is responsible for a rise in the mitochondrial production of ROS, targeting antioxidants to mitochondria and increasing their overall antioxidant potential is expected to ameliorate diabetic symptoms [34]. In order to enhance the antioxidant capacity of the body, we must increase the exogenous intake of antioxidants or stimulate the endogenous synthesis of antioxidants such as superoxide dismutase and reduced glutathione. The inhibition of AR and advanced glycation end products formation is yet another mode for diabetes treatment, which is not dependent on the control of blood glucose, and would be useful in prevention of certain diabetic complications [35].

2. Therapeutic agents

Both synthetic compounds and natural products have been evaluated as inhibitors against the formation of advanced glycation end products (AGEs). The synthetic AGEs inhibitors so far discovered are divided into three classes: (a) carbonyl trapping agents which attenuate carbonyl stress; (b) metal ion chelators, which suppress glycoxidations; and (c) cross-link breakers that reverse AGE cross-links [9]. However, despite of their inhibitory capacities against the formation of AGEs, many synthetic inhibitors of AGEs formation were withdrawn from clinical trials due to relatively low efficacies, poor pharmacokinetics, and unsatisfactory safety [36, 37].

For exaple, aminoguanidine (AG), a nucleophilic hydrazine compound which prevents the formation of AGEs was withdrawn from the crucial phase III of clinical trials because of safety concerns and apparent lack of efficacy [38]. On the other hand natural products have been proven relatively safe for human consumption and many plant extracts have been tested for their ability to prevent AGEs formation [39]. Moreover, a number of plant-derived products have been shown to possess hypoglycemic, hypolipidemic as well as antioxidant properties [40]. Some important compounds such as phenolics [41, 42], oligo- and polysaccharides [43, 44], carotenoids [45, 46], unsaturated fatty acids [45, 46] and many others have been reported to possess anti-glycating activity. Thus, the daily consumption of dietary components, mainly from plant sources which have an antioxidant effect, is considered to be of potential benefit for prevention of diabetes and diabetic complications [47]. For example ethanol fractions of *Melissa officinalis, L* (Lemon balm) were shown to possess high inhibitory effect on the formation of advanced glycation end products in the late stage of the glycation process [48]. Green tea consumption (drinking) also significantly reduced the advanced glycation, the accumulation of AGEs and the cross-linking of tail tendon collagen in diabetes [49]. Moreover, phenolics, particularly flavonoids, are responsible for the anti-glycation activity of herbal infusions [50]. Another beverage consumed worldwide, the coffee, is also rich in phenolic compounds, mainly caffeoylquinic, p-coumaroylquinic, feruoylquinic and dicaffeoylquinic acids which inhibit protein glycation and dicarbonyl compounds formation [51]. Except polyphenols which constitute a major group of plant-derived compounds with anti-glycation activity, some amino acids [52-54], triterpenes and saponins [55, 56], polysaccharides and oligosaccharides [43, 44, 57, 58] were shown to decrease the AGEs formation. Taurine, a sulfur amino acid, was shown to reduce acrylamide production in potato chip models suggesting its potential use in food processing to decrease acrylamide formation [53]. Another amino acid, arginine, was shown to have an immunomodulatory effect and to inhibit AGEs formation in *in vitro* studies [54].

3. Polyphenols

The anti-glycation capacity of numerous medicinal herbs and dietary plants was comparable with [59], or even stronger than that of aminoguanidine [42, 60, 61-63]. Several studies have demonstrated that the anti-glycation activity correlates significantly with the phenolic content of the tested plant extracts [5, 50, 59, 64, 65]. Polyphenols are the most abundant dietary antioxidants, being common constituents of fruits, vegetables, cereals, seeds, nuts, chocolate, and beverages, such as coffee, tea, and wine. They have been shown to lead to many health benefits, such as prevention of cancer [66], neurodegenerative diseases [67], cardiovascular diseases [68] and diabetes [69].

Although polyphenols are chemically characterized as compounds with phenolic structural features, this group of natural products is highly diverse and contains several sub-groups of phenolic compounds. The diversity and the wide distribution of polyphenols in plants have led to different ways of categorizing these naturally occurring compounds. Polyphenols have been classified by the source of origin, biological function, and chemical structure. Also, the majority of polyphenols in plants exist as glycosides with different sugar units and with sugars acylated at different positions of the polyphenol skeletons [70]. According to the chemical structure of the aglycones, polyphenols are subdivided into the following groups:

1. **Phenolic acids** are among the most important non-vitamin antioxidant phytochemicals naturally present in almost all vegetables and fruits. Their biological activity is related to their lipophilicity and is influenced by the presence of ring substituent hydroxyl groups and, in the case of polyhydroxylated phenolic esters, by the length of the ester moiety [71]. Phenolic acids are non-flavonoid polyphenolic compounds which can be divided into two main types, benzoic acid and cinnamic acid derivatives based on C1–C6 and C3–C6 backbones (Figure 1) [70].

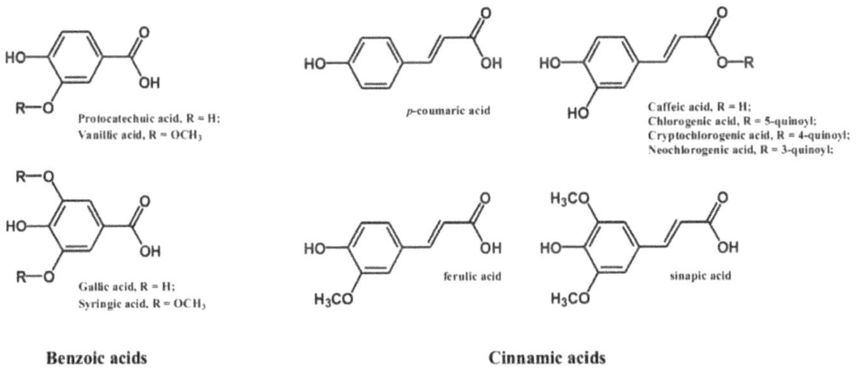

Figure 1. Phenolic acids: Left, Benzoic acids; right, Cinnamic acids. Tsao R (2010).

Cinnamic acids and derivatives

Caffeic acid is a naturally occurring cinnamic acid, found in many vegetables and herbs, e.g. coffee, pear, basil, oregano and apple [72]. In 2009 Gugliucci et al. demonstrated that caffeic acid in *Ilex paraguariensis* extracts inhibits the generation of fluorescent AGEs in *in vitro* experiments [65]. Moreover, extracts from two *Chrysanthemum* species (*C. morifolium* R. and *C. indicum* L.) demonstrated marked inhibition of the formation of AGEs and CML in *in vitro* model systems [42]. The plant extracts inhibited the formation of total AGEs after one week of incubation in BSA/glucose and BSA/fructose systems. Furthermore, the inhibitory effect of the *Chrysanthemum* extracts at a concentration of 5.0 $mg.ml^{-1}$ was stronger than that of AG at a concentration of 1 mM as a positive control. The active components in these plants were characterized by liquid chromatography-diode array detector-atmospheric pressure chemical ionization/mass spectrometry, which showed that *C. indicum* L. conatins large amounts of caffeic acic, as well as luteolin and kaemferol. The other *Chrysanthemum* species (*C. morifolium* R.) contains chlorogenic acid, flavonoid glucoside varieties, and apigenin [42]. **Chlorogenic acids**, esters formed between certain *trans* cinnamic acids and (-)-quinic acid, are the major phenolic compounds in coffee, strawberries, pineapple, apple, and sunflower. **5-caffeoylquinic acid (5-CQA)** is the only chlorogenic acid commercially available and has been extensively studied due to its antioxidant activity. Chlorogenic acids are free radical and metal scavengers, and along with other biological activities they may interfere with glucose absorption and have been shown to modulate gene expression of antioxidant enzymes [73]. Coffee fractions, in which chlorogenic acids are the main compounds, have been shown to inhibit the formation of CML in a concentration-dependent manner. In addition polyphenols such as *caffeoylquinic, p-coumaroylquinic, feruloylquinic* and *dicaffeoylquinic* acids contributed to about 70% of the antioxidant capacity of the coffee fractions [51]. Notably, *Ilex paraguariensis*, like coffee, contains a high concentration of caffeic acid, mostly esterified as chlorogenic acids [65]. Large amounts of chlorogenic acid were identified also in *Chrysanthemum morifolium* R. [42].

Recently, Jang et al. (2010) isolated three **quinic acid derivatives** from ethyl acetate soluble extract of the leaves and stems of *Erigeron annuus*. The structures of these compounds were identified as **3-caffeoylquinic acid, 3,5-di-O-caffeoylquinic acid methyl ester, and 3,5-di-O-caffeoyl-epi-quinic acid.** The last compound exhibited the most potent inhibitory activity against AGEs formation (IC_{50} value of 6.06 μM *vs.* 961 μM for AG) and prevented opacification of rat lenses, while **3-caffeoylquinic acid** (a **monocaffeoylquinic acid**) was not effective. **Two caffeoyl erigerosides and a sucrose ester** also were more effective AGEs inhibitors than AG. This is the first report on **3,5-di-O-caffeoyl-epi-quinic acid** as an inhibitor of RLAR (rat lens aldose reductase), AGEs formation, AGEs-BSA cross-linking, and cataractogenesis [62].

Ferulic acid (FA) is another cinnamic acid that occurs naturally and which is present in drinks and foods, e.g., rice, wheat, oats and some fruits and vegetables [58]. In 2002 Kikuzaki et al. reported that ferulic acid possesses free radical scavenging properties toward hydroxyl radicals, peroxynitrite and oxidized low-density lipoprotein *in vitro* [74]. It has been also shown that ferulic acid can bind human serum albumin (HSA) to form complexes [75]. This interaction led to a significant reduction of the HSA α-helix structure and caused structural changes to the protein providing unusual protective effects against protein oxidation. Silván et al. (2011) reported that the addition of ferulic acid reduces the formation of CML and fluorescent AGEs *in vitro* by nearly 90% [76]. It was shown that the presence of ferulic acid in samples containing proteins and fructose prevents the blocking of free amino groups by about 15% and 30% in soy glycinin and BSA glycation model systems, respectively. Based on previously published results, as well as on the latest findings regarding ferulic acid, Silván et al. (2011) concluded that FA might prevent AGE formation by some of the following ways: acting as an antioxidant, binding amino groups, and inhibiting sugar autoxidation and early Maillard Reaction Products (MRP) degradation [76]. However, the exact mechanism of anti-glycation by ferulic acid demands further investigations.

In 2011 Miroliaei et al. reported the presence of **rosmarinic acid,** a dimmer of caffeic acid, in *Melissa officinalis* L. extract. They demonstrated that treatment of BSA with this herb resulted in a profound prevention of structural changes caused by D-glucose keeping the protein molecule close to its native polar conformation. The extract has the potential to arrest changes in the α-conformers by concealing the glycation sites and lowering the extent of the solvent-accessible surface area, thereby producing barriers for cross β-structure formation. The behavior of the balm extract in this respect resembles that of molecular chaperones which block the hydrophobic surfaces of substrate proteins. Moreover, when albumin molecules were treated with glucose in the presence of balm extract, a lower affinity of glycated BSA for RAGE receptors was observed. Based on the above experimental data, researchers concluded that the herb extract, possessing chaperone-like activity, would afford a protective effect against AGE-induced toxicity by suppression of receptor signaling pathways (e.g. RAGE antagonists) [48]. Also, Ma et al. (2011) reported that rosmarinic acid, isolated from *Salvia miltiorrhiza Bge,* has a more potent inhibitory effect against the formation of AGEs in α-glucosidase (IC_{50} 0.04 μM) than the positive control (AG with IC_{50} of 0.11 μM) [63].

Benzoic acid derivatives

Three derivatives of gallic acid: **ethyl gallate, pentagalloyl glucose,** and **protocatechuic acid** were isolated from ethyl acetate fraction of *Rhus verniciflua* extracts [77]. These gallic acid derivatives have been shown to inhibit recombinant human aldose reductase as well as the accumulation of advanced glycation endproducts in BSA-glucose model system.

In 2007 Ardestani et al. reported that the *Cyperus rotundus* extract (CRE) has a potent antioxidant activity and chelating properties. CRE inhibits high fructose-induced oxidative damage to protein in a dose-dependent manner by decreasing protein carbonyl (PCO) formation and preserving protein thiols from oxidation [59]. Recently, RP-HPLC analysis of *C. rotundus* revealed the presence of phenolic compounds such as **gallic acid**, **p-coumaric acid** (a typical cinnamic acid), and **epicatechin** (flavanol) [78]. Accordingly, the potent inhibitory activity of *C. rotundus* on AGEs formation and protein oxidation might be related to its polyphenolic content.

A new gallic acid-derivative, 7-*O*-galloyl-D-sedoheptulose (GS), was identified in *Cornus officinalis* (2007). This polyphenolic compound showed beneficial effect on the early stage of the diabetic kidney disease [79]. GS reduced renal glucose, AGE formation, and oxidative stress in diabetic rats. Moreover, GS did not show any toxicity at 20 and 100 mg, and reduced Maillard reaction-induced CML *via* the marked inhibition of mitochondrial lipid peroxidation. It also effectively ameliorates the increases in serum creatinine and urinary protein to nearly normal levels [80].

2. Flavonoids

Flavonoids have the C6–C3–C6 general structural backbone in which the two C6 units (Ring A and Ring B) are of phenolic nature (Figure 2)

Due to the hydroxylation pattern and variations in the chromane ring (Ring C), flavonoids can be further allocated to different sub-groups such as **anthocyanins, flavan-3-ols, flavones, flavanones and flavonols.** While the vast majority of the flavonoids have their Ring B attached to the C2 position of Ring C, in some flavonoids such as **isoflavones and neoflavonoids**, Ring B is connected at the C3 and C4 position of Ring C, respectively.

Since beans, particularly soybean, are a major constituent of the diet in many cultures, the role of isoflavones has, thus, great impact on human health [70]. Flavonoids are present in various kinds of vegetables, tea, and red wine [81]. Numerous flavonoids are well-known antioxidants, effective in trapping free radicals, and in such a way participating in maintaining the overall plant cell redox homeostasis [82]. The primary structure of flavonoids (three benzene rings with one or more hydroxyl groups) is the key factor determining their antioxidation capacity. The antioxidation activity may involve the ability of flavonoids to scavenge free radicals, chelation of transition metal ions, sparing of LDL associated antioxidants, and binding to macromolecules or interaction with other kinds of antioxidants [83].

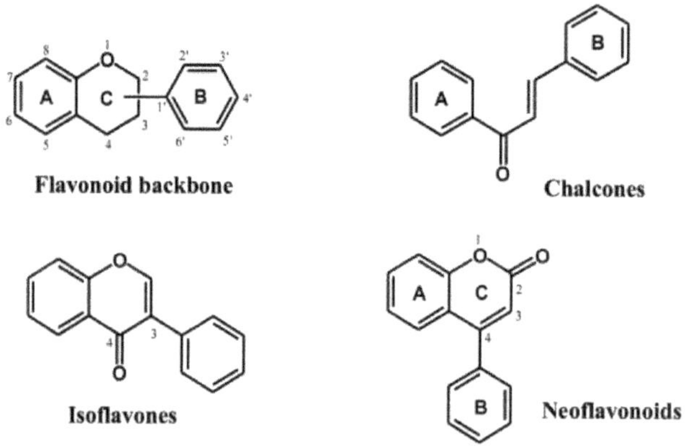

Figure 2. Flavonoid structures. Tsao R (2010).

The antioxidant activity of flavonoids has been demonstrated in many lipid systems. Therefore, they are speculated to have potential in atherosclerosis prevention [81]. The antioxidant capacity of flavonoids depends on both their structure and glycosylation pattern. *Cuminum cyminum* (CC), commonly known as Jeera, was found to contain 51.87% w/w **flavonoids**, which were proposed to be responsible for its antiglycation property. Researchers have shown that treatment of streptozotocin-diabetic rats with CC reduced the renal oxidative stress and AGEs accumulation by increasing the antioxidant defense and reducing the free radical induced lipid peroxidation. The antioxidant activity of superoxide dismutase, catalase and reduced glutathione, increased upon CC treatment. Further experiment revealed that the antihyperglycemic effect of CC may be due to protection of surviving pancreatic β cells, and increase in insulin secretion and glycogen storage [84].

Flavonoids are also abundant in honeys, and honeys rich in flavonoids, such as buckwheat honey, exhibited higher antioxidant activity than flavonoid-poorer honeys such as acacia honey [85]. Raw Millefiori honey, for example, is packed full of antioxidants [86-88]. In addition to the direct contribution to the radical scavenging activity, the polyphenolic content also influences the honey color. Conjugated systems of double bonds, such as those present in flavonoids, terpenes, isoprene units and long chain phenolic acids present in honey, constitute chromophores that absorb photons of visible light giving rise to colors ranging from yellow to brown. Several reports point out the positive and highly significant correlation between honey color, phenolic content and antioxidant activity [85, 87, 88].

Chalcones, though lacking the heterocyclic Ring C, are still categorized as members of the flavonoid family [70]. The chalcone **butein** isolated from ethyl acetate fraction of *Rhus verniciflua* proved to be a potent inhibitor of Recombinant Human ALR2 (rhALR2) with an IC_{50} value of 0.7 μM. Butein also strongly inhibits the advanced glycation end products accumulation *in vitro*. It has been reported that the hydroxyl groups at the 3′-, 4′-, 5-, and 7-positions of flavones increase their AGE inhibitory activities, whereas the methylation or glycosylation of the 3′- or 4′-hydroxyl group reduces this activity [89]. In agreement with this report, the open ring form of 3-, 4-, 2′-, and 4′-tetrahydroxylated flavone, butein, has an increased AGE inhibitory activity. From these results the conclusion could be drawn that the inhibitory effect of *Rhus verniciflua*, and especially of the ethyl acetate fraction on rhALR2, is possibly exerted by butein acting as an active component.

In apple trees (*Malus domestica*), the major sub-family of flavonoids is represented by **dihydrochalcones**, which are found in large amounts (up to 5% of dry weight) in leaves and in immature fruits [90]. Among the known dihydrochalcones, **phloridzin** and its aglycone, **phloretin**, are simple forms [91] and their biosynthesis in Malus has been recently described [92]. Bernonville et al. (2010), demonstrated the presence of **phloridzin** alone or in combination with two additional dihydrochalcones, identified as **sieboldin** and **trilobatin** [60]. **Phloridzin** was shown to inhibit glucose intestinal absorption and renal resorption, resulting in normalization of blood glucose and overall diminution of glycaemia in animal models [93]. Based on the results of the antioxidant assays and the fact that **sieboldin** was several folds more efficient than **phloridzin** in inhibiting AGE formation (10-fold lower IC_{50} than phloridzin and 40-fold lower IC_{50} (0.2 mM) than AG), a role for **sieboldin** in inhibiting the formation of intermediate glycation products is suggestive [60].

Isoflavones

Isoflanones have their ring B attached to the C3 position of ring C. They are mostly found in the leguminous family of plants [70]. Soy products and soybeans are particularly abundant sources of **isoflavones** which have both antioxidant and phytoesterogenic activities that may contribute to their potential anticarcinogenic and cardioprotective effects [94-96]. High soybean consumption has been implicated in the longevity of the Japanese [97]. Genistein and daidzein are the two main isoflavones in soy along with glycetein, biochanin A and formononetin [98]. In 2009 Hsieh et al. reported that soy isoflavones supplementation significantly and dose-dependently decreased the concentration of protein carbonyls in the liver, kidney and brain in D-galactose treated mice. Soy isoflavones administration effectively attenuate oxidative damage and improve parameters related to aging and Alzheimer's disease [99].

Puerarin (daidzein-8-C-glucoside) is an isoflavone glycoside isolated from the root of *Pueraria lobata* and has various pharmacological effects, including anti-hyperglycemic and anti-allergic properties [100-102]. Additionally, puerarin has been reported to effectively inhibit advanced glycation end products formation which is one of the typical risk factors for diabetic complications [103]. In 2010 Kim et al. reported that puerarin administration to mouse mesanglial cells increased heme oxygenase-1(HO-1) protein levels in a dose-dependent manner [104]. This enzyme participates in conversion of heme to biliverdin, which is rapidly metabolized to bilirubin, a potent antioxidant [105]. Moreover, puerarin treatment was able to enhance the phosphorylation of protein kinase C δ-subunit which primarily regulates the expression of HO-1, which in turn inhibited AGE-induced inflammation in mouse mesanglial cells.

Flavones, Flavonols, Flavanones and Flavanonoles

These flavonoid subgroups are the most common and almost ubiquitous throughout the plant kingdom. (Figure 3)

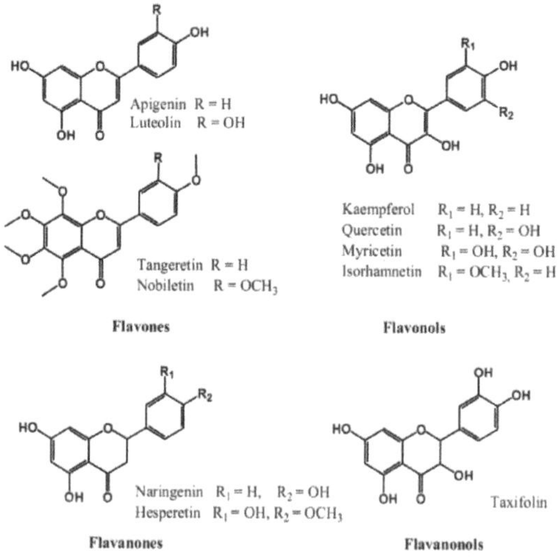

Figure 3. Flavones, Flavonols, Flavanones and Flavanonoles. Tsao R (2010).

Flavones and their 3-hydroxy derivatives flavonols, including their glycosides, methoxides, and other acylated products, make this the largest subgroup among all polyphenols [70]. The most common flavonol aglicones, *quercetin and kaemferol,* alone have at least 279 and 347 different combinations, respectively [106-108].

Kaempferol is a well known anti-oxidant flavonol aglycone that possesses anti-inflammatory properties resulting from its ability to diminish the formation of reactive species (RS) [109]. **Kaempferol** was detected in plant extracts of two *Chrysanthemum* species [42] and *Erigeron annuus* [110]. *Kaempferol* causes the inhibition of the inducible nitric oxide synthase and cyclooxygenase-2 and the down-regulation of NF-κB pathway [111]. In 2010 Kim et al. demonstrated that the short-term feeding of aged rats with kaempferol modulated both AGE accumulation and RAGE expression which are dependent on NF-κB transcriptional activity. Furthermore, kaempferol suppressed age-related NF-κB activation and its pro-inflammatory genes through the suppression of AGE-induced NADPH oxidase activation [112]. **Quercetin** is another example of flavonol aglycone found in citrus fruit, buckwheat and onions. Many researchers examined the ability of *quercitrin* to protect against protein damage (AGEs formation) using *in vitro* model systems [77, 113]. Both *quercetin* and *kaempferol* together with *kaempferol-3-O-rutinoside* were reported as the main polyphenols present in crude extracts of aerial parts of *Cassia auriculata*. Ethyl acetate fraction of this medicinal plant showed radical scavenging activity and inhibition of lipid peroxidation. The guava leaves extracts also are a very good source of phenolic compounds such as *gallic acid, ferrulic acid, quercetin and quercetin derived glycosides*. The phenolic compounds of guava leaf extracts significantly decreased fasting blood glucose levels in streptozotocin-induced diabetic rats, decreased glycation products, lipid peroxidation and improved the antioxidant status in a dose-dependent manner [114]. Thus, the effect of guava leaves on glycation may be due to the different composition of the phenolic compounds. The latter also showed strong inhibitory effects on the glycation of albumin, especially quercetin exhibited over 95% inhibitory effect at a concentration of $100 \mu g.ml^{-1}$. The anti-glycation activity of the aqueous extracts of guava was higher than that of AG and green tea polyphenols [115]. More interestingly, flavonoids with the 3',4'-dihydroxy group (i.e., **quercetin** and **quercitrin**) demonstrated the highest inhibitory activity against AGEs formation after incubation at 37°C for 14 days, with IC_{50} values much lower than that of AG [111]. Besides *quercetin,* the flavonoid glycosides *isoquercitrin* (quercetin-3-β-glucopyranoside) and *hyperin* (quercetin-3-D-galactoside) are well-known antioxidants [116-118]. *Isoquercitrin* showed outstanding antioxidant activity in yeast cells by increasing the activity of superoxide dismutase (SOD) [118]. It also demonstrated free radical scavenging properties [119]. Marzouk et al. (2006) reported that hyperin strongly inhibited the formation of 1,1-diphenyl-2-picrylhydrazyl (DPPH) free radicals [117] and lipid peroxidation, as well as the hydroxyl radical and superoxide anion generation [120]. Also, hyperin inhibits lipopolysaccharide (LPS)-induced nitric oxide (NO) production [120]. It is worth mentioning that *isoquercitrin* and *hyperin* showed a dose-dependent inhibitory activity against the formation of AGEs which was stronger than that of the AG positive control. Thereby, hyperin demonstrated a greater effect by inhibiting AGEs formation by 92% as compared to isoquercitrin, which inhibited the accumulation of AGEs by 89.6%.

In 2011 Manaharan et al. reported the presence of **quercetin-3-O-β-D-galactopyranoside** in ethanolic extracts of *Peltophorum pterocarpum* leaves as the major bioactive compound. *Peltophorum pterocarpum* leaf and bark extracts were shown to inhibit aldose reductase far better than the pure compound quercetin [121]. The plant leaf and bark extracts were found to be about 28-fold and 56-fold more effective than quercetin, respectively in inhibiting aldose reductase which points to their potential use in hyperglycemia treatment. Also, the HPLC profiles of the active ethyl acetate fraction from *Nelumbo nucifera* leaves indicated the presence of **quercetin 3-O-b-D-glucopyranoside** and **quercetin 3-O-b-D-**

glucuronopyranoside. In terms of *N. nucifera's* antioxidant effect, the leaf extract exhibited potent antioxidant capacities in the DPPH and total ROS assay. The leaf extracts also showed remarkable inhibitory activities on RLAR and AGE formation [35].

Quercetin-3-O-rutinoside (Rutin), a common dietary flavonoid is an established antioxidant. It is found in fruits, vegetables and plant-derived beverages such as tea and wine [122]. Gut microflora in the large intestine metabolize rutin to a variety of compounds that include *quercetin* and phenol derivatives such as 3,4-dihydroxyphenylacetic acid (DHPAA), 3,4-dihydroxytoluene (DHT), 3-hydroxyphenylacetic acid (HPAA), and 4-hydroxy-3-methoxyphenylacetic acid (homovanillic acid, HVA) [122-124]. Rutin metabolites, particularly those that include vicinal hydroxyl groups in their structure such as 3,4-dihydroxyphenylacetic acid (DHPAA) and 3,4-dihydroxytoluene (DHT), are powerful inhibitors of the formation of CML and fluorescent derivatives (370-440 nm and 335-385 nm) in histone H1 caused by ADP-ribose. The plasma concentrations of these rutin metabolites are expected to effectively neutralize the reported plasma concentrations of glyoxal and methylglyoxal [125]. Rutin was also found to inhibit the formation of glycation products in collagen type I induced by glucose *in vitro* [126] and to be an effective inhibitor of lipoprotein glycation by increasing the resistance of LDL to HG/Cu (II)-mediated oxidation [127].

In 2009 Tsuji-Naito et al. reported that **apigenin** (*4',5, 7-trihydroxyflavone*) in *C. indicum* L. is a minor flavonoid aglycone, although *apigenin* in *C. morifolium* R. is the main component of the plant extract which inhibits AGEs accumulation. Large amounts of another flavone – **luteolin** (*2-(3,4-dihydroxyphenyl)-5,7-dihydroxy-4-chromenone)* were found in *C. indicum* L [42]. **Apigenin, luteolin, apigenin-7-O- β -D-glucuronide methyl ester** and **apigenin-7-O- β -D-glucuronide** were indentified in the ethyl acetate-soluble extract of flowers of *Erigeron annuus* [110]. This is the first report on apigenin-7-*O*-*β*-D-glucuronide and apigenin-7-*O*-*β*-D-glucuronide methyl ester having significant inhibitory activity towards aldose reductase and AGEs formation, which makes it worth to further study their potential for treatment of diabetic complications. The presence of *luteolin* together with *maysin* was also reported in the silk of *Zea mays*. These flavonoids are abundant in corn silk and *in vitro* glycation studies demonstrated their role in inhibiting AGE formation with **luteolin** being exceptionally active [61].

Hispidulin is a flavone compound from *Artemisia campestris ssp.* glutinosa [128], while **vitexin** (*apigenin-8-C-glucoside*) and **isovitexin** (*apigenin-6-C-glucoside*) are flavone C-glucosides, which have been identified in mung bean extract. The use of *A. campestris* has been recommended in Tunisian folk medicine for their antivenom [129], anti-inflammatory, antirheumatic and antimicrobial activities [130]. Recent studies provide for the first time data on the effect of an ethyl-acetate fraction from *A. capillaries* on the oxidative stress and antioxidant enzymes in high-fat diet induced obese and type 2 diabetic mice [131]. In 2010 Sefi et al. demonstrated that administration of an aqueous extract of *A. campestris* to diabetic rats increased significantly serum insulin levels, reduced serum glucose level by 60% ($p < 0.001$) and tended to bring the glucose value to near normal after 21 days. The ability of *A. campestris* extracts to reduce the blood glucose level could be attributed to a stimulation of langerhans islets, to an improvement of the peripheral sensitivity to remnant insulin, and to the strong antioxidant properties of the plant compounds [132].

Vitexin and isovitexin have been identified in other plants, such as bamboo leaves [133] and pigeonpea leaves [134]. Their anti-glycation activity could be attributed to their free radical scavenging and/or metal ion trapping activities, as they failed to directly trap reactive carbonyl species, such as methylglyoxal [64].

The number of **flavanones**, and their **3-hydroxy derivatives (flavanonols, which are also referred to as dihydroflavonols)** identified in the last 15 years has significantly increased. Some **flavanones** have unique substitution patterns, e.g., prenylated flavanones, furanoflavanones, pyranoflavanones, benzylated flavanones, giving a large number of substituted derivatives of this subgroup [135]. A new compound that was designated as **4`-O-[β -D-apiosyl (1→2)] - β -D-glucopyranosyl] - 5-hydroxyl-7-O-sinapylflavanone** was isolated from *Viscum album* (European Mistletoe). This compound together with previously identified compounds in *V. album*, **5,7-dimethoxy-4`-O-β -D-glucopyranoside flavanone, 4`,5-dimethoxy-7-hydroxy flavanone** and **5,7-dimethoxy-4`-hydroxy flavanone**, showed a potent anti-glycation activity, *i.e.* 72.5% (IC_{50} = 199.85 ± 0.067 mM) as well as superoxide anion scavenging capacity. The antioxidant potential of **4`,5-dimethoxy-7-hydroxy flavanone** (IC_{50} = 58.36 ± 2.9m M) was determined to be greater than that of *rutin* used as a standard [41]. Recently, Jang et al. (2010) isolated from the flowers of *E. annuus* a novel **2,3-dioxygenated flavanone**, **erigeroflavanone**, which was also shown to possess a strong anti–AGE activity [110]. In addition, the presence of **fustin** (2-(3, 4-dihydroxyphenyl)-3,7-dihydroxy-2,3-dihydrochromen-4-one), a flavanonol together with quercetin, morin (flavonol), and butein was reported in an ethyl-acetate fraction of *Rhus verniciflua*. All these flavonoids, especially those with hydroxyl groups at the 3′-,4′-,5′-, and 7-positions have shown a significant inhibitory activity against AGE in *in vitro* experiments [77].

In 2007 two **dihydroflavonol glycoside, engeletin** and **astilbin**, were isolated from an ethyl acetate extract of the leaves of *Stelechocarpus cauliflorus* R.E. Fr. The inhibitory activity of **engeletin** against a recombinant human aldose reductase (IC_{50} value = 1.16 μM) was twice that of **quercetin** used as a positive control (2.48 μM), and 23 times greater than that of **astilbin** (26.7 μM). **Engeletin** was shown to inhibit the aldose reductase uncompetitively. On the other hand, in contrast to its inhibitory activity against AR, **astilbin** was more potent than **engeletin** in suppressing AGE formation. Moreover**, astilbin** was almost as potent as the positive control, quercetin, in inhibiting advanced glycation end-products accumulation. Interestingly, the only structural difference between **engeletin** and **astilbin** is the number of hydroxyl groups in the B ring. Of both compounds only **astilbin** has the catechol orientation. Both **astilbin** and **taxifolin (2, 3-dihydro quercetin)**, its aglycone, have been demonstrated to protect against oxidative damage [136]. Therefore, the antioxidant flavonoids such as **engeletin** and **astilbin** are potentially useful for therapeutic prevention of diabetic complications resulting from AGEs accumulation [137].

Flavanols and procyanidins

Flavanols or flavan-3-ols are often called catechins (Figure 4). They differ from most flavonoids in that they do not have a double bond between C2 and C3, and there is no C4 carbonyl in ring C [70]

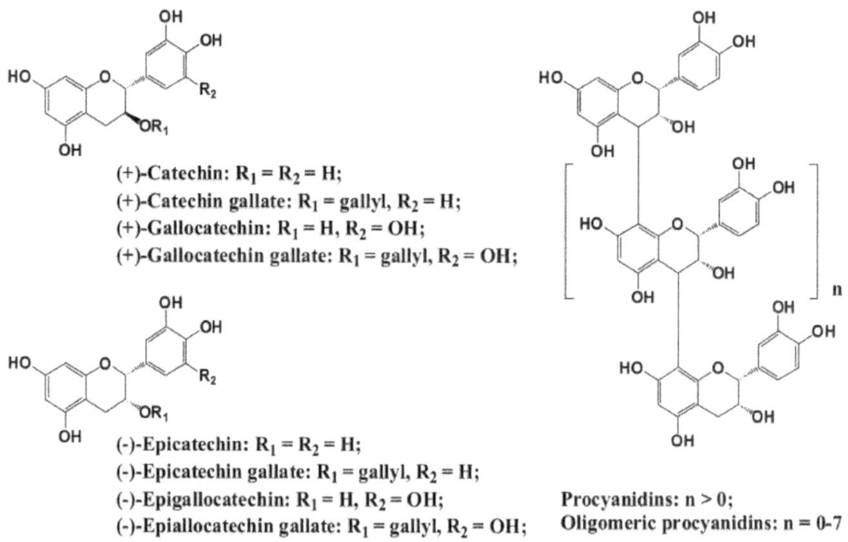

Figure 4. Flavanols and procyanidins. Tsao R (2010).

Catechin is the isomer with *trans* configuration and **epicatechin** is the one with *cis* configuration. Each of these two configurations has two steroisomers, *i.e.*, **(+)-catechin, (–)-catechin, (+)-epicatechin and (–)-epicatechin. (+)-Catechin and (–)-epicatechin** are the two isomers often found in food plants. Catechin and epicatechin can form polymers, which are often referred to as **proanthocyanidins** because an acid-catalyzed cleavage of the polymeric chains produces **anthocyanidins** [70]. The presence of **catechins** was reported in green tea, which is an excellent source of many polyphenol antioxidants [49]. The most important catechins of green tea are **(-)-epicatechin (EC), (-)-epicatechin- 3-gallate (ECG), (-)-epigallocatechin (EGC) and (-)-epigallocatechin-3-gallate (EGCG)** [138]. Nearly 80% of the extract of green tea is a mixture of **catechins** namely **epigallocatechin** (EGC), **epicatechin (EC), epigallocatechin-3-gallate (EGCG)** and **epicatechin-3-gallate (ECG)**. The sum of **EGC** and **EGCG** weighed more than 70% of the catechin mixture in the green tea extract [49]. It was shown that green tea treatment of diabetic rats significantly reduced the blood glucose level. This antihyperglycemic effect may be linked to enhanced basal and insulin-stimulated glucose uptake in rat adipocytes [139], inhibition of the intestinal glucose transporter [140], and decreased expression of genes that control gluconeogenesis [141]. Green tea supplementation also reduced the accumulation of AGEs in diabetic rats as indicated by decreased collagen linked fluorescence [49]. In 2009 Rasheed et al. reported that EGCG significantly decreased AGE-stimulated gene expression and the production of TNFα and matrix metalloproteinase-13 (MMP-13) in human chondrocytes. The inhibitory effect of EGCG on the AGE-BSA-induced expression of TNFα and MMP-13 was mediated at least in part *via* suppression of p38-MAPK and activation of JNK. In addition, EGCG inhibited the phosphorylating activity of IKKβ kinase in an *in vitro* assay and also the AGE-mediated activation and DNA binding activity of NF-κB by suppressing the degradation of its inhibitory protein IκBα in the cytoplasm [142]. EGCG has also been shown to prevent intracellular AGEs formation and the production of proinflammatory cytokines in monocytes under hyperglycemic conditions [143]. EGCG is capable of trapping reactive dicarbonyl species, such as methylglyoxal and glyoxal, as demonstrated by Ho and coworkers in 2007 [144]. Data from HPLC-DAD demonstrated that EGCG was able to bind lipoproteins and to enhance the antioxidant and antiglycation properties of LDL [145].

Proanthocyanidins are traditionally considered to be condensed tannins and depending on the interflavanic linkages, oligomeric proanthocyanidins can have A-type structure in which monomers are linked through C2–*O*–C7 or C2–*O*–C5 bonding, or B-type in which C4–C6 or C4–C8 bonds are common (Figure 4) [70]. *Catechin and epicatechin procyanidin B2 (a dimer-type proanthocyanidin)* isolated from cinnamon bark extract have been shown to possess a significant MGO trapping activitiy, with *procyanidin B2* demonstrating the strongest inhibition of AGE formation among *proanthocyanidins* isolated from cinnamon bark [146]. All these flavonoids potently inhibited (more than 50%) the formation of pentosidine and CML [147] with *procyanidin B2* showing the highest inhibition capacity (almost 80%) on the CML formation. **Proanthocyanidins** could further abate the MGO mediated formation of cross-links in creatine kinase in a dose-dependent manner. They also exerted various protective effects on glucose consumption impaired by high MGO concentrations through potential interaction with proteins involved in insulin signaling pathways [147]. **Proanthocyanidin B-4** and two more **nitrogen containing flavonoids** were identified for the first time in *Actinidia arguta* [148]. The N-containing flavonoids comprise a very small class of natural products, which have been only rarely isolated from natural sources. The ^{1}H- and ^{13}C-NMR spectral data revealed that these newly isolated compounds are **6- and 8-(2- pyrrolidinone-5-yl)-(-)-epicatechins,** which may be produced by condensation between (-)-epicatechin and 5-hydroxypyrrolidin-2-one under acidic conditions. **Proanthocyanidin B-4** and the two novel compounds showed a significant activity against AGEs formation with IC_{50} values for **proanthocyanidin B-4** of 10.1 μM, which is lower than that of the well known glycation inhibitor AG [148].

Geraniin, an **ellagitannin**, is the major **tannin** in *Geranium thunbergii*. It was also identified as the major bioactive compound in an ethanolic *Nephelium lappaceum* L. rind extract [149]. Previous studies have shown that *N. lappaceum* rind extract exhibits high anti-oxidant activity [150]. In 2011 Palanisamy et al. reported the ability of *geraniin* to scavenge free radicals and to possess *in vitro* hypoglycemic activity [149]. *Geraniin* is also an excellent inhibitor of carbohydrate hydrolysing enzymes (α-glucosidase and α-amylase) – superior to the positive control acarbose (carbohydrate hydrolysis inhibitor). It was far more effective in preventing polyol and advanced glycation endproducts formation as compared to the positive controls quercetin and green tea which reveals geraniin as an ideal candidate for the management of hyperglycemia in diabetic individuals [149].

4. Other phenolic compounds

A stilbene glucoside - *2,3,5,4′-tetrahydroxystilbene 2-O-β-D-glucoside (THSG)* is a natural compound with strong antioxidative and anti-inflammatory properties, which has been reported as the major bioactive compound from *Polygonum multiflorum* Thunb., a traditional Chinese herbal tea [151,152]. It was shown to efficiently inhibit the formation of AGEs in a dose-dependent manner by trapping reactive MGO under physiological conditions (pH 7.4, 37°C) [153]. More than 60% of MGO was trapped by THSG within 24 hours and THSG was much more effective than resveratrol and its methylated derivative pterostilbene (two major bioactive stilbenes) [153]. In 2011 **Chompoo et al.** isolated two previously described compounds from *Alpinia zerumbet,* namely **5,6-dehydrokawain (DK)** and **dihydro-5,6-dehydrokawain (DDK)** which are kawalactones. DK and DDK were present in all six different parts of the plant but rhizomes had higher inhibitory activity against AGEs formation than the other parts [154]. A previous study also provided data about the antioxidant acitivities of DDK present in leaves and rhizomes of *A. zerumbet* [155]. Among the compounds isolated from *Alpinia zerumbet* rhizomes, DK had the strongest inhibitory activity against BSA glycation with IC_{50} value of 15.9 μM. DK has been also shown to inhibit human platelet aggregation and to possess anti-inflamatory and cancer chemoprotective therapeutic properties [156].

5. Terpenes, carotenoids and polyunsaturated fatty acids

A terpene, 8(17),12-Labdadiene-15,16-dial (labdadiene) was isolated for the first time from the rhizome of *Alpinia zerumbet* together with **5,6-dehydrokawain (DK)** and **dihydro-5,6-dehydrokawain (DDK)** [154]. In contrast to DK which strongly inhibited AGEs formation in BSA **labdadiene** markedly suppressed the fructosamine adduct formation with IC_{50} = 51.1 μg/mL. **Labdadiene** was also more efficient than DK in inhibiting glycation-induced protein oxidation and the formation of α-dicarbonyl compounds, at first place preventing glyoxal accumulation. It is possible that the aldehyde groups of labdadiene have a significant role in inhibiting AGEs formation. These aldehyde groups may compete with sugars for Schiff's bases formation and/or limit the amount of amines available for glucose attachment. The fructosamine assay revealed that **labdadiene** has strong activity when compared to **rutin** and **quercetin** used as positive controls, although the inhibitory mechanism of **labdadiene** is likely to differ from that of **rutin** and **quercetin** [154].

The ability of microalgal extracts to inhibit AGEs formation differs from that of many other plant species and is not promoted by phenolic compounds. There is a weak correlation between the antiglycative activity and the total phenolic content of several microalgae as demonstrated by the very small correlation coefficient. For example, in total AGEs inhibition, the value of R^2 was 0.035 for ethyl acetate fractions of green microalgae *Chlorella* [45]. In microalgae, a wide range of antioxidants can be produced such as **carotenoids, polyunsaturated fatty acids** and **polysaccharides** [157]. HPLC and gas chromatography (GC) analysis revealed that **carotenoids,** especially **lutein** in *Chlorella* and **unsaturated fatty acids**, mainly of **linoleic acid, arachidonic acid** and **eicosapentaenoic acid** in *Nitzschia laevis* contributed to the strong antiglycative capacities of these species [45]. The green microalga *Chlorella zofingiensis* accumulates primary carotenoids such as **lutein** and **β-carotene** to protect the cells from oxidative damage [158]. Results showed that **lutein** and some **unsaturated fatty acids** effectively inhibited the formation of both total AGEs and specific AGEs *in vitro* in a dose-dependent manner. For **lutein**, at the concentration of 0.8 $mg.ml^{-1}$, the inhibitory efficacy was comparable to or even higher than the effect of 1 mM AG solution [45]. It is noteworthy that if the amount of primary carotenoids is not enough, secondary carotenoids (i.e. **astaxanthin, canthaxanthin and adonixanthin**) are generated to diminish the excessive oxidative stress. The green microalga *Chlorella zofingiensis* is known as a natural source of **astaxanthin,** a red ketocarotenoid that is a potent anti-oxidant. It acts as the major secondary carotenoid and over 90% of the **astaxanthin** is in the form of mono- and di-esters [158]. The antioxidant activity of **astaxanthin** is an order of magnitude higher than that of other carotenoids such as zeaxanthin, **lutein**, canthaxanthin and β-carotene, and 100 times higher than the antioxidant activity of a-tocopherol [159]. It was shown that under heterotrophic conditions the colour of *C. zofingiensis* gradually changed from green to red, indicating the accumulation of astaxanthin within algal cells. *C. zofingiensis* extracts and especially the red one are suggested to scavenge hydroxyl radicals and/or to chelate transition metals. Seven major fractions were obtained from astaxanthin-rich extract of *C. zofingiensis*. HPLC results revealed that they are astaxanthin diester, astaxanthin monoester, adonixanthin ester, free astaxanthin, free adonixanthin, lutein and zeaxanthin, and neoxanthin. The **astaxanthin diester** was found to be the most potent antiglycative compound among all fractions [46].

Several unsaturated fatty acids showed inhibitory activity against AGEs formation. Although palmitoleic and oleic acid were involved in the inhibition of pentosidine formation, the main contributors were linoleic acid, arachidonic acid, and eicosapentaenoic acid [45].

Hydrophobic compounds which also display antioxidant and hypoglycemic activities, such as oleanolic and ursolic acid, were observed in non-polar extracts from medicinal herbs [160]. Ursolic acid and its isomer, oleanolic acid, are triterpenoid compounds found across the vegetal kingdom that have anti-inflammatory, anti-arthritic, cytostatic, antiproliferative, and hepato-protective effects in mice, as well as membrane stabilizing properties [161-163]. One of the major components of yerba maté (*Ilex paraguariensis*) is oleanolic acid (OA) [65]. It has been reported that oleanolic acid has hypoglycemic and hypolipidemic effects in diabetic rats [164]. Oleanolic acid or ursolic acid (UA) intake at 0.1 or 0.2% increased the content of both acids in the kidney, dose-dependently decreased plasma glucose, HbA1c, renal N^{ε}-(carboxymethyl)lysine, urinary glycated albumin and urinary albumin levels. OA or UA intake significantly reduced renal pentosidine and decreased AR activity. The triterpens have been shown also to elevate plasma insulin levels and renal creatinine clearance as well as to decrease renal sorbitol and fructose concentrations [55].

Arjunolic acid (2,3,23-trihydroxyolean-12-en-28-oic acid, AA), a natural pentacyclic triterpenoid saponin isolated from the bark of *Terminalia arjuna,* is well known to display various biological functions, including antioxidative [165], antifungal [166], hepatoprotective [167], and antibacterial activities [168]. AA plays a protective role against hepatotoxicity induced by environmental toxins such as drugs and chemicals. AA was shown to be effective in preventing the formation of reactive oxygen species (ROS), reactive nitrogen species (RNS), HbA1c, AGEs, and oxidative stress signaling cascades. AA also has been reported to protect against poly (ADP-ribose) polymerase (PARP)-mediated DNA fragmentation. Treatment with AA both before and after diabetes, on the other hand, prevented the NO signaling pathways and thereby brought the affected organs back to their physiological state. AA treatment was effective in preventing the phosphorylation of IκBα and NF-κB p65. Aslo, treatment with AA could prevent the hyperglycemia-induced phosphorylation of extracellular signal-regulated kinase (ERK) and p38. It was observed that the antidiabetic as well as antioxidant properties of AA were comparable to those of insulin [56].

6. Polysaccharides

In recent years a lot of attention has been paid to **polysaccharides** because of their unique biological activities [169]. Yang et al. (2009) have confirmed the presence of high quantity of polysaccharides in longan pericarp tissues (*Dimocarpus longan* Lour.) Polysaccharides of longan fruit pericarp (PLFP) have been found to be strong radical scavengers [170]. It is hypothesized that there is a link between the antioxidant and anti-glycative properties of PLFP. On the other hand, polysaccharides are composed of monosaccharides, which can compete with glucose for binding free amino groups in proteins thus lowering the effective concentration of glycation targets in proteins. This might be another pathway in which PLFP inhibits the formation of advanced glycation end products [44]. Moreover, it has been found that the molecular weight of the polysaccharide chain and the antioxidant activity of PLFP can be modified by ultrasonic treatment [44].

In 2009 Chen et al. reported on the ability of *Ganoderma lucidum* **polysaccharides** (GLP) to reduce lipid peroxidation and blood glucose levels in diabetic rats [171]. Administration of middle or high doses of GLP in diabetic mice significantly decreased blood glucose and HbA_{1c} [43]. Blood cholesterol and triglyceride levels were also improved which could be ascribed to the reduced blood glucose levels [172]. GLP administration to diabetic rats further influenced the myocardial hydroxyproline as well as the soluble and insoluble myocardial collagen. After 16-week treatment of diabetic mice with GLP, the cross-linked and non-cross-linked collagens tended to decrease. The AGEs formation was also significantly reduced upon GLP treatment. In addition, the activities of antioxidant enzymes such as SOD, GSH-Px, and CAT from streptozotocin-treated diabetic rats were significantly enhanced after GLP administration [43]. Another study showed that all **polysaccharides from pumpkin** *(Cucurbita moschata)* (PPs) inhibited the formation of dicarbonyl compounds. Two of the ethanolic fractions isolated from *pumpkin*, PPIII and PPII, were proposed to be stronger inhibitors than AG. PPIII was shown to have 65% inhibitory effect at a concentration of 50μM. PPs also inhibited the aldose reductase in a dose-dependent manner [57]. Least but not last, the anti-glycative activity of PPs increased with a decrease in their molecular weight.

7. Other anti-glycative compouds

The major biochemical constituents of *Withania somnifera* roots are steroidal alkaloids and steroidal lactones from the class called **withanolides** [173]. To date, up to 19 **withanolide derivatives** have been isolated from *Withania* roots [174]. Recently, *Withania* and its active components were shown to scavenge free radicals and to inhibit lipid peroxydation [175]. *Withania* have been reported to suppress AGE linked fluorescence of rat's tail tendon collagen, which was explained by its antioxidant and free radical scavenging effect [176].

Melanoidins are another class of antioxidants receiving attention in recent years. These polymeric brown compounds formed in the last stage of the Maillard reaction were supposed to be involved in the color and flavor development of thermally-processed foods. They are present in food and beverages such as coffee, beer, traditional balsamic vinegar, cocoa and bread [177-179]. It seems that during coffee roasting phenolic compounds are involved in the Maillard reaction to partially form the brown, water soluble polymers known as coffee melanoidins. Physiological studies indicated that some of the coffee effects arise not from caffeine but from melanoidins [51, 180]. The high molecular weight compound (HMWC) fraction of coffee was shown to inhibit BSA glycation by acting as radical scavenger and Fe-chelator in the post-Amadori phase of the reaction and by inhibiting the production of dicarbonyl reactive compounds during glucose autoxidation [51]. Verzelloni et al. (2011) noted that this fraction is rich in melanoidins and concluded that melanoidins could be mainly responsible for the anti-glycative activity of the HMWC fraction. Also, the presence of proteins and chlorogenic acids, incorporated in the melanoidins' structure, has been reported [51].

Interestingly, the total phenolic content and honey color are predictive markers of the antioxidant activity in honey [85]. The radical-scavenging activity of honey was higher in honeys with high phenolic content and of darker color [181]. On the other hand, the color of honey may also result from non-enzymatic browning (the Maillard reaction) [182]. The brown, carbohydrates based melanoidin polymers have been shown to possess strong antioxidant activity [183]. It is thought that one of the possible mechanisms by which MRPs may act as both antioxidants and antibacterial agents, is their metal-chelating activitiy [184].

In 2010 Ye et al. uncovered the inhibitory effect of **fermentation byproducts** on AGE formation [185]. Japanese distilled spirit can be prepared from starchy substances, such as rice, barley and sweet potato. Recycled distilled residues (DRs) of rice and barley spirit as well as their vinegars inhibited the formation of N^{ε}(carboxymethyl)lysine (CML), a major

AGE in BSA model system. The high protein levels together with free lysines and arginines present in DRs of rice and barley spirits raise the possibility that these proteinaceous ingredients inhibit AGEs formation by competing with BSA for the glycation reaction. However, the low protein content in DRs of sweet potato spirit, which is accompanied by strong anti-glycation activity, argues against the above suggestion. Distilled residues and their derived vinegars are extremely complex mixtures containing caffeic acid as the dominant phenolic constituent and hence further investigations are required to elucidate the anti-glycative mechanism of DRs [185].

Vitamins and some trace elements such as zinc and selenium are also a part of the human antioxidant defence system which must be delivered by diet. For example, treatment with vitamin E prevents renal hypertrophy in streptozotocin diabetic rats [186]. Similarly, a combination of vitamins E and C administered for 12 weeks decreased lipid peroxidation and augmented the activities of antioxidant enzymes in the kidneys of stretozotocin diabetic rats. The treatment also reduced urinary albumin excretion, decreased kidney weight and reduced the thickness of the glomerular basement membrane [187]. Tocotrienol, a component of vitamin E that may accumulate more effectively in membranes than α-tocopherol [188], was shown to ameliorate experimental nephropathy in STZ-diabetic rats [189]. Pyridoxamine (PM), one of three natural forms of vitamin B_6, was also reported to inhibit the Maillard reaction and PM application in diabetic nephropathy has now progressed to a phase III clinical trials. PM inhibits post-Amadori steps of the Maillard reaction by sequestering catalytic metal ions and blocking the oxidative degradation of Amadori intermediates [190]. Besides vitamin E and PM, **α-lipoic acid** (ALA) also possesses a strong antioxidant properties [191]. In rat soleus muscle, inhibition of glycogen synthesis and acceleration of glucose oxidation, have been correlated with the uncoupling effect of this acid. Thus, ALA may regulate glucose metabolism in muscles in a way which does not mimic the action of insulin. It has been reported that after long-term incubation in cell culture, ALA behaves as an antioxidant, whereas after short-term incubation and quick uptake by cultured cells it may act as a pro-oxidant. By acting as an antioxidant, ALA reduces oxidative stress and the formation of AGEs and improves insulin sensitivity in skeletal muscles and liver.

Recently, the effect of citric acid on the pathogenesis of diabetic complications has been reported. Citrate, a natural, dietary chelator found in citrus fruits [192], is widely used in food products as a preservative and to enhance tartness. Oral administration of citric acid to diabetic rats delayed the development of cataracts, inhibited the accumulation of AGEs such as N^{ε}-(carboxyethyl) lysine (CEL) and CML in lens proteins. Citric acid also inhibited the development of nephropathy (albuminuria) and significantly reduced ketonemia in diabetic rats [193]. On the other hand, the administration of citric acid did not affect blood glucose or HbA_{1c} but decreased the concentration of AGEs in lens. Since citrate did not directly inhibit the formation of CEL from acetol, most probably the inhibition of CEL formation by citric acid is secondary to the inhibition of ketogenesis.

AGEs and carbonyl accumulation have been shown to decrease in Zn-co-incubated samples containing BSA and glucose [194]. Zn also inhibited glycation and β-aggregation in BSA-containing samples. It has been suggested that during the glycation reaction, Zn prevented the β-sheet formation in albumin by promoting the native α-sheet conformation. The protection of thiol groups, which has been observed in Zn-containing samples, could be explained through one of the following three mechanisms: (1) direct binding of Zn to the thiols, (2) steric hindrance as a result of binding to other protein sites in close proximity to the thiol group or (3) a conformational change resulting from binding to another site on the protein [195].

8. Conclusion

Considering that AGEs are believed to act as major pathogenic propagators in many human diseases, and especially in diabetes and its complications, it is of great interest to identify anti-glycative substances and to examine their mode of action. The current review provides examples for the anti-glycation activity of plant-derived substances which target the essential stages of glycation through i) antiglycemic or hypoglycemic action, ii) inhibition of Amadori products formation or intervention in the post-Amadori phase of the reaction, iii) inhibition of the formation of AGEs precursors (oxidation products of sugars and early MRPs), and iv) reduction of AGEs cross-linking. This anti-glycation activity correlates with the phenolic content of the plant extracts although there is a wide range of others, nonphenolic compounds such as terpens, carotenoids, polyunsaturated fatty acids, polysaccharides, withanolides, and melanoidins which demonstrate a high potential to reduce the non-enzymatic protein glycosylation. The plant-derived anti-glycative comounds appear attractive candidates for the development of new generation therapeutics for treatment of diabetic complications and prophylaxis of aging, and point to the importance of an antioxidant-rich diet, as part of the overall diabetes management strategy.

Author details

Mariela Odjakova*, Eva Popova and Merilin Al Sharif
Department of Biochemistry, Faculty of Biology, Sofia University, Bulgaria

Roumyana Mironova
Roumen Tsanev Institute of Molecular Biology at the Bulgarian Academy of Sciences, Bulgaria

Acknowledgement

This work was supported by grant DID02-31/09 from National Science Fund.

9. References

[1] Baynes JW, Watkins NG, Fisher CI, Hull CJ, Patrick JS, Ahmed MU, Dunn JA, Thorpe SR (1989) The Amadori product on protein: structure and reactions. Prog Clin Biol Res 304: 43–67

* Corresponding Author

[2] Bry L, Chen PC, Sacks DB (2001) Effects of hemoglobin variants and chemically modified derivatives on assays for glycohemoglobin. Clin Chem 47(2): 153–163
[3] Neglia CI, Cohen HJ, Garber AR, Ellis PD, Thorpe SR, Baynes JW (1983) 13C NMR investigation of nonenzymatic glucosylation of protein. Model studies using RNase A. J Biol Chem. 258:14279-14283
[4] Lapolla A, Traldi P, Fedele D (2005) Importance of measuring products of non-enzymatic glycation of protein. Clin Biochem. 38:103-115
[5] Hsieh CL, Yang MH, Chyau CC, Chiu CH, Wang HE, LinYC, ChiuWT, Peng RY (2007) Kinetic analysis on the sensitivity of glucose- or glyoxal-induced LDL glycation to the inhibitory effect of Psidium guajava extract in a physiomimic system. BioSystems 88(1-2): 92-100
[6] Kikuchi S, Shinpo K, Takeuchi M, Yamagishi S, Makita Z, Sasaki N, Tashiro K (2003) Glycation-a sweet temper for neuronal death. Brain Res. Rev. 41:306-323
[7] Simm A, Wagner J, Gursinsky T, Nass N, Friedrich I, Schinzel R, Czeslik E, Silber RE, Scheubel RJ (2007) Advanced glycation endproducts: A biomarker for age as an outcome predictor after cardiac surgery? Exp Gerontol. 42(7): 668-675
[8] Brownlee M, Cerami A, Vlassara H (1988) Advanced glycosylation end products in tissue and biochemical basis of diabetic complications. New Engl. J. Med. 318(20):1315-1321
[9] Reddy VP, Beyaz A (2006) Inhibitors of the Maillard reaction and AGE breakers as therapeutics for multiple diseases. Drug Discov Today 11(13-14): 646-654
[10] Ahmed N (2005) Advanced glycation endproducts-role in pathology of diabetic complications. Diabetes Res Clin Pract. 67(1): 3-21
[11] Stitt AW (2005) The maillard reaction in eye diseases. *Ann N Y Acad Sci* 1043:582-597
[12] Forbes JM, Yee LT, Thallas V, Lassila M, Candido R, Jandeleit-Dahm KA, Thomas MC, Burns WC, Deemer EK, Thorpe SR, Cooper ME, Allen TJ (2004) Advanced Glycation End Product Interventions Reduce Diabetes-Accelerated Atherosclerosis. Diabetes 53(7): 1813-1823
[13] Yamamoto Y, Doi T, Kato I., Shinohara H, Sakurai S, Yonekura H, Watanabe T, Myint KM, Harashima A, Takeuchi M, Takasawa S, Okamoto H, Hashimoto N, Asano M, Yamamoto H (2005) Receptor for advanced glycation end products is a promising target of diabetic nephropathy. Ann. N.Y. Acad. Sci. 1043:562-566
[14] Jono T, Kimura T, Takamatsu J, Nagai R, Miyazaki K, Yuzuriha T, Kitamura T, Horiuchi S (2002) Accumulation of imidazolone, pentosidine and N(epsilon)-(carboxymethyl)lysine in hippocampal CA4 pyramidal neurons of aged human brain. Pathol. Int. 52(9): 563-571
[15] Boušová I, Vukasović D, Juretić D, Palička V, Dršata J (2005) Enzyme activity and AGE formation in a model of AST glycoxidation by D-fructose *in vitro.* Acta Pharm. 55: 107-114
[16] Ames JM (1998) Applications of the Maillard reaction in the food industry. Food Chem 62(4): 431-439
[17] Hardy J, Parmentier M, Fanni J (1999) Functionality of nutrients and thermal treatments of food. Proc Nutr Soc 58(3): 579-585

[18] Chao PC, Hsu CC, Yin MC (2009) Analysis of glycative products in sauces and sauce-treated foods. Food Chem 113: 262-266
[19] Delgado-Andrade C, Seiquer I, Navarro P, Morales, FJ (2007) Maillard reaction indicators in diets usually consumed by adolescent population. Mol Nutr Food Res 51: 341-351
[20] Zhao W, Devamanoharan PS, Varma SD (2000) Fructose-mediated damage to lens α-crystallin: prevention by pyruvate. Biochim Biophys. Acta 1500(2): 161-168
[21] Kawai T, Takei I, Tokui M, Funae O, Miyamoto K, Tabata M, Hirata T, Saruta T, Shimada A, Itoch H (2010) Effects of epalrestat, an aldose reductase inhibitor, on diabetic peripheral neuropathy in patients with type 2 diabetes, in relation to suppression of N_{ε}- carboxymethyl lysine. J Diabetes Complications 24(6):424-432
[22] Ramana KV, Srivastava SK (2010) Aldose Reductase: A Novel Therapeutic Target for Inflammatory Pathologies. Int J Biochem Cell Biol 42(1): 17-20
[23] Boulanger E, Wautier MP, Wautier JL, Boval B, Panis Y, Wernert N, Danze PM, Dequiedt P (2002) AGE binds to mesothelial cells via RAGE and stimulate VCAM-1 expression. Kidney Int 61(1):148-156
[24] Boulanger E, Grossin N, Wautier MP, Taamma R, Wautier JL (2007) Mesothelial RAGE activation by AGEs enhances VEGF release and potentiates capillary tube formation. Kidney Int 71(2):126-133
[25] Schmidt AM, Yan SD, Yan SF, Stern DM (2001) The multiligand receptor RAGE as a progression factor amplifying immune and inflammatory responces. J. Clin. Invest 108(7):949-955
[26] Naziroğlu M, Butterworth P (2005) Protective effects of moderate exercise with dietary vitamin C and E on blood antioxidative defense mechanism in rats with streptozotocin-induced diabetes. Can J Appl Physiol. 30(2): 172-185
[27] Zhang Q, Ames JM, Smith RD, Baynes JW, Metz TO (2009) A perspective on the Maillard reaction and the analysis of protein glycation by mass spectrometry: probing the pathogenesis of chronic disease. J Proteome Res 8(2):754-769
[28] Wang AL, Yu A, Qi H, Zhu XA, Tso M (2007) AGEs mediates expression and secretion of TNF alpha in rat retinal microglia. Exp Eye Res 84(5):905-913
[29] Halliwell B (2009) The wanderings of free radical. Free Radic Biol Med 46(5):531-542
[30] Li L, Renier G (2006) Activation of nicotinamide adenine dinucleotide phosphate (reduced form) oxidase by advanced glycation end products links oxidative stress to altered retinal vascular endothelial growth factor expression. Metabolism 55(11): 1516-1523
[31] Inoguchi T, Li P, Umeda F, Yu HY, Kakimoto M, Imamura M, Aoki T, Etoh T, Hashimoto T, Naruse M, Sano H, Utsumi H, Nawata H (2000) High glucose level and free fatty acid stimulate reactive oxygen species production through protein kinase C-dependent activation of NAD(P)H oxidase in cultured vascular cells. Diabetes 49(11): 1939-1945
[32] Xia L, Wang H, Munk S, Kwan J, Goldberg HJ, Fantus IG, Whiteside CI (2008) High glucose activates PKC-zeta and NADPH oxidase through autocrine TGF-beta 1 signaling in mesanglial cells. Am. J. Physiol. 295(6): F1705-F1714

[33] Lee EY, Chung CH, Kim JH, Joung HJ, Hong SY (2006) Antioxidants ameliorate the expression of vascular endothelial growth factor mediated by protein kinase C in diabetic podocytes. Nephrol. Dial. Transplant. 21(6): 1496-1503

[34] Edeas M (2009) Anti-oxidants, controverses et perspectives: comment expliquer l'echec. J. Soc Biol 203(3):271-280

[35] Jung HA, Jung YJ, Yoon NY, Jeong da M, Bae HJ, Kim DW, Na DH, Choi JS (2008) Inhibitory effects of Nelumbo nucifera leaves on rat lens aldose reductase, advanced glycation end products formation, and oxidative stress. Food Chem Toxicol 46(12): 3818-3826

[36] Kawanishi K, Ueda H, Moriyasu M (2003) Aldose reductase inhibitors from the nature. Curr Med Chem. 10(15): 1353–1374

[37] Manzanaro S, Salva J, de la Fuente JA (2006) Phenolic marine natural products as aldose reductase inhibitors. J Nat Prod. 69(10): 1485–1487

[38] Thornalley PJ (2003) Use of aminoguanidine (Pimagedine) to prevent the formation of advanced glycation endproducts. Arch Biochem Biophys 419(1): 31–40

[39] Lee GY, Jang DS, Lee YM, Kim JM, Kim JS (2006) Naphthopyrone glucosides from the seeds of *Cassia tora* with inhibitory activity on advanced glycation end products (AGEs) formation. Arch Pharm Res. 29(7):587-590

[40] Vasu VT, Modi H, Thaikoottathil JV, Gupta S (2005) Hypolipidaemic and antioxidant effect of Enicostemma littorale Blume aqueous extract in cholesterol fed rats. J. Ethnopharmacol. 101(1-3): 277–282

[41] Choudhary MI, Maher S, Begum A, Abbaskhan A, Ali S, Khan A, Shafique-ur-Rehman, Atta-ur-Rahman (2010) Characterization and antiglycation activity of phenolic constituents from *Viscum album* (European Mistletoe). Chem Pharm Bull (Tokyo) 58(7):980-982

[42] Tsuji-Naito K, Saeki H, Hamano M (2009) Inhibitory effects of Chrysanthemum species extracts on formation of advaced glycation end products. Food Chem. 116(4):854-859

[43] Meng G, Zhu H, Yang S, Wu F, Zheng H, Chen E, Xu J (2011) Attenuating effects of *Ganoderma lucidum* polysaccharides on myocardial collagen cross-linking relates to advanced glycation end product and antioxidant enzymes in high-fat-diet and streptozotocin-induced diabetic rats. Carbohydrate Polymers 84(1):180-185

[44] Yang B, Zhao M, Jiang Y (2009) Anti-glycated activity of polysaccharides of longan (Dimocarpus longan Lour.) fruit pericarp treated by ultrasonic wave Food Chem 114 (2): 629-633

[45] Sun Z, Peng X, Liu J, Fan K-W, Wang M, Chen F (2010) Inhibitory effect of microalgal extracts on the formation of advanced glycation endproducts (AGEs). Food Chem. 120(1):261-267

[46] Sun Z, Liu J, Zeng X, Huangfu J, Jiang Y, Wang M, Chen F(2011) Astaxanthin is responsible for antiglycoxidative properties of microalga Chlorella zofingiensis. Food Chem. 126(4):1629-1635

[47] Yazdanparast R, Ardestani A, Jamshidi S (2007) Experimental diabetes treated with *Achillea santolina*: Effect on pancreatic oxidative parameters.J. Ethnopharmacol. 112(1):13-18

[48] Miroliaei M, Khazaei S, Moshkelgosha S, Shirvani M (2011) Inhibitory effects of Lemon balm (Melissa officinalis, L.) extract on the formation of advanced glycation end products. Food Chem. 129(2):267-271

[49] Babu PV, Sabitha KE, Shyamaladevi CS (2008) Effect of green tea extract on advaced glycation and cross-linking of tail tendon collagen in streptozotocin induced diabetic rats. Food Chem Toxicol. 46(1): 280-285

[50] Ho SC, Wu SP, Lin SM, Tang YL (2010) Comparison of anti-glycation capacities of several herbal infusions with that of green tea. Food Chem. 112(3):768-774

[51] Verzelloni E, Tagliazucchi D, Rio D, Calani L, Conte A (2011) Antiglycative and antioxidative properties of coffee fractions. Food Chem. 124(4):1430-1435

[52] Huang JS, Chuang LY, Guh JY, Yang YL, Hsu MS (2008) Effect of taurine on advanced glycation end products-induced hypertrophy in renal tubuar epithelial cells. Toxicol Appl Pharmacol. 233(2):220-226

[53] Shin DC, Kim CT, Lee YC, Choi WJ, Na YJ (2010) Reduction of acrylamide by taurine in aqueous and potato chip model systems. Food Res Int 43(5):1356-1360

[54] Huang KH, Pai MH, Wu CH, Liu JJ, Yeh SL(2010) Supplemental dietary arginine reduces renal RAGE expression and oxidative damage in rats with streptozotocin-induced type 2 diabetes.e-SPEN 5(2): e77-e84. Available:http://dx.doi.org/10.1016/j.eclnm.2010.02.001

[55] Wang ZH, Hsu CC, Huang CN, Yin MC (2010) Anti-glycative effect of oleanolic acid and ursolic acid in kidney of diabetic mice. Eur J Pharmacol. 628(1-3):255-260

[56] Manna P, Das J, Ghosh J, Sil PC (2010) Contribution of type 1 diabetes to rat liver dysfunction and cellular damage via activation of NOS, PARP, IkappaBalpha/NF-kappaB, MAPKs, and mitochondria-dependent pathways: Prophylactic role of arjunolic acid. Free Radic Biol Med. 48(11):1465-1484

[57] Wang X, Zhang LS, Dong LL (2010) Inhibitory effect of polysaccharides from pumpkin on advanced glycation end-products formation and aldose reductase activity. Food Chem. 120(1):261-267

[58] Wang J, Sun B, Cao Y, Tian Y (2009) Protein glycation inhibitory activity of wheat bran feruoyl oligosaccharides. Food Chem. 112(2):350-353

[59] Ardestani A, Yazdanparast R (2007) *Cyperus rotundus* suppresses AGE formation and protein oxidation in a model of fructose-mediated protein glycoxidation. Int J Biol Macromol. 41(5):572-578

[60] Dugé de Bernonville T, Guyot S, Paulin JP, Gaucher M, Loufrani L, Henrion D, Derbré S, Guilet D, Richomme P, Dat JF, Brisset MN (2010) Dihydrochalcones: Implication in resistance to oxidative stress and bioactivities against advanced glycation end-products and vasoconstriction. Phytochemistry 71(4):443-452

[61] Farsi DA, Harris CS, Reid L, Bennett SA, Haddad PS, Martineau LC, Arnason JT (2008) Inhibition of Non-enzymatic Glycation by Silk Extracts from a Mexican Land Race and Modern Inbred Lines of Maize (*Zea mays*). Phytother. Res. 22:108-112

[62] Jang DS, Yoo NH, Kim NH, Lee YM, Kim CS, Kim J, Kim JH,Kim JS (2010) 3,5-Di-O-caffeoyl-epi-quinic acid from the leaves and stems of Erigeron annuus inhibits protein glycation, aldose reductase, and cataractogenesis. Biol Pharm Bull 33(2):329-333

[63] Ma HY, Gao HY, Sun L, Huang J, Xu XM, Wu LJ (2011) Constituents with α-glucosidase and advanced glycation end-product formation inhibitory activities from *Salvia miltiorrhiza* Bge. J Nat Med. 65(1):37-42
[64] Peng X, Zheng Z, Cheung KW, Shan F, Ren GX, Chen SF, Wang M (2008) Inhibitory effect of mung bean extract and its constituents vitexin and isovitexin on the formation of advanced glycation endproducts. Food Chem. 106(2):457-481
[65] Gugliucci A, Bastos DH, Schulze J, Souza MF (2009) Caffeic and chlorogenic acids in Ilex paraguariensis extracts are the main inhibitors of AGE generation by methylglyoxal in model proteins. Fitoterapia 80(6):339-344
[66] Landis-Piwowar KR, Huo C, Chen D, Milacic V, Shi G, Chan TH, Dou QP (2007) A Novel Prodrug of the Green Tea Polyphenol (−)-Epigallocatechin-3-Gallate as a Potential Anticancer Agent Canser Res. 67: 4303-4310
[67] Mandel S, Weinreb O, Amit T, Youdim BH (2004) Cell signaling pathways in the neuroprotective actions of the green tea polyphenol (-)-epigallocatechin-3-gallate: implications for neurodegenerative diseases. J. Neurochem 88(6):1555-1569
[68] Vinson JA, Proch J, Bose P, Muchler S, Taffera P, Shuta D, Samman N, Agbor GA (2006) Chocolate Is a Powerful ex Vivo and in Vivo Antioxidant, an Antiatherosclerotic Agent in an Animal Model, and a Significant Contributor to Antioxidants in the European and American Diets. J Agric Food Chem 54: 8071-8076
[69] Kowluru RA, Kanwar M (2007) Effects of curcumin on retinal oxidative stress and inflammation in diabetes. Nutr Metab. 4:8
[70] Tsao R (2010) Chemistry and Biochemistry of Dietary Polyphenols. Nutrients 2: 1231-1246
[71] Moridani My, Sconbie H, Jamshidzadeh A, Salehi P, O' Brien PJ (2001) Caffeic acid, chlorogennic acid, and dihydrocaffeic acid metabolism: glutathione conjugate formation. Drug Metab Dispos 29:1432-1439
[72] Clifford MN (1999) Chlorogenic acids and other cinnamates-nature, occurrence and dietary burden. J Sci Food Agric 79(3):362-372
[73] Fiuza SM, Gomes C, Teixeira LJ, Girão da Cruz MT, Cordeiro MN, Milhazes N, Borges F, Marques MP (2004) Phenolic acid derivatives with potential anticancer properties--a structure-activity relationship study. Part 1: methyl, propyl and octyl esters of caffeic and gallic acids. Bioorg Med Chem. 12(13):3581-3589
[74] Kikuzaki H, Hisamoto M, Hirose K, Akiyama K, Taniguchi H (2002) Antioxidant properties of ferulic acid and its related compounds. J Agric Food Chem 50(7):2161-2168
[75] Kang J, Liu Y, Xie MX, Li S, Jiang M,Wang YD (2004). Interactions of human serum albumin with chlorogenic acid and ferulic acid. Biochim Biophys Acta 1674(2): 205-214
[76] Silván JM, Assar SH, Srey C, Dolores del Castillo M, Ames JM (2011) Control of the Maillard reaction by ferulic acid. Food Chem 128(1):208-213
[77] Lee EH, Song DG, Lee JY, Pan CH, Um BH, Jung SH (2008) Inhibitory Effect of the Compounds Isolated from *Rhus verniciflua* on Aldose Reductase and Advanced Glycation Endproduct. Biol Pharm Bull. 31(8):1626-1630

[78] Proestos C, Chorianopoulos N, Nychas GJE, Komaitis M (2005) RP-HPLC analysis of the phenolic compounds of plant extracts. Investigation of their antioxidant capacity and antimicrobial activity. J. Agric. Food Chem.53: 1190-1195.

[79] Yamabe N, Kang KS, Matsuo Y, Tanaka T, Yokozawa T (2007) Identification of Antidiabetic Effect of Iridoid Glycosides and Low Molecular Weight Polyphenol Fractions of Corni Fructus, a Constituent of Hachimi-jio-gan, in Streptozotocin-Induced Diabetic Rats. *Biol. Pharm. Bull.* 30(7): 1289—1296

[80] Yamabe N, Kang KS, Park CH,Tanaka H,Yokozawa T (2009) 7-O-Galloyl-D-sedoheptulose Is a Novel Therapeutic Agent against Oxidative Stress and Advanced Glycation Endproducts in the Diabetic Kidney. Biol. Pharm. Bull. 32(4): 657—664

[81] Terao J, Kawai Y, Murota K (2008) Vegetable flavonoids and cardiovascular disease. Asia Pac. J. Clin. Nutr. 17(S1): 291–293

[82] Hernández I, Alegre L, Van Breusegem F, Munné-Bosch S (2009) How relevant are flavonoids as antioxidant in plants? Trends Plant Sci 14 (3):125-132

[83] Fraga CG (2007) Plant polyphenols : how to translate their in vitro antioxidant actions to in vivo conditions. IUBMB Life 59(4-5):308-315

[84] Jagtar AG, Patil PB (2010) Antihyperglycemic activity and inhibition of advanced glycation end product formation by Cuminum cyminum in streptozotocin induced diabetic rats. Food Chem Toxicol 48(8-9):2030-2036

[85] Bertoncelj J, Doberšek U, Jamnik M, Golob T (2007). Evaluation of the phenolic content, antioxidant activity and colour of Slovenian honey. Food Chem. 105(2): 822–828.

[86] Blasa M, Candiracci M, Accorsi A, Piacentini MP, Albertini MC, Piatti E (2006). Raw Millefiori honey is packed full of antioxidants. Food Chem 97: 217–222

[87] Estevinho L, Pereira AP, Moreira L, Dias LG, Pereira E (2008). Antioxidant and antimicrobial effects of phenolic compounds extracts of Northeast Portugal honey. Food Chem Toxicol 46(12): 3774–3779

[88] Gheldof N, Engeseth NJ (2002). Antioxidant capacity of honeys from various floral sources based on the determination of oxygen radical absorbance capacity and inhibition of in vitro lipoprotein oxidation in human serum samples. J Agric Food Chem 50(10): 3050–3055

[89] Matsuda H, Wang T, Managi H, Yoshikawa M (2003) Structural requirements of flavonoids for inhibition of protein glycation and radical scavenging activities. Bioorg. Med. Chem. 11: 5317-5323

[90] Pontias I, Treutter D, Paulin JP, Brisset MN (2008) *Erwinia amylovora* modifies phenolic profiles of susceptible and resistant apple through its type III secretion system. Physiol. Plant. 132(3): 262-271

[91] Williams AH (1964) Dihydrochalcones; their occurrence and use as indicators in plant chemical taxonomy. Nature 202: 824–825

[92] Gosh C, Halbwirth H, Kuhn J, Miosic S, Stich K (2009) Biosynthesis of phloridzin in apple (Malus domestica Borkh). Plant Sci. 176(2): 223-231

[93] Ehrenkranz JR, Lewis NG, Kahn CR, Roth J (2005) Phlorizin: a review. Diabetes Metab. Res. Rev. 21(1): 31–38

[94] Messina MJ, Loprinzi CL (2001) Soy for breast cancer survivors: a critical review of the literature. J. Nutr. 131 (11), 30955–31085

[95] Omoni AO, Aluko RE (2005) Soybean foods and their benefits: potential mechanisms of action. Nutr. Rev. 63(8): 272–283

[96] Rimbach G, Boesh-Saadatmandi C, Frank J, Fuchs D, Wenzel U, Daniel H, Hall WL, Weinberg PeterD (2008) Dietary isoflavones in the prevention of cardiovascular disease – A molecular perspective. Food Chem. Toxicol. 46(4): 1308–1319

[97] Sho H (2001) History and characteristics of Okinawan longevity food. Asia Pac J Clin Nutr. 10(2): 159-164

[98] Mazur WM, Duke JA, Wahala K, Rasku S, Adlercreutz H (1998) Isoflavonoids and lignans in legumes: Nutritional and heath aspects in Humans. *J.Nutr. Biochem* 9:193-200

[99] Hsieh HM, Wu WM, Hu ML (2009) Soy isoflavones attenuate oxidative stress and improve parameters related to aging and Alzheimer's disease in C57BL/6J mice treated with D-galactose. Food Chem Toxicol. 47(3):625-632

[100] Hsu FL, Lin IM, Kuo DH, Chen WC, Su HC, Cheng JT (2003) Antihyperglycemic effect of puerarin in streptozotocin-induced diabetes rats. J. Nat. Prod. 66(6): 788–792

[101] Xiong FL, Sun XH, Gan L, Yang XL, Xu HB (2006) Puerarin protects rat pancreatic islets from damage by hydrogen peroxide. Eur. J. Pharmacol. 529(1-3): 1–7

[102] Xu XH (2003) Effects of puerarin on fatty superoxide in aged mice induced by D-galactose. Chin. J. Chin. Master. Med. 28(1): 66–69

[103] Kim J., Lee YM, Lee GY, Jang DS, Bae KH, Kim JS (2006) Constituents of the Root of Pueraria lobata inhibit formation of advanced glycation end products (AGEs) Arch. Pharm. Res. 29(10): 821-825

[104] Kim KM, Jung DH, Jang DS, Kim YS, Kim JM, Kim HN, Surh YJ, Kim JS (2010). Puerarin suppresses AGEs-induced inflammation in mouse mesangial cells: A possible pathway through the induction of heme oxygenase-1 expression. Toxicol Appl Pharmacol 244(2): 106–113

[105] Alam, Cook JL (2003) Transcriptional regulation of the heme oxygenase-1 gene via the stress response element pathway. *Curr. Pharm. Des.* 9(30): 2499-2511

[106] Tsao R, McCallum J (2009) Chemistry of flavonoids. In: de la Rosa L, Alvarez-Parrilla E, González-Aguilar G, editors. *Fruit* and *Vegetable Phytochemicals: Chemistry, Nutritional Value and Stability*. Blackwell Publishing: Ames, IA, USA, pp131-153;

[107] Valant-Vetschera KM, Wollenweber E (2006) Flavones and Flavonols. In: Andersen ØM, Markham KR, editors. *Flavonoids: Chemistry, Biochemistry and Applications.* CRC Press/Taylor & Francis Group: Boca Raton, FL, USA, pp 617-748

[108] Williams C (2006) Flavone and flavonol O-glycosides. In: Andersen ØM, Markham KR, editors. *Flavonoids: Chemistry, Biochemistry and Applications.* CRC Press/Taylor & Francis Group: Boca Raton, FL, USA, pp. 749-856

[109] Blonska M, Czuba ZP, Krol W (2003) Effect of flavone derivatives on interleukin-1beta (IL-1beta) mRNA expression and IL-1beta protein synthesis in stimulated RAW 264.7 macrophages. Scand J Immunol 57(2):162–166

[110] Yoo NH, Jang DS, Yoo JL, Lee YM, Kim YS, Cho JH, Kim JS (2008) Erigeroflavanone, a Flavanone Derivative from the Flowers of *Erigeron annus* with Protein Glycation and Aldose Reductase Inhibitory Activity. J Nat Prod. 71:713-715

[111] García-Mediavilla V, Crespo I, Collado PS, Esteller A, Sánchez-Campos S, Tuñón MJ, González-Gallego J (2007) The anti-inflammatory flavones quercetin and kaempferol cause inhibition of inducible nitric oxide synthase, cyclooxygenase-2 and reactive C-protein, and down-regulation of the nuclear factor kappaB pathway in Chang Liver cells. Eur J Pharmacol 557(2-3):221–229

[112] Kim JM, Lee EK, Kim DH, Yu BP, Chung HY (2010) Kaempferol modulates pro-inflammatory NF-κB activation by suppressing advanced glycation endproducts-induced NADPH oxidase. American Aging Association 32(2):197–208

[113] Kim HY, Lee JM, Yokozawa T, Sakata K, Lee S (2011) Protective activity of flavonoid and flavonoid glycosides against glucose-mediated protein damage. Food Chem 126:892-895

[114] Soman S, Rauf AA, Indira M, Rajamanickam C (2010) Antioxidant and Antiglycative Potential of Ethyl Acetate Fraction of *Psidium guajava* Leaf Extract in Steptozotocin-Induced Diabetic Rats. Plant Foods Hum Nutr 65(4):386-391

[115] Wu JW, Hsieh CL, Wang HY, Chen HY (2009) Inhibitory effect of guava (Psidium guajava L.) leaf extracts and its active compounds on the glycation process of protein. Food Chem 113:78-84

[116] Marzouk MS, Moharram FA, El Dib RA, El-Shenawy SM, Tawfike AF (2009) Polyphenolic profile and bioactivity study of Oenothera speciosa Nutt. aerial parts. Molecules 14(4): 1456–1467

[117] Marzouk MS, Moharram FA, Haggag EG, Ibrahim MT, Badary OA(2006) Antioxidant flavonol glycosides from *Schinus molle*. Phytother Res. 20(3):200-205

[118] Silva C., Raulino RJ, Cerqueira DM, Mannarino SC, Pereira MD, Panek AD, Silva JFM, Menezes FS, Eleutherio ECA (2009). In vitro and in vivo determination of antioxidant activity and mode of action of isoquercitrin and Hyptis fasciculate. Phytomedicine 16(8): 761–767

[119] Panda S, Kar A (2007). Antidiabetic and antioxidative effects of Annona squamosa leaves are possibly mediated through quercetin-3-O-glucoside. BioFactors 31(3-4): 201–210

[120] Lee S, Park HS, Notsu Y, Ban HS, Kim YP, Ishihara K, Hirasawa N, Jung SH, Lee YS, Lim SS, Park EH, Shin KH, Seyama T, Hong J, Ohuchi K (2008). Effects of hyperin, isoquercitrin and quercetin on lipopolysaccharide-induced nitrite production in rat peritoneal macrophages. Phytother Res. 22(11): 1552–1556

[121] Manaharan T, Teng LL, Appleton D, Ming CH, Masilamani T, Palanisamy UD (2011) Antioxidant and antiglycemic potential of Peltophorum pterocarpum plant parts. Food Chem 129(4):1355-1362

[122] Winter J, Moore LH, Dowell VR, Jr, Bokkenheuser VD (1989) C-ring cleavage of flavonoids by human intestinal bacteria. Appl. Environ. Microbiol. 55(5):1203–1208

[123] Braune A, Gütschow M, Engst W, Blaut M (2001) Degradation of quercetin and luteolin by Eubacterium ramulus. Appl Environ Microbiol 67(12):5558-5567

[124] Schneider H, Simmering R, Hartmann L, Pforte H, Blaut M (2000) Degradation of quercetin-3-glucoside in gnotobiotic rats associated with human intestinal bacteria. J. Appl. Microbiol. 89(6):1027–1037

[125] Pashikanti S, de Alba DR, Boissonneault GA, Cervantes-Laurean D (2010) Rutin metabolites: Novel inhibitors of nonoxidative advanced glycation end products. Free Radic Biol Med 48(5): 656–663

[126] Cervantes-Laurean D, Schramm DD, Jacobson EL, Halaweish I, Bruckner GG, Boissonneault GA(2006) Inhibition of advanced glycation end product formation on collagen by rutin and its metabolites. J. Nutr. Biochem. 17(8): 531–540

[127] Wu CH, Lin JA, Hsieh WC, Yen GC (2009) Low-Density-Lipoprotein (LDL)-Bound Flavonoids Increase the Resistance of LDL to Oxidation and Glycation under Pathophysiological Concentrations of Glucose in Vitro. J Agric Food Chem. 57(11):5058-5064

[128] Hurabielle M, Eberle J, Paris M (1982) Flavonoids of Artemisia campestris, ssp. glutinosa. Planta. Med. 46(2): 124–125

[129] Memmi A, Sansa G, Rjeibi I, EI Ayeb M, Srairi-Abid N, Bellasfer Z, Fekhih A (2007) Use of medicinal plants against scorpionic and ophidian venoms. *Arch. Inst. Pasteur. Tunis.* 84(1-4): 49-55

[130] Behmanesh B, Heshmati GA, Mazandarani M, Rezaei MB, Ahmadi AR, Ghaemi EO, Bakhshandeh Nosrat S (2007) Chemical composition and antibacterial activity from essential oil of *Artemisia sieberi* Besser subsp. *Sieberi* in North of Iran. Asian J. Plant. Sci. 6(3): 562-564

[131] Hong JH, Lee IS (2009) Effects of *Artemisia capillaris* ethyl acetate fraction on oxidative stress and antioxidant enzyme in high-fat diet induced obese mice. Chem. Biol. Interact. 179(2-3): 88-93

[132] Sefi M, Fetoui H, Makni M, Zeghal N (2010) Mitigating effects of antioxidant properties of Artemisia campestris leaf extract on hyperlipidemia, advanced glycation end products and oxidative stress in alloxan-induced diabetic rats. Food Chem Toxicol. 48:1986-1993

[133] Zhang Y, Bao B, Lu B, Ren Y, Tie X, Zhang Y (2005) Determination of flavone C-glucosides in antioxidant of bamboo leaves (AOB) fortified foods by reversed-phase high-performance liquid chromatography with ultraviolet diode array detection. *Journal of Chromatography* A 1065: 177-185

[134] Fu Y, Zu Y, Liu W, Hou C, Chen L, Li S, Shi X, Tong M (2007). Preparative separation of vitexin and isovitexin from pigeonpea extracts with macroporous resins. *Journal of Chromatography* A 1139(2): 206-213

[135] Grayer RJ, Veitch NC (2006) Flavanones and dihydroflavonols. In: Andersen ØM, Markham KR, editors. *Flavonoids: Chemistry, Biochemistry and Applications.* CRC Press/Taylor & Francis Group: Boca Raton, FL, USA, pp. 918-1002.

[136] Haraguchi H, MochidaY, Sakai S, Masuda H, Tamura Y, Mizutani K, Tanaka O, Chou WH (1996) Protection against oxidative damage by dihydroflavonols in Engelhardtia chrysolepis. Biosci. Biotechnol. Biochem. 60(6): 945–948

[137] Wirasathien L, Pengsuparp T, Suttisri R, Ueda H, Moriyas M, Kawanishi K (2007) Inhibitors of aldose reductase and advanced glycation end-products formation from the leaves of Stelechocarpus cauliflorus R.E.Fr. Phytomedicine 14:546-550

[138] McKay DL, Blumberg JB (2002) The role of tea in human health: an update. J. Am. Coll. Nutr. 21 (1): 1–13

[139] Wu LY, Juan CC, Ho LT, Hsu YP, Hwang LS(2004) Effect of green tea supplementation on insulin sensitivity in Sprague-Dawley rats. J. Agr. Food Chem. 52(3): 643–648

[140] Kobayashi Y, Suzuki M, Satsu H, Arai S, Hara Y, Suzuki K, Miyamaoto Y, Shimizu M(2000) Green tea polyphenols inhibit the sodium dependent glucose transporter of intestinal epithelial cells by a competitive mechanism. J. Agr. Food Chem. 48(11):5618–5623

[141] Waltner-Law ME, Wang XL, Law BK, Hall RK, Nawano M(2002) Epigallocatechin gallate, a constituent of Green tea represses hepatic glucose production. J. Biol. Chem. 277(38): 34933–34940

[142] Rasheed Z, Anbazhagan AN, Akhtar N, Ramamurthy S, Voss FR, Haqqi TM (2009). Green tea polyphenol epigallocatechin-3-gallate inhibits advanced glycation end product-induced expression of tumor necrosis factor-α and matrix metalloproteinase-13 in human chondrocytes. *Arthritis Res Ther* 11(3):R71

[143] Wu CH, Wu CF, Huang HW, Jao YC, Yen GC (2009). Naturally occurring flavonoids attenuate high glucose-induced expression of proinflammatory cytokines in human monocytic THP-1 cells. Mol Nutr Food Res 53(8): 984–995

[144] Sang S, Shao X, Bai N, Lo CY, Yang CS, Ho CT (2007). Tea polyphenol (-)-epigallocatechin-3-gallate: A new trapping agent of reactive dicarbonyl species. Chem Res Toxicol 20(12): 1862–1870

[145] Wu CH, Yeh CT, Yen GC (2010) Epigallocatechin gallate (EGCG) binds to low-density lipoproteins (LDL) and protects them from oxidation and glycation under high-glucose conditions mimicking diabetes. Food Chem 121(3):639-644

[146] Peng X, Cheng KW, Ma J, Chen B, Ho CT, Lo C, Chen F, Wang M (2008) Cinnamon Bark Proanthocyanidins as Reactive Carbonyl Scavengers To Prevet Formation of Advanced Glycation Endproducts. J Agric Food Chem 56:1907-1911

[147] Peng X, Ma J, Chao J, Sun Z, Chang RCC, Tse I, Li ETS, Chen F, Wang M (2010) Beneficial Effects of Cinnamon Proanthocyanidins on the Formation of Specific Advanced Glycation Endproducts and Methylglyoxal-Induced Impairment on Glucose Consumption. J Agric Food Chem 58:6692-6696

[148] Jang DS, Lee GY, Lee YM, Kim YS, Sun H, Kim DH, Kim JS (2009) Flavan-3-ols Having a γ-Lactam from the Roots of Actinidia arguta Inhibit the Formation of Advanced Glycation End Products in Vitro. Chem Pharm Bull. 57(4):397-400

[149] Palanisamy U, Ling L, Manaharan T, Appleton D (2011) Rapid isolation of geraniin from Nephelium lappaceum rind waste and its anti-hyperglycemic activity. Food Chem 127:21-27

[150] Palanisamy U, Cheng HM, Masilamani T, Subramaniam T, Ling L T, Radhakrishnan AK (2008) Rind of the rambutan, Nephelium lappaceum, a potential source of natural antioxidants. Food Chem 109(1): 54–63

[151] Wang X, Zhao L, Han T, Chen S, Wang J (2008) Protective effects of 2,3,5,4'-tetrahydroxystilbene-2-O-β-D-glucoside, an active component of Polygonum multiflorum Thunb, on experimental colitis in mice. Eur. J. Pharmacol. 578(2-3): 339–348

[152] Lv L, Gu X, Tang J, Ho CT (2007) Antioxidant activity of stilbene glycoside from Polygonum multiflorum Thunb in vivo. Food Chem. 104(4): 1678–1681

[153] Lv L, Shao X, Wang L, Huang D, Ho CT, Sang S (2010) Stilbene Glucoside from Polygonum multiflorum Thunb.: A Novel Natural Inhibitor of Advanced Glycation End Product Formation by Trapping of Methylglyoxal. J. Agric. Food Chem. 58(4): 2239–2245

[154] Chompoo J, Upadhyay A, Kishimoto W, Makise T, Tawata S (2011) Advanced glycation end products inhibitiors from Alpinia zerumbet rhizomes. Food Chem 129:709-715

[155] Elzaawely AA, Xuan TD, Tawata S (2007). Essential oils, kava pyrones and phenolic compounds from leaves and rhizomes of Alpinia zerumbet (Pers.) B.L. Burtt & R.M.Sm. and their antioxidant activity. Food Chem 103(2): 486–494

[156] Jantan I, Raweh SM, Sirat .M, Jamil S, MohdYasin, YH, Jalil J, Jamal JA (2008) Inhibitory effect of compounds from Zingiberaceae species on human platelet aggregation. *Phytomedicine* 15(4): 306-309

[157] Chen F (1996) High cell density culture of microalgae in heterotrophic growth. Trends in Biotechnology 14(11): 421–426

[158] Bar E, Rise M, Vishkautsan M, Arad S (1995). Pigment and structural changes in Chlorella zofingiensis upon light and nitrogen stress. J Plant Pysiol 146(4): 527–534

[159] Miki W (1991). Biological functions and activities of animal carotenoids. IJPAC 63: 141–146

[160] Hsieh CL, Peng CH, Chyau CC, Lin YC, Wang HE, Peng RY (2007) Low-density lipoprotein, collagen, and thrombin models reveal that Rosemarinus officinalis L. exhibits potent antiglycative effects. J Agric Food Chem 55(8):2884-2891

[161] Saraswat B, Visen PK, Agarwal DP (2000) Ursolic acid isolated from Eucalyptustereticornis protects against ethanol toxicity in isolated rat hepatocyes. Phytother Res14(3):163-166

[162] Martin-Aragón S, de Las Heras B, Sanchez-Reus MI, Benedi J (2001) Pharmacological modification of endogenous antioxidant enzymes by ursolic acid on tetrachloride-induced liver damage in rats and primary cultures of rat hepatocytes. Exp Toxicol Pathol 53(2-3):199-206

[163] Saravanan R, Viswanathan P, Kodukkur VP (2006) Protective effect of ursolic acid on ethanol-mediated experimental liver damage in rats. Life Sci 78(7):713-718

[164] Gao D, Li Q, Li Y, Liu Z, Fan Y,Han Z, Li J, Li K(2007) Antidiabetic potential of oleanolic acid from Ligustrum lucidum Ait. Can J PhysiolPharmacol. 85(11): 1076–1083

[165] Manna P, Sinha M, Pal P, Sil PC (2007) Arjunolic acid, a triterpenoid saponin, ameliorates arsenic-induced cyto-toxicity in hepatocytes. Chem. Biol. Interact. 170(3):187–200

[166] Masoko P, Mdee LK, Mampuru LJ, Eloff JN (2008) Biological activity of two related triterpenes isolated from Combretum nelsonii (Combretaceae) leaves. Nat. Prod. Res. 22(12):1074–1084

[167] Manna P, Sinha M, Sil PC (2007) Protection of arsenic-induced hepatic disorder by arjunolic acid. Basic Clin. Pharmacol. Toxicol. 101(5):333–338

[168] Djoukeng JD, Abou-Mansour E, Tabacchi R, Tapondjou AL, Bouda H, Lontsi D (2005) Antibacterial triterpenes from Syzygium guineense (Myrtaceae). J. Ethnopharmacol. 101(1-3):283–286

[169] Schepetkin IA, Quinn MT (2006) Botanical polysaccharides: Macrophage immunomodulation and therapeutic potential. Int Immunopharmacol 6(3): 317–333

[170] Yang B, Zhao MM, Shi J, Cheng GP, Ruenroengklin N, Jiang YM (2008). Variations in water-soluble saccharides and phenols in longan fruit pericarp after drying. *Journal of Food Process Engineering* 31(1): 66-77

[171] Chen XP, Chen Y, Li SB, Chen YG, Lan JY, Liu L P (2009). Free radical scavenging of Ganoderma lucidum polysaccharides and its effect on antioxidant enzymes and immunity activities in cervical carcinoma rats. Carbohydrate Polymers 77(2): 389–393

[172] Jain SK, Rains JL, Croad JL (2007). Effect of chromium niacinate and chromium picolinate supplementation on lipid peroxidation, TNF-alpha, IL-6, CRP, glycated hemoglobin, triglycerides, and cholesterol levels in blood of streptozotocin-treated diabetic rats. *Free Radic Biol Med* 43(8): 1124-1131

[173] Ganzera M, Choudhary MI, Khan IA (2003) Quantitative HPLC analysis of withanolides in Withania somnifera. Fitotherapia 74 (1-2):68–76

[174] Zhao J, Nakamura N, Hattori M, Kuboyama T, Tohda C, Komatsu K (2002) Withanolide derivatives from the roots of Withania somnifera and their neurite outgrowth activities. Chem. Pharm. Bull. (Tokyo) 50(6):760–765

[175] Jayaprakasam B, Strasburg GA, Nair MG (2004) Potent lipid peroxidation inhibitors from Withania somnifera fruits. Tetraherdron 60 (13): 3109-3121

[176] Babu VP, Gokulakrishnan A, Dhandayuthabani R, Ameethkhan D, Kumar CV, Ahamed MI (2007) Protective effect of Withania somnifera (Solanaceae) on collagen glycation and cross-linking. Comp Biochem Physiol. Part B147:308-313

[177] Borrelli RC, Fogliano V (2005). Bread crust melanoidins as potential prebiotic ingredients. Mol Nutr Food Res 49(7):673–678

[178] Rufian-Henares JA, Morales FJ (2007). Angiotensin-I converting enzyme inhibitory activity of coffee melanoidins. J Agric Food Chem 55(4): 1480–1485

[179] Tagliazucchi D, Verzelloni E, Conte A (2008). Antioxidant properties of traditional balsamic vinegar and boiled must model systems. Eur Food Res Technol 227(3): 835–843

[180] Cao Z, Chen X, Chang E, Du J (2009) Effective Components of Chinese Herbal Compound Decoction and Maillard reaction. Clin J Integr Med. 15(3):224-228

[181] Brudzynski K, Miotto D (2011) The relationship between the content of Maillard reaction-like products and bioactivity of Canadian honeys. Food Chem 124(3): 869–874

[182] Turkmen N, Sari F, Poyrazoglu ES, Velioglu YS (2006) Effects of prolonged heating on antioxidant activity and color of honey. Food Chem 95(4): 653–657
[183] Morales FJ, Jiménez-Perez SJ (2004) Peroxyl radical scavenging activity of melanoidins in aqueous systems. Eur Food Res Technol 218(6): 515–520
[184] Einarsson H, Snygg BG, Eriksson C (1983) Inhibition of bacterial growth by Maillard reaction products. J Agric Food Chem 31(5): 1043–1047
[185] Ye XJ, Ng TB, Nagai R (2010). Inhibitory effect of fermentation byproducts on formation of advanced glycation end-products. Food Chem 121(4): 1039–1045
[186] Kim SS, Gallaher DD, Csallany AS (2000) Vitamin E and probucol reduce urinary lipophilic aldehydes and renal enlargement in streptozotocin-induced diabetic rats. Lipids 35(11): 1225–1237
[187] Kedziora-Kornatowska K, Szram S, Kornatowski T, Szadujkis-Szadurski L, Kedziora J, Bartosz G (2003) Effect of vitamin E and vitamin C supplementation on antioxidative state and renal glomerular basement membrane thickness in diabetic kidney. Nephron. Exp. Nephrol. 95(4):e134–e143
[188] Yoshida Y, Saito Y, Jones LS, Shigeri Y (2007) Chemical reactivities and physical effects in comparison between tocopherols and tocotrienols: physiological significance and prospects as antioxidants. J. Biosci. Bioeng. 104(6): 439–445
[189] Kuhad A, Chopra K (2009) Attenuation of diabetic nephropathy by tocotrienol: involvement of NFkB signaling pathway. Life Sci. 84(9-10): 296–301
[190] Voziyan PA, Hudson BG (2005) Pyridoxamine: The many virtues of a Maillard Reaction inhibitor. Ann N Y Acad Sci 1043:807–816
[191] Dicter N, Madar Z, Tirosh O (2002) Alpha-lipoic acid inhibits glycogen synthesis in rat soleus muscle via its oxidative activity and the uncoupling of mitochondria. J. Nutr. 132(10):3001-3006
[192] Penniston KL, Nakada SY, Holmes RP, Assimos, DG (2008) Quantitative assessment of citric acid in lemon juice, lime juice, and commercially-available fruit juice products. J Endourol. 22(3): 567–570
[193] Nagai R, Nagai M, Shimasaki S, Baynes JW, Fujiwara Y (2010) Citric acid inhibits development of cataracts, proteinuria and ketosis in streptozotocin (type 1) diabetic rats. Biochem Biophys Res Commun. 393(1): 118–122
[194] Tupe RS, Agte VV (2010) Role of zinc along with ascorbic acid and folic acid during long-term in vitro albumin glycation. Br J Nutr 103(3):370–377
[195] Powell SR (2000) The antioxidant properties of zinc. J Nutr 130(5S Suupl): 1447S–1454S

Chapter 10

Alpha-1-Acid Glycoprotein (AGP) as a Potential Biomarker for Breast Cancer

Kevin D. Smith, Jennifer Behan,
Gerardine Matthews-Smith and Anthony M. Magliocco

Additional information is available at the end of the chapter

1. Introduction

The majority of plasma proteins are modified by the addition of oligosaccharide chains (glycans) to their surface. The process, known as glycosylation, is responsible for introducing huge structural variation and is important in the determination of functional properties expressed by the overall glycoprotein. Glycan composition and structure can vary widely, potentially showing disease- specificity, unlike the underlying polypeptide sequence [1-3]. Therefore analysis of their structures could offer more accurate and condition-specific markers which is critical when early treatment and monitoring is crucial in tackling diseases. The monomeric units of glycans are the monosaccharides, linked together via glycosidic bonds between the –OH reducing group of C1 and any other –OH of adjacent residues (a condensation reaction). The bonds exist in either an α or β anomeric configuration, depending on the orientation of the bond, providing further opportunity to generate structural variability.

Unlike the structure of the polypeptide backbone, glycosylation is not directly encoded by specific genes, but is reliant upon the concerted action of a series of highly specific enzymes (glycosyltransferases and glycosidases), which are resident in the ER and Golgi of cells, to initiate and processing glycan chain precursors. Any alterations in the expression of the genes encoding these enzymes due to, for example a pathophysiological condition, will affect the structure and composition of the glycans generated. During the synthesis and processing of glycans, they are transferred by the enzyme oligosaccharyltransferase from a lipid-linked oligosaccharide donor to polypeptide chains, either at asparagine (N-linked) or Serine/Threonine (O-linked) residues. N-linked glycans share a common pentasaccharide core ($Man_3GlcNAc_2$), however the overall structures can vary widely, primarily differing by the sequence and quantity of monosaccharides of which they are composed. Glycan synthesis then proceeds by the sequential addition/removal of individual monosaccharide units.

Glycan expression is cell and tissue specific and dependent on the presence of (patho)-physiological conditions [4]. Abnormal expression of even a single enzyme participating in this process may alter the subsequent steps and give rise to aberrant oligosaccharide structures. The cell types containing these biomolecules determine the enzymes expressed and therefore the glycans – variation reflects these source cells or tissue and the physiological and biochemical conditions present [3-4]. For example, abnormal or unusual glycan moieties are often detected on the surface of tumour cells, as well as among secreted glycoproteins. Many well known tumour-associated antigens are glycolipids or glycoproteins that contain aberrant glycan structures [5].

The post translational modification of proteins with oligosaccharide chains offers several advantages as a potential source of cancer biomarkers [5-6]. While protein biomarkers frequently occur in low abundance, distinctive cellular glycan structures, including those that are tumour-specific, are typically fairly abundant. Such structures are usually synthesised in multiple copies on a single glycoprotein molecule. In addition, the same tumour-specific glycan structures may be present on several different glycoproteins. Importantly, many currently used biomarker molecules, such as prostate specific antigen, CA125, and carcinoembryonic antigen, are monitored solely based on protein levels, not on their tumour-specific glycan moieties.

Glycoproteins with modified glycosylation resulting from the presence of cancer may be identifiable in body fluids that are typically amenable for clinical testing [6]. Alpha – 1 (or $\alpha 1$) – acid glycoprotein (AGP, or orosomucoid) is an important example of a naturally occurring N-linked plasma glycoprotein secreted by liver parenchymal cells. It was first isolated and characterised in the 1950s and although its specific biological role/function has yet to be clearly defined it is considered a natural anti-inflammatory and immunomodulatory agent [7-9], the concentration and/or 'normal' glycosylation of which may change under various physiological and pathophysiological conditions. It is a highly glycosylated (45%) protein and therefore investigations into alterations in its glycan profile during different diseases has been, and is still, of great interest (Figure 1). Sialic acid, in the form of N-acetylneuraminic acid (NeuAc), accounts for 12% of the total number of monosaccharides, and is responsible for giving AGP its low isoelectric point of 2.7.

Alterations in the protein levels of AGP have been well documented for numerous physiological and pathophysiological conditions including lung and breast cancer [10] and malignant mesothelioma [11]. It has been suggested that a marked increase in AGP concentration may limit adverse reactions such as inflammation by providing a form of negative feedback [12]. Increased concentrations are associated with expression of the ORM-1 and ORM-1 genetic variants of AGP; however the proportions of each do not necessarily differ to that of a healthy individual [10]. In terms of plasma concentration, increased levels of AGP have been detected in the plasma of patients with breast cancer [13] and have also been shown to increase with disease progression [14] but to return to normal upon treatment with tamoxifen [15]. This indicates that the prognosis of breast cancer appears to be linked to the APR in general and AGP levels in particular. Not only is there a dearth of studies determining the levels of AGP in breast cancer but more importantly there has been no investigations of the glycosylation of AGP in breast cancer, the extent that it is altered and/or its significance as a diagnostic marker.

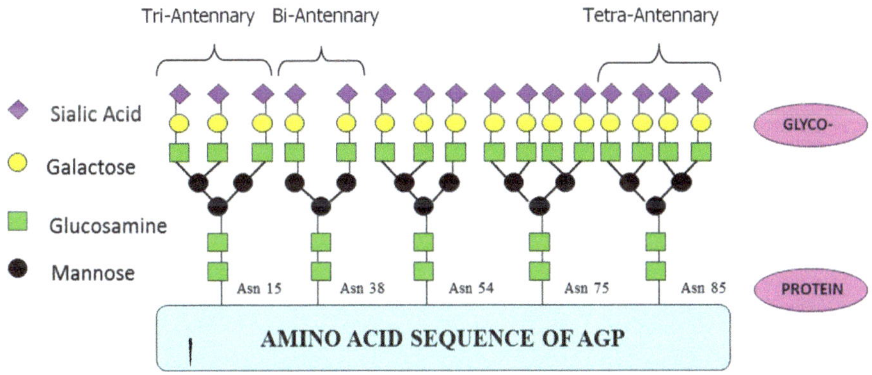

Figure 1. General schematic illustrating the structure of alpha-1-acid glycoprotein.

Although the concentration of AGP alone is not diagnostic for a particular pathological condition, the altered glycosylation of AGP (microheterogeneity) in different diseases, provides an alternative biomarker target. These alterations have the potential to be markers for particular diseases and also disease progression. The major form of heterogeneity, type I, is associated with a reduction in the number of branches on the oligosaccharide chains [4, 16-17]; the minor form (type II) concentrates on the composition of the oligosaccharide chains namely the extent of fucosylation and sialylation of the five oligosaccharide chains [18].

There is potential for the existence of 10^5 different glycoforms of AGP due to the huge structural variability provided mainly by the presence of glycans. However in normal, non-pathological conditions there are only 12-20 expressed, each exhibiting various degrees of branching, fucosylation and sialylation [18]. The number of feasible glycoforms is reduced because the asparagines (Asn) residues are selective for the type of glycans they express in terms of the degree of branching. The first and second Asn sites (Asn 15 and 38) prefer to harbour bi-antennary glycans whilst Asn 15 will not bind a tetra-antennary chain and Asn 38 never binds fucosylated glycans. Conversely, Asn 75 and Asn 85 prefer more branched glycans; in fact site Asn 75 never carries bi-antennary chains and Asn 85 usually expresses the greatest degree of α1,3-fucosylation [16]. This increased branching could explain why only these two sites potentially carry a tetra-antennary chain with more than one fucose residue [8]. It has been found that the majority of AGP glycans have a tri/tetra-antennary structure (85-90%) and the remaining 10-15% is bi-antennary glycans [19]. Sialic acid can be bound α2-3 or α2-6 on terminal galactose residues; fucose can be bound α1-3 to an external N-acetlyglucosamine and also by α1-6 and α1-2 to core N-acetlyglucosamine and galacose residues respectively [8]. The extent of fucosylation in a healthy population varies with 30% [8] to 40% [20] containing no fucose.

Altered glycosylation of AGP has been widely studied in a number of pathophysiological conditions. Decreased branching of the five glycans of AGP has been demonstrated in acute inflammation while an increase is associated with chronic inflammation [21-23]. Chronic inflammatory states such as those found in patients with RA are characterised by an increase in branching [24, 25], increased fucosylation of the chains [26-31] and increased sialic acid content [32-33]. The expression of the antigen sialyl lewis X (SLe^X) has been found to increase on AGP oligosaccharides [20, 25-26] at any one of the 5 sites of glycosylation [27]. SLe^X is normally expressed on white blood cells and aids through binding to the endothelial ligand, E-Selectin, the extravasation of leukocytes into tissues to mount their inflammatory response. De Graaf and colleagues [20] proposed that the increased expression of this antigen on AGP represents a negative feedback response during inflammatory conditions. Further work by Jørgensen *et al.* [26] using a microtitre cell-protein binding assay discovered that SLe^X-containing AGP expressed in patients with RA was able to inhibit binding of SLe^X presenting cells to E-Selectin. Thus, during inflammatory states such as those in RA, the abnormally glycosylated form of AGP has the ability to competitively bind to E-Selectin and inhibit leukocyte extravasation exerting an overall anti-inflammatory effect. Highly branched structures have also been noted in individuals with liver disease, along with increased fucosylation of these chains [34-36]. Increased concentrations of AGP were found in the cerebrospinal fluid (CSF) of patients with multiple sclerosis (MS) compared to disease free individuals [37] and in the plasma/serum of patients with burns injuries [38]. It is known that various factors act to influence AGP glycosylation including the congentital disorders [28], pregnancy [29, 33], drug or glucocorticoid use [39], and oral contraceptive use [40].

Investigations into the heterogeneity of AGP in cancer have found an increase in bi-antennary glycan content with increased fucosylation and sialylation. An early study by [41] reported that the major microheterogeneity of AGP from malignant, benign and normal groups was significantly different from each other. When compared with normal controls, sera from patients with inflammatory lung disease (benign group) exhibited an increase in tri-and tetra-antennary glycans and in sera from lung cancer patients (malignant group) there was an increase in bi-antennary expression. This work was contradicted the following year in an investigation [42] which repeated the study including an evaluation of the microheterogeneity of AGP in colorectal cancer and reported that there was no significant difference in the microheterogeneity of AGP from malignant and non-malignant inflammatory disease. Evidence to support the cancer-induced expression of bi-antennary glycans on AGP detailed by Hansen and colleagues was provided by [43] who reported a similar finding on AGP from the ascitic fluid of liver cancer patients.

Reports of changes in the minor microheterogeneity of AGP from studies of the oligosaccharide content of the chains in cancer have been detailed. Early studies [44-45] observed an increase in the sialic acid content of cancer AGP glycans compared to normal controls. A more recent study by Hashimoto *et al.* [46] on AGP branching and fucosylation in cancer concluded that changes in microheterogeneity could be used as a marker of carcinoma progression and prognosis. Patients with advanced stages of the disease who displayed highly branched glycans with a high fucose content for a significant duration post-surgery were associated with poor prognosis whereas, patients who did not display this increase in branching and fucosylation were likely to have a good, more reassuring prognosis. When highly branched and fucosylated glycans were present a prolonged period after surgery, a poor prognosis was associated. Similar analyses can be associated with surgery and other traumas in general [47].

2. Problem statement

The majority of studies [reviewed in 7-9] have concluded that, although increased levels of AGP have been detected in the plasma of patients with breast cancer, this concentration is largely unrelated to disease progression. The prognostic value of AGP glycosylation in breast cancer is still largely unproven. Although the limited research to date has discovered structural differences in the glycosylation in AGP isolated from different cancers and early and advanced stages of the same cancer, very few studies have looked at breast cancer samples. We hypothesise that alpha-1-acid glycoprotein (AGP), which is a common constituent in all blood, could be a diagnostic marker for early breast cancer.

3. Application area

There is a critical need for information regarding a reliable biomarker that could be used to predict the onset, severity, progression and prognosis of breast cancer. The identification of alterations in AGP glycosylation specific to individual stages of breast cancer could, through being altered to the "normal" healthy AGP glycosylation pattern, result in the development of a diagnostic test based on AGP glycosylation for the onset, progression and/or prognosis of breast cancer.

4. Aims of study

To determine whether specific alterations in the glycosylation pattern of alpha-1-acid glycoprotein (AGP) could be diagnostic for the for the onset, progression and/or prognosis of breast cancer.

5. Study design and methodology (Figure 2)

5.1. Patient samples

Breast cancer can occur as a non-malignant or malignant tumour. Non-malignant cancers can be further subdivided into ductal carcinoma *in situ* (DCIS) and lobular carcinoma *in situ* (LCIS). In situ refers to the state and location of the breast epithelial cells in that they have gone through malignant transformation and are now proliferating but remain at the site of origin and do not penetrate the basement membrane into surrounding tissues. DCIS and LCIS are not at risk of metastatic spread as there are no blood vessels or lymphatics in the epithelial layer of the breast [48]. The two most common types of invasive or malignant breast cancer are Invasive ductal carcinoma (IDC) and Invasive lobular carcinoma (ILC). Approximately 75% of breast cancers are IDC with ILC accounting for a further 10%. The remaining 15% of diagnosed invasive breast cancers are made up of rare types such as mucinous, medullary, tubular, and papillary. Blood samples were taken from non-fasting patients with consent by Calgary Laboratory Services. The blood samples were drawn into 5mL Red Top tubes and processed within 30 minutes. During processing, the blood samples were spun for 30 minutes to separate them into components. The samples were stored at -80 degrees Celsius. The patient details are given in Table 1.

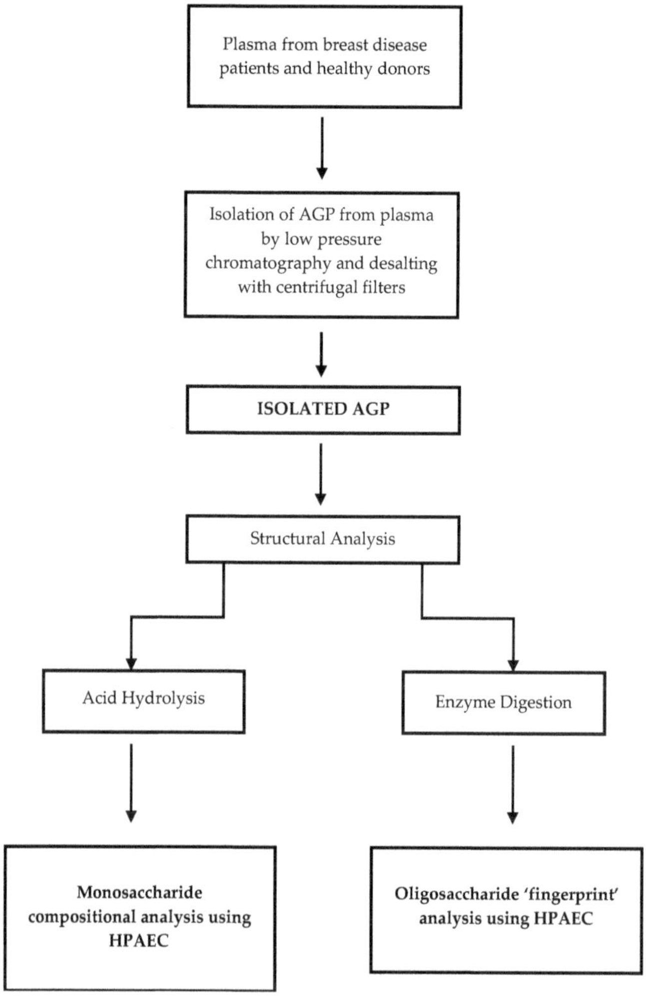

Figure 2. The design of the study to determine the glycosylation patterns of AGP in breast cancer

No	Age	Description	Her2	Malignant.	Grade	ER
1	62	Benign proliferative breast disease	N/A	No	.	.
2	78	DCIS (+ invasive ductal carcinoma)	neg.	Yes	3.00	.
3	48	Benign proliferative breast disease	N/A	No	.	.
4	77	Invasive duct carcinoma	neg.	Yes	1.00	1.00
5	51	Invasive duct carcinoma	neg.	Yes	2.00	1.00
6	35	Invasive duct carcinoma	neg.	Yes	1.00	1.00
7	39	Benign proliferative breast disease	N/A	No	.	.
8	35	Malignant phyllodes tumor.	neg.	Yes	.	.
9	53	DCIS (+ invasive ductal carcinoma)	neg.	Yes	.	.
10	49	Invasive duct carcinoma	pos.	Yes	2.00	1.00
11	76	Invasive duct carcinoma	neg.	Yes	2.00	1.00
12	65	Invasive duct carcinoma	neg.	Yes	3.00	1.00
13	32	Invasive ductal carcinoma	pos.	Yes	.	.
14	43	Invasive duct carcinoma	neg.	Yes	3.00	1.00
15	40	Biphasic fibroepithelial lesion	N/A	No	.	.
16	48	Benign proliferative breast disease	N/A	No	.	.
17	50	Invasive duct carcinoma	pos.	Yes	3.00	1.00
18	69	Invasive duct carcinoma	neg.	Yes	2.00	.
19	45	Invasive duct carcinoma	pos.	Yes	3.00	.00

Table 1. Demographic and clinical details of patients used in study

5.2. Materials

5.2.1. AGP Isolation

Polyethylene glycol (PEG 3350), the low pressure chromatographic material (Cibacron Blue 3GA, Q-sepharose and Red Sepharose CL-6B), potassium thiocyanate, sodium acetate, sodium chloride and Trizma base were purchased from Sigma (Poole, UK). Bio-Rad, Hemel Hampstead, UK, provided the poly-prep disposable 10ml columns. HPLC-grade water and the Centricon YM10 centrifugal filter device were purchased from Rathburn Chemicals (Walkerburn, UK) and Millipore (Bedford, USA) respectively.

5.2.2. High pH anion exchange chromatography monosaccharide analysis

HPAEC was carried out on a DX600 system supplied by Dionex (Camberley, UK), consisting of a GP50 gradient pump and an ED40 electrochemical detector, controlled with PeakNet software via a Dell OptiPlex GX110 personal computer. Separation of monosaccharides was achieved using a CarboPac PA-100 (250x4mm) and guard column (50x4mm). HPLC grade water was purchased from Rathburn Chemicals, Walkerburn, UK, and sodium hydroxide 50% w/v was obtained from BDH, Poole, UK. The monosaccharide standards and Dowex-50W were purchased from Sigma, Poole, UK. HPLC grade trifluoroacetic acid was purchased from Perbio Science, UK Ltd, Talenhall, UK.

5.2.3. Oligosaccharide analysis

PNGase F and buffers (NP40 and NE buffer G7) were supplied by New England BioLabs (New England, USA). HPAEC was carried out on a DX500 system supplied by Dionex (Camberley, UK), consisting of a GP40 gradient pump and an ED40 electrochemical detector, controlled with PeakNet software via a Vtech personal computer. Separation of monosaccharides was achieved using a CarboPac PA-100 (250x4mm) and guard column (50x4mm). HPLC grade water was purchased from Rathburn Chemicals, Walkerburn, UK, and sodium hydroxide 50% w/v was obtained from BDH, Poole, UK. Sodium acetate was purchased from Sigma (Poole, UK).

5.3. Methods

5.3.1. Polyethylene Glycol (PEG) precipitation

First, PEG 3350 was added to each plasma sample to a final concentration of 40% w/v in a microcentrifuge tube. This was then mixed vigorously for ten minutes before being stored overnight at 4ºC. Subsequently the mixture was centrifuged at 14,000 rpm for 30minutes. The supernatant was removed to a clean eppendorf and frozen until required, with the pellet being discarded.

5.4. Low pressure chromatography to isolate AGP

A disposable BioRad low-pressure chromatography column (10ml) was packed to a bed volume of approximately 6ml with the appropriate chromatographic resin. Several volumes of the appropriate Elution buffer (Table 2) was washed through the column using a Pharmacia LKB peristaltic pump at a flow rate of 0.5ml/min, with the change in UV absorbance at 280nm being detected using a Pharmacia LKB optical unit. Peak fractions were collected in 15ml disposable centrifuge tubes and dried under vacuum to approximately 2ml between stages, using a centrifugal evaporator. 2-3ml of the PEG-precipitated serum sample is first applied to a Cibacron blue column, to which the bilirubin site of albumin binds. The remaining soluble proteins are eluted with 0.05M Tris/0.1M potassium chloride/0.02% Azide (pH7.0) elution buffer while the albumin is removed from the column with 0.5M potassium thiocyanate. One peak is collected upon elution. After drying the peak is next applied to the ion-exchange column Q-sepharose. This divides the eluate from the previous column into three fractions; α_1-antitrypsin, AGP and transferrin. A column gradient of 0.075-0.5M sodium chloride (pH 6.5) is utilised, and the second peak, which contains AGP, is collected. The column is regenerated between samples by washing the column with 0.075M NaCl buffer until a flat baseline is observed. The AGP is then, after again drying to 2-3ml, applied to a column containing the Red sepharose resin in order to remove any trace proteins that may remain. The AGP in the sample is eluted with a 0.03M sodium acetate elution buffer and collected, while the bound proteins are removed with 1M sodium chloride. The AGP peak is then dried to 2ml in preparation for desalting.

	Buffer	
Chromatography resin	**Elution**	**Desorption**
Cibacron Blue	50mM Trizma/0,1mM KCl/0.02% Azide at pH 7.0	500mM KSCN/50mM Trizma/0.1mM KCl/0.02% Azide pH 7.0
Q-sepharose fast flow	20mM Trizma/75mM NaCl pH6.5	**Equilibrium buffer**: 20mM Trizma/75mM NaCl pH 6.5
Red sepharose CL-6B	30mM NaOAc	1M NaCl/30mM NaOAc

Table 2. Buffers for each stage of low-pressure chromatography

5.5. Desalting

In The AGP peak from red sepharose was subsequently desalted using centrifugal ultrafiltration. 2mL of the sample was added to a Centricon centrifugal filter device. This removes any salt in the solution and also acts to concentrate the AGP. The centricons were spun at 4000 rpm until the filtrate had passed entirely through the membrane into the filtrate vial. To ensure complete removal of salt from the sample 1ml of HPLC water was added to the sample reservoir and spun as before. The centricon was then inverted and spun for 10 minutes at 1000 rpm, transferring the concentrate into the collection vial. The desalted glycoprotein is then transferred to an eppendorf and dried down to completion using a centrifugal evaporator.

5.6. High pH Anion Exchange Chromatography (HPAEC)

5.6.1. Preparation of monosaccharides: Acid Hydrolysis of AGP

Approximately 50µg of isolated AGP was hydrolysed using 100µl 2M trifluoroacetic acid (TFA) and 50µl 4M hydrochloric acid in a reacti-vial then sealed with a Teflon disc. The reacti-vial was then placed on a heating block pre-heated to 100ºC for four hours in order to reach optimum hydrolysis. The monosaccharides were separated from the peptide fragments by eluting the hydrolysate down a Dowex-50W cation exchange column. A Pasteur pipette was plugged with glass wool and loaded with the resin to a bed volume of 1ml. Each AGP sample was added to a separate column and washed through with HPLC water. The neutral monosaccharides were washed through and collected, then dried to completion in preparation for analysis.

5.6.2. Monosaccharide analysis

Monosaccharide analysis was carried out using a DX600™ system. Separation of the monosaccharides, as their oxyanions, was performed at pH 13 on a CarboPac PA-100 column. The monosaccharides were resolved using an isocratic elution with 30mM NaOH, at a flow rate of 0.5ml/min for 35 minutes. The column was regenerated after each run with 0.5M NaOH for 5 minutes, prior to equilibration with 20mM NaOH for a further

10 minutes. Detection of monosaccharides was by pulsed amperometric detection (PAD) at the following pulse potentials: 0 sec: E=0.05V; 0.29sec: E=0.05V; 0.49sec: E=0.05V; 0.50sec: E=0.05V; 0.51sec: E=0.6V; 0.6sec: E=0.6V; 0.61sec: E=-0.6V; 0.65sec: E=-0.6V; 0.66sec: E=0.05V (Figure 3).

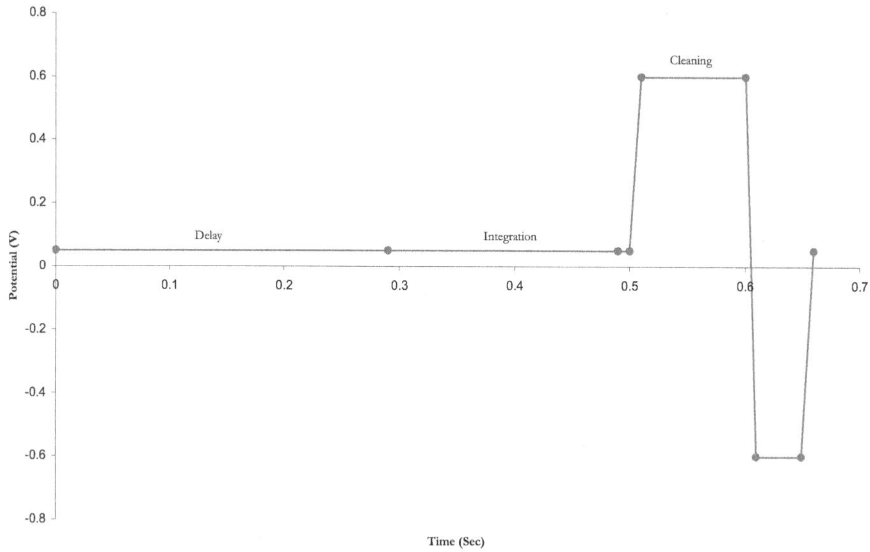

Figure 3. The triple pulsed waveform used in pulsed amperometric detection

An internal standard (IS) of 2-deoxy-D-galactose was utilised to determine the elution position of the monosaccharides by dividing the elution time of the IS by the time of the unknown peak. The calculated ratio can then be compared to known standards.

5.6.3. HPAEC-PAD oligosaccharide analysis

The required amount (50-100μg) of AGP was reconstituted in 100 μL of HPLC grade water. The AGP solution was subsequently denatured through boiling for one hour. To the denatured AGP 10 μL of NP-40, 10 μL of NE Buffer G7 and 100U of PNGase F were added. PNGaseF cleaves the oligosaccharide chains from the AGP peptide moiety. The enzymatic reaction was allowed to proceed overnight while incubated at 37C. Following incubation the solution was subjected to cold ethanol precipitation. 1ml of cold ethanol was added to each sample and then centrifuged at 10000rpm for 10 minutes which pelleted the peptide fractions and left the soluble oligosaccharides in the supernatant. Thus the supernatant was removed and dried to completion.

Analysis of the enzymatically released oligosaccharides was performed using a DX500™ system. Similar to monosaccharide analysis, oligosaccharides are eluted as their oxyanions, formed in the highly alkaline environment. Resolution is achieved using a CarboPac PA-100 column packed with a micro-pellicular resin. The column was equilibrated for 10 minutes with 10% 1M sodium hydroxide, 5% 1M sodium acetate and 85% HPLC grade water. Following equilibration a linear gradient was developed over 40 minutes to a final eluant composition of 10% sodium hydroxide, 20% sodium acetate and 70% water. These conditions were maintained for 5 minutes before the column was regenerated for ten minutes with 50% sodium hydroxide and 50% water. The pulsed amperometry was performed as described for the monosaccharide analysis.

6. Results

This investigation studied AGP from 19 breast cancer sufferers with either malignant or non malignant forms of breast cancer in term of the monosaccharide composition of the oligosaccharide chains and the oligosaccharide fingerprint of the intact chains themselves. Under alkaline conditions, the hydroxyl groups of a monosaccharide, either individually or as part of a sequence, are ionised to varying degrees such that they exist as negatively charged oxyanions. The varying location of the OH groups results in slight differences in the pKa value (ranging from 12-14) of individual monosaccharides under these conditions. These differences in ionisation, which are unique both with respect to individual monosaccharides and also cumulatively in terms of a sequence of monosaccharides, can be exploited in chromatographical terms using the anion exchange stationary phases and the electrochemical detection (based on pulsed potentials) components of high pH anion exchange chromatography (HPAEC). For monosaccharide analysis, an alkaline environment of 0.03M sodium hydroxide ionises monosaccharides in the hydrolysate and results in the formation of oxyanions unique to individual monosaccharides. The interaction of the charged oxyanions with the strong anion-exchange CarboPac ™ column and the slight differences in relative pKa values aids the chromatographic separation of individual monosaccharides. The monosaccharide compositions were determined, after acid hydrolysis, using high pH anion exchange chromatography (HPAEC). The results obtained are summarised in Table 3 and a typical HPAEC profile is shown in Figure 4.

The monosaccharide composition of the AGP oligosaccharide chains was found to differ between normal, non-malignant and malignant groups. Of all the samples containing fucose, the malignant breast cancer group displayed the highest average level of this monosaccharide compared to the normal and non malignant breast disease group. Only the malignant breast cancer population had N-acetylgalactosamine present; this monosaccharide is not usually found in AGP and was completely absent from the normal samples analysed in the study. The graphs in Figures 5-7 illustrate the statistical significance of the fucose and Nacetylgalactosamine composition of AGP with respect to malignancy.

Patient	**Description**	Galactose*	Fucose[1]	GalNAc[1]
1	Benign proliferative breast disease	14.12	1.22	.00
2	DCIS (+ invasive ductal carcinoma)	15.58	1.78	1.25
3	Benign proliferative breast disease	14.57	.98	.00
4	Invasive duct carcinoma	16.22	3.12	2.87
5	Invasive duct carcinoma	17.33	4.12	3.55
6	Invasive duct carcinoma	16.47	2.95	1.14
7	Benign proliferative breast disease	14.29	1.04	.00
8	Malignant phyllodes tumor.	18.29	4.56	3.22
9	DCIS (+ invasive ductal carcinoma)	15.98	2.55	.98
10	Invasive duct carcinoma	17.78	4.01	3.68
11	Invasive duct carcinoma	17.56	3.93	3.44
12	Invasive duct carcinoma	19.91	6.02	4.45
13	Invasive ductal carcinoma	18.84	5.10	4.06
14	Invasive duct carcinoma	19.76	5.77	4.89
15	Biphasic fibroepithelial lesion	11.75	.76	.00
16	Benign proliferative breast disease	13.95	1.31	.00
17	Invasive duct carcinoma	19.82	5.52	4.36
18	Invasive duct carcinoma	17.43	4.19	3.49
19	Invasive duct carcinoma	19.64	5.59	4.80

* units in mol/mol AGP

Table 3. The monosaccharide compositions of oligosaccharide chains from AGP, isolated from healthy volunteers, patients with non malignant breast cancer and patients with malignant breast cancer, determined using high pH anion exchange chromatography

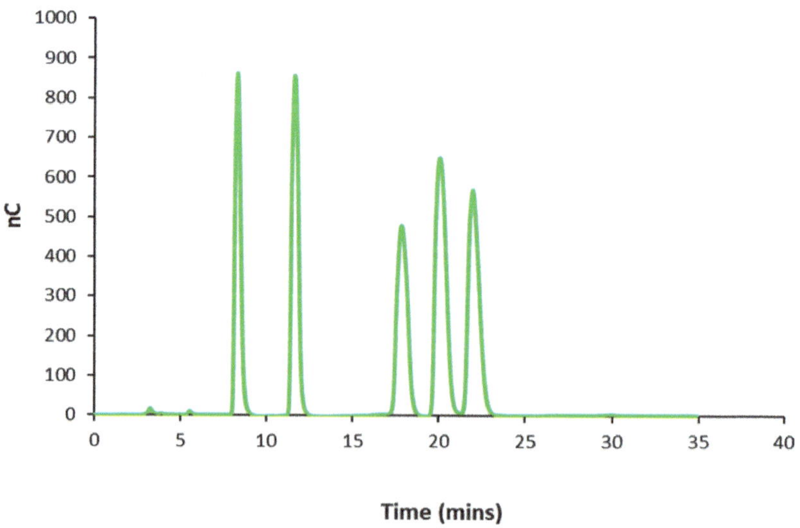

Figure 4. The separation of a mixture of monosaccharides by high pH anion exchange chromatography

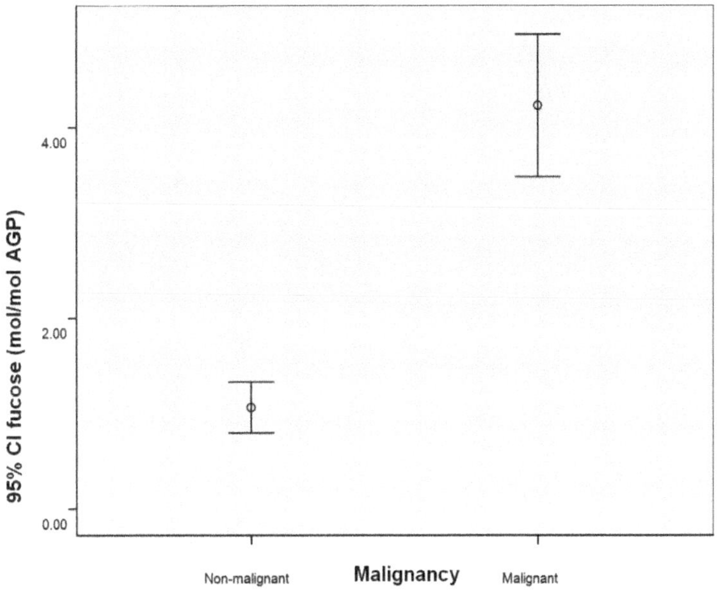

Figure 5. Fucose Composition of AGP with Respect to Malignancy

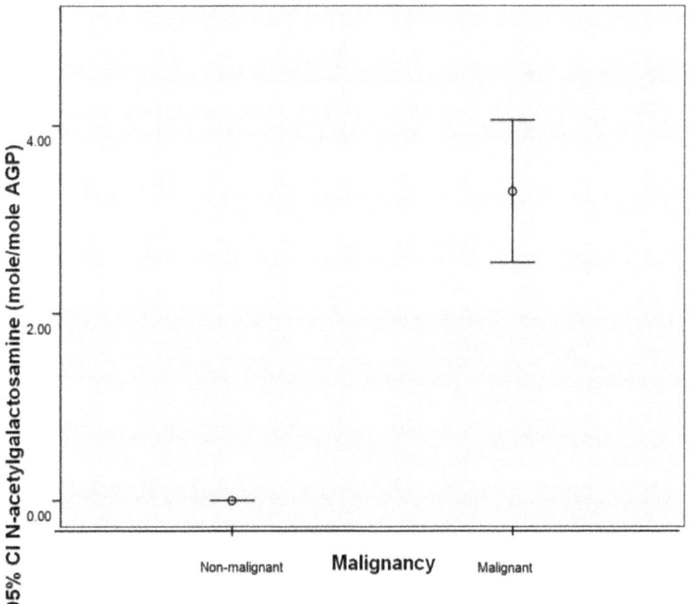

Figure 6. N-acetylgalactosamine Composition of AGP with Respect to Malignancy

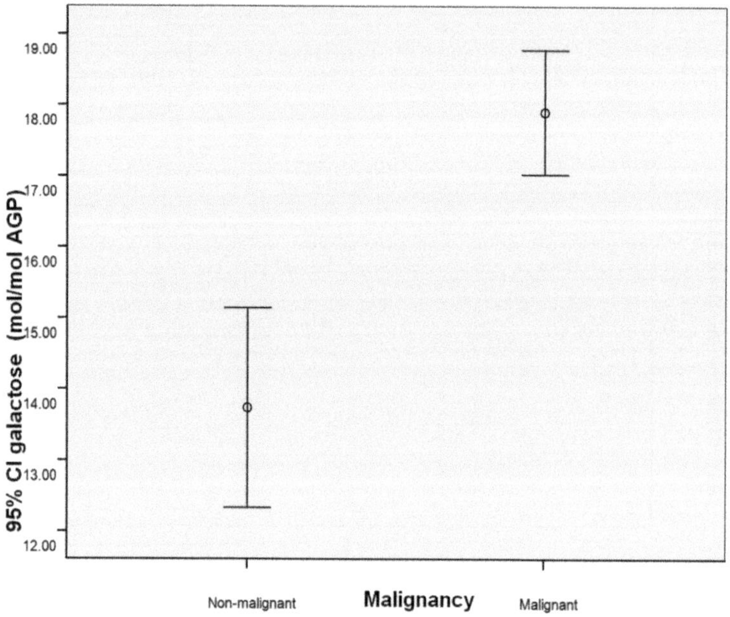

Figure 7. Galactose Composition of AGP with Respect to Malignancy

The monosaccharides fucose and N-acetylgalactosamine are not normally found on AGP oligosaccharide chains; however, both were found to be present in plasma from breast cancer patients. Most importantly, the presence of N-acetylgalactosamine appears to correlate with the presence of malignancy (Figure 6). While fucose appears to be present in small levels in patients with benign proliferative breast disease and in increased levels in malignant breast disease (Figure 5), N-acetylgalactosamine appears to be present only in ductal carcinoma in-situ (DCIS) and invasive carcinoma. Each AGP molecule has five oligosaccharide chains, and each chain can have two, three, or four branches. Given that each branch contains a galactose molecule (refer to Figure 1), a higher degree of branching can be identified by a higher level of galactose. Therefore, increased galactose levels appear to be indicative of increased Malignancy (Figure 7). The AGP oligosaccharide profiles differed between the normal, non-malignant and malignant breast cancer groups.

The flexibility of HPAEC is illustrated by the fact that it also allows the efficient separation of oligosaccharide chains on the basis of their size, formal charge, monosaccharide composition and intra-chain linkages [49]. A major advantage over other techniques is that the HPAEC of oligosaccharides requires no prior derivatisation and can detect to picomole sensitivity. The separation of oligosaccharide structures using HPAEC is based primarily on the negative charge present on the oligosaccharides due to the presence of terminal sialic acid residues. The greater the number of sialic acid residues, the greater the negative charge. This leads to a stronger interaction between the oligosaccharide and the resin in the HPAEC analytical column. The application of a sodium acetate gradient causes the displacement of oligosaccharides, thus residues with a greater charge require a higher acetate concentration to cause displacement. Therefore, for example, a trisialylated oligosaccharide structure is retained longer than a bisialylated structure. Thus there is a positive correlation between the degree of negative charge and the elution time. Within charge bands, structures can be further separated on the basis of size i.e. a biantennary, bisialylated structure will be retained less than a triantennary, bisialylated oligosaccharide.

On a more subtle level the separation of oligosaccharides within each charge band is determined by the sequence of monosaccharides which form the chain and also the linkages between them. The involvement of hydroxyl groups in glycosidic bonds between monosaccharides can alter the charge of the sequence. Therefore oligosaccharides with the same monosaccharides but differing in one linkage can be separated using this technique. The analysis of oligosaccharides is qualitative, unlike monosaccharide analysis which is both qualitative and quantitative, allowing the pattern of oligosaccharides to be determined [49-50]. The separation of a standard library of the oligosaccharide chains of AGP is shown in Figure 8. In the profile, each peak corresponds to a single oligosaccharide chain. The sialylated portion of oligosaccharide chains,such as those found in AGP, are separated into distinct charge bands based on the number of terminal sialic acids.

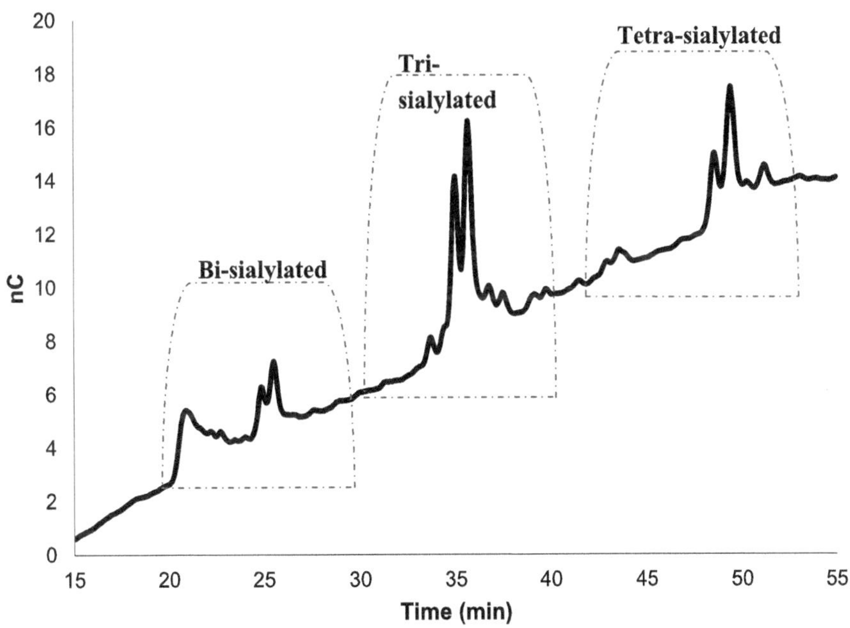

Figure 8. A HPAEC separation of a commercially available library of AGP oligosaccharides illustrating their separation into charged bands based on the number of sialic acid residues present at the chain terminus.

The greater the number of sialic acids, the greater the overall negative charge resulting in a longer retention time. Typically bisialylated oligosaccharides elute between 20 and 30 minutes; trisialylated, 30 to 40 minutes and tetrasialylated 40-50 minutes (Figure 8). The oligosaccharide profiles for the AGP oligosaccharides from the clinical samples analysed in this study are shown in Figure 9.

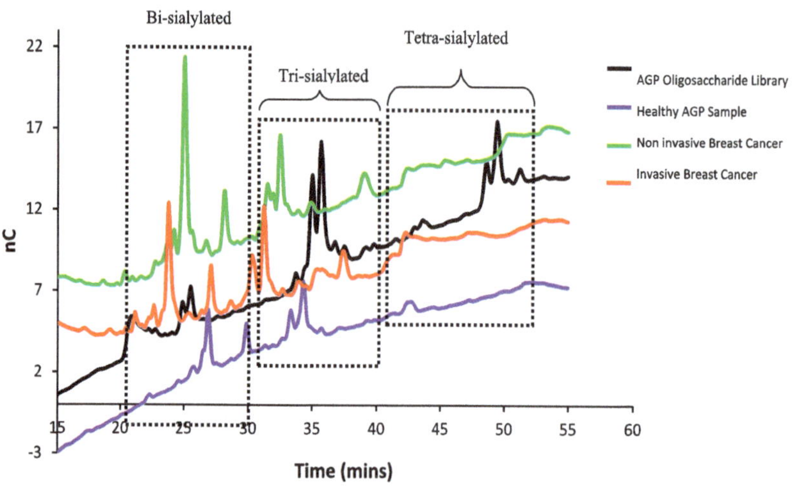

Figure 9. Representative oligosaccharide profiles of AGP, isolated from healthy volunteers (purple line), non malignant breast cancer (green line), malignant breast cancer (red line) were separated using high pH anion exchange chromatography and compared with a standard library of AGP oligosaccharides (black line).

The oligosaccharide profiles of AGP oligosaccharides from healthy volunteers, non malignant breast cancer and malignant breast cancer are comparable with the standard library (Figure 8) in that they both display peaks in the bi-, tri- and tetra-sialylated regions. Although qualitatively similar, the profiles for the breast cancer populations are noticeable for the earlier elution of the comparable peak regions. This reduction in retention is indicative of the presence of fucose on the oligosaccharide chains [49-50] which reinforces the results of the monosaccharide compositional analysis. Additionally, the two cancer populations appear to have fewer tetra-sialylated chains (elute 40-50 minutes) even if the presence of fucosylation is taken into account.

7. Further research

The oligosaccharide chains of each AGP in this study were released and analysed by high pH anion exchange chromatography [Figure 8] and showed differences that could be diagnostic and the basis of future research. Monosaccharide compositional analysis using HPAEC has highlighted the differences that exist between normal and breast disease populations. Specifically the presence of fucose, galactose and N-acetylgalactosamine appears to be correlated with breast cancer malignancy. The presence of N-acetylgalactosamine in the malignant population, together with its absence from the healthy and non-malignant populations suggests that the modification could form the basis of a serum biomarker for breast cancer malignancy with potential prognostic utility. HPAEC also provides useful information about the oligosaccharide structures present on AGP and allows the development of an oligosaccharide 'fingerprint' for each cohort of breast cancer patients.

8. Conclusion

Glycosylated molecules play major roles in oncogenesis, but their potential as cancer biomarkers remains unclear. Historically the reasons were twofold. Firstly, the large heterogeneity of structures arising from the cellular biosynthesis of glycans were not able to be detected and resolved by the analytical technology available at that point in time. Secondly the definitive diagnosis of cancerous disease normally required invasive techniques. These two hindrances have been addressed through the development of techniques such as high pH anion exchange chromatography (HPAEC) which now allows the sensitive (to picomole level) determination of glycosylated structures including linkage isomers of oligosaccharide chains. The solution to the second problem could lie in the development of tests based on the presence, in biological fluids, of glycosylated molecules which are disease-specific markers.

Alpha-1-acid glycoprotein is a prominent example of a molecule in which alterations in the structure of the surface oligosaccharides has provided some potential in the diagnosis and prognosis of physiological and pathophysiological conditions and determination of the effectiveness of treatments. There have been a vast number of studies concerned with quantifying the levels and characterising the microheterogeneity of AGP in a number of diseases [Reviewed in 7-9]. Variation in the branching of the chains, levels of fucose, sialic acid and the presence or absence of antigens such as sialyl Lewis X are commonly reported and represent the potential exploitation of AGP heterogeneity as prognostic and diagnostic markers of disease [7].

In spite of major advances in detection and treatment, deaths from breast cancer are still high therefore there is a very definite requirement for the identification of a breast cancer specific biomarker to indicate the onset of the disease. Breast cancer is the most common cancer in women and currently diagnosis is reliant upon mammography and invasive techniques such as biopsy. There is no serum biomarker currently used in clinical practice for the detection and diagnosis of breast cancer. The discovery of a breast cancer-specific biomarker whose presence and/or altered expression in the serum precedes the appearance of a malignant mass would allow the development of a non-invasive serologic test and provide an invaluable screening tool in the assessment of high risk individuals.

The current study has identified that the quantitative and qualitative changes in AGP glycosylation observed among patients with malignant and non malignant breast cancer may in fact, provide the basis of a serum biomarker with potential prognostic utility. The monosaccharide and oligosaccharide composition of the AGP glycosylation was found to differ between normal, non-invasive and invasive groups. From our results, fucose, N-acetylgalactosamine and galactose could be used as a screening test for early detection of breast cancer, and also as markers of breast carcinoma progression or "tumour load", as changes in the levels of the substances correlate with disease progression. Most importantly, N-acetylgalactosamine could be used as a marker for the presence of malignancy in the breast, as N-acetylgalactosamine is present only in the plasma of patients with malignant breast lesions. The implications of the use of N-acetylgalactosamine as a marker for the presence of breast malignancy is profound; a definite diagnosis of breast carcinoma could be carried out by simple blood tests.

The current standard method of diagnosing early stages of breast cancer is mammography which, along with other imaging technology, requires expensive instrumentation. The use of blood tests to diagnose breast cancer would obviously have numerous advantages over mammography, and would prove to be an invaluable tool in the fight against breast cancer. The use of blood tests based on differences in AGP glycosylation for diagnosing breast cancer could be done inexpensively using ELISA. Mammography requires radiologists to review images, while screening for AGP markers can be performed by laboratory technicians with a high degree of automation. Mammography suffers from a relatively high degree of false diagnoses, while screening for N-acetylgalactosamine appears to diagnose breast cancer with 100% accuracy. Breast cancer has a high rate of survival if it is detected in its early stages. Unfortunately, large segments of the at-risk female population do not receive mammograms regularly, as a result of lack of access to mammography facilities, high cost, discomfort caused by the procedure, and potential or perceived risks of radiation exposure. The factors preventing women from receiving mammograms regularly would not be present with blood tests. Small blood samples could be taken from patients with minimal discomfort and cost in local medical centres. These samples could then be sent to large, centralised testing facilities for processing. The improved accessibility and convenience of this strategy over mammography would result in a greater segment of the at-risk female population receiving regular breast cancer screening, which would result in more early diagnoses, and therefore higher survival rates.

The use of blood tests that employ AGP alterations as markers of breast cancer has obvious commercial applications. These blood tests would likely become routine diagnostic procedures like the prostate-specific antigen (PSA) test. The potential market value of a breast cancer blood test would be comparable to that of the PSA test, which is about $1.5 billion (USD) per year in the United States alone. Clearly, on an international scale, blood tests for diagnosing breast cancer would have a huge market value. In the event that AGP alterations are found to be markers of other cancers, the international market value of AGP-based blood tests would likely double or triple, reaching as high as $20 billion (USD) per year.

Author details

Kevin D. Smith* and Jennifer Behan
School of Life, Sport and Social Sciences,
Edinburgh Napier University, Sighthill Campus, Edinburgh, EH11 4BN, Scotland

Gerardine Matthews-Smith
School of Nursing, Midwifery and Social Care,
Edinburgh Napier University, Sighthill Campus, Edinburgh, EH11 4BN, Scotland

Anthony M. Magliocco
Department Chair Anatomic Pathology,
Esoteric Laboratory Services, H Lee Moffitt Cancer, Tampa, Florida, USA

Acknowledgement

Kevin Smith acknowledges funding from Friends for an Earlier Breast Cancer Test and Professor David George, University of Glasgow. Jennifer Behan was a funded by a Caledonian Scholarship from the Carnegie Trust for the Universities of Scotland.

9. References

[1] Kolarich D, Lepenies B, Seeberger PH. (2012) Glycomics, glycoproteomics and the immune system. Curr Opin Chem Biol. Accessed 2012 Jan 3. [Epub ahead of print].
[2] Hennet T. (2012) Diseases of glycosylation beyond classical congenital disorders of glycosylation. Biochim Biophys Acta. Accessed 2012 Feb 9. [Epub ahead of print]
[3] Dall'Olio F, Malagolini N, Trinchera M, Chiricolo M. (2012) Mechanisms of cancer-associated glycosylation changes. Front Biosci. 1;17:670-99.
[4] van Dijk W., Turner G A. and Mackiewicz A. (1994) Changes in glycosylation of acute-phase proteins in health and disease: occurrence, regulation and function. Glycosylation and Disease. 1, 5-14.
[5] Adamczyk B, Tharmalingam T, Rudd PM. (2011) Glycans as cancer biomarkers. Biochim Biophys Acta. Accessed Dec 9. [Epub ahead of print].
[6] Narimatsu H, Sawaki H, Kuno A, Kaji H, Ito H, Ikehara Y. (2010) A strategy for discovery of cancer glyco-biomarkers in serum using newly developed technologies for glycoproteomics. FEBS J. 277(1):95-105.
[7] Ceciliani F, Pocacqua V. (2007) The acute phase protein α1-acid glycoprotein: a model for altered glycosylation during diseases. Current Protein and Peptide Science. 8, 91-108.
[8] Fournier T, Medjoubi-N N , Porquet D. (2000) Alpha-1-acid glycoprotein. Biochimica et Biophysica Acta. 1482, 157-171.
[9] Israili Z H, Dayton G. (2001) Human alpha-1-glycoprotein and its interactions with drugs. Drug Met. Revs. 33, 161-235

* Corresponding Author

[10] Duché J., Urien, S., Simon N, Malaurie E, Monnet I, Barré J. (2000) Expression of the genetic variants of human alpha-1-acid glycoprotein in cancer. Clin. Biochem. 33, 197-202.

[11] Hervé F, Duché J, Jaurand M. (1998) Changes in expression and microheterogeneity of the genetic variants of human α1-acid glycoprotein in malignant mesothelioma. J. Chrom. B. 715, 111-23.

[12] Crestani B, Rolland C, Lardeux B, Fournier T, Bernuau D, Poüs C, Vissuzaine C, Li L, Aubier M. (1998) Inducible Expression of the α_1-Acid Glycoprotein by Rat and Human Type II Alveolar Epithelial Cells. J. Immunol. 160, 4596-4605.

[13] Thompson DK, Haddow JE, Smith DE, Ritchie RF (1983). Elevated serum acute phase protein levels as predictors of disseminated breast cancer. Cancer; 51 2100-2104.

[14] Kailajarva M, Ahokoski O, Virtanen A, Salminen E, Irjala K. (2000). Early effects of adjuvant tamoxifen therapy on serum hormones, proteins and lipids. Anticancer Research, 20, 1323-1327.

[15] Dorssers LC, vander Flier S, Brinkman S, van Agthoven T, Veldscholte J, Berns EM, Klijn JG, Beex LV, Foekens JA (2001). Tamoxifen resistance in breast cancer: elucidating mechanisms. Drugs; 61, 1721-1733.

[16] Treuheit MJ, Costello CE, Halsall HB. (1992) Analysis of the five glycosylation sites of human α1-acid glycoprotein. Biochem. J.1992, 283, 105-112.

[17] Bayard B, Kerckaert JP (1980) Evidence for uniformity of the Carbohydrate Chains in Individual Glycoprotein Molecular Variants. Biochem Biophys Res Commun 95, 777-784.

[18] Albani JR. (1997) Binding effect of progesterone on the dynamics of α1-acid glycoprotein. Biochimica et Biophysica Acta 1336, 349-359.

[19] Perkins SJ, Kerckaert J-P, Loucheux-lefebvre M. (1985) The shapes of biantennary and tri/tetraantennary α_1-acid glycoprotein by small angle neutron and X-ray scattering. European J Biochem 147, 151-171.

[20] De Graaf TW, Van der Stelt ME, Anbergen MG, van Dijk W. (1993) Inflammation-induced expression of Sialyl Lewis X-containing glycan structures on α1-acid glycoprotein (orosomucoid) in human sera. J. Exp. Med. 177, 657-666.

[21] Fassbender K, Zimmerli W, Kissling R, Sobieska M, Aeschlimann A, Kellnar M, Müller W. (1991) Glycosylation of α1-acid glycoprotein in relation to duration of disease in acute and chronic infection and inflammation. Clinica Chimica Acta. 203, 315-328.

[22] Kratz E, Poland DCW, van Dijk W, Kątnik-Prastowska I. (2003) Alterations of branching and differential expression of sialic acid on alpha-1-acid glycoprotein in human seminal plasma. Clinica chimica acta: international journal of clinical chemistry. 331, 87-95.

[23] van den Heuvel MM, Poland DCW, De Graaf CS, Hoefsmit ECM, Postmus PE, Beelen RHJ, van Dijk W. (2000) The degree of branching of the glycans of α1-acid glycoprotein in asthma. American Journal of Respiratory and Critical Care Medicine. 161, 1972-1978.

[24] Smith KD, Pollacchi A, Field M, Watson J. (2002) The heterogeneity of the glycosylation of alpha-1-acid glycoprotein between the sera and synovial fluid in rheumatoid arthritis. Biomed. Chromatogr. 16, 261-266.

[25] Elliott MA, Elliott HG, Gallagher K, McGuire J, Field M, Smith KD. (1997) An investigation into the Concanavalin A reactivity, fucosylation and oligosaccharide microheterogeneity of α_1-acid glycoprotein expressed in the sera of rheumatoid arthritis patients. J. Chrom., Biomed. Appl. 688, 229-237.
[26] Jorgenson HG, Priest R, Smith K D. (1998) Modulation of sialyl Lewis X dependent cell binding to E-selectin by glycoforms of alpha-1-acid glycoprotein expressed in rheumatoid arthritis. Biomed. Chromatogr. 12 : 1-7.
[27] Dage JL, Ackermann B, and Halsall HB. (1998) Site localization of sialyl Lewis^X antigen on α_1-acid glycoprotein by high performance liquid chromatography-electrospray mass spectrometry. Glycobiology. 8, 755-760.
[28] van Dijk W, Koeleman C, Van het Hof B, Poland D, Jakobs C, Jaeken, J. (1998) Increased α3-fucosylation of α_1-acid glycoprotein in patients with congenital disorder of glycosylation type IA (CDG-Ia) Trends Glycosci Glycotecnol.10, 235-245.
[29] Havenaar EC, Hoff RC, van den Eijnden DH, van Dijk W. Severe rheumatoid arthritis prohibits the pregnancy-induced decrease in α3-fucosylation of α_1 – acid glycoprotein. Glycoconjugate J 1998, 15, 389-395.
[30] Poland DC, Schalkwijk CG, Stehouwer CD, Koeleman C.A, van het Hof B, van Dijk W. (2001) Increased alpha3-fucosylation of alpha1-acid glycoprotein in Type I diabetic patients is related to vascular function. Glycoconjugate Journal. 2001, 18, 261-268.
[31] Rydén I, Påhlsson P, Lundblad A., Skogh T. (2002) Fucosylation of α1-acid glycoprotein (orosomucoid) compared with traditional biochemical markers of inflammation in recent onset rheumatoid arthritis. Clinica Chimica Acta. 317, 221-229.
[32] Elliott MA, Jørgensen HG, Smith KD. (1998) Hypersialylation of α_1-acid glycoprotein in rheumatoid arthritis. Pharmacy and Pharmacology Communications. 4, 545-54.
[33] Pawiłowicz, M., Hirnle, L. and Kątnik-Prastowska, I. (2006) Alterations of N-glycan branching and expression of sialic acid on amniotic fluid alpha-1-acid glycoprotein derived from second and third trimesters of normal and prolonged pregnancies. Clinica Chimica Acta. 367, 86-92.
[34] Anderson N, Pollacchi A, Hayes P, Therapondos G., Newsome P, Boyter A, and Smith K. (2002) A preliminary evaluation of the differences in the glycosylation of alpha-1-acid glycoprotein between individual liver diseases. Biomed. Chromatogr. 16, 365-372.
[35] Rydén I, Påhlsson P, Lindgren S. (2002) Diagnostic accuracy of alpha (1)-acid glycoprotein fucosylation for liver cirrhosis in patients undergoing hepatic biopsy. Clinical chemistry. 48, 2195-2201.
[36] Mooney P, Hayes P, Smith K. (2006) The putative use of α-1-acid glycoprotein as a non-invasive marker of fibrosis. Biomed. Chromatogr. 20, 1351-1358.
[37] Adam P, Sobek O, Táborsk L, Hildebrand T, Tutterová, Žáček P. (2003) CSF and serum orosomucoid (α-1-acid glycoprotein) in patients with multiple sclerosis: a comparison among particular subgroups of MS patients. Clinica Chimica Acta. 334, 107-110.
[38] French D, Watson J, McCahill B, Taggart I, Smith KD. (2002) A preliminary evaluation of the functional significance of alpha-1-acid glycoprotein glycosylation on wound healing. Biomedical Chromatography. 16, 412-419.

[39] Pos O, van Dijk W, Ladiges N, Linthorst C, Sala M, van Tiel D, Boers W. (1998) Glycosylation of four acute-phase glycoproteins secreted by rat liver cells in vivo and in vitro. Effects of inflammation and dexamethasone Eur J Cell Biol 46, 121-128.
[40] Brinkman-Van der Linden ECM, Havenaar EC, Van Ommen ECR, Van Kamp GJ, Gooren LJG, van Dijk W. (1996) Oral estrogen treatment induces a decrease in expression of sialyl Lewis x on α1-acid glycoprotein in females and male-to-female transsexuals. Glycobiology. 6, 407-412.
[41] Hansen JS, Larsen VA, Bøg-Hansen TC. (1984) The microheterogeneity of α1-acid glycoprotein in inflammatory lung disease, cancer of the lung and normal health. Clinica Chimica Acta. 138, 41-47.
[42] Bleasby AJ, Knowles JC, Cooke NJ. (1985) Microheterogeneity of α1-acid glycoprotein: lack of discrimination between benign and malignant inflammatory disease of the lung. Clinica Chimica Acta. 150, 231-235.
[43] Fujii M, Takahashi N, Hayashi H, Furusho T, Matsunaga K, Yoshikumi C. (1988) Comparative study of α_1-acid glycoprotein molecular variants in ascitic fluid of cancer and non-cancer patients. Anticancer Research. 8, 303-306
[44] Moule SK, Peak M, Thompson S, Turner GA. (1987) Studies of the sialylation and microheterogeneity of human serum α1-acid glycoprotein in health and disease. Clinica Chimica Acta. 166, 177-185.
[45] Turner GA, Skillen AW, Buamah P, Guthrie D, Welsh J, Harrison J, Kowalski A. (1985) Relation between raised concentrations of fucose, sialic acid and the acute phase proteins in serum from patients with cancer: choosing suitable serum glycoprotein markers. Journal of Clinical Pathology. 38, 588-592.
[46] Hashimoto S, Asao T, Takahashi J, Yagihashi Y, Nishimura T, Saniabadi A R, Poland DCW, van Dijk W, Kuwano H, Kochibe N Yazawa S. (2004) α1-acid glycoprotein fucosylation as a marker of carcinoma progression and prognosis. Cancer. 101, 2825-2836.
[47] van Dijk W, Havenaar EC, Brinkman-van der Linden ECM. (1995) α1-acid glycoprotein (orosomucoid): pathophysiological changes in glycosylation in relation to its function. Glycoconjugate J. 12, 227-233.
[48] Sakorafas GH, Farley DR (2003) Optimal management of ductal carcinoma in situ of the breast. Surg Oncol 2003: 12:221-240.
[49] Behan J, Smith, K.D. (2011) The analysis of glycosylation: A continued need for high pH anion exchange chromatography. Biomed. Chrom. 25, 39-46.
[50] Smith KD. (1997) Structural Elucidation of the N-linked Oligosaccharides of Glycoproteins using High pH Anion Exchange Chromatography. Advances in Macromolecular Carbohydrate Research 1, 65-91.

Section 4

Cell Biology of Glycosylation

Chapter 11

Impact of Yeast Glycosylation Pathway on Cell Integrity and Morphology

Anna Janik, Mateusz Juchimiuk, Joanna Kruszewska, Jacek Orłowski, Monika Pasikowska and Grażyna Palamarczyk

Additional information is available at the end of the chapter

1. Introduction

The subject of protein N- and O-glycosylation in the yeast *S.cerevisiae* has already been well described in many excellent reviews [1-3]. However, less is known about the occurrence of these processes in the yeast *Candida albicans*, an opportunistic human pathogen.

In the *N*-glycosylation reactions catalysed by the Alg enzymes, dolichyl phosphate (DolP) serves as a lipid acceptor of sugar residues. The Alg enzymes are present both on the cytosolic side of the ER: Alg7, Alg1, Alg2 and Alg11) and on its lumenal face(Alg3, Alg5, Alg6, Alg8-10 and Alg12 [3-6]. Cytosolic Alg enzymes, necessary for viability of yeast and mammalian cells in culture, produce DolPP-GlcNAc$_2$Man$_5$ from DolP, UDPGlcNAc and GDPMan. There is also circumstantial evidence that DolPP-GlcNAc$_2$Man$_5$ is delivered into the lumen of the ER by a putative flippase (Rft1p) [7]. Within the ER the DolPP-GlcNAc$_2$Man$_5$ product is converted to DolPP-GlcNAc$_2$Man$_9$Glc$_3$ using DolPMan and DolPGlc as sugar donors [3]. As it was mentioned already DolPMan is also a substrate for protein *O*-glycosylation, where it serves as a donor of the first mannose to be attached to hydroxyl groups of serine and threonine. The second and subsequent mannose residues are transferred directly from GDPMan [1, 2]. DolPMan is also involved in the synthesis of the sugar part of glycosylphosphatidyl inositol anchor in yeast and other eukaryotes [2]. Moreover, a large group of cell wall glycoproteins is attached to the glucan polymers via a GPI-remnant structure.

On the other hand, there is a growing literature describing the involvement of cell wall carbohydrates in fungus-host interactions (for review see [8]), as well as in the maintenance of cell wall integrity [9]. Thus, one can predict a functional link between N-glycosylation and O-mannosylation of cell wall proteins, cell wall integrity and/or fungus–host interactions. Nonetheless, open questions remain concerning the regulatory mechanisms of early events of protein glycosylation and their impact on the synthesis of outer glycan chain in cell surface glycoproteins of *S. cerevisiae* and *C. albicans*.

A model of the *S. cerevisiae* cell wall has been described. It contains four classes of interacting components: chitin, β1,3 and β1,6 glucan and mannoproteins [10,11]. This structure is similar to that of the *C. albicans* cell wall [11-13].

2. Contribution of the mevalonate pathway to protein glycosylation and cell wall integrity

Mevalonate pathway in yeasts is important not only for ergosterol biosynthesis but also for the production of nonsterol molecules, deriving from farnesyl diphosphate. Formation of cell wall proteins, i.e. the glycosyl- phosphatidylinositol (GPI) anchored (GPI-CWP) and proteins with internal repeats (pir-CWP,) requires, as an initial intermediate, DolP synthesized together with the other isoprenoid lipids in the mevalonate pathway.

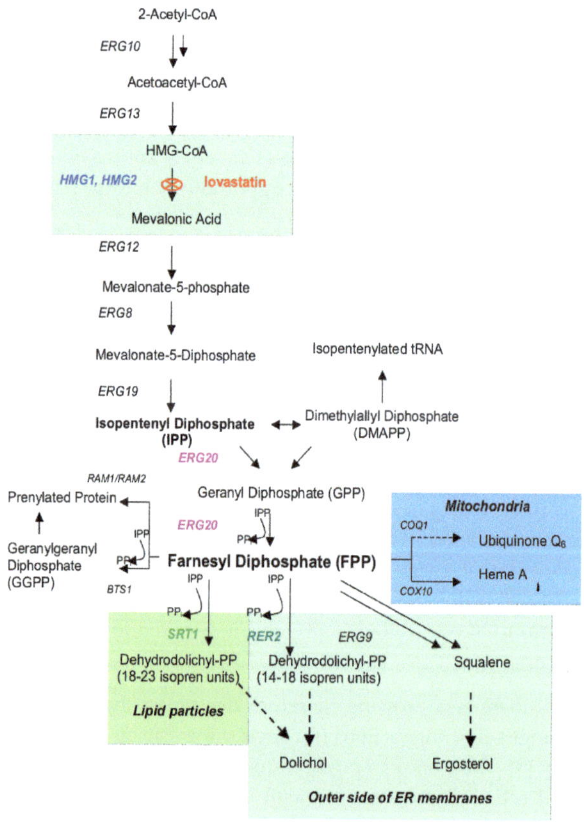

Figure 1. Isoprenoid (mevalonate) pathway in the yeast *Saccharomyces cerevisiae*.

Dolichol biosynthetic pathway has multiple levels of regulation. Thus, its cellular level is amenable to alterations affecting protein glycosylation and, in consequence, the cell wall structure. The pathway is shared with other isoprenoid lipids (Fig.1). The most abundant branch of the pathway, leading to sterol biosynthesis, is one of the main targets for anti-fungal drugs, which exploit the differences in the pathways and the end product - ergosterol in fungal cells and cholesterol in animals. The mevalonate pathway diverges after the synthesis of farnesyl diphosphate by farnesyl diphosphate synthase (FPPS), encoded by the *ERG20* gene.

Products of the subsequent reactions are shown together with the names of the genes, in capital letters, encoding the enzymes catalyzing them. Enzyme affected by lovostatin, an inhibitor of isoprenoid pathway, is indicated (K. Grabinska PhD thesis, IBB 2002)

Using a yeast based two hybrid system, we have identified the Yta7 protein interacting with FPPS, and showed that it was membrane-associated and localised both to the nucleus and the endoplasmic reticulum (ER). In order to assess the importance of the mevalonate pathway for cell wall synthesis and its role in cell-wall integrity (Fig 2), we investigated the effects of *YTA7* deletion at the transcriptional level. Our data [14] show that loss of *YTA7* function leads to activation of the genes implicated in cell wall integrity pathway (*CRZ1, FKS2,* and *KNR4)* and highlight a possible link between dolichol metabolism and cell wall synthesis.

Moreover, farnesol, which is likely to be derived from dephosphorylation of FPP, inhibits growth of *S. cerevisiae* and *C. albicans.* This inhibition is concomitant with a significant loss of intracellular diacylglycerol (DAG) [15], which is an activator of the PKC1-signaling pathway involved in the maintenance of cell wall integrity (described later, compare Fig.3)

De novo synthesis of dolichol starts with the 1'-4 condensation of farnesyl diphosphate (FPP) with 11-15 isopentenyl pyrophosphate units to form polyprenyl diphosphate (dehydrodolichol diphosphate) (Fig. 2). This reaction is catalysed by *cis*-prenyltransferase (*cis*-Ptase) encoded in *S. cerevisiae* by the *RER2* and *SRT1* genes [16].

Our results indicate that FPP or its derivatives regulate the transcription of *RER2* and *SRT1* genes as well as of *DPM1,* which encodes Dpm1p [17]. To enter the glycosylation pathways the *RER2* and *SRT1* gene products (dehydrodolichol diphosphate) need to be dephosphorylated, reduced and phosphorylated again by CTP-dependent dolichyl kinase, Sec59p [18-20].

Biosynthesis of isopentenyldiphosphate (IPP) from mevalonate diphosphate is catalysed by mevalonate pyrophosphate decarboxylase (Fig.1). Subsequently IPP is converted to dimethylallyl diphosphate (DMAPP) and DMAPP (C_5) is condensed with farnesyl diphosphate, composed of 3 isoprene units (C_{15}). This reaction is catalysed by Rer2 and Srt1p. Further elongation of the isoprenol chain occurs, by step wise addition of the 5-carbon isoprene units to reach species-specific chain length. Poliprenol diphosphate is dephosphorylated and reduced by alfa saturase (Dfg10p) to form dolichol. To enter glycosylation pathway dolichol is phosphorylated by CTP-dependent dolichol kinase.

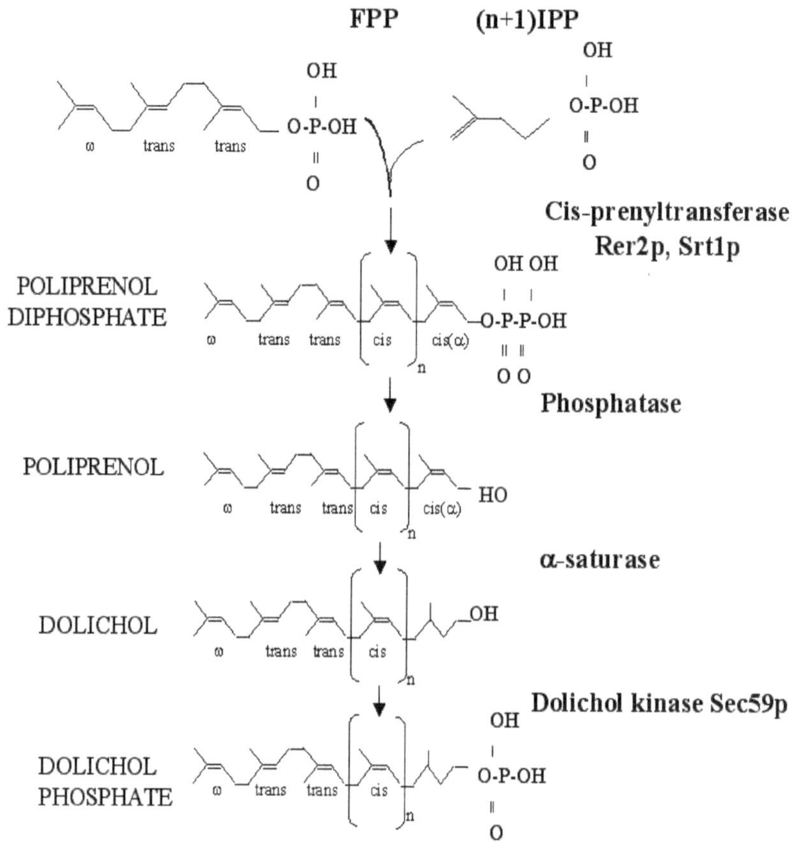

n=10-14 isoprene units for Rer2p and 14-23 for Srt1p products. (Adopted from [21])

Figure 2. Dolichyl phosphate biosynthesis *de novo* in yeast.

Our data indicate that in the rapidly dividing yeast cells "de novo" biosynthesis of dolichyl phosphate described above is a main source of its supply.

However, it is thought that in non dividing cells dolichyl phosphate might be derived mainly from recycling after each cycle of protein glycosylation, either by transfer of the oligosaccharide from DolPPGlcNAc2Man9Glc3 onto acceptor protein or from the transfer across the ER membranes of single glucose (Glc) and mannose (Man) residues from DolPMan or DolPGlc [22].

The length of dolichol molecules is species-specific and in yeast contains 14-18 isoprene units [23]. Although a great deal of progress has been made in the understanding of the enzymatic steps responsible for polyprenyl chain length termination and conversion of dehydrodolichol to dolichol, some open questions still remain. *In vitro,* cell extracts and membrane fractions from *S. cerevisiae* catalyse the biosynthesis of dolichol backbone i.e. dehydrodolichol [24] whereas in *vivo* yeast synthesise dolichols. On the base of the results obtained so far it was assumed that in *S. cerevisiae* polyprenyl diphosphates synthesised *in vitro* undergo immediate dephosphorylation and reduction of its alpha residue [24, 25]. A similar conclusion was reached for the rat liver system [26] and in general for metazoan eukaryotic cells [22].

3. Cell wall alteration resulting from the defect in dolichol and dolichyl phosphate formation

It has been shown that depletion of GDPMan pyrophosphorylase activity in *C.albicans* [27] leads to cell lysis, inefficient cell separation, impaired bud growth, clumping and flocculation, as well as increased sensitivity to a wide range of antifungal drugs. GDPMan pyrophosphorylase (*Ca SRB*1 encoded) catalyses the transfer of Man-1P to GDP and is responsible for the biosynthesis of GDPMan, the major sugar donor in yeast and filamentous fungi. On the other hand, overexpression of GDPMan pyrophosphorylase encoding gene (*mpg1*) in the filamentous fungus *Trichoderma reesei* resulted in a two-fold increase in GDPMan level, overglycosylation of secreted proteins, increased transcription of *dpm1*, DolPMan-synthase encoding gene, and increased enzyme activity [28]. Thus, it was assumed that the cellular level of GDPMan is one of the factors playing a regulatory role in protein glycosylation.

In this work we have concentrated on the assembly of mono- and oligo-saccharide lipid carrier (DolP), which is another substrate in protein glycosylation, and on it's effect on cell wall integrity and cell morphology.

In *C.albicans,* which unlike *S.cerevisiae* is an obligatory diploid, we initially cloned the genes involved in dolichol biosynthesis (*CaRER2 CaSRT1*) and phosphorylation (*CaSEC59*). To construct mutants defective in Rer2p or Sec59p activities, we used "URA-blaster" cassette [29] to delete one chromosomal copy of the gene. Since the genes under investigation were either essential or generated mutants with strong phenotypes, we adopted a conditional mutant approach and the second copy of the gene was placed under the control of a regulatable promoter.

By growing the strains in repressive conditions we were able to demonstrate that the defect in dolichol backbone synthesis or its phosphorylation, resulted in the aberrant cell wall structure and increased sensitivity to some antifungal drugs. Moreover, the normal morphogenesis of the fungus, e.g. hyphae formation, was prevented (Juchimiuk et al. 2011, in preparation).

Recently, we have cloned an ortholog of *S.cerevisiae DFG10* from *C.albicans* (orf 19.7841). *S.cerevisiae DFG10* encodes a protein with 3-oxo-5-alpha-steroid 4-dehydrogenase activity, lately postulated to be involved in dehydrodolichol reduction [30]. The latter was based on the finding that the mutant, named *dfg10-100,* isolated in a genetic screen for the strains defective in filamentous growth [31], was shown to be defective in N-linked glycosylation.

This defect was conferred by orf encoding the human SRD5A3 protein, involved in the reduction of dehydrodolichol diphosphate into dolicholdiphosphate. Thus, it was assumed that SRD5A3 is the human ortholog of the yeast *DFG10* gene encoding dehydrodolichyl diphosphate reductase. The product of *DFG10* shows 25% amino acid identity and 43% similarity with the human SRD5A3 protein, forming with others SRD proteins the steroid 5a-reductase family [30].

Double deletion of the *C.albicans* orf19.7841 rendered viable *Cadfg10* mutant strain. Its biochemical analysis revealed that the strain produces 70% of dehydro- and 30% of dolichols, whereas in the wild type, parental strain only dolichols are synthesized. The mutant strain was also oversensitive to tunicamycin, an inhibitor of N-glycosylation. Indeed, a defect in protein glycosylation was confirmed by assessing the degree of glycosylation of the marker glycoprotein hexosaminidase. Moreover, the *Cadfg10* mutant exhibits abnormal hyphal growth and increased resistance to some antifungal agents, thus indicating alterations of the *C.albicans* cell wall (Janik et al., manuscript in preparation).

However, based on our results, only 30% of dehydrodolihol is reduced to dolichol in *Cadfg10* null mutant. Thus, it is possible that *Ca*Dfg10p is not the only protein involved in dehydrodolichol reduction. Another likely candidate is an ortholog of *S.cerevisiae TSC 13*, encoding steroid reductase, i.e. putative enoyl reductase, encoded in *C.albicans* by orf 19.3293. However, the possible role of orf 19.3293 in dolichol synthesis is now under investigation.

We have also studied alterations in cell wall composition and integrity in *S. cerevisiae dpm1-1* mutant impaired in DolPMan formation and in the *sec59* mutant impaired in dolichol kinase activity. In our earlier work we were able to demonstrate that overexpression of the *S.cerevisiae DPM1* gene, encoding DolPMan synthase in *T.reesei* and *Aspergillus nidulans* led to increased secretion of fully glycosylated proteins, concomitant with alteration of the cell wall ultrastructure (*T.reesei*) or with accumulation of overproduced glycoproteins in the periplasmic space (*A.nidulans*) [32, 33].

For the *S. cerevisiae dpm1-1* as well as for the *sec59-1* mutants we observed an increased sensitivity to Calcofluor White (CFW) and an upregulated chitin level. Both mutated strains were also oversensitive to a variety of the external agents including antifungal drugs [34].

We have shown that the *sec59* and *dpm1* mutants are affected in cell wall composition [35]. A search for multicopy suppressors of the mutant phenotype, resulted in the isolation of the *RER2* gene , which as described above, is involved in the synthesis of the dolichol backbone and enhances protein glycosylation. In addition, the *sec 59-1* phenotype could be rescued by overexpression of the *ROT1* gene, encoding the endoplasmic reticulum Rot1 protein. The latter is implicated in cell wall biogenesis and acts as a chaperone for misfolded proteins [36, 37]. Recently, we have shown also that Rot1 interacts with Ost3, one of the nine subunits of the oligosaccharyltransferase complex, the key enzyme of *N*-glycosylation. Deletion of *OST3* in the *rot1-1* mutant causes a temperature sensitive phenotype as well as sensitivity towards compounds interfering with cell wall biogenesis, such as Calcofluor White, caffeine, Congo Red and hygromycin B. Oligosaccharyltransferase activity determined *in vitro* in membranes from *rot1-1ost3Δ* cells was found to be decreased to 45% compared to wild-type membranes, and model glycoproteins of *N*-glycosylation, like carboxypeptidase CPY, Gas1 or DPAP B, displayed an underglycosylation pattern. A physical interaction between Rot1 and Ost3 was demonstrated by affinity chromatography. Moreover, Rot1 was also found to be involved in the *O*-mannosylation process, as glycosylation of distinct glycoproteins of this type was affected as well. Altogether, it can be assumed that Rot1 acts also as a chaperone required to ensure proper glycosylation [38].

As already mentioned, a defect in protein O-mannosylation in fungi results in impaired cell wall integrity [9]. This process is initiated at the luminal side of the ER. The key enzyme of O-mannosylation is protein -O-mannosyltransferase (Pmtp) catalysing direct transfer of Man from DolPMan into the serine/threonine OH group in acceptor protein. This is followed by the addition of a short linear glycan, composed of mannosyl residues, directly from GDPMan. Whereas O-mannosylation is initiated in the ER, further modifications of the glycan chain occur in the Golgi apparatus. In *S. cerevisiae* the Pmt proteins are encoded by seven *PMT* genes [39]. A family of *PMT* genes was also identified in *C. albicans* [40] and in another fungal pathogen *Cryptococcus neoformans* [41]. Phylogenetic analysis of Pmt proteins revealed that they can be grouped in three subfamilies, Pmt1p, Pmt2p and Pmt4p [42]. The functions of Pmt proteins were studied, and it was demonstrated that in *S.cerevisiae* O-mannosylation affects protein stability, localisation and transport from the ER [43]. In *C.albicans* O-mannosylation is important for morphogenesis, adherence to host cells and virulence [44]. Characterization of the *PMT* gene family in *C. neoformans* revealed that Pmt proteins play a crucial role in maintaining cell morphology and cell wall integrity [41]. Similarly, in the filamentous fungus *T.reesei* diminished overall activity of Pmt proteins resulted in decreased O- as well as N-glycosylation and aberrant cell wall composition [45].

Studies of the effect of tunicamycin revealed the effect of the dolichol dependent protein N-glycosylation on *C. albicans* biofilm development. In its normal niche *C.albicans* forms biofilms that are attached to cell surfaces. In addition to the mucosal surfaces, such biofilms are often formed on implanted medical devices [46, 47]. Fully mature *C.albicans* biofilms consist of a complex of yeast, hyphae and pseudohyphae and exhibit increased resistance to antifungal drugs [48, 49]. Tunicamycin is a nucleoside antibiotic, produced by *Streptomyces lysosuperficus* that blocks the transfer of GlcNAc-1-P from UDPGlcNAc to DolP. This reaction, catalysed by the Alg7 protein, initiates the formation of DolPP-oligosaccharide and hence the whole N-glycosylation process. It has been demonstrated that physiological concentrations of tunicamycin display a significant inhibitory effect on biofilm development and maintenance, without affecting overall cell growth or morphology. Based on the above, conclusion was reached underlying the role of N-glycosylation in the developmental stages of biofilm formation [50]

4. Cell morphology in Golgi glycosylation mutants

A number of data indicates that a defect in glycosylation process occurring in the Golgi stack might affect cell morphology and virulence.

Initial steps of protein N- and O-glycosylation described so far occur in the ER. "Dolichol-dependent" glycosylation ends with the formation of DolPP-oligosaccharide (DolPPGlcNAc2Man9Glc3) and subsequent transfer of the oligosaccharide to the beta amido group of asparagine within the N-glycosylation site (Asn/X/Ser). Such a glycosylated peptide undergoes partial trimming, which is species-specific and in yeast involves removal of the three Glc and one Man residues. Partially processed glycopeptide is transported to the Golgi stack where the saccharide part (GlcNAc2Man8) undergoes further processing and maturation. Whereas glycosylation reactions occurring in the ER are well conserved in the eukaryotic cells, N-glycan processing in the Golgi is greatly diverse. In yeast, the core structure of glycoproteins is hypermannosylated (up to 200 Man residues), forming a backbone made up of alpha 1,6 linked residues, branched by alpha 1,2- and alpha 1,6 mannosyl residues. In addition, another mature core structure occurs in the Golgi, i.e. GlcNAc2Man8-13 [51]. A number of Golgi mannosyltransferases is involved in the synthesis of the sugar backbone and branching. It offers a vast majority of modifications of the sugar structure. The *OCH1* gene encodes an alpha 1,6 mannosyltransferase initiating addition of the outer sugar chain (www.yeastgenome.org). Deletion or depletion of *OCH1* causes lethality or slow-growth. The *och1* mutants cannot form high mannose oligosaccharides (~50+ Man residues) so they have decreased levels of cell wall mannoproteins, causing weakness and defects in bud formation and hypersensitivity to agents that attack the cell wall (calcofluor white, hygromycin B and SDS). Weakening of cell walls in hypotonic solution can be partially suppressed by the addition of osmotic stabilizers such as salt or sorbitol.

Mannan branching involves Mnn1p, one of five *S. cerevisiae* proteins of the MNN1 family. It is alpha-1,3-mannosyltransferase, an integral membrane glycoprotein of the Golgi complex, required for addition of alpha1,3-mannose linkages to N- and O-linked oligosaccharides,. Mnn1 mutant exhibits decreased sensitivity to hygromycin B [52]. It has been demonstrated that a double deletion of the *OCH1* and *MNN1* genes renders a strain which is unable to grow at non-permissive temperature and is defective in cell cytokinesis. On the other hand, no clear effect on the cell integrity was observed [53]. However, the role of Mnn5 protein (an iron regulated alpyha-1,2 mannosyltransferase) from *C.albicans* in cell wall integrity maintenance was demonstrated [54]. A *mnn5* delta mutant contains a reduced amount of cell wall mannan and hypersensitivity to the cell wall damaging agents. Disruption of the *C. albicans OCH1* homologue resulted in a temperature-sensitive growth defect and cellular aggregation. Outer chain elongation of *N*-glycans was absent in the null mutant, demonstrated by the lack of the alpha 1,6-linked polymannose backbone and the underglycosylation of *N*-acetylglucosaminidase. Moreover, the null mutant was hypersensitive to a range of cell wall perturbing agents. These mutants had near normal growth rates *in vitro* but exhibited decreased virulence in a murine model of systemic infection. Based on this the importance of *N*-glycan outer chain epitopes to the host-fungal interaction and virulence was assumed [55, 56].

5. Evidence for the role of glycan in cell defence mechanisms

The fungal cell wall is a highly dynamic structure, essential for the shape and stability of the cell. Thus in yeast and fungi cell wall integrity is tightly controlled by the activation of the protein kinase C –dependent MAP-kinase pathway [57] (Fig. 3).

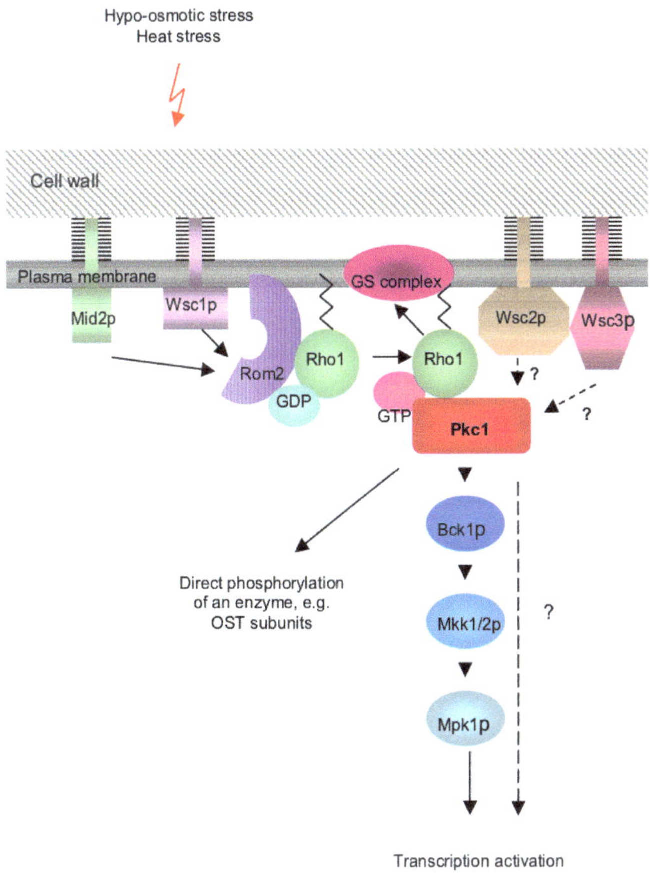

Figure 3. Cell wall integrity pathway (CWIP) in *Saccharomyces cerevisiae*

The sensors of the cell wall damage located in the plasma membrane (Mid2p, Wsc1-3p) pass the signal thorough the Rom1p to Rho1p, which in turn activates the Pkc1p- dependent cascade of MAP kinases (Bck1p, Mkk1/2p) Mpk1 kinase activates its main targets Swi4/6p and Rlm1p, transcription factors that activate expression of the cell wall genes (adopted from [57]).

Efficiency of this process depends, among others, on the glycosylation status of the receptor proteins located in the plasma membrane [58]. It has been demonstrated that the plasma membrane protein Mid2, a putative mechanosensor, responds to cell wall stresses and changes in the cell morphology induced by pheromone treatment. The response is related to the glycosylation status of the Mid2 protein, which is highly O-glycosylated and contains two potential glycosylation sites, one of them at Asn-35, carrying N-linked sugars. It was demonstrated that O-glycosylation is responsible for the stability of the protein, whereas the presence of the N-linked sugar chain is a prerequisite for its function as a sensor of external stimuli.

Our data [35] indicate that impairement of dolichol kinase (Sec59p) activity is concomitant with a defect in plasma membrane/cell wall Gas1p glycosylation and the activation of the CWIP. These results were further confirmed by the analysis of the cell wall composition of the *sec59-1* mutant as compared to the parental wild type. The cell wall of the mutant strain contains a significantly increased amount of chitin and beta-1,6 glucan (2.7 and 1.7 fold, respectively). Simultaneously we observed an activation of the CWIP in *sec59-1*, as compared to the wild type cells. All these differences were abolished by overexpression of the suppressor *RER2* gene, leading to increased synthesis of the oligosaccharide carrier (DolP) and protein glycosylation.

An activation of the cell defence mechanisms was also observed in *C.albicans OCH1* mutants, affected in outer sugar chain elongation (compare above).

Together, based on the knowledge acquired so far, it can be assumed that the glycosylation pathway in yeast and fungi offers many levels of regulation which might influence the final quality and quantity of cell wall glycoproteins and consequently cell surface immunogeneity and the fungal-host interaction. This includes also the processes occurring in the ER i.e. dolichol biosynthesis and glycosylation steps involving dolichol.

Author details

Anna Janik, Mateusz Juchimiuk, Joanna Kruszewska,
Jacek Orłowski, Monika Pasikowska and Grażyna Palamarczyk*
Institute of Biochemistry and Biophysics, Polish Academy of Sciences, Laboratory of Fungal Glycobiology, Warsaw, Poland

Acknowledgement

The experimental work was supported by the grant N N303 577238 from Ministry of Science and Higher Education, Poland, to G.P.

6. References

[1] Tanner W, Lehle L. Protein glycosylation in yeast. . Biochim Biophys Acta 1987; 906, :81-99.
[2] Herscovics A., Orlean P Glycoprotein biosynthesis in yeast. FASEB J. 71993; 540-550
[3] Burda P., Aebi, M. The dolichol pathway of N-linked glycosylation, Biochim. Biophys. Acta 1999;1426, 239-257

* Corresponding Author

[4] Cippolo J.F., Trimble R.B., Chi J.H., Yan Q., Dean N. The yeast *ALG11* gene specifies addition of the terminal alpha 1,2 Man to $Man_5GlcNAc_2PPDolichol$ intermediate formed on the cytosolic side of the endoplasmic reticulum J. Biol. Chem 2001; 276, 4267-4277

[5] Huffaker T., Robbins PW. Yeast mutants deficient in protein glycosylation, Proc. Natl. Acad Sci USA 1983; 80, 7466-7470

[6] Reiss G., te Heesen S., Zimmerman J., Robbins P.W., Aebi M. Isolation of the ALG6 locus of *Saccharomyces cerevisiae* required for glucosylation of the N-linked glucosylation pathway. Glycobiology.1996;6:493-498

[7] Hellenius A., Aebi M. Intracellular functions of N-linked glycans 2001; Science 291, 2364-2369

[8] Masuoka J. Surface glycans of *Candida albicans* and other pathogenic fungi: physiological roles, clinical uses and experimental challenges. Clin. Microbiol Rev 2004; 17, 281-310

[9] Strahl-Bolsinger, S., Gentzsch, M., Tanner W. Protein O-mannosylation. Biochim. Biophys. Acta 1999; 1426, 297-307.

[10] De Nobel H., Sietsma J.H., Van Den Ende, H., Klis, F. Model organisation and construction of the fungal cell wall in The Mycota VIII. 2001; Biology of the Fungal Cell , Howard/Gow Eds. Springer-Verlag, Berlin Heidelberg pp181-200.

[11] Klis, F. M., P. De Groot, and K. Hellingwerf Molecular organisation of the cell wall of *Candida albicans*. Med. Mycol 2001; 39,S1,1–8.

[12] Mormeneo S., Marcilla A., Iranzo M., Sentandreu R. Structural mannoproteins released by βelimination from *Candida albicans* cell wall. FEMS Microbiol. Lett 1994; 123, 131-136.

[13] Kapteyn, J. C., L. L. Hoyer, J. E. Hecht, W. H. Müller, A. Andel, A. J. Verkleij, M. Makarow, H. Van Den Ende, and F. M. Klis The cell wall architecture of *Candida albicans* wild-type cells and cell wall-defective mutants. Mol. Microbiol 2000; 35:601–611.

[14] Kuranda K, Grabinska K, Berges T, Karst F, Leberre V, Sokol S, François J, Palamarczyk G The *YTA7* gene is involved in the regulation of the isoprenoid pathway in the yeast *Saccharomyces cerevisiae*. FEMS Yeast Res 2009; 9:381-390.

[15] Machida K, Tanaka T Farnesol-induced generation of reactive oxygen species dependent on mitochondrial transmembrane potential hyperpolarization mediated by F(0)F(1)-ATPase in yeast. FEBS Lett.1999; 462(1-2):108-12

[16] Sato, M., Sato, K., Nishikawa, Si., Hirata, A., Kato, Ji., Nakano, A The yeast *RER2* gene, identified by endoplasmic reticulum protein localization mutations encodes *cis*-prenyltransferase, a key enzyme in dolichol biosynthesis. Mol Cell Biol 1999; 19: 471–483

[17] Grabińska K, Sosińska G, Orłowski J, Swiezewska E, Berges T, Karst F, Palamarczyk G Functional relationships between the Saccharomyces cerevisiae cis-prenyltransferases required for dolichol biosynthesis. Acta Biochim Pol 2005; 52: 221-32.

[18] Adair W.L., Cafmeyer N Cell cycle dependence of dolichol phosphate biosynthesis. Archiv. Biochem. Biophys 1987; 258, 491-497

[19] Burton W.A., Scher M.G., Waechter C.J. Enzymatic phosphorylation of dolichol in central nervous tissues. J. Biol. Chem 1979; 254, 7129-7136

[20] Szkopińska A., Nowak L, Swiezewska E., Palamarczyk G CTP-dependent lipid kinases of yeast. Archiv. Biochem. Biophys 1988; 266, 124-131

[21] Grabinska K, Palamarczyk G Dolichol biosynthesis in the yeast *Saccharomyces cerevisiae*: an insight into the regulatory role of farnesyl diphosphate synthase. FEMS Yeast Res 2002; 2:259–265.

[22] Hartley MB and Imperiali B., At the membrane frontier: A prospectus on the remarkable evolutionary conservation of polyprenols and polyprenyl-phosphates. Archiv. Biochem. Biophys. 2012; 517, 83-97.).

[23] Jung, P. and Tanner, W Identification of the lipid intermediate in yeast mannan biosynthesis. Eur. J. Biochem 1973; 37, 16-21

[24] Szkopin.ska, A., Karst, F. and Palamarczyk G Products of S. cerevisiae cis-prenyltransferase activity in vitro. Biochimie 1996; 78, 225-235

[25] Sagami, H., Igarashi, Y., Tateyama, S., Ogura, K., Roos, J. and Lennarz, W.J Enzymatic formation of dehydrodolichal and dolichal, new products related to yeast dolichol biosynthesis. J. Biol. Chem. 1996; 271, 9560-9566.

[26] Sagami, H.A., Kurisaki, A. and Ogura, K Formation of dolichol from dehydrodolichol is catalysed by NADPH-dependent reductase localised in microsomes of rat liver. J. Biol. Chem 1993; 268,10109-10113

[27] Warit S, Zhang N, Short A, Walmsley RM, Oliver SG, Stateva LI Glycosylation deficiency phenotypes resulting from depletion of GDP-mannose pyrophosphorylase in two yeast species. Mol Microbiol 2000 ; 36:1156-66.

[28] Zakrzewska A, Palamarczyk G, Krotkiewski H, Zdebska E, Saloheimo M, Penttilä M, Kruszewska JS Overexpression of the gene encoding GTP:mannose-1-phosphate guanyltransferase, mpg1, increases cellular GDP-mannose levels and protein mannosylation in *Trichoderma reesei.* Appl Environ Microbiol. 2003; 69:4383-4389.

[29] Fonzi,W.A, Irwin M.Y 1993; Isogenic strain construction and gene mapping in Candida albicans. Genetics. 134:717-28.

[30] Cantagrel V, Lefeber DJ, Ng BG, Guan Z, Silhavy JL, Bielas SL, Lehle L, Hombauer H, Adamowicz M, Swiezewska E, De Brouwer AP, Blümel P, Sykut-Cegielska J, Houliston S, Swistun D, Ali BR, Dobyns WB, Babovic-Vuksanovic D, van Bokhoven H, Wevers RA, Raetz CR, Freeze HH, Morava E, Al-Gazali L, Gleeson JG SRD5A3 is required for converting polyprenol to dolichol and is mutated in a congenital glycosylation disorder 2010; Cell. 142 (2):203-217.

[31] Mosch HU and Fink GR Dissection of filamentous growth by transposon mutagenesis in Saccharomyces cerevisiae 1997; *Genetics* 145(3):671-84

[32] Kruszewska, J., Butterweck, A., Kurzątkowski, W., Migdalski, A., Kubicek, C.P. and Palamarczyk, G. Overexpression of the *Saccharomyces cerevisiae* Mannosylphosphodolichol synthase -encoding gene in *Trichoderma reesei* results in the increased level of protein secretion and abnormal cell ultrastructure. Appl. Environ. Microbiol 1999; 65, 2382-2387.

[33] Perlińska-Lenart U, Kurzatkowski W, Janas P, Kopińska A, Palamarczyk G, Kruszewska JS.. Protein production and secretion in an Aspergillus nidulans mutant impaired in Glycosylation, Acta Biochim Pol. 2005; 52: 195-206.

[34] Juchimiuk, M., Pasikowska, M., Zatorska, E., Laudy, A., Smoleńska-Sym, G., Palamarczyk, G. Defect in dolichol-dependent glycosylation increases sensitivity of *Saccharomyces cerevisiae* towards anti-fungal drugs. Yeast 2010; 27: 637-645.
[35] Orłowski, J., Machula, K., Janik, A., Zdebska, E., Palamarczyk,G Dissecting the role of dolichol in cell wall assembly in the yeast mutants impaired in early glycosylation reactions. Yeast 2007; 24: 239-252.
[36] Takeguchi, M., Kimala, Y., Kohno, K. *Saccharomyces cerevisiae* Rot1 is an essential molecular chaperone in the endoplasmic reticulum. Mol Biol Cell.; 2008; 19: 3514-3525.
[37] Machi K, Azuma M, Igarashi K, Matsumoto T, Fukuda H, Kondo A, Ooshima H Rot1p of Saccharomyces cerevisiae is a putative membrane protein required for normal levels of the cell wall 1,6-beta-glucan. Microbiology 2004; 150:3163-3173.
[38] Pasikowska M., Palamarczyk G, and Lehle L The essential endoplasmic reticulum chaperone Rot1 is required for protein *N*- and *O*-glycosylation in yeast 2012; Glycobiology, accepted
[39] Girrbach V, Zeller T, Priesmeier M, Strahl-Bolsinger S Structure-function analysis of the dolichyl phosphate-mannose: protein O-mannosyltransferase ScPmt1p. 2000; J Biol Chem. 275:19288-19296.
[40] Ernst JF, Prill SK. O-glycosylation 2001; Med Mycol. 39 Suppl 1:67-74.
[41] Willger SD, Ernst JF, Alspaugh JA, Lengeler KB. Characterization of the PMT gene family in Cryptococcus neoformans. PLoS One. 2009; 4(7):e6321.
[42] Gentzsch M, Tanner W. Protein-O-glycosylation in yeast: protein-specific mannosyltransferases Glycobiology 1997; 7, 481-4866.
[43] Harty C, Strahl S, Römisch K O-mannosylation protects mutant alpha-factor precursor from endoplasmic reticulum-associated degradation. Mol Biol Cell 2001; 12:1093-10101.
[44] Prill SK, Klinkert B, Timpel C, Gale CA, Schröppel K, Ernst JF. PMT family of Candida albicans: five protein mannosyltransferase isoforms affect growth, morphogenesis and antifungal resistance. Mol Microbiol 2005; 55(2):546-560.
[45] Górka-Nieć W, Kania A, Perlińska-Lenart U, Smoleńska-Sym G, Palamarczyk G, Kruszewska JS Integration of additional copies of *Trichoderma reesei* gene encoding protein O-mannosyltransferase I results in a decrease of the enzyme activity and alteration of cell wall composition.. Fungal Biol. 2011; 115:124-32.
[46] Chandra J, Kuhn DM, Mukherjee PK, Hoyer LL, McCormick T, Ghannoum MA. Biofilm formation by the fungal pathogen Candida albicans: development, architecture, and drug resistance. J Bacteriol. 2001; 183(18):5385-94.
[47] Ramage G, Wickes BL, López-Ribot JL. Inhibition on Candida albicans biofilm formation using divalent cation chelators (EDTA). Mycopathologia. 2007;164(6):301-6. Epub 2007 Oct 2.
[48] Ramage G, López-Ribot JL. Techniques for antifungal susceptibility testing of Candida albicans biofilms Methods Mol Med. 2005; 118:71-79.
[49] Kojic EM, Darouiche RO. 2004. Candida infections of medical devices. Clin Microbiol Rev.;17(2):255-67.
[50] Uppuluri P, Pierce CG, López-Ribot JL.2009. Candida albicans biofilm formation and its clinical consequences. Future Microbiol. 2009; 4(10):1235-7

[51] Silberstein S, Gilmore R. 1996 Biochemistry, molecular biology, and genetics of the oligosaccharyltransferase. FASEB J.;10(8):849-58.
[52] Dean, N. 1995. Yeast glycosylation mutants are sensitive to aminoglycosides. Proc Natl Acad Sci U S A 92:1287–1291.
[53] Zhou J, Zhang H, Liu X, Wang PG, Qi Q. 2007. Influence of N-glycosylation on Saccharomyces cerevisiae morphology: a golgi glycosylation mutant shows cell division defects. Curr Microbiol.;55(3):198-204.
[54] Bai C, Xu XL, Chan FY, Lee RT, Wang Y. 2006. MNN5 encodes an iron-regulated alpha-1,2-mannosyltransferase important for protein glycosylation, cell wall integrity, morphogenesis, and virulence in Candida albicans.. Eukaryot Cell.;5 (2):238-247.
[55] Buurman ET, Westwater C., Hube B, Brown A.J., Odds FC, Gow N.A. 1998 Molecular analysis of CaMnt1p, a mannosyltransferase important for adhesion and virulence of *Candida albicans*. Proc. Natl. Acad. Sci. USA 95, 7670-7675.
[56] Bates S., MacCallum, D.M., Bertram, G., Munro, C.A., Hughes, H.B., BuurmanE.T., Brown A.J.P., Odds, F.C. and Gow, N.A.R. 2005 *Candida albicans* Pmr1p: a secretory pathway Ca^{2+}/Mn^{2+} P-type ATPase is required for glycosylation and virulence. J. Biol. Chem. 2005 ;280(2):1051-60.
[57] Levin DE. 2005. Cell wall integrity signaling in Saccharomyces cerevisiae.. Microbiol Mol Biol Rev. 69(2):262-91.
[58] Hutzler F, Gerstl R, Lommel M, Strahl S. 2008 Protein N-glycosylation determines functionality of the Saccharomyces cerevisiae cell wall integrity sensor Mid2p. Mol Microbiol. 68(6):1438-49. Epub 2008 Apr 11.

Chapter 12

The Role of O-Linked β-N-Acetylglucosamine (GlcNAc) Modification in Cell Signaling

Paula Santoyo-Ramos, María Cristina Castañeda-Patlán
and Martha Robles-Flores

Additional information is available at the end of the chapter

1. Introduction

The modification of serine or threonine residues of nuclear and cytoplasmic proteins by the monosaccharide, β-D-Nacetylglucosamine was discovered in the early 1980s by Torres and Hart [1]. However, relative development in this field has remained sluggish for nearly two decades, mainly due to the lack of tools and techniques for the identification and quantification of O-GlcNAc modification in proteins.

O-GlcNAc is a unique type of intracellular glycan attachment. Protein glycosylation traditionally refers to the covalent attachment of complex oligosaccharides to proteins in intraluminal compartments or cellular membranes, or in proteins that are destined for secretion. In contrast, O-GlcNAcylation is abundant in cytoplasmic and nuclear proteins, which are modified with a single β-*N*-acetylglucosamine monosaccharide moiety through an *O*-β-glycosydic attachment to serine and/or threonine side chains of the polypeptide backbone [2].

Because O-GlcNAc modification in protein occurs at serine/threonine residues, the potential for interplay between serine/threonine phosphorylation and O-GlcNAc modification has been realized very early on. However, unlike phosphorylation, which is regulated by hundreds of kinases and phosphatases, O-GlcNAc cycling has only two mediators: the enzymes O-GlcNAc transferase (OGT), and O-GlcNAc amidase (OGA) [2]. The addition and removal of O-GlcNAc is mediated by the concerted action of these two enzymes. Uridine diphosphate N-acetylglucosamine (UDP-GlcNAc) is the sugar donor for the O-GlcNAc modification, for glycophosphatidylinositol lipids synthesis, N-glycosylation, and other cellular processes.

2. Synthesis of UDP-GlcNAc from glucose via the hexosamine biosynthetic pathway

The synthesis of UDP-GlcNAc from glucose occurs via the hexosamine biosynthetic pathway [3], as depicted in Figure 1. When glucose enters a cell, it becomes phosphorylated by hexokinase, and can be redirected from the main glycolytic/glycogen pathways to secondary pathways. About 2- 5% of intracellular glucose enters the hexosamine biosynthetic pathway (HBP); thus, the amount of protein GlcNAcylation is considered to be sensitive to the nutrient (i.e., glucose and/or glutamine). Glutamine/fructose aminotransferase (GFAT) commits glucose to this pathway, and represents the access to the HBP. This pathway links glycolytic metabolism with the amino acid metabolism via the requirement of glutamine to produce glucosamine-6-phosphate. The HBP pathway culminates with the formation of UDP-GlcNAc, the high-energy donor substrate for the O-GlcNAc transferase. As can be observed in Figure 1, the biosynthesis of UDP-GlcNAc is affected and regulated by nearly every metabolic pathway in the cell, and because OGT-catalyzed O-GlcNAcylation is sensitive to insulin, to nutrients, and to cellular stress, it has been proposed that O-GlcNAcylation serves primarily to modulate cellular signaling and transcription regulatory pathways in response to nutrients and stress [3- 6].

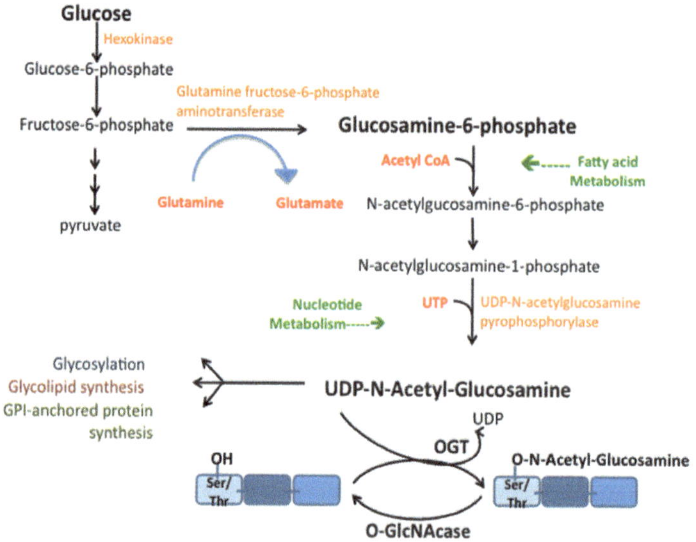

Figure 1. The synthesis of Uridine diphospho-β-N-acetylglucosamine (UDP-GlcNAc) from glucose (Hexosamine biosynthetic pathway [HBP]) and the O-GlcNAcylation cycle. Glucose is metabolized through the HBP to form the high-energy intermediate UDP-GlcNAc, which serves as the three-sugar donor for O-linked N-acetylglucosaminyl transferase (OGT). This enzyme catalyzes the transfer of GlcNAc from UDP-GlcNAc to the OH of serine or threonine residues of a protein substrate through β-glycosydic attachment. O-GlcNAcase hydrolyzes the glycosydic linkage to generate free GlcNAc and naked protein. The enzymes and the biosynthetic pathways involved in the process are depicted.

3. Enzymes that regulate O-GlcNAcylation cycle

The enzymes that catalyze the addition and removal of O-GlcNAc have been cloned and characterized. Similar to phosphorylation, O-linked -N-acetylglucosamine (O-GlcNAc) modification of nuclear and cytosolic proteins is an abundant, dynamic, and inducible posttranslational modification. However, in mammals, only one enzyme attends to the transference of O-GlcNAc: the Uridine diphospho-N-acetylglucosamine polypeptide β-N-acetylglucosaminyltransferase (OGT). This enzyme is regulated by binding partners as well as by posttranslational modification and alternative splicing. OGT interacts with several proteins that appear to target it to different locations inside the cell [4]. OGT is conserved from *Caenorhabditis elegans* to humans and it has been demonstrated that it is required for life because knock-out of the gene encoding OGT is lethal in embryonic mammalian stem cells [4]. The removal of O-GlcNAc is catalyzed by a neutral β-N-acetylglucosaminidase (O-GlcNAcase). The main characteristics of the enzymes involved in the O-GlcNAcylation cycle are the following:

3.1. OGT

The gene encoding OGT has been cloned from several organisms and is well conserved. There is 99% identity between rat and human OGT, and 61% identity between rat and *Caenorhabditis elegans* OGT. In mammalian cells, OGT is encoded by a single copy X-linked gene, while in plants, there are two homologs: spy, and secret agent [7-9].

OGT is expressed in all tissues studied, although it appears to be particularly rich in pancreas, brain, and thymus. OGT is a unique glycosyltransferase because it is a soluble protein rather than a type II membrane protein and has little or no homology to other glycosyltransferases. OGT contains two domains, an N-terminal Tetratricopeptide repeat (TPR) domain, and a C-terminal catalytic domain [7-9]. These domains are separated by a bipartite Nuclear localization sequence (NLS), thus, the enzyme is located predominantly within the nucleus. In mammalian cells, splicing of OGT mRNA leads to alternative transcripts, of which two are well characterized: mitochondrial OGT (mOGT), and nucleocytoplasmic OGT (ncOGT). mOGT has nine TPR repeats and an alternative N-terminus, which contains a mitochondrial targeting sequence.

ncOGT is the most widely studied of the known splice variants. It is a 1,037-amino-acid protein (110 kDa) and was thought to exist predominantly either as a homotrimer comprised of identical 110-kDa subunits or as a heterotrimer comprised of two identical 110-kDa subunits coupled with a 78-kDa subunit. ncOGT was originally isolated from rat liver cytosol, has a pH optimum near 6 (active pH range, 6–7.5), and, unlike many glycosyltransferases, it is not dependent on divalent cations [7].

3.2. O-GlcNAcase

O-GlcNAcase is a soluble, cytosolic β-N-acetylglucosaminidase expressed in all tissues examined and predominantly in brain. O-GlcNAcase is well conserved in mammals, with

97.8% identity between the human and the mouse gene and with 29% identity between the human and the C. elegans gene [10]. In contrast to lysosomal hexosaminidases (A and B), O-GlcNAcase is specific for β-N-acetylglucosamine, is not inhibited by N-acetylgalactosamine, and has a neutral optimum pH of 5.5–7 (lysosomal hexosaminidases have pH ~3.5–5.5). O-GlcNAcase co-purifies with a complex of proteins, is probably regulated by its interactions with other proteins, and is phosphorylated, suggesting an additional mechanism for its regulation [10]. O-GlcNAcase is efficiently inhibited by O-(2-acetamido-2-deoxy-D-glucopyranosylidene) amino-N-phenylcarbamate (PUGNAc; Ki 54 nm). It is a 917-amino-acid protein, with at least two functional domains: an N-terminal hexosaminidase domain, and a C-terminal histone acetyltransferase (HAT) domain [11]. It has been shown that the C-terminal domain of O-GlcNAcase acetylates both free-core histones and nucleosomal histone proteins. Consistent with these data, OGT is known to associate with histone deacetylase complexes and to promote transcriptional silencing and dynamic modification of transcriptional complexes by O-GlcNAc and histone [11].

4. O-GlcNAc-phosphorylation interplay

Glycomic analyses have shown that O-GlcNAcylation has extensive cross-talk with phosphorylation, where it serves as a nutrient/stress sensor to modulate signaling, transcription, and cytoskeletal functions. Nearly all *O*-GlcNAc-modified proteins are also phosphorylated, supporting the notion that interplay exists between the two modifications [3]. Indeed, it has been reported that *O*-GlcNAcylation and phosphorylation both occur in serine and threonine residues in many instances, and that the two modifications can compete for occupancy at the same site or at adjacent residues [12]. Therefore, it has been proposed that a Yin-Yang relationship does exist at the global level as well as at specific sites in several proteins. YinOYang prediction residues refer to S/T residues that can be either phosphorylated or modified by O-GlcNAc (http://www.cbs.dtu.dk/services/YinOYang/). A recently developed Website contains the most up-to-date list of published O-GlcNAc modification sites and an algorithm to predict whether a site might be O-GlcNAcylated (http://cbsb.lombardi.georgetown.edu/hulab/OGAP.html) [3]. However, subsequent studies have identified increasing numbers of O-GlcNAc modification sites that are adjacent or even distal to phosphorylation sites. In addition, pharmacologically inhibiting dephosphorylation does not decrease *O*-GlcNAcylation in all proteins, and inhibition of OGA or overexpression of OGT can produce increases in phosphorylation [13]. Thus, these data suggest a complex interplay between the two modifications. Therefore, although regulatory roles of O-GlcNAc interplay with phosphorylation occur, the binary Yin-Yang model is extremely simplistic and it is noteworthy that different types of interplay between these two posttranslational modifications can take place [12, 13].

Our understanding of the regulatory mechanisms involved in O-GlcNAc modification and its interplay with serine/threonine phosphorylation in proteins remains elusive. This is probably due to the fact that O-GlcNAcylation remained undetected until the early 1980s by commonly used analytical protein methods, including gel electrophoresis and the majority of forms of High-pressure liquid chromatography (HPLC) [3]. For example, addition of the sugar does not generally affect migration of a polypeptide in gel electrophoresis, or upon isoelectric focusing, or even in high-resolution, two-dimensional gels. In addition, the sugar modification is rapidly hydrolyzed by cellular hexosaminidases upon cellular damage or during protein isolation if countermeasures are not employed [3]. A major breakthrough in the detection and site mapping of O-GlcNAc occurred first with the development of Fourier transform mass spectrometers capable of Electron capture dissociation (ECD), and subsequently with the development of ion-trap mass spectrometers, which could perform Electron transfer dissociation-Mass spectrometry (ETD-MS). These recent successes in O-GlcNAc modification-site mapping in proteins have revealed two important clues: first, nearly all O-GlcNAc modified proteins are known phospho-proteins, and second, the prevalence of tyrosine phosphorylation among O-GlcNAc- modified proteins is exceptionally higher (~68%) than its normal occurrence (~2%) alone [2]. In this respect, Mishra S et al [14] recently reported that tyrosine phosphorylation interacts with O-GlcNAc modification, a phenomenon not previously known [14]. Subsequently, two additional articles were published showing that O-GlcNAc modification of Insulin receptor substrate 1 (IRS1) occurs in close proximity of tyrosine phosphorylation sites and affects the tyrosine phosphorylation-dependent function of IRS1 [15,16].

The extensive cross-talk between O-GlcNAcylation and phosphorylation represents a new paradigm for cellular signaling. Accurate delineation of the complex cross-talk between O-GlcNAcylation and phosphorylation could elucidate the global complex regulatory mechanism vital for cellular functions, such as regulation of cell cycle and cell growth, proliferation, and apoptosis.

5. O-GlcNAcylation in cell signaling and cellular stress

A generalization with respect to the roles of O-GlcNAcylation in cellular signaling has emerged during the past two decades. The primary function of O-GlcNAcylation appears to be the modulation of cellular processes in response to nutrients and to cellular stress [3, 17]. Cells dynamically induce O-GlcNAc protein modification in response to numerous forms of cellular stress, and this appears to be a protective response of cells. For example, blocking the hexosamine biosynthetic pathway ablates glucose-mediated cellular protection [18]. Raising O-GlcNAc levels by either inhibition of O-GlcNAcase or overexpression of OGT rendered cells more tolerant to several forms of stress, such as exposition to ultraviolet (UV) irradiation, ethanol, and sodium arsenite [18].

Many transcription factors are modified by O-GlcNAcylation in response to physiological stimuli, cell cycle stage, and developmental stage, and this modification can modulate their function in different ways [19]. O-linked GlcNAc moieties on transcription factors may be recognized by several components of the transcriptional machinery, serve as a nuclear localization signal, antagonize the action of protein kinases by masking potential serine and threonine sites for phosphorylation, modulate DNA binding activity or half-life, and increase the stability of transcription factors in the cell [19]. Although quantitatively the majority of O-GlcNAc occurs in chromatin proteins, many cytosolic enzymes, including kinases, glycolytic enzymes, the majority of cytoskeleton regulatory proteins, and the cytoskeleton proteins themselves are also modified [3]. In most cells, OGT is found mainly within the nucleus, and O-GlcNAcase is found mainly within the cytosol. In the following section, we will describe how O-GlcNAcylation affects cellular signaling, using as examples the following four proteins involved in carcinogenesis:

5.1. O-GlcNAc modification of protein kinase C (PKC)

PKC comprises a family of 10 lipid-dependent serine/threonine kinases that play key roles in proliferation and cell cycle progression, differentiation, tumorigenesis, apoptosis, cytoskeletal remodeling, ion-channel modulation, and secretion. Activation of PKC isozymes is dependent on tyrosine-kinase receptors and G-protein-coupled receptors. Based on differences in sequence homology and in biochemical properties, PKC isozymes have been classified into three subfamilies: the conventional PKC isozymes (cPKC) α, βI, βII, and γ, which are dependent on diacylglycerol (DAG) and Ca^{2+} for their activity and which respond to phorbol esters; the novel PKC (nPKC) δ, ε, η, and θ, which are insensitive to Ca^{2+} but DAG-dependent, and the atypical PKC (aPKC) ι/λ and ζ, which neither require Ca^{2+} nor respond to DAG or phorbol esters [20].

It has been reported that increased flux through hexosamine biosynthetic pathway affects the activity and translocation of certain PKC isoforms [21]. We reported for the first time that conventional, novel, and atypical PKC isozymes are all posttranslationally modified by O-GlcNAc [20]. Activation of PKC is generally correlated with its serine-theonine phosphorylation and translocation to cell membranes, but it has been also shown that other posttranslational modifications (PTM), such as tyrosine phosphorylation or tyrosine nitration, may modulate PKC activity [22, 23]

We have investigated the posttranslational modifications induced on PKC isozymes as result of their activation upon exposure of cells to a direct PKC activator (the phorbol ester Tetradecanoyl phorbol acetate [TPA]), or to an extracellular ligand known to activate PKC-dependent pathways [20]. Using freshly isolated rat hepatocytes, we studied the effect of epinephrine on PKC isoforms via activation of α1-adrenergic receptors in comparison with acute TPA treatment. The cells were incubated for 5 min in the absence (vehicle) or presence of 1 μM TPA or 10 μM epinephrine plus 10 μM propranolol (β-adrenergic antagonist). PKC isozymes were immunoprecipitated from cell extracts, and the posttranslational modifications produced on them were analyzed by Western blotting utilizing antibodies that recognizes O-linked GlcNAc in proteins (Affinity Bioreagents, Inc).

We found that posttranslational modifications other than Ser/Thr phosphorylation, such as tyrosine nitration and tyrosine phosphorylation, are commonly present in all PKC isozymes and that they change rapidly and dynamically in response to extracellular stimuli [20]. Our data demonstrated for the first time that all PKC isozymes are also dynamically modified by O-linked β-N-acetylglucosamine (O-GlcNAc); the presence of this modification was confirmed in part by Fourier transformation-Ion cyclotron resonance (FT-ICR) mass spectrometry analysis, and interestingly, the O-GlcNAc modified Ser or Thr were mapped at similar positions in several PKC isozymes.

General analysis of the collected data showed several interesting things [20]. First, it was noteworthy that O-GlcNAc modification appeared at similar positions in all PKC isoforms, such as the middle part of the molecule. Second, despite the percentage of sequence identity varied between PKC isoforms, the O-GlcNAc-modification was found in different peptides with similar sequences shared by several isozymes, and third, in many instances, the probably modified residue found agreed with the predicted O-GlcNAc potential sites (http://www. cbs.dtu.dk/services/YinOYang/), with YinOYang predictions (S/T residues that can be either phosphorylated or modified by O-GlcNAc), and in the case of PKCε, matched exactly with T710 located at 'turn motif', which is known to be an autophosphorylation 'priming site'. In other instances, the modified residue found did not exactly match the predicted ones, but were mapped near the phosphorylation 'priming sites': T517 in PKC epsilon, located immediately prior to the beginning of 'activation loop', T689 and S690, present at the end of the 'hydrophobic motif' of PKCγ, or S670, present at 'turn motif' in PKCθ, immediately prior to the S676 'priming site'. Therefore, these findings suggest that O-GlcNAcylation and phosphorylation may modulate each other.

The biochemical meaning of these posttranslational modifications for PKC alpha and PKCδ activity was investigated [20]. The results obtained suggested that while Ser/Thr phosphorylation at C terminus and tyrosine phosphorylation status appear to regulate the activation states of both PKCα and PKCδ isozymes, only PKCα activity appears to be regulated as well, in this case, in a negative manner, by tyrosine nitration and O-GlcNAc posttranslational modifications. Thus, our data indicated that phosphorylation status, both in Ser/Thr and in Tyr residues, may regulate the activity of all PKC isozymes, but the biochemical consequences of tyrosine nitration and O-GlcNAc modifications may be different for each PKC isozyme.

5.2. O-GlcNAc modification of c-Myc

c-Myc (v-myc myelocytomatosis viral oncogene homolog), a helix-loop-helix leucine zipper transcription factor, contributes to the development of several malignancies. In a normal resting cell, c-Myc levels are very low. However, upon mitogenic stimulation, c-myc expression levels are highly increased, as well as upon canonical Wnt signaling stimulation.

C-myc heterodimerizes with Max to regulate transcription of proliferation and cell differentiation genes [24]. Cell proliferation requires the coordinated activity of cytosolic and mitochondrial metabolic pathways to provide ATP and building blocks for DNA, RNA, and protein synthesis. Many metabolic pathway genes are targets of the c-myc oncogene and cell-cycle regulator. Morris F et al [25] demonstrated that Myc expression also increased global O-linked N-acetylglucosamine protein modification, and inhibition of hexosamine biosynthesis selectively reduced growth of Myc-expressing cells, suggesting its importance in Myc-induced proliferation.

c-Myc can be phosphorylated by casein kinase II [26] and by Mitogen-activated protein kinase (MAP kinase) [24]. It also has been reported that can be modified by O-GlcNAc [28, 29]. Phosphorylation at Thr-58 and/or Ser-62 in the N-terminal transcription activation/malignant transformation domain (TAD) of c-Myc may modulate transactivation and co-transformation by c-Myc [29].

The O-GlcNAc modification of c-Myc has been identified both in mammalian (rabbit reticulocyte lysate and Chinese hamster ovary [CHO]) cell line and in insect cell systems [19]. This modification of c-Myc was shown by three different methods as follows: (i) demonstration of lectin- binding to in vitro translated protein using a protein-protein interaction mobility-shift assay; (ii) glycosidase or glycosyltransferase treatment of in vitro translated protein analyzed by lectin-affinity chromatography, and (iii) direct characterization of the sugar moieties on purified recombinant protein overexpressed in either insect cells or CHO cells [29]. O-GlcNAc-modified sites within c-Myc were originally found to be located near the transcriptional activation domain [28]. A later study identified Thr-58, an in vivo phosphorylation site in the transactivation domain, as the major site of O-GlcNAc modification [29]. This suggests a mutually exclusive modification of c-Myc by either phosphorylation or O-GlcNAc modification. The transactivation domain of c-Myc associates with the tumor suppressor Retinoblastoma protein Rb and the Rb-related protein p107 in vitro [19, 30]. Thus, the presence of O-GlcNAc modification on c-Myc may result in altered interaction with Rb and Rb-related protein p107, thereby interfering with transactivation by c-Myc. Interestingly, Thr-58 is located within a mutational hot spot in lymphomas, suggesting that this region is associated with increased tumorigenicity [19].

5.3. O-GlcNAc modification of β-catenin

The Wnt pathway plays an important role in development and in regulation of adult stem-cell systems. Deregulation of Wnt signaling causes developmental defects and cancer. Canonical Wnt signaling operates through regulating the phosphorylation and degradation of the transcription co-activator β-catenin [31]. Without stimulation by Wnt, β-catenin is assembled into the so-called "destruction complex" in which Adenomatous polyposis coli (APC) plays a central role, and includes Axin, Glycogen synthase kinase 3 (GSK3)β, and Casein kinase 1 (CK1). This complex directs a series of phosphorylation events on β-catenin that targets it for ubiquitination and subsequent proteolysis via the proteasome. Stimulation by Wnt leads to inhibition of β-catenin breakdown, allowing β-catenin to accumulate, enter the nucleus, and activate a Wnt target gene program [31, 32].

The phosphorylation of β-catenin is key in the regulation of its intracellular levels, hence in its transcriptional activity. Nuclear β-catenin is the hallmark of activated canonical Wnt signaling. Thus, nuclear import/export of β-catenin represents a crucial step in regulating signaling competent β-catenin levels and serves as an attractive target for pharmacological interventions in cancer and other diseases associated with altered Wnt signaling. However, the mechanisms that regulate the nuclear localization of β-catenin remain unclear. β-catenin contains no recognizable NLS; thus, it has been proposed that it is imported by a piggy-back mechanism. However, it has been demonstrated that β-catenin nuclear import can occur in the absence of transport factors, such as importins or the Ran GTPase [33]. Moreover, β-catenin was found to compete with importin-β for docking to components of the nuclear pore complex. Indeed, the central arm repeats are required for β-catenin import, and these are structurally related with the importin-β HEAT (Huntington, Elongation factor 3, PR65/A, TOR) repeats that bind the nuclear pore complex [34]. With respect to the β-catenin nuclear export, although the precise mechanism is not completely understood, several distinct β-catenin nuclear export pathways have been reported to date.

Sayet et al [35] demonstrated that O-GlcNAcylation of β-catenin negatively regulates its levels in the nucleus. TCF reporter plasmid (TOPflash) reporter assays and mRNA expression of β-catenin's target genes indicated that O-GlcNAcylation of β-catenin results in a decrease in its transcriptional activity. The authors showed also that this novel modification of β-catenin regulates its nuclear localization and transcriptional function. Their results indicated that O-GlcNAcylation of β-catenin is inversely related to its nuclear localization and transcriptional function. This finding highlights O-GlcNAcylation as a new level of regulation of β-catenin transcriptional function.

5.4. O-GlcNAc modification of p53

The tumor suppressor protein p53 is a cell cycle regulator, and mutations in p53 cause cancer. This protein is considered the major 'gatekeeper' of genomic stability. Under conditions of cellular stress, environmental damage or genetic catastrophe, p53 expression and stability increase, followed by the induction of genes that promote cell-cycle arrest, apoptosis, and autophagy [13].

Owing to the pivotal role of p53 in maintaining genomic integrity, p53 is tightly regulated by proteolytic degradation. Normally, p53 levels are low due to continuous degradation by ubiquitin-dependent proteolysis. In unstressed cells, p53 interacts with Mdm2, which acts as an ubiquitin ligase, leading to degradation of p53 by the proteasome [36, 37]. The stability of p53 is also affected by phosphorylation. The amino-terminal domain of the tumor suppressor contains the transactivation domain and several known phosphorylation sites. Phosphorylation at Ser18 and Ser23 promotes p53 stability and tumor suppression, and phosphorylation of p53 at Thr155 (which resides in the DNA-binding domain) promotes p53 degradation by the COP9 signalosome [38].

The mechanism by which O-GlcNAc modification enhances p53 stability has been recently established by the identification of the O-GlcNAc residue within p53 employing mass spectrometry (MS). Activation of p53 by stress involves the O-GlcNAc modification of Ser-149, and modification of this site interferes with phosphorylation at Thr-155 by means of the COP9-associated kinases. Lowering phosphorylation at Thr-155 weakens the interaction of p53 with Mdm2 and decreases p53 ubiquitination/proteolysis, resulting in higher stability of the p53 protein [39]. However, mutating Ser-149 to Ala does not significantly decrease O-GlcNAc modification of p53 [39], suggesting that in addition to Ser-149, there are other residues within p53 that become O-GlcNAc-modified.

6. O-GlcNAcylation and cancer

Altered O-linked GlcNAc modification has been linked with several human diseases, including cardiovascular disease, neurodegenerative disorders, diabetes mellitus, and cancer [13]. Given the extensive cross-talk between O-GlcNAcylation and phosphorylation and the known roles of phosphorylation in mechanisms underlying cancer, it is not surprising that O-GlcNAcylation is also involved in the etiology of cancer.

The unique metabolism of tumors was described many years ago by Otto Warburg, who identified tumor cells with increased glycolysis and decreased mitochondrial activity [40]. Warburg found that normal tissues used mitochondrial oxidation to account for 90% of ATP production, with glycolysis accounting for 10%. However, tumors used less of the highly efficient oxidative phosphorylation, producing < 50% of ATP from oxidation and >50% from glycolysis. This shift was thought to occur even though there was sufficient oxygen to support mitochondrial function and is called "aerobic glycolysis" [40].

Aerobic glycolysis is just one component of the metabolic transformation. In order to engage in replicative division, a cell must duplicate its genome, proteins and lipids and assemble the components into daughter cells; in short, it must become a factory for macromolecular biosynthesis. Thus, enhanced biosynthetic capacity is a key feature of the metabolic transformation of tumor cells. Synthesis of nucleotides and fatty acids and consumption of glucose and glutamine are widespread among tumors and tumor cell lines. It appears likely that these activities, particularly the use of glutamine as a source of both reductive power and anaplerosis, are general characteristics of tumor cell growth and proliferation [41]. Glutamine contributes essentially to every core metabolic task of proliferating tumor cells: it participates in bioenergetics, supports cell defenses against oxidative stress, and complements glucose metabolism in the production of macromolecules [41]. Interest in glutamine metabolism has been further heightened by recent findings that c-myc controls glutamine uptake and degradation, and that glutamine itself exerts influence over a number of signaling pathways that contribute to tumor growth [13, 41]. Because HBP flux and UDP-GlcNAc availability are directly affected by many different nutrients, such as glucose, fatty acids, and amino acids, it is possible that in cancer cells, the HBP, together with O-GlcNAc modification, serves as a glucose/nutrient sensor that links metabolism to the activation of many oncogenic signaling pathways within the cell.

Western blot methods showed decreased *O*-GlcNAc levels in some tumor samples compared with controls. However, other studies suggested that *O*-GlcNAc levels increase in some cancers. There are potential sources of confusion from these studies. Crucial tumorigenic roles of specific O-GlcNAcylated proteins might be masked on analyzing only global cellular O-GlcNAcylation. In addition, cellular levels of O-GlcNAc, OGT, and OGA have a tight relationship with multiple cellular activities. Despite conflicting data in primary tumor samples, increased O-GlcNAcylation does appear to be a general characteristic of cancer cells. Histological sections from breast, lung, and colon tumors demonstrated increased O-GlcNAcylation compared with matched adjacent tissue [42]. In the case of the lung and colon tissue, both OGT and OGA expression appeared to increase. Similarly, in patients with chronic lymphocytic leukemia, the O-GlcNAcylation of proteins was increased when compared with normal lymphocytes [43]. Furthermore, breast cancer cells in which OGT was knocked down and that were transplanted into nude mice formed fewer tumors compared with controls [13].

Chemotherapeutics targeting OGT or OGA inhibition could potentially alter tumor function or could render tumors more susceptible to other chemotherapeutic agents; however, which adverse affects OGT or OGA inhibition might have on normal cells is unknown.

7. Conclusions

Many cytoplasmic and nuclear proteins are dynamically modified by O-linked β-N-acetylglucosamine (O-GlcNAc). However, the precise function that it carries out on each protein remains unknown. Interestingly, O-GlcNAc shares many common traits with O-phosphate. They are both dynamic modifications processed by specific enzymes that modify serine/threonine residues and that rapidly respond to extracellular stimuli. Posttranslational modifications are essential devices to generate the tremendous diversity, complexity, and heterogeneity of gene products. Determination of the in vivo roles that they play is one of the main challenges in proteomics and signal transduction research.

Author details

Paula Santoyo-Ramos, María Cristina Castañeda-Patlán and Martha Robles-Flores*
Department of Biochemistry, Faculty of Medicine, Universidad Nacional Autónoma de México (UNAM), Mexico City, Mexico

8. References

[1]. Torres CR, Hart GW (1984). Topography and polypeptide distribution of terminal N acetylglucosamine residues on the surfaces of intact lymphocytes. Evidence for O-linked GlcNAc. J Biol Chem. 259(5): 3308-17.
[2] Mishra S, Ande SR, Salter NW (2011). O-GlcNAc modification: why so intimately associated with phosphorylation? Cell Communication and Signaling 9:1-4.
[3] Hart GW, Slawson C, Ramirez-Correa G, and Lagerlof O (2011). Cross talk between O-GlcNAcylation and phosphorylation: roles in signaling, transcription, and chronic disease. Annu. Rev. Biochem 80:825–58.
[4] Ngoh GA, Jones SP (2008). New insights into metabolic signaling and cell survival: the role of beta-O-linkage of N-acetylglucosamine. J Pharmacol Exp Ther. 327(3):602-9.
[5] Walgren JL, Vincent TS, Schey KL, Buse MG (2003). High glucose and insulin promote O-GlcNAc modification of proteins, including alpha-tubulin. Am J Physiol Endocrinol Metab. 284(2):E424-E434.

* Corresponding Author

[6] Housley MP, Rodgers JT, Udeshi ND, Kelly TJ, Shabanowitz J, Hunt DF, Puigserver P, Hart GW (2008). O-GlcNAc regulates FoxO activation in response to glucose. J Biol Chem. 283(24):16283-92.

[7] Nolte D, Müller U (2002). Human O-GlcNAc transferase (OGT): genomic structure, analysis of splice variants, fine mapping in Xq13.1. Mamm Genome. 13(1):62-4.

[8] Kreppel LK, Blomberg MA, Hart GW (1997). Dynamic glycosylation of nuclear and cytosolic proteins. Cloning and characterization of a unique O-GlcNAc transferase with multiple tetratricopeptide repeats. J Biol Chem. 272(14):9308-15.

[9] Lubas WA, Frank DW, Krause M, Hanover JA (1997). O-Linked GlcNAc transferase is a conserved nucleocytoplasmic protein containing tetratricopeptide repeats. J Biol Chem. 272(14):9316-24.

[10] Wells L, Gao Y, Mahoney JA, Vosseller K, Chen C, Rosen A, Hart GW (2002). Dynamic O-glycosylation of nuclear and cytosolic proteins further characterization of the nucleocytoplasmic beta-N-acetylglucosaminidase. J. Biol. Chem. 277:1755-61.

[11] Toleman C, Paterson AJ, Whisenhunt TR, Kudlow JE (2004). Characterization of the histone acetyltransferase (HAT) domain of a bifunctional protein with activable OGlcNAcase and HAT activities. J Biol Chem 279: 53665-53673.

[12] Hart GW, Housley MP, Slawson C (2007). Cycling of O-linked beta-N-acetylglucosamine on nucleocytoplasmic proteins. Nature 446 (7139):1017-22.

[13] Slawson C, Hart GW (2011). O-GlcNAc signaling: implications for cancer cell biology. Nat Rev Cancer 11: 678-684.

[14] Ande SR, Moulik S, Mishra S (2009). Interaction between O-GlcNAc modification and tyrosine phosphorylation of prohibitin: implication for a novel binary switch. PLoS ONE, 4(2):e4586.

[15] Whelan SA, Dias WB, Lakshmanan T, Lane MD, Hart GW (2010). Regulation of insulin receptor 1 (IRS-1)/AKT kinase mediated insulin signaling by O-linked-{beta}-N-acetylglucosamine (O-GlcNAc) in 3T3-L1 adipocytes. J Biol Chem 285: 5204-11.

[16] Klein A, Berkaw MN, Buse MG, Ball LE (2009). O-GlcNAc modification of insulin receptor substrate-1 (IRS-1) occurs in close proximity to multiple SH2 domain binding motifs. Mol Cell Proteomics 8:2733-45.

[17] Kazemi Z, Chang H, Haserodt S, McKen C, Zachara NE (2010). O-linked beta-N-acetylglucosamine (O-GlcNAc) regulates stress-induced heat shock protein expression in a GSK-3beta-dependent manner. J Biol Chem. 285 (50):39096-107.

[18] Chatham JC, Marchase RB (2010). Protein OGlcNAcylation: a critical regulator of the celular response to stress. Curr Signal Transduct Ther. 5(1):49-59.

[19] Özcan S, Andrali SS, Cantrell JEL (2010). Modulation of transcription factor function by O-GlcNAc modification. Biochim Biophys Acta 1799: 353–364

[20] Robles-Flores M, Meléndez L, García W, Mendoza-Hernández G, Lam TT, Castañeda-Patlán C, González-Aguilar H (2008). Posttranslational modifications on protein kinase C isozymes. Effects of Epinephrine and phorbol esters. BBA Mol Cell Res 1783: 695-712

[21] Matthews JA, Acevedo-Duncan M, Potter RL (2005). Selective decrease of membrane-associated PKC-α and PKC-ε in response to elevated intracellular O-GlcNAc levels in transformed human glial cells, Biochim. Biophys. Acta 1743: 305–315.

[22] Konishi H, Tanaka M, Takemura Y, Matsuzaki H, Ono Y, Kikkawa U, Nishizuka Y. (1997). Activation of protein kinase C by tyrosine phosphorylation in response to H_2O_2, Proc. Natl. Acad. Sci. U. S. A. 94: 11233–11237.

[23] Balafanova Z, Bolli R, Zhang J, Zheng Y, Pass JM, Bhatnagar A, Tang XL, Wang O, Cardwell E, Ping P (2002). Nitric oxide (NO) induces nitration of protein kinase Cε, facilitating PKCε translocation via enhanced PKCε-RACK2 interactions, (PKCε), facilitating PKCε J Biol Chem 277 (17) 15021-27.

[24] Meyer N., Penn LZ (2008). Reflecting on 25 years with MYC, Nat Rev Cancer 8: 976–990.

[25] Morrish F, Isern N, Sadilek M, Jeffrey M, Hockenbery DM (2009). c-Myc activates multiple metabolic networks to generate substrates for cell-cycle entry. Oncogene 28, 2485–2491

[26] Luscher B, Kuenzel EA, Krebs EG, Eisenman RN (1989). Myc oncoproteins are phosphorylated by casein kinase II, EMBO J. 8: 1111–1119.

[27] Seth A, Alvarez E, Gupta S, Davis RJ (1991). A phosphorylation site located in the NH2-terminal domain of c-Myc increases transactivation of gene expression, J Biol Chem 266: 23521–23524.

[28] Chou TY, Dang CV, Hart GW (1995). Glycosylation of the c-Myc transactivation domain, Proc. Natl. Acad. Sci. U. S. A. 92: 4417–4421.

[29] Chou TY, Hart GW and Dang CV (1995). C-Myc is glycosylated at Threonine 58, a known phosphorylaltion site and a mutational hot spot in lymphomas. J Biol Chem 270(32): 18961-18965.

[30] Gu W, Bhatia K, Magrath IT, Dang CV, Dalla-Favera R (1994). Binding and suppression of the Myc transcriptional activation domain by p107, Science 264: 251–254.

[31] Cadigan KM and Peifer M (2009). Wnt signaling from development to disease: insights from model systems. (Abstract) Cold Spring Harb Perspect Biol 1:a002881

[32] Wend P, Holland JD, Ziebold U, Birchmeier W (2010). Wnt signaling in stem and cancer stem cells. Semin Cell Dev Biol 21: 855–863.

[33] Fagotto F, Gluck U and Gumbiner BM (1998). Nuclear localization signal-independent and importin/karyopherin-independent nuclear import of β-catenin. Curr Biol 8:181-190.

[34] Kulay U, Izaurralde E, Bischoff FR, Mattaj IW and Gorlich D (1997). Dominant-negative mutants of importin-β block multiple pathways of import and export through the nuclear pore complex. EMBO J 16: 1153-1163.

[35] Sayat R, Leber B, Grubac V, Wiltshire L, Persad S (2008). O-GlcNAc-glycosylation of β-catenin regulates its nuclear localization and transcriptional activity. Exp Cell Res 314: 2774-2787.

[36] Haupt Y, Maya R, Kazaz A, Oren M (1997). Mdm2 promotes the rapid degradation of p53, Nature 387: 296–299.

[37] Kubbutat MH, Jones SN, Vousden KH (1997). Regulation of p53 stability by Mdm2, Nature 387: 299–303.

[38] Bech-Otschir D, Kraft R, Huang X, Henklein P, Kapelari B, Pollmann C, Dubiel W (2001). COP9 signalosome-specific phosphorylation targets p53 to degradation by the ubiquitin system, EMBO J. 20: 1630–1639.

[39] Yang WH, Kim JE, Nam HW, Ju JW, Kim HS, Kim YS, Cho JW (2006). Modification of p53 with O-linked N-acetylglucosamine regulates p53 activity and stability. Nat. Cell Biol. 8: 1074–1083.

[40] Warburg O (1956). On respiratory impairment in cancer cells. Science 124: 269–270.

[41] DeBerardinis RJ, Sayed N, Ditsworth D, and Thompson CB (2008). Brick by brick: metabolism and tumor cell growth. Curr Opin Genet Dev. 18(1): 54–61.

[42] Mi W, Gu Y, Han C, Liu H, Fan Q, Zhang X, Cong Q, Yu W (2011). O-GlcNAcylation is a novel regulator of lung and colon cancer malignancy. Biochim Biophys Acta 1812: 514–519

[43] Shi Y, Tomic J, Wen F, Shaha S, Bahlo A, Harrison R, Dennis JW, Williams R, Gross BJ, Walker S, Zuccolo J, Deans JP, Hart GW, Spaner DE (2010). Aberrant O-GlcNAcylation characterizes chronic lymphocytic leukemia. Leukemia 24: 1588-1598

Chapter 13

The Role of Glycans in Apical Sorting of Proteins in Polarized Epithelial Cells

Kristian Prydz, Gro Live Fagereng and Heidi Tveit

Additional information is available at the end of the chapter

1. Introduction

Epithelia consist of highly differentiated cells that form tight layers lining the inner cavities and the outer surface of the body. In classical cell biology, carbohydrate groups are attached to proteins (called glycoproteins or proteoglycans, depending on the carbohydrate structure) and lipids (glycolipids) - facing towards the exterior from the outer lipid leaflet of the cell membrane. These carbohydrate structures were previously thought to constitute a protective water-binding glycocalix, but were also known early on to be attachment sites for viruses, parasites, bacteria and bacterial toxins. Today, the glycan moieties of glycoproteins and proteoglycans are known to participate in protein sorting, transport and signaling processes in eukaryotic cells. Such processes are more complex in the epithelial cell, which possesses two distinct plasma membrane areas; the apical membrane domain, directed towards the inner cavities of the body, and the basolateral membrane domain facing the bloodstream. These two opposite membrane domains are segregated and differ in protein and lipid composition, and to maintain this polarity, vectorial transport of newly synthesized protein and lipid molecules is required. Proteins destined for the plasma membrane must traverse the secretory pathway. Such proteins emerge from ribosomes that dock onto endoplasmic reticulum (ER) membranes, due to recognition of a signal sequence in the protein itself. During co-translational import into the ER lumen, the starting point of the secretory pathway, the majority of translocated proteins receive carbohydrate modifications of the N-glycan type on asparagine residues, when a mannose-rich branched glycan structure with three terminal glucose residues is transferred *en bloc* from the lipid carrier dolichol to a proper modification site by an oligosaccharyl transferase. Such N-glycan structures are utilized as recognition units on glycoproteins by the protein folding quality control system which either allows further transport along the secretory pathway or diverts misfolded proteins to proteasomal degradation in the cytoplasm (1).

Properly folded proteins exit the ER in membrane carriers that are transported to the intermediate compartment (IC), and from there further to the Golgi apparatus. While N-glycans are modified to become more complex structures by enzymes in the Golgi apparatus, are other classes of glycans attached to alternative glycosylation sites in available protein cores. O-glycosylation on serine and threonine residues, starting with N-acetyl-galactosamine often results in mucin-type branched glycans, while glycosaminoglycan modification of serines with a neighboring glycine results in proteoglycans. Proteins and proteoglycans exiting from the *trans* side of the Golgi apparatus – the *trans*-Golgi network - are either transported in vesicles moving directly to the cell surface or to an endosomal compartment, from where the cell surface can be reached via indirect routes. Besides being the main organelle for carbohydrate synthesis and modification, the Golgi apparatus has been regarded as the main sorting center for polarized delivery of proteins in epithelial cells. In the trans-Golgi network, molecules are sorted and packaged into vesicles destined either for the apical or the basolateral cell surface domain. Although the *trans*-Golgi network has been established as a main protein sorting center (2), both pre- and post-Golgi polarized sorting has been reported for epithelial cells (3).

Molecular mechanisms governing polarized protein sorting in epithelial cells have been studied for to decades. Adapter proteins, tethering factors and fusion mediators have been assigned to different organelles and transport steps. Some proteins are shown to be required for particular routes although the molecular role played, is uncertain. Sorting signals have been localized to cytoplasmatic protein tails, transmembrane regions, specific globular protein domains, as well as post-translational modifications like lipid anchors and carbohydrate structures. The definition of a protein sorting signal, is that this molecular feature works after transfer to a previously non-sorted or oppositely sorted protein. In studies of glycans as sorting mediators, this requires manipulation and transfer of glycan modification sites. Such studies have indicated that both N- and O-glycans can mediate apical sorting, as well as long unbranched glycosaminoglycan chains of the chondroitin sulfate type. In this chapter we emphasize how different post-translationally added carbohydrate structures participate in polarized sorting of proteins. Although there are many examples that glycan structures and glycosylation sites in proteins are central to protein sorting, the underlying molecular mechanisms and their intracellular sites of action have not been thoroughly described.

2. Polarized epithelial cells

Single layers of epithelial cells coat the inner cavities of the body, like the intestinal tract, lungs, and kidney tubules. Attachment to the underlying tissues is mediated by proteins in the basal membrane domain that bind firmly to extracellular matrix components, while an apical region of the plasma membrane, with microvillus protrusions faces the lumenal environment. The apical membrane domain of the cell surface is segregated from the lateral and basal regions, together termed the basolateral domain, by tight junctions, protein complexes consisting of transmembrane anchored adhesion proteins, like claudins and occuldins. The tight junctions fulfill several functions. Together with other junctional comlexes, they connect the individual cells in an epithelial layer by closing the gap between all neighboring cells in the monolayer. At the same time, the tight junctions introduce a restriction to lateral movement of proteins and lipids from the apical domain of the plasma membrane to the basolateral domain, and *vice versa* (4). In addition, tight junctions contribute to the fence function of epithelia, by restricting the intercellular route for passage of macromolecules. The tightness of epithelia is variable with the physiological functions of the tissue in question, some epithelia allow passage of small proteins, while some even restrict the passage of water molecules, like the epithelium of the distal kidney tubules where reabsorption of water molecules takes place (Figure 1).

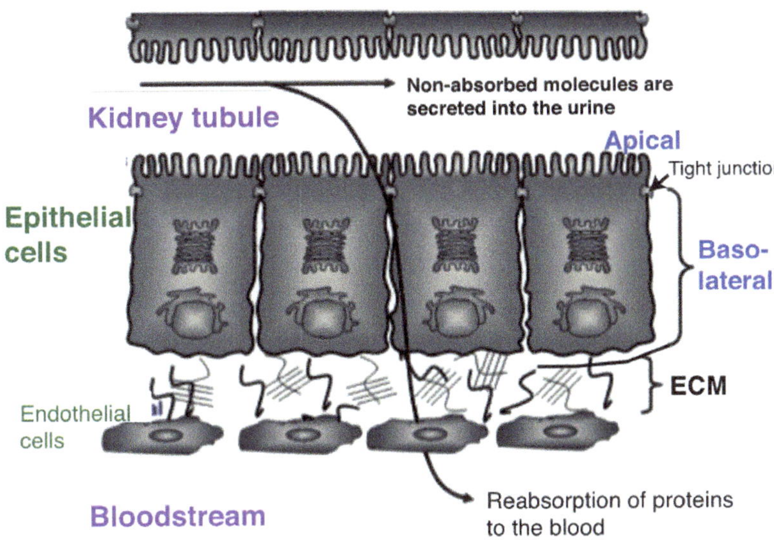

Figure 1. Polarized kidney epithelial cells. A single kidney epithelium monolayer consists of polarized cells with two surface domains segregated by tight junctions. The apical domain facing the kidney tubule lumen, and the basolateral domain attached to the extracellular matrix (ECM), facing the bloodstream.

The apical membrane domain takes on specialized tissue functions, while the basolateral domain fulfills more general functions, like uptake of nutrients from the blood, building blocks like amino acids, lipids, and monosaccharides for synthesis of macromolecules. The functional differences of the two cell surface domains requires some differences in protein and lipid composition of these two opposite membrane areas, which again requires sorting mechanisms that direct newly synthesized proteins and lipids to the proper plasma membrane region. Receptors and transporters for binding of ligands like lipoproteins and growth factors and sugars, amino acids, sulfate and phosphate must be sorted basolaterally, together with attachment proteins, while several proteins involved in ion secretion to the body cavities must be sorted apically. A number of epithelia are immunologically active tissues that engage in uptake and transcytosis of immunologically active molecules, like immunoglobulins. Thus, it is of the outmost importance that the corresponding receptors are correctly sorted and transported.

2.1. General aspects of cell polarity

Most proteins destined for the cell surface in animal cells are modified with carbohydrate structures during their passage through the secretory pathway. Such carbohydrate structures are functionally important to the individual protein at the final cellular destination, but may also play a role in guiding the protein to its correct site of action. Thus, glycan modifications can contribute to the generation and maintenance of proper cellular organization in several ways, by positioning the protein correctly, and by fine-tuning the activity. Cell polarity is not restricted only to epithelial cells. Endothelial cells also possess tight junctions and apical and basolateral membrane domains similar to those observed in epithelia, and even in neurons there are restrictions to lateral protein mobility in the plasma membrane (5). In the hepatocyte, the apical membrane is divided into several separate domains that form bile canaliculi together with corresponding domains of neighboring hepatocytes. In these cells, the basolateral surface domain is continuous and constitutes the non-canalicular (sinusoidal) regions of the plasma membrane, covering approximately 85 % of the total cell surface area. Studies of protein sorting in hippocampal neurons have indicated that the apical membrane of epithelial cells and the hippocampal axons share some characteristics with respect to protein sorting (6, 7), suggesting also that dendritic and basolateral transport mechanisms share common features.

Much of our knowledge on polarized sorting and transport of proteins has been derived from studies with epithelial cell lines, such as Madin-Darby canine kidney (MDCK) cells and the colon cancer cell line $CaCo^{-2}$. These, and other epithelial cell lines, have been cultured on filters with an appropriate pore size and allowed to form confluent monolayers with a measurable electrical resistance. MDCK cells can grow on almost any type of support, since these cells produce their own matrix components needed for subsequent attachment and differentiation. The relative tightness of epithelial monolayers may in many instances be determined from trans-epithelial resistance values that are easily measured when the epithelium has achieved a sufficient tightness. The ease by which kidney epithelial cells are grown on filters, and the tightness of the monolayers formed, grants differential experimental access to the apical and basolateral membrane domains and their corresponding medium reservoirs. This allows for both biochemical studies and various modes of microscopy. Apical and basolateral medium samples can be collected without mixing, and the apical and basolateral membrane domains are differentially accessible to biotinylation reagents, antibodies and reagents for chemical modification. Biochemical studies of polarity are not carried out with similar ease for neuronal cells and hepatocytes. Although hepatocytes sometimes are classified as epithelial cells, the apical membrane normally forms perpendicularly to the growth substratum in culture. Thus, polarity studies with hepatocytes are more rare than for other epithelial cell types. While some conclusions from studies carried out with the MDCK cell line might have transition value to hepatocytes and neurons, there are also evident differences. In MDCK cells, proteins destined for the apical membrane are mostly delivered directly to this membrane domain, without transient basolateral appearance. In hepatocytes, however, most newly synthesized apical proteins are first transported to the basolateral membrane domain, followed by endocytosis and transcytosis across to the apical, bile canalicular membrane (8). Many polarized cell types possess both direct and indirect transport routes for newly synthesized proteins to the cell surface, observed both in epithelial cell lines like $CaCo^{-2}$ (9) and in neurons (10).

3. The secretory pathway

Access to the secretory pathway is normally granted to polypeptides that emerge from ribosomes with an N-terminal signal sequence that is recognized by the signal recognition particle (SRP) in the cytoplasm followed by docking onto the ER membrane. However, signals elsewhere in a protein may also mediate uptake into the ER lumen, sometimes giving rise to a different protein topology. After ER entry, and during further transport through the secretory pathway, a large number of different types of post-translational modifications may occur. Proteolytic removal of the N-terminal signal sequence is a frequent event, while proteolysis at the C-terminus occurs more seldom, when a single trans-membrane domain is cleaved to transfer the remaining protein onto a glycosylphosphatidyl inositol (GPI) lipid anchor. It has been estimated that approximately 250 proteins are prone to transfer onto GPI-anchors (11). The oxidative environment in the ER lumen promotes the stability of disulfide bridges, which are enzymatically formed by protein disulfide isomerase. An important aspect of protein assembly is the formation of multimeric proteins from single polypeptide subunits. Excess subunits often expose their more hydrophobic interaction surfaces and are thus recognized in the same way as misfolded proteins. A functional relationship between different classes of post-translational events, like formation of oligomers and glycosylation has been reported to influence apical sorting of some proteins (12-14), and will be discussed later.

The altogether most common modification of proteins is addition of N-glycans in the ER. A pre-assembled glycan moiety, rich in mannoses and with terminal glucose units is built onto the isoprene lipid dolichol and transferred to an appropriate asparagine-X-serine/threonine site by an oligosaccharyl transferase, as the polypeptide moves through the Sec 61 translocon into the ER lumen. Glycans added in this manner are used as handles for protein folding investigations carried out by chaperones, and the status of their terminal sugar residues also indicates when and where to depart from the ER (1, 15). The N-glycans acquired and subjected to initial trimming in the ER can function as mediators of forward transport in the early secretory pathway for several proteins. L-type lectins, like ERGIC-53, VIP-36 and VIP-like lectins bind immature N-glycan groups in a Ca^{2+} dependent manner (16) and mediate efficient transport of glycoproteins towards the intermediate compartment (IC) and *cis*-Golgi region in this way (17). In the Golgi apparatus, N-glycans are trimmed further by removal of more mannose units, before addition of novel sugar units takes place. A variable number of *antennae* receive N-acetyl-glucosamine, galactose, and sialic acid units, or only some of these sugars. The glycans may also receive fucose units and more rare modifications, variable from tissue to tissue. N-glycans may appear at the cell surface as hybrid structures, where some *antennae* have maintained a mannose-rich branch, while others have been rebuilt to a more complex variant. Some glycoproteins arrive at the plasma membrane with only high-mannose branches in their N-glycans. In some cases, at least, this correlates to a Golgi-independent transport route to the cell surface (18), but a good correlation between N-glycan structures at the cell surface and the transport route followed to get there has not been established. From a standard high-mannose N-linked glycan structure in the ER, an almost unlimited number of carbohydrate structures may be generated after departure from the ER, during the subsequent passage through the Golgi apparatus. In fact, the glycan modification of a particular protein depends among other factors on the tissue where expression takes place and the developmental stage of the organism.

The Golgi apparatus is localized in a central, perinuclear position in vertebrate cells, where stacks of membrane-limited *cisternae* are joined together in a membrane ribbon. In other animals, like *Drosophila melanogaster* (19), and in plant cells, Golgi stacks are spread in the cytoplasm without ribbon-forming interconnections. Still, general mechanisms of protein exit from the ER and transport towards and through the Golgi apparatus are very similar. Glycan modification of proteins and lipids, and their further transport through the secretory pathway share similar principles across the plant and animal worlds, although the glycans acquired are structurally different. Significant attention has been given to the *trans*-most *cisterna* or region of the Golgi apparatus, the *trans*-Golgi network, as an important site for segregation of newly synthesized proteins on their way to apical and basolateral membrane domains of polarized epithelial cells (2). Early studies of epithelial polarity were performed with the MDCK cell line (20-23), where viral proteins, like influenza hemeagglutinin (HA) and the vesicular stomatitis virus (VSV) G protein, were used as markers for apical and basolateral protein delivery, respectively. Initially, experiments were carried out with double infection of epithelial cells by viruses with opposite budding polarity. The polarized epithelial cells could be studied in a subsequent time window until the intracellular transport machinery had been shut down (20, 21). By electron microscopy, both influenza HA and VSV G protein were observed in the *trans*-Golgi network of the same cell, contributing, together with biochemical assays, to the conclusion that apical and basolateral proteins travel together through the Golgi apparatus and are only sorted and segregated upon exit from the *trans*-Golgi network (2). Later, as molecular biology techniques developed, epithelial cells were transfected to express the viral envelope proteins, including manipulated variants that could shed light on where in the proteins sorting information is localized (24-26). The early studies of polarized sorting of viral proteins did not indicate a role for glycans as apical sorting information, but rather implicated that sorting was mediated by the transmembrane and cytoplasmic regions of apically transported proteins like (27, 28, 25). A similar sorting outcome as that reported for epithelial MDCK cells was reported for hippocampal neurons, where influenza HA localized mainly to the axon, while the VSV G protein localized to the dendrites (6). The GPI-linked protein Thy-1 confirmed the expectation of being sorted exclusively to the axon (29), in line with the view that the axon resembles the apical surface of the epithelium, since a large number of GPI-linked proteins were observed to localize predominantly apically in MDCK cells (30).

4. Sorting sites in polarized cells

The *trans*-Golgi network is, as previously noted, not equally important as a sorting site of the secretory pathway in all polarized cell types, since hepatocytes deliver most apical proteins via the basolateral surface domain (9). Eventually, endosomes have been found to be transit compartments between the *trans*-Golgi network and the cell surface both in the secretory (31-33) and transcytotic pathways (34-37), also in epithelial MDCK cells. Thus, both direct and indirect routes towards the cell surface generally exist for apical and basolateral cargo transport, and numerous types of carriers have been implicated in post-Golgi transport (38, 39). Clearly, the *trans*-Golgi network is observed as an important sorting site for apical and basolateral cargo molecules, based on studies of fluorescently labeled model proteins in living cells (40). However, the observation of tubulo–saccular membrane structures emanating from the *trans*-Golgi network, carrying incompletely segregated apical and basolateral proteins, supports the existence of additional post-Golgi sorting stations (41, 42). The choice of model proteins for apical and basolateral cargo transport routes, evidently influences the observations made.

More recently, a number of studies have suggested that protein sorting in the secretory pathway is not exclusively a late-and post-Golgi affair. For instance, the *trans*-most Golgi *cisterna* is not the only one in a Golgi stack that displays clathrin coats, indicating the possibility of departure from the Golgi apparatus earlier than the *trans*-Golgi network. These coats may drive the formation of vesicles that are targeted for the endosomal/ lysosomal pathway, but also for the basolateral membrane (43, 44). Other coat types have been observed even at the rim of the *cisternae* preceding the clathrin-coated *cisternae* (45). Such coats were shown to engage in transport towards the plasma membrane. Thus, several *cisternae* could constitute the exit interface of Golgi membranes (46). In mammalian cells, the intermediate compartment (IC), consisting of vacuolar and tubular membrane domains and functionally positioned between the ER and the Golgi apparatus, is also stably localized in a perinuclear position at the cell center, close to the Golgi *cisternae*. Although the *trans* region of the Golgi apparatus has been regarded as the primary sorting area for proteins and lipids in the secretory pathway, the IC has been suggested as a protein and lipid sorting station operating prior to entry into the Golgi apparatus (47-49). In fact, lateral segregation of proteins in the secretory pathway can occur already in the ER (50-53, 3, 18).

Where in the secretory pathway sorting takes place is an interesting question in relation to the role played by glycans, since the input into and output from the Golgi apparatus of glycoproteins, proteoglycans (PGs) and glycolipids is very different with respect to glycan structures. As far as we know today, the glycan structures of glycoproteins entering the Golgi apparatus are identical or very similar, while the action of Golgi glycosidases, glycosyltransferases, epimerases and sulfotransferases generate the astonishing diversity observed for glycan structures, even on individual proteins. This is important to consider when discussing possible targeting information in N-glycan structures, since the terminal sugars of N-glycans are likely to be too diverse to serve as general recognition units, while the high-mannose structures existing early in the secretory pathway have proven to serve as secretory cargo ligands for protein lectins like ERGIC-53 and VIP-36 (54, 55).

In addition to N-linked glycosylation, another important protein glycosylation mechanism exists that is initiated with the addition of an N-acetyl-galactosamine (GalNAc) unit to a serine or threonine in an acceptor protein core early in the Golgi apparatus, or just before entry into the *cis*-Golgi lumen. Further extension by the action of several glycosyltransferases results in a branched O-glycan, sometimes sulfated.. Some O-glycosylated secretory proteins are classified as mucins. These are heavily glycosylated and are mainly secreted apically from different epithelial tissues to constitute the mucus layer lining the body cavities. Mucins have attracted much attention because their glycan structures are variable from tissue to tissue and change during transformation of normal epithelial cells into cancer cells (56, 57).

The third of the major glycosylation mechanism in metazoans is the polymerization of long, linear glycosaminoglycan (GAG) chains onto protein acceptor cores, resulting in proteoglycans (PGs). The major classes of GAG chains that decorate PGs are chondroitin sulfate (CS)/dermatan sulfate (DS) and heparan sulfate (HS)/heparin. These polymers are all connected to a protein core via a linker tetrasaccharide consisting of xylose-galactose-galactose-glucuronic acid, extending from –serine-glycine-sites in the polypeptide, with the xylose attached to the serine. While the initiation of the linker region might take place before entry into the Golgi apparatus, GAG chains are shown to be enzymatically polymerized in the Golgi apparatus itself and subsequently modified to a variable extent by epimerization and sulfation. Keratan sulfate PGs are less abundant and extend from N- or O-linked glycans. In addition, hyaluronic acid is classified as a GAG, but is not attached to a protein core (58). Some PGs are defined as hybrids, since the protein core may carry more than one type of GAG. In addition GAGs may co-exist together with N-glycans and O-glycans on the same PG protein core.

5. Glycans and protein sorting

Early speculations on protein sorting in epithelial cells suggested that apical protein transport would most likely require sorting signals, since the apical surface is the specialized domain of epithelial plasma membranes. Basolateral transport could then occur by default, since this surface domain bears resemblance to non-polarized cells in which anterograde sorting was not discussed (59). Since a number of GPI-linked proteins had been detected at the apical surface of epithelial MDCK cells (30) such lipid anchors were proposed to possess apical sorting information. The subsequent discovery of basolateral sorting signals in the cytoplasmic domain of several transmembrane protein receptors (60-66) indicated that polarized basolateral transport also requires sorting. Later it was discovered that basolateral targeting does not have a monopoly on cytoplasmic protein domain sorting signals (67-69). The situation for GPI-linked proteins was later shown to be quite complex. The GPI-anchor alone is not sufficient to direct a protein to the apical surface. There is usually an additional or a sole requirement for N-glycans (12, 70) or also for additional oligomerization (71).

5.1. N-glycans and apical sorting

N-glycans were introduced as apical sorting signals in the 1990-ies. When polarized MDCK cells were transfected to express human erythropoietin, the glycoprotein that possesses three sites for N-glycosylation, was secreted from the apical membrane. Mutagenesis studies demonstrated, however, that only one of the N-glycosylation sites was critically important for apical sorting (72). In another study, the non-glycosylated protein rat growth hormone (rGH) was expressed in polarized MDCK cells and the secretion pattern was compared to that of rGH variants with one or two N-glycosylation sites. Addition of glycosylation sites stimulated apical secretion compared to the random secretion pattern observed for unglycosylated rGH (73). Such examples of apical targeting mediated by N-glycan groups lead to a search for sorting lectins that could mediate apical sorting at the required site in the Golgi apparatus, the *trans*-Golgi network. A promising candidate was VIP36, a protein lectin which was extracted in the detergent insoluble fraction, proposed to contain lipid raft associated proteins that are destined for the apical membrane (74). Later studies revealed that VIP36 in fact is localized more to the early secretory pathway (75), fitting with the observed binding affinity for high-mannose glycans (54). Still, manipulation of the expression level of VIP36 affected the apical sorting stringency of gp80 (or clusterin), the major endogenous glycoprotein secreted from MDCK cells. An early indication of the importance of N-glycans for apical secretion came from a study of this protein, which was secreted randomly from polarized MDCK cells in the presence of the glycosylation inhibitor tunicamycin (76). Enhanced expression of VIP36 in MDCK cells directed gp80 more strictly towards the apical medium (77). The fact that VIP36 and other lectins in the secretory pathway recognize high-mannose glycan structures (54, 16), indicates that their site of action is early in the pathway, or that their substrates are proteins that maintain a high mannose structure throughout the pathway (78). Thus, the role of VIP36 and related lectins in polarized sorting in epithelial cells requires more research to become fully understood.

Cholesterol and glycosphingolipid-rich lipid "rafts" have been proposed to function as sorting platforms in the TGN for vectorial delivery of proteins and lipids to the apical membrane of epithelial cells. The lipid raft concept describes regions of a membrane that are stable in time due to a higher content of cholesterol and more saturated fatty acids than surrounding membrane areas, with more freely moving lipids (79, 80). Since direct experimental contact with intracellular membranes is difficult, much of the evidence has been generated by indirect methods. Cholesterol depletion from cells by synthesis inhibition and extraction from cellular membranes has been shown to reduce apical transport efficiency of influenza HA and gp80 (81). A criterion for a particular protein to be regarded as raft associated has been an observable insolubility to treatment with cold detergent. In agreement with this view has cholesterol depletion been shown to both reduce apical transport efficiency and detergent insolubility (81). A reduction in apical transport is, however, not necessarily specifically interfering with apical sorting mechanisms, but could mediate a general reduction of apical transport capacity, since cholesterol depletion also reduces apical secretion of non-sorted proteins (82). A class of proteins proposed to require lipid rafts to reach the apical surface, is constituted by GPI-linked proteins (83, 84). However, a GPI anchor as such is not sufficient to drive the protein exclusively in the apical direction. This has been shown in many cases to require the presence of N-glycans (12, 71, 70) and also oligomerization of the apically sorted protein (71, 13). An additional level of complexity was added, when it also was shown that the structure of the GPI-anchor itself influences whether the GPI-linked protein is transported apically or basolaterally (85, 86). Some GPI-linked proteins are transported basolaterally, presumably because their GPI-anchors do not favour oligomerization (85). The GPI-linked PG glypican carries both heparan sulfate (HS) chains and N-glycans, and is mainly transported basolaterally (87). However, when the sites for HS attachment are removed, glypican is mainly transported apically. Since the GPI-anchor and the N-glycans are the same in both cases, this would indicate that the presence of HS chains reduces the ability of glypican to oligomerize. Alltogether, the complexity of this field underlines one important fact: it is easier to study secretory proteins than proteins with membrane attachment.

Apically targeted proteins that are not associated with lipid rafts, meaning they are not associated with detergent-resistant microdomains, do not seem to require oligomerization to reach the apical surface (13). Still some clustering mechanism can be required. A group of small lectins with one or two carbohydrate binding domains are called galectins. Although galectins are synthesized in the cytoplasm and are devoid of known signals for membrane translocation, several members of this family of proteins have been shown to be of the outmost importance for epithelial cell differentiation and glycoprotein sorting mechanisms (88-92). The galectins are to a variable extent able to cross intracellular membranes or the plasma membrane to gain access to the lumen of intracellular compartments like endosomes, or to the outside of the cell, where (in both cases) the glycans of glycoproteins and glycolipids are localized. The translocation process has been shown to require glycan counter-receptors at the opposite side of the plasma membrane in the case of galectin-1 translocation (93). This might also be a requirement for other galectins and underscores that galectin translocation could be a highly regulated process, because the proper counter-receptors might be required. Galectin-3 has been shown to promote apical sorting of glycoproteins that are not raft-associated. Depletion of galectin-3 from MDCK cells resulted in missorting of non-raft glycoproteins, but did not affect the apical sorting of raft-lipid associated proteins (88). The effect on sorting seems to be that apical transport requires galectin-3 mediated glycoprotein clustering to be efficient (94). In polarized epithelial HT-29 cells, galectin-4 recruits N-linked glycoproteins to detergent resistant membrane fractions and further to the apical membrane (89). Knock-down of galectin-4 did not inhibit Golgi exit of glycoproteins, but reduced their surface appearance. These, and other indications point to endosomes as the site of the galectin-mediated glycoprotein clustering required for efficient apical transport. Galectin-9 has affinity for a class of glycosphingolipids at the apical surface of epithelial MDCK cells and is essential to the organisation of the apical membrane (90). Whether galectin-9 also contributes to apical transport of these glycolipids is not yet known.

5.2. O-glycans

O-glycans are branched and diverse, tissue-specific glycan structures initiated by the addition of an N-acetyl-galactosamine unit to a serine or threonine residue in an acceptor protein core (95). Although some bioinformatic tools have been developed for the purpose, these sites are not easily predicted from analysis of the amino acid sequence. As described for N-linked glycans, O-linked sugars may also direct apical sorting in polarized MDCK cells, as demonstrated for the neurotrophin receptors, where the stalk region rich in O-glycans was required for apical transport (96). O-glycans are also important for apical sorting of the human intestinal sucrase-isomaltase in MDCK cells (97). The sorting domain could induce apical sorting after transfer to rat growth hormone (rGH) in MDCK cells (98). In some cases, both N-linked and O-linked sugars play essential roles in apical targeting of a protein, as shown for bovine enteropeptidase (99, 100). The heavily glycosylated transmembrane mucin glycoprotein MUC1 is efficiently delivered to the apical membrane of epithelial MDCK cells in a glycan-dependent manner (101). The apical targeting was not affected by galectin-3 knockdown, but was blocked when O-glycan synthesis was inhibited. Thus, the mechanistic basis for apical sorting of MUC1 is in all likelihood a glycan recognition event, but this remains to be characterized.

5.3. Glycosaminoglycans (GAGs) – The glycans decorating proteoglycans (PGs)

Proteoglycans are expressed in vertebrate cells, in *Drosophila melanogaster* (102-105), in *C. elegans* (106, 107) and in simple organisms as the starlet sea anemone *Nematostella vectensis* (108). Synthesis, sorting and transport of proteoglycans can therefore be studied in a number of different model organisms. When arriving at the cell surface, proteoglycans can participate in binding and uptake of signalling molecules, such as fibroblast growth factors. Proteoglycans at the cell surface may be endocytosed and pass via endosomes to lysosomes (109), but have also been observed in the nucleus (110-112), co-localising with growth factors. The route taken from endosomes to the nucleus is essentially unknown. In rat hepatocytes, heparan sulfate proteoglycans are transported from the *trans*-Golgi network to the cell surface in vesicles different from those that transports serum albumin, apolipoprotein E and fibrinogen (113, 114). Frequently, proteoglycans are sorted to the regulated secretory pathway, being released together with other contents of storage granules upon a proper stimulus (115-117)The negatively charged proteoglycans may bind small, positively charged molecules, such as histamine (118-120) and proteases (121, 122).

The definition of a proteoglycan is a protein core that is modified with one or several glycosaminoglycan (GAG) chains in the Golgi apparatus. Three different classes of protein-linked GAG chains are classified as heparan sulfate (HS)/heparin, chondroitin sulfate (CS)/dermatan sulfate (DS) or keratan sulfate (KS). The classification is based on the sugar units that constitute the repeating disaccharides in each type of the long, linear GAG chains (58). The most widespread GAGs are those of the CS/DS type and the HS/heparin type.

The synthesis of these GAG chains is initiated by the stepwise enzymatic addition of one xylose, two galactoses and a glucuronic acid, which makes up the linker region of the GAG chain. The fifth sugar added decides whether a GAG chain enters the HS/heparin pathway or the CS/DS track. The xylose is thought to be added to a serine with a neighboring glycine residue already in the ER by xylosyl transferase I or II (123, 124). The two galactoses are subsequently added by galactosyl transferase I and II, which are enzymes localized to the *cis*-Golgi region (125). The addition of glucuronic acid takes place in the *medial/trans* Golgi (126). A protein core that carries several GAG chain binding sites can be modified with both CS and HS chains and would then be called a hybrid proteoglycan (127, 128). The polymerization is catalyzed by enzyme complexes adding disaccharides of hexosamine and hexuronic acid nature. In the case of CS and DS, the hexosamine is N-acetyl-galactosamine, while in the case of HS and heparin it is N-acetyl-glucosamine. The hexuronic acid added is in all cases glucuronic acid, but this may subsequently be epimerized to iduronic acid in DS, HS and heparin by specific epimerases. All these classes of GAG chains are modified by sulfotransferases in various positions on the disaccharides and to a variable extent (58). All in all, although linear structures, the GAG chains may contain a variety of negatively charged domains that have affinity for proteins with clusters of positively charged amino acids.

Epithelial cells transport proteoglycan components of the extracellular matrix (ECM) and receptor proteins that attach the basolateral membrane domain of the epithelium to the ECM by vectorial basolateral routes. Basolateral localization of syndecan-1 in MDCK cells requires a signal positioned in the 12 most distal amino acids of its cytoplasmic tail (129), while basolateral secretion of the major basement membrane heparan sulfate (HS) PG occurs by a pH-dependent sorting mechanism (130). The fact that different types of proteoglycans are secreted apically and basolaterally indicates that these molecules are actively sorted. While HSPGs are mainly secreted basolaterally from MDCK cells, CSPGs are mainly secreted apically (131). The observation that endogenously synthesized CSPGs of high molecular mass and individual hexyl-β-D xylosides that have acquired CS chains, are mainly secreted apically in MDCK cells, indicated that CS chains may contain essential apical sorting determinants (132). Still, the interpretation of the CS data is not straightforward, since the presence of a CS chain in amyloid precursor-like protein 2 did not alter the predominant basolateral secretion observed for the CS-deficient variant (133). Addition of a CS chain did also not alter the basolateral sorting pattern of the H1 subunit of the asialoglycoprotein receptor, when it was expressed in MDCK cells (134). Since the H1 subunit is a trans-membrane protein, the lack of effect of the CS chain could be explained by the fact that glycan signals often are recessive to signals in the protein core, as observed for transmembrane receptors like the polymeric IgA receptor (135) and the low density lipoprotein receptor (64). HS chains, on the other hand, could possess basolateral sorting information. The GPI-linked HSPG glypican was detected mainly at the basolateral surface of both CaCo-2 and MDCK cells (87). A variant of glypican lacking sites for HS attachment was transported predominantly to the apical surface in MDCK cells, presumably directed there by the GPI membrane anchor and/or the N-linked glycan groups. However, the latter question was not addressed by removal of the N-glycans. HS chains could either promote basolateral sorting of glypican, or simply interfere with the recognition of the apical sorting information present in the molecule, which comes into play when the HS chains have been removed. As discussed above, the effect of the HS chains could merely be blocking of oligomerization.

A number of studies have been carried out in polarized MDCK cells with the small CSPG serglycin as a model protein. Serglycin was expressed with green fluorescent protein fused to the C-terminus. The expressed serglycin-GFP fusion proteoglycan obtained mainly CS chains and was mainly secreted into the apical medium (Figure 2) of epithelial MDCK cells (53).

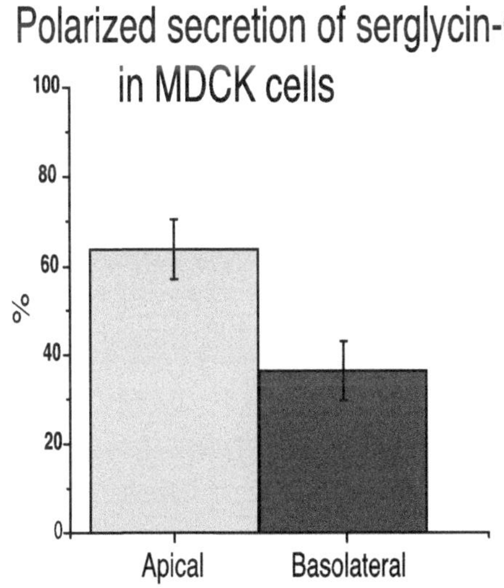

Figure 2. Polarized secretion of serglycin-GFP. A fusion protein of serglycin and GFP was expressed in polarized, filter-grown MDCK cells. Secretion of serglycin-GFP to the apical and basolateral medium was quantified from Western blots and expressed as % distribution in the apical and basolateral medium.

However, the minor fraction secreted into the basolateral medium was several times more intensely sulfated and showed a different sensitivity to the sulfation inhibitor chlorate (136). A sulfation intensity difference was also observed for the linker region sugars, indicating that the apical and basolateral serglycin-GFP molecules were segregated at an early stage of the secretory pathway, where the linker region is formed. A fraction of the serglycin constructs studied had the ability to bypass the Golgi apparatus and appear at the cell surface as a variant without GAG chains (137). Also this variant was secreted predominatly apically, indicating that a sorting event had already occurred prior to Golgi entry. This is in line with the observation that apical and basolateral glycoproteins in MDCK cells possess differential detergent extractability already in their high-mannose form (52), and raises the question whether glycans are signals directing a protein into the right pathway or merely a product of the pathway followed, due to a signal present in the immature protein or proteoglycan. The difference in sulfation intensity observed for apically and basolaterally secreted serglycin-GFP was essentially abolished upon neutralization of the secretory pathway with the macrolide Bafilomycin A_1. The effects of neutralization were by far most prominent in the apical route (138). The differences in sulfation intensity in the apical and basolateral secrtory route were observed also for endogenous proteins and could be overcome by transfection leading to over-expression of the PAPS-transporter (139). This indicates that the difference observed in the sulfation intensity in the apical and basolateral secretory routes was due to a reduced concentration of nucleotide sulfate (PAPS) in the apical route, a difference that was subsequently counteracted by expression of more PAPS transporter. In a recent study, the GAG binding domain of serglycin, which has the ability to carry several GAG chains, was transferred to the non-glycosylated model protein rat growth hormone (rGH). The GAG domain still gave rise to CS chains in the new protein context, and redirected the randomly secreted protein rGH more towards the apical surface (Figure 3) of MDCK cells (140). This is the first published example of transplantable sorting information in a GAG binding site with chondroitin sulfate chains. An interesting aspect of this study was that, although the secretion polarity of the serglycin GAG domain was maintained in the new protein environment provided by rGH, the sulfation intensity was no longer different in the apical and basolateral secretory pathways, indicating that different routes have been followed, or different enzymatic regimes have been recruited for GAG synthesis (140).

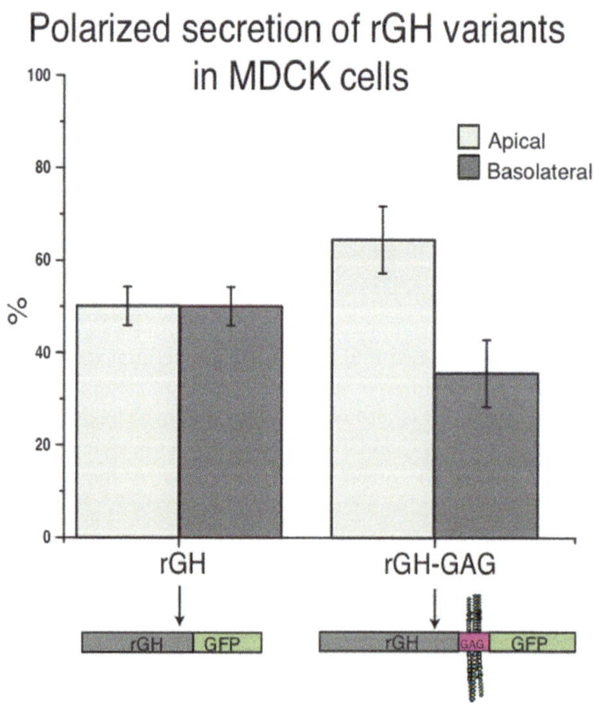

Figure 3. Polarized secretion of rat growth hormone (rGH) variants. Secretion from transfected polarized MDCK cells of a fusion protein of rGH with GFP was compared to that of an rGH-GFP fusion protein with the glycosaminoglycan (GAG) domain of serglycin inserted between rGH and GFP (rGH-GAG). Secretion of rGH and rGH-GAG to the apical and basolateral medium was quantified from Western blots and expressed as % distribution in the apical and basolateral medium.

6. Glycans are not the whole story

Although there evidently are many examples of glycan-mediated sorting to the apical surface of epithelia, there are many examples of apical sorting not requiring glycans. In fact, quite a few proteins are transported apically in a glycan-independent manner (141), like CD3-epsilon which has no N-glycan modification (142-146). It is evident that both proteins with (99) and without (142) N-linked sugars are transported to the apical surface, while some proteins carrying N-glycans are transported basolaterally (147). This could depend on variability in the structure of the N-glycans on different proteins, on the surrounding context in the protein (72), or the glycans could be recessive to other sorting signals in the molecule. The lectins localized to the early secretory pathway has affinity for high-mannose ligands, while the galectins generally has the better affinity for galactose containing glycans, found in both glycoproteins and glycolipids. While terminal processing of N-glycan structures seemed important for endolyn, gp114 and DPPIV (100, 148), there was little missorting of glycoproteins in a mutant MDCK cell line deficient in UDP-galactose delivery to the Golgi apparatus (149). A detailed investigation of the structure of the N-glycans attached to a dually glycosylated variant of rGH revealed only minor differences in these glycan structures after apical and basolateral transport and secretion, indicating that, although the glycosylated variant was mainly secreted apically, there were no observable differences that could mediate sorting based on terminal glycan structure (150).

Reduction in the sulfation level of GAG chains and attempts to generate shorter GAG chains have not resulted in alterations in the transport polarity of PGs. As long as the sorting machineries have not been outlined, the role of CS chains as apical sorting mediators and HS chains as basolateral sorting mediators remains incompletely understood.

Two classes of lectins have been implicated as mediators of apical sorting of N-glycanated proteins in polarized epithelial cells, the high mannose binding lectin VIP36, and possibly some related lectins, and the galectin family. While the former are mostly localized to the early secretory pathway, the latter family members are found in the lumen of endosomes. Thus, any suggestion for a glycan based sorting mechanism in the *trans*-Golgi network, a major sorting site in the epithelial MDCK cell line, lacks a description at the molecular level. How and when glycans influence the sorting and transport of their host glycoproteins and proteoglycans is therefore open to future investigations.

Author details

Kristian Prydz*, Gro Live Fagereng and Heidi Tveit
Department of Molecular Biosciences,
University of Oslo, Oslo, Norway

* Corresponding Author

7. References

[1] Helenius A, Aebi M (2004) Roles of N-linked glycans in the endoplasmic reticulum. Annu Rev Biochem.73:1019-49.

[2] Griffiths G, Simons K (1986) The trans Golgi network: sorting at the exit site of the Golgi complex. Science.234(4775):438-43.

[3] Prydz K, Dick G, Tveit H (2008) How many ways through the Golgi maze? Traffic.9(3):299-304.

[4] Matter K, Balda MS (2003) Signalling to and from tight junctions. Nat Rev Mol Cell Biol.4(3):225-36.

[5] Winckler B, Forscher P, Mellman I (1999) A diffusion barrier maintains distribution of membrane proteins in polarized neurons. Nature.397(6721):698-701.

[6] Dotti CG, Simons K (1990) Polarized sorting of viral glycoproteins to the axon and dendrites of hippocampal neurons in culture. Cell.62(1):63-72.

[7] Dotti CG, Banker G (1991) Intracellular organization of hippocampal neurons during the development of neuronal polarity. Journal of cell science Supplement.15:75-84.

[8] Schell MJ, Maurice M, Stieger B, Hubbard AL (1992) 5'nucleotidase is sorted to the apical domain of hepatocytes via an indirect route. J Cell Biol.119(5):1173-82.

[9] Bartles JR, Hubbard AL (1988) Plasma membrane protein sorting in epithelial cells: do secretory pathways hold the key? Trends Biochem Sci.13(5):181-4.

[10] Lasiecka ZM, Winckler B (2011) Mechanisms of polarized membrane trafficking in neurons -- focusing in on endosomes. Mol Cell Neurosci.48(4):278-87.

[11] Nosjean O, Briolay A, Roux B (1997) Mammalian GPI proteins: sorting, membrane residence and functions. Biochim Biophys Acta.1331(2):153-86.

[12] Benting JH, Rietveld AG, Simons K (1999) N-Glycans mediate the apical sorting of a GPI-anchored, raft-associated protein in Madin-Darby canine kidney cells. J Cell Biol.146(2):313-20.

[13] Paladino S, Sarnataro D, Tivodar S, Zurzolo C (2007) Oligomerization is a specific requirement for apical sorting of glycosyl-phosphatidylinositol-anchored proteins but not for non-raft-associated apical proteins. Traffic.8(3):251-8.

[14] Imjeti NS, Lebreton S, Paladino S, de la Fuente E, Gonzalez A, Zurzolo C (2011) N-Glycosylation instead of cholesterol mediates oligomerization and apical sorting of GPI-APs in FRT cells. Mol Biol Cell.22(23):4621-34.

[15] Freeze HH, Aebi M (2005) Altered glycan structures: the molecular basis of congenital disorders of glycosylation. Curr Opin Struct Biol.15(5):490-8.

[16] Kamiya Y, Kamiya D, Yamamoto K, Nyfeler B, Hauri HP, Kato K (2008) Molecular basis of sugar recognition by the human L-type lectins ERGIC-53, VIPL, and VIP36. J Biol Chem.283(4):1857-61.

[17] Mitrovic S, Ben-Tekaya H, Koegler E, Gruenberg J, Hauri HP (2008) The cargo receptors Surf4, endoplasmic reticulum-Golgi intermediate compartment (ERGIC)-53, and p25 are required to maintain the architecture of ERGIC and Golgi. Mol Biol Cell.19(5):1976-90.

[18] Prydz K, Tveit H, Vedeler A, Saraste J (2012) Arrivals and departures at the plasma membrane: direct and indirect transport routes. Cell Tissue Res.

[19] Yano H, Yamamoto-Hino M, Abe M, Kuwahara R, Haraguchi S, Kusaka I, et al. (2005) Distinct functional units of the Golgi complex in Drosophila cells. Proc Natl Acad Sci U S A.102(38):13467-72.

[20] Rindler MJ, Ivanov IE, Plesken H, Rodriguez-Boulan E, Sabatini DD (1984) Viral glycoproteins destined for apical or basolateral plasma membrane domains traverse the same Golgi apparatus during their intracellular transport in doubly infected Madin-Darby canine kidney cells. J Cell Biol.98(4):1304-19.

[21] Fuller SD, von Bonsdorff CH, Simons K (1985) Cell surface influenza haemagglutinin can mediate infection by other animal viruses. Embo J.4(10):2475-85.

[22] Rindler MJ, Ivanov IE, Plesken H, Sabatini DD (1985) Polarized delivery of viral glycoproteins to the apical and basolateral plasma membranes of Madin-Darby canine kidney cells infected with temperature-sensitive viruses. J Cell Biol.100(1): 136-51.

[23] Rodriguez-Boulan E, Musch A, Le Bivic A (2004) Epithelial trafficking: new routes to familiar places. Current opinion in cell biology.16(4):436-42.

[24] Compton T, Ivanov IE, Gottlieb T, Rindler M, Adesnik M, Sabatini DD (1989) A sorting signal for the basolateral delivery of the vesicular stomatitis virus (VSV) G protein lies in its luminal domain: analysis of the targeting of VSV G-influenza hemagglutinin chimeras. Proc Natl Acad Sci U S A.86(11):4112-6.

[25] Scheiffele P, Roth MG, Simons K (1997) Interaction of influenza virus haemagglutinin with sphingolipid-cholesterol membrane domains via its transmembrane domain. Embo J.16(18):5501-8.

[26] Lin S, Naim HY, Rodriguez AC, Roth MG (1998) Mutations in the middle of the transmembrane domain reverse the polarity of transport of the influenza virus hemagglutinin in MDCK epithelial cells. J Cell Biol.142(1):51-7.

[27] Kundu A, Avalos RT, Sanderson CM, Nayak DP (1996) Transmembrane domain of influenza virus neuraminidase, a type II protein, possesses an apical sorting signal in polarized MDCK cells. J Virol.70(9):6508-15.

[28] Bonilha VL, Marmorstein AD, Cohen-Gould L, Rodriguez-Boulan E (1997) Apical sorting of influenza hemagglutinin by transcytosis in retinal pigment epithelium. J Cell Sci.110 (Pt 15):1717-27.

[29] Dotti CG, Parton RG, Simons K (1991) Polarized sorting of glypiated proteins in hippocampal neurons. Nature.349(6305):158-61.

[30] Lisanti MP, Sargiacomo M, Graeve L, Saltiel AR, Rodriguez-Boulan E (1988) Polarized apical distribution of glycosyl-phosphatidylinositol-anchored proteins in a renal epithelial cell line. Proc Natl Acad Sci U S A.85(24):9557-61.

[31] Futter CE, Connolly CN, Cutler DF, Hopkins CR (1995) Newly synthesized transferrin receptors can be detected in the endosome before they appear on the cell surface. J Biol Chem.270(18):10999-1003.

[32] Leitinger B, Hille-Rehfeld A, Spiess M (1995) Biosynthetic transport of the asialoglycoprotein receptor H1 to the cell surface occurs via endosomes. Proc Natl Acad Sci U S A.92(22):10109-13.

[33] Ang AL, Taguchi T, Francis S, Folsch H, Murrells LJ, Pypaert M, et al. (2004) Recycling endosomes can serve as intermediates during transport from the Golgi to the plasma membrane of MDCK cells. J Cell Biol.167(3):531-43.

[34] Orzech E, Cohen S, Weiss A, Aroeti B (2000) Interactions between the exocytic and endocytic pathways in polarized Madin-Darby canine kidney cells. J Biol Chem.275(20):15207-19.

[35] Polishchuk R, Di Pentima A, Lippincott-Schwartz J (2004) Delivery of raft-associated, GPI-anchored proteins to the apical surface of polarized MDCK cells by a transcytotic pathway. Nat Cell Biol.6(4):297-307.

[36] Lock JG, Hammond LA, Houghton F, Gleeson PA, Stow JL (2005) E-cadherin transport from the trans-Golgi network in tubulovesicular carriers is selectively regulated by golgin-97 Rab11 in recycling endosomes regulates the sorting and basolateral transport of E-cadherin. Traffic.6(12):1142-56.

[37] Lock JG, Stow JL (2005) Rab11 in recycling endosomes regulates the sorting and basolateral transport of E-cadherin. Mol Biol Cell.16(4):1744-55.

[38] Folsch H (2005) The building blocks for basolateral vesicles in polarized epithelial cells. Trends Cell Biol.15(4):222-8.

[39] Farr GA, Hull M, Mellman I, Caplan MJ (2009) Membrane proteins follow multiple pathways to the basolateral cell surface in polarized epithelial cells. J Cell Biol.186(2):269-82.

[40] Keller P, Toomre D, Diaz E, White J, Simons K (2001) Multicolour imaging of post-Golgi sorting and trafficking in live cells. Nat Cell Biol.3(2):140-9.

[41] Polishchuk RS, Polishchuk EV, Marra P, Alberti S, Buccione R, Luini A, et al. (2000) Correlative light-electron microscopy reveals the tubular-saccular ultrastructure of carriers operating between Golgi apparatus and plasma membrane. J Cell Biol.148(1):45-58.

[42] Polishchuk EV, Di Pentima A, Luini A, Polishchuk RS (2003) Mechanism of constitutive export from the golgi: bulk flow via the formation, protrusion, and en bloc cleavage of large trans-golgi network tubular domains. Mol Biol Cell.14(11):4470-85.

[43] Deborde S, Perret E, Gravotta D, Deora A, Salvarezza S, Schreiner R, et al. (2008) Clathrin is a key regulator of basolateral polarity. Nature.452(7188):719-23.

[44] Gonzalez A, Rodriguez-Boulan E (2009) Clathrin and AP1B: key roles in basolateral trafficking through trans-endosomal routes. FEBS Lett.583(23):3784-95.

[45] Ladinsky MS, Mastronarde DN, McIntosh JR, Howell KE, Staehelin LA (1999) Golgi structure in three dimensions: functional insights from the normal rat kidney cell. J Cell Biol.144(6):1135-49.

[46] Patterson GH, Hirschberg K, Polishchuk RS, Gerlich D, Phair RD, Lippincott-Schwartz J (2008) Transport through the Golgi apparatus by rapid partitioning within a two-phase membrane system. Cell.133(6):1055-67.

[47] Sannerud R, Marie M, Nizak C, Dale HA, Pernet-Gallay K, Perez F, et al. (2006) Rab1 defines a novel pathway connecting the pre-Golgi intermediate compartment with the cell periphery. Mol Biol Cell.17(4):1514-26.
[48] Marie M, Dale HA, Sannerud R, Saraste J (2009) The function of the intermediate compartment in pre-Golgi trafficking involves its stable connection with the centrosome. Mol Biol Cell.20(20):4458-70.
[49] Saraste J, Dale HA, Bazzocco S, Marie M (2009) Emerging new roles of the pre-Golgi intermediate compartment in biosynthetic-secretory trafficking. FEBS Lett.583(23):3804-10.
[50] Muniz M, Morsomme P, Riezman H (2001) Protein sorting upon exit from the endoplasmic reticulum. Cell.104(2):313-20.
[51] Morsomme P, Prescianotto-Baschong C, Riezman H (2003) The ER v-SNAREs are required for GPI-anchored protein sorting from other secretory proteins upon exit from the ER. J Cell Biol.162(3):403-12.
[52] Alfalah M, Wetzel G, Fischer I, Busche R, Sterchi EE, Zimmer KP, et al. (2005) A novel type of detergent-resistant membranes may contribute to an early protein sorting event in epithelial cells. J Biol Chem.280(52):42636-43.
[53] Tveit H, Dick G, Skibeli V, Prydz K (2005) A proteoglycan undergoes different modifications en route to the apical and basolateral surfaces of Madin-Darby canine kidney cells. J Biol Chem.280(33):29596-603.
[54] Hara-Kuge S, Ohkura T, Seko A, Yamashita K (1999) Vesicular-integral membrane protein, VIP36, recognizes high-mannose type glycans containing alpha1-->2 mannosyl residues in MDCK cells. Glycobiology.9(8):833-9.
[55] Nyfeler B, Zhang B, Ginsburg D, Kaufman RJ, Hauri HP (2006) Cargo selectivity of the ERGIC-53/MCFD2 transport receptor complex. Traffic.7(11):1473-81.
[56] Hanisch FG, Reis CA, Clausen H, Paulsen H (2001) Evidence for glycosylation-dependent activities of polypeptide N-acetylgalactosaminyltransferases rGalNAc-T2 and -T4 on mucin glycopeptides. Glycobiology.11(9):731-40.
[57] Brockhausen I (2006) Mucin-type O-glycans in human colon and breast cancer: glycodynamics and functions. EMBO reports.7(6):599-604.
[58] Prydz K, Dalen KT (2000) Synthesis and sorting of proteoglycans. J Cell Sci.113 Pt 2:193-205.
[59] Simons K, Wandinger-Ness A (1990) Polarized sorting in epithelia. Cell.62(2):207-10.
[60] Mostov KE, de Bruyn Kops A, Deitcher DL (1986) Deletion of the cytoplasmic domain of the polymeric immunoglobulin receptor prevents basolateral localization and endocytosis. Cell.47(3):359-64.
[61] Mostov KE, Breitfeld P, Harris JM (1987) An anchor-minus form of the polymeric immunoglobulin receptor is secreted predominantly apically in Madin-Darby canine kidney cells. J Cell Biol.105(5):2031-6.
[62] Casanova JE, Apodaca G, Mostov KE (1991) An autonomous signal for basolateral sorting in the cytoplasmic domain of the polymeric immunoglobulin receptor. Cell.66(1):65-75.

[63] Hunziker W, Harter C, Matter K, Mellman I (1991) Basolateral sorting in MDCK cells requires a distinct cytoplasmic domain determinant. Cell.66(5):907-20.
[64] Matter K, Hunziker W, Mellman I (1992) Basolateral sorting of LDL receptor in MDCK cells: the cytoplasmic domain contains two tyrosine-dependent targeting determinants. Cell.71(5):741-53.
[65] Reich V, Mostov K, Aroeti B (1996) The basolateral sorting signal of the polymeric immunoglobulin receptor contains two functional domains. J Cell Sci.109 (Pt 8): 2133-9.
[66] Distel B, Bauer U, Le Borgne R, Hoflack B (1998) Basolateral sorting of the cation-dependent mannose 6-phosphate receptor in Madin-Darby canine kidney cells. Identification of a basolateral determinant unrelated to clathrin-coated pit localization signals. J Biol Chem.273(1):186-93.
[67] Marzolo MP, Yuseff MI, Retamal C, Donoso M, Ezquer F, Farfan P, et al. (2003) Differential distribution of low-density lipoprotein-receptor-related protein (LRP) and megalin in polarized epithelial cells is determined by their cytoplasmic domains. Traffic.4(4):273-88.
[68] Takeda T, Yamazaki H, Farquhar MG (2003) Identification of an apical sorting determinant in the cytoplasmic tail of megalin. Am J Physiol Cell Physiol.284(5):C1105-13.
[69] Hodson CA, Ambrogi IG, Scott RO, Mohler PJ, Milgram SL (2006) Polarized apical sorting of guanylyl cyclase C is specified by a cytosolic signal. Traffic.7(4):456-64.
[70] Pang S, Urquhart P, Hooper NM (2004) N-glycans, not the GPI anchor, mediate the apical targeting of a naturally glycosylated, GPI-anchored protein in polarised epithelial cells. J Cell Sci.117(Pt 21):5079-86.
[71] Paladino S, Sarnataro D, Pillich R, Tivodar S, Nitsch L, Zurzolo C (2004) Protein oligomerization modulates raft partitioning and apical sorting of GPI-anchored proteins. J Cell Biol.167(4):699-709.
[72] Kitagawa Y, Sano Y, Ueda M, Higashio K, Narita H, Okano M, et al. (1994) N-glycosylation of erythropoietin is critical for apical secretion by Madin-Darby canine kidney cells. Exp Cell Res.213(2):449-57.
[73] Scheiffele P, Peranen J, Simons K (1995) N-glycans as apical sorting signals in epithelial cells. Nature.378(6552):96-8.
[74] Fiedler K, Simons K (1996) Characterization of VIP36, an animal lectin homologous to leguminous lectins. J Cell Sci.109 (Pt 1):271-6.
[75] Fullekrug J, Scheiffele P, Simons K (1999) VIP36 localisation to the early secretory pathway. J Cell Sci.112 (Pt 17):2813-21.
[76] Urban J, Parczyk K, Leutz A, Kayne M, Kondor-Koch C (1987) Constitutive apical secretion of an 80-kD sulfated glycoprotein complex in the polarized epithelial Madin-Darby canine kidney cell line. J Cell Biol.105(6 Pt 1):2735-43.
[77] Hara-Kuge S, Ohkura T, Ideo H, Shimada O, Atsumi S, Yamashita K (2002) Involvement of VIP36 in intracellular transport and secretion of glycoproteins

in polarized Madin-Darby canine kidney (MDCK) cells. J Biol Chem.277(18):16332-9.

[78] Hara-Kuge S, Seko A, Shimada O, Tosaka-Shimada H, Yamashita K (2004) The binding of VIP36 and alpha-amylase in the secretory vesicles via high-mannose type glycans. Glycobiology.14(8):739-44.

[79] Simons K, Ikonen E (1997) Functional rafts in cell membranes. Nature.387(6633):569-72.

[80] Surma MA, Klose C, Simons K (2011) Lipid-dependent protein sorting at the trans-Golgi network. Biochim Biophys Acta.

[81] Keller P, Simons K (1998) Cholesterol is required for surface transport of influenza virus hemagglutinin. J Cell Biol.140(6):1357-67.

[82] Prydz K, Simons K (2001) Cholesterol depletion reduces apical transport capacity in epithelial Madin-Darby canine kidney cells. Biochem J.357(Pt 1):11-5.

[83] Lisanti MP, Caras IW, Davitz MA, Rodriguez-Boulan E (1989) A glycophospholipid membrane anchor acts as an apical targeting signal in polarized epithelial cells. J Cell Biol.109(5):2145-56.

[84] Lisanti MP, Le Bivic A, Saltiel AR, Rodriguez-Boulan E (1990) Preferred apical distribution of glycosyl-phosphatidylinositol (GPI) anchored proteins: a highly conserved feature of the polarized epithelial cell phenotype. J Membr Biol.113(2):155-67.

[85] Paladino S, Lebreton S, Tivodar S, Campana V, Tempre R, Zurzolo C (2008) Different GPI-attachment signals affect the oligomerisation of GPI-anchored proteins and their apical sorting. J Cell Sci.121(Pt 24):4001-7.

[86] Puig B, Altmeppen HC, Thurm D, Geissen M, Conrad C, Braulke T, et al. (2011) N-glycans and glycosylphosphatidylinositol-anchor act on polarized sorting of mouse PrP(C) in Madin-Darby canine kidney cells. PLoS One.6(9):e24624.

[87] Mertens G, Van der Schueren B, van den Berghe H, David G (1996) Heparan sulfate expression in polarized epithelial cells: the apical sorting of glypican (GPI-anchored proteoglycan) is inversely related to its heparan sulfate content. J Cell Biol.132(3):487-97.

[88] Delacour D, Jacob R (2006) Apical protein transport. Cell Mol Life Sci.63(21):2491-505.

[89] Stechly L, Morelle W, Dessein AF, Andre S, Grard G, Trinel D, et al. (2009) Galectin-4-regulated delivery of glycoproteins to the brush border membrane of enterocyte-like cells. Traffic.10(4):438-50.

[90] Mishra R, Grzybek M, Niki T, Hirashima M, Simons K (2010) Galectin-9 trafficking regulates apical-basal polarity in Madin-Darby canine kidney epithelial cells. Proc Natl Acad Sci U S A.107(41):17633-8.

[91] Boscher C, Dennis JW, Nabi IR (2011) Glycosylation, galectins and cellular signaling. Current opinion in cell biology.23(4):383-92.

[92] Rondanino C, Poland PA, Kinlough CL, Li H, Rbaibi Y, Myerburg MM, et al. (2011) Galectin-7 modulates the length of the primary cilia and wound repair in polarized kidney epithelial cells. Am J Physiol Renal Physiol.301(3):F622-33.

[93] Seelenmeyer C, Wegehingel S, Tews I, Kunzler M, Aebi M, Nickel W (2005) Cell surface counter receptors are essential components of the unconventional export machinery of galectin-1. J Cell Biol.171(2):373-81.
[94] Delacour D, Greb C, Koch A, Salomonsson E, Leffler H, Le Bivic A, et al. (2007) Apical sorting by galectin-3-dependent glycoprotein clustering. Traffic.8(4):379-88.
[95] Jensen PH, Kolarich D, Packer NH (2010) Mucin-type O-glycosylation--putting the pieces together. FEBS J.277(1):81-94.
[96] Yeaman C, Le Gall AH, Baldwin AN, Monlauzeur L, Le Bivic A, Rodriguez-Boulan E (1997) The O-glycosylated stalk domain is required for apical sorting of neurotrophin receptors in polarized MDCK cells. J Cell Biol.139(4):929-40.
[97] Alfalah M, Jacob R, Preuss U, Zimmer KP, Naim H, Naim HY (1999) O-linked glycans mediate apical sorting of human intestinal sucrase-isomaltase through association with lipid rafts. Curr Biol.9(11):593-6.
[98] Spodsberg N, Alfalah M, Naim HY (2001) Characteristics and structural requirements of apical sorting of the rat growth hormone through the O-glycosylated stalk region of intestinal sucrase-isomaltase. J Biol Chem.276(49):46597-604.
[99] Zheng X, Lu D, Sadler JE (1999) Apical sorting of bovine enteropeptidase does not involve detergent-resistant association with sphingolipid-cholesterol rafts. J Biol Chem.274(3):1596-605.
[100] Alfalah M, Jacob R, Naim HY (2002) Intestinal dipeptidyl peptidase IV is efficiently sorted to the apical membrane through the concerted action of N- and O-glycans as well as association with lipid microdomains. J Biol Chem.277(12):10683-90.
[101] Kinlough CL, Poland PA, Gendler SJ, Mattila PE, Mo D, Weisz OA, et al. (2011) Core-glycosylated mucin-like repeats from MUC1 are an apical targeting signal. J Biol Chem.286(45):39072-81.
[102] Campbell AG, Fessler LI, Salo T, Fessler JH (1987) Papilin: a Drosophila proteoglycan-like sulfated glycoprotein from basement membranes. J Biol Chem.262(36):17605-12.
[103] Cambiazo V, Inestrosa NC (1990) Proteoglycan production in Drosophila egg development: effect of beta-D-xyloside on proteoglycan synthesis and larvae motility. Comp Biochem Physiol B.97(2):307-14.
[104] Graner M, Stupka K, Karr TL (1994) Biochemical and cytological characterization of DROP-1: a widely distributed proteoglycan in Drosophila. Insect biochemistry and molecular biology.24(6):557-67.
[105] Spring J, Paine-Saunders SE, Hynes RO, Bernfield M (1994) Drosophila syndecan: conservation of a cell-surface heparan sulfate proteoglycan. Proc Natl Acad Sci U S A.91(8):3334-8.
[106] Rogalski TM, Williams BD, Mullen GP, Moerman DG (1993) Products of the unc-52 gene in Caenorhabditis elegans are homologous to the core protein of the mammalian basement membrane heparan sulfate proteoglycan. Genes & development.7(8):1471-84.

[107] Schimpf J, Sames K, Zwilling R (1999) Proteoglycan distribution pattern during aging in the nematode Caenorhabditis elegans: an ultrastructural histochemical study. The Histochemical journal.31(5):285-92.
[108] Feta A, Do AT, Rentzsch F, Technau U, Kusche-Gullberg M (2009) Molecular analysis of heparan sulfate biosynthetic enzyme machinery and characterization of heparan sulfate structure in Nematostella vectensis. Biochem J.419(3):585-93.
[109] Egeberg M, Kjeken R, Kolset SO, Berg T, Prydz K (2001) Internalization and stepwise degradation of heparan sulfate proteoglycans in rat hepatocytes. Biochim Biophys Acta.1541(3):135-49.
[110] Fedarko NS, Conrad HE (1986) A unique heparan sulfate in the nuclei of hepatocytes: structural changes with the growth state of the cells. J Cell Biol.102(2):587-99.
[111] Ishihara M, Fedarko NS, Conrad HE (1986) Transport of heparan sulfate into the nuclei of hepatocytes. J Biol Chem.261(29):13575-80.
[112] Tumova S, Hatch BA, Law DJ, Bame KJ (1999) Basic fibroblast growth factor does not prevent heparan sulphate proteoglycan catabolism in intact cells, but it alters the distribution of the glycosaminoglycan degradation products. Biochem J.337 (Pt 3): 471-81.
[113] Nickel W, Huber LA, Kahn RA, Kipper N, Barthel A, Fasshauer D, et al. (1994) ADP ribosylation factor and a 14-kD polypeptide are associated with heparan sulfate-carrying post-trans-Golgi network secretory vesicles in rat hepatocytes. J Cell Biol.125(4):721-32.
[114] Barthel A, Nickel W, Tonko C, Soling HD (1995) Sorting and budding of constitutive secretory vesicles in hepatocytes and hepatoma cells. Adv Enzyme Regul.35:283-92.
[115] Abrink M, Grujic M, Pejler G (2004) Serglycin is essential for maturation of mast cell secretory granule. J Biol Chem.279(39):40897-905.
[116] Kolset SO, Prydz K, Pejler G (2004) Intracellular proteoglycans. Biochem J.379(Pt 2):217-27.
[117] Henningsson F, Hergeth S, Cortelius R, Abrink M, Pejler G (2006) A role for serglycin proteoglycan in granular retention and processing of mast cell secretory granule components. FEBS J.273(21):4901-12.
[118] Brion C, Miller SG, Moore HP (1992) Regulated and constitutive secretion. Differential effects of protein synthesis arrest on transport of glycosaminoglycan chains to the two secretory pathways. J Biol Chem.267(3):1477-83.
[119] Grimes M, Kelly RB (1992) Sorting of chromogranin B into immature secretory granules in pheochromocytoma (PC12) cells. Annals of the New York Academy of Sciences.674:38-52.
[120] Castle AM, Castle JD (1998) Enhanced glycosylation and sulfation of secretory proteoglycans is coupled to the expression of a basic secretory protein. Mol Biol Cell.9(3):575-83.
[121] Lutzelschwab C, Pejler G, Aveskogh M, Hellman L (1997) Secretory granule proteases in rat mast cells. Cloning of 10 different serine proteases and a carboxypeptidase A

from various rat mast cell populations. The Journal of experimental medicine.185(1): 13-29.

[122] Huang C, Sali A, Stevens RL (1998) Regulation and function of mast cell proteases in inflammation. Journal of clinical immunology.18(3):169-83.

[123] Esko JD, Stewart TE, Taylor WH (1985) Animal cell mutants defective in glycosaminoglycan biosynthesis. Proc Natl Acad Sci U S A.82(10):3197-201.

[124] Gotting C, Prante C, Kuhn J, Kleesiek K (2007) Proteoglycan biosynthesis during chondrogenic differentiation of mesenchymal stem cells. The Scientific World Journal. 7:1207-10.

[125] Esko JD, Weinke JL, Taylor WH, Ekborg G, Roden L, Anantharamaiah G, et al. (1987) Inhibition of chondroitin and heparan sulfate biosynthesis in Chinese hamster ovary cell mutants defective in galactosyltransferase I. J Biol Chem.262(25):12189-95.

[126] Silbert JE, Sugumaran G (2002) A starting place for the road to function. Glycoconj J.19(4-5):227-37.

[127] Rapraeger A, Jalkanen M, Endo E, Koda J, Bernfield M (1985) The cell surface proteoglycan from mouse mammary epithelial cells bears chondroitin sulfate and heparan sulfate glycosaminoglycans. J Biol Chem.260(20):11046-52.

[128] Sugahara K, Mizuno N, Okumura Y, Kawasaki T (1992) The phosphorylated and/or sulfated structure of the carbohydrate-protein-linkage region isolated from chondroitin sulfate in the hybrid proteoglycans of Engelbreth-Holm-Swarm mouse tumor. European journal of biochemistry / FEBS.204(1):401-6.

[129] Miettinen HM, Edwards SN, Jalkanen M (1994) Analysis of transport and targeting of syndecan-1: effect of cytoplasmic tail deletions. Mol Biol Cell.5(12):1325-39.

[130] Caplan MJ, Stow JL, Newman AP, Madri J, Anderson HC, Farquhar MG, et al. (1987) Dependence on pH of polarized sorting of secreted proteins. Nature. 329(6140):632-5.

[131] Svennevig K, Prydz K, Kolset SO (1995) Proteoglycans in polarized epithelial Madin-Darby canine kidney cells. Biochem J.311 (Pt 3):881-8.

[132] Kolset SO, Vuong TT, Prydz K (1999) Apical secretion of chondroitin sulphate in polarized Madin-Darby canine kidney (MDCK) cells. J Cell Sci.112 (Pt 11):1797-801.

[133] Lo AC, Thinakaran G, Slunt HH, Sisodia SS (1995) Metabolism of the amyloid precursor-like protein 2 in MDCK cells. Polarized trafficking occurs independent of the chondroitin sulfate glycosaminoglycan chain. J Biol Chem.270(21):12641-5.

[134] Kobialka S, Beuret N, Ben-Tekaya H, Spiess M (2009) Glycosaminoglycan chains affect exocytic and endocytic protein traffic. Traffic.10(12):1845-55.

[135] Mostov KE, Deitcher DL (1986) Polymeric immunoglobulin receptor expressed in MDCK cells transcytoses IgA. Cell.46(4):613-21.

[136] Vuong TT, Prydz K, Tveit H (2006) Differences in the apical and basolateral pathways for glycosaminoglycan biosynthesis in Madin-Darby canine kidney cells. Glycobiology.16(4):326-32.

[137] Tveit H, Akslen LK, Fagereng GL, Tranulis MA, Prydz K (2009) A secretory Golgi bypass route to the apical surface domain of epithelial MDCK cells. Traffic.10(11):1685-95.

[138] Grondahl F, Tveit H, Prydz K (2009) Neutralization of endomembrane compartments in epithelial MDCK cells affects proteoglycan synthesis in the apical secretory pathway. Biochem J.418(3):517-28.

[139] Dick G, Grondahl F, Prydz K (2008) Overexpression of the 3'-phosphoadenosine 5'-phosphosulfate (PAPS) transporter 1 increases sulfation of chondroitin sulfate in the apical pathway of MDCK II cells. Glycobiology.18(1):53-65.

[140] Hafte TT, Fagereng GL, Prydz K, Grondahl F, Tveit H (2011) Protein core-dependent glycosaminoglycan modification and glycosaminoglycan-dependent polarized sorting in epithelial Madin-Darby canine kidney cells. Glycobiology.21(4):457-66.

[141] Rodriguez-Boulan E, Gonzalez A (1999) Glycans in post-Golgi apical targeting: sorting signals or structural props? Trends Cell Biol.9(8):291-4.

[142] Alonso MA, Fan L, Alarcon B (1997) Multiple sorting signals determine apical localization of a nonglycosylated integral membrane protein. J Biol Chem.272(49):30748-52.

[143] Marzolo MP, Bull P, Gonzalez A (1997) Apical sorting of hepatitis B surface antigen (HBsAg) is independent of N-glycosylation and glycosylphosphatidylinositol-anchored protein segregation. Proc Natl Acad Sci U S A.94(5):1834-9.

[144] Bravo-Zehnder M, Orio P, Norambuena A, Wallner M, Meera P, Toro L, et al. (2000) Apical sorting of a voltage- and Ca2+-activated K+ channel alpha -subunit in Madin-Darby canine kidney cells is independent of N-glycosylation. Proc Natl Acad Sci U S A.97(24):13114-9.

[145] Meerson NR, Bello V, Delaunay JL, Slimane TA, Delautier D, Lenoir C, et al. (2000) Intracellular traffic of the ecto-nucleotide pyrophosphatase/phosphodiesterase NPP3 to the apical plasma membrane of MDCK and Caco-2 cells: apical targeting occurs in the absence of N-glycosylation. J Cell Sci.113 Pt 23:4193-202.

[146] Trischler M, Koch-Brandt C, Ullrich O (2001) Apical transport of osteopontin is independent of N-glycosylation and sialylation. Mol Membr Biol.18(4):275-81.

[147] Su T, Cariappa R, Stanley K (1999) N-glycans are not a universal signal for apical sorting of secretory proteins. FEBS Lett.453(3):391-4.

[148] Potter BA, Ihrke G, Bruns JR, Weixel KM, Weisz OA (2004) Specific N-glycans direct apical delivery of transmembrane, but not soluble or glycosylphosphatidylinositol-anchored forms of endolyn in Madin-Darby canine kidney cells. Mol Biol Cell. 15(3):1407-16.

[149] Le Bivic A, Garcia M, Rodriguez-Boulan E (1993) Ricin-resistant Madin-Darby canine kidney cells missort a major endogenous apical sialoglycoprotein. J Biol Chem. 268(10):6909-16.

[150] Moen A, Hafte TT, Tveit H, Egge-Jacobsen W, Prydz K (2011) N-Glycan synthesis in the apical and basolateral secretory pathway of epithelial MDCK cells and the influence of a glycosaminoglycan domain. Glycobiology.21(11):1416-25.

Chapter 14

The Role of Glycosylation in Receptor Signaling

Brian J. Arey

Additional information is available at the end of the chapter

1. Introduction

Glycosylation is an important and highly regulated mechanism of secondary protein processing within cells. It plays a critical role in determining protein structure, function and stability. Structurally, glycosylation is known to affect the three dimensional configuration of proteins. This is of particular importance when considering protein-protein interactions such as those that occur between protein ligands and their cognate receptors or in the creation of other large macromolecular complexes. Many secreted proteins, such as hormones or cytokines, are glycosylated and this has been shown to impact in determining their activity when bound to receptors. Changes in these complexes result in alterations in how they recruit, interact and activate signaling proteins (e.g. G proteins). Additionally, signaling proteins are also glycosylated and this has distinct effects on their function. Ultimately, these effects help determine which signaling pathways are activated within the cell (Figure 1). Thus, glycosylation plays a key role in determining the cellular response to exogenous factors. This chapter will provide an overview of how glycosylation of ligands, their receptors, and signaling proteins affects signal transduction in mammalian cells by discussing specific examples of how receptor signaling is regulated by glycosylation.

2. Role of glycosylation in protein function

Although carbohydrates added to proteins are known to be highly flexible and mobile within the constraints of the glycoprotein, they are known to provide a key stabilizing force for proteins within their microenvironments. Of particular importance is the role that carbohydrates play in achieving the proper three dimensional conformation of glycoproteins(1, 2). As the carbohydrates are added to the nascent protein within the endoplasmic reticulum, carbohydrates (monosaccharides) are added to the protein on specific amino acid residues. Glycosylation has been reported on 8 different amino acids with the most common residue for carbohydrate addition being asparagine (N-glycosylation). This process can aid in the final protein product folding correctly into its three dimensional, biologically active conformation. However, this is not the case for all glycoproteins although it has been noted for a significant number (2). Interestingly the mechanism of adding these sugar residues is complex and not fully understood but is known to require several enzymes and is physiologically regulated. This suggests that glycosylation, as well as other secondary protein processing, is vital to the biological function of these proteins.

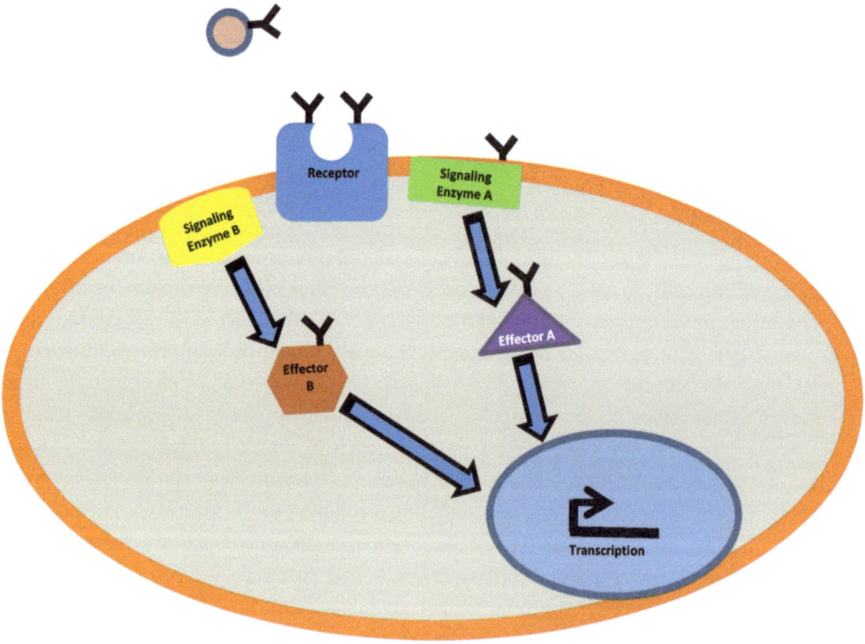

Figure 1. Regulation of receptor-ligand signaling by glycosylation. Glycosylation occurs at every level of the receptor signaling mechanism and therefore can impact the biological responses induced by receptor-ligand binding. Glycosylation (black lines) can occur on the ligand itself, the receptor, as well as on key signaling enzymes and effector proteins (hexagon,triangle). All of which play an improtant role in driving the biological responses in the cell.

In addition to its effects on driving correct folding of glycoproteins, glycosylation also has other effects on the physicochemical properties of these proteins. These effects help to determine the glycoprotein's overall energy and this can affect many of the biological functions that the protein performs (for a more detailed review see (3)). For example, glycosylation is well known to play a role in modulating thermostability of proteins as well as the overall charge. Of particular interest to the development of new therapies is the role that glycosylation plays in affecting protein-protein interactions. Intermolecular association that occur between protein ligands and their cognate receptors or between activated receptors and their intracellular signaling machinery have been shown to be modulated by the presence of glycosylation (4). Many examples exist that suggest that glycosylation of either a receptor or its ligand aids in determining the resulting biological responses. The primary mechanism for these effects lies in the ability of carbohydrates to modulate the overall energy state of the protein (2, 3).

In terms of thermostability, studies of various glycoproteins have focused on the thermodynamics of select placement or displacement of glycosylation on proteins (2). These studies have revealed that addition of even a single monosaccharide to a protein can significantly impact the fluctuation of that protein between folded and unfolded states (3). Through detailed NMR evaluation and use of statistical tools it has been found that certain commonalities exist for the placement of carbohydrates on protein and predict a functional role for these sites in stabilization of protein structures. One such study has found that glycosylation can occur on almost any part of a protein's structure but that bends or turns in the structure are preferred. Similarly, it has been found that glycosylation of proteins has a greater impact at less structured regions of a protein highly suggest that glycosylation plays a key role in protein stabilization (2, 3). In addition to aiding the stablization of proteins in a microenvironment, glycosylation has also been found to play a key role in stabilizing glycoproteins in the macroenvironment through alteration in half-life. There are numerous reports that the presence of polysaccharides added as secondary protein processing prolongs the half-life of these proteins including antibodies, hormones and cytokines (5).

3. Role of glycosylation in receptor function

3.1. Viral coat proteins

One of the best studied glycoproteins is the HIV viral coat protein, GP120. The description of the role of viral glycoproteins in host-virus interactions have been studied extensively and reviewed in detail elsewhere (5). However, this important interaction deserves mention. The GP120 protein is integral to the initiation of contact between the HIV virus particle and its host cell by mediating the adhesion of the viral particle to the host cell surface. It is a heavily glycosylated protein owing nearly half of its mass to the presence 27 glycosylated residues (5). This protein acts as part of a co-receptor complex with the host cell CD4 protein. Association between CD4 and GP120 leads to conformational changes in these proteins that ultimately lead to membrane fusion between the host cell and the virus particle. The presence of this glycosylation acts as a natural barrier to defending immune cells and antibodies such that it is difficult for the natural immune system to recognize and target the HIV virus for elimination.

3.2. Interleukins

Interleukins are secreted glycoproteins of the immune system that communicate both positive and negative regulatory signals to the various cellular and components that make up the innate and acquired immune responses. Interleukins and other cytokines exert their actions on their target cells through interactions with specific receptors. Most cytokines like interleukins are found in their mature state as glycosylated proteins. In the case of these important glycoprotein modulators, both N-linked and O-linked glycosylation has been described (6). The role of glycosylation in affecting cytokine function has been of interest from both the protein ligand and the receptor perspectives. Due to the large number of different cytokines, for the purposes of this chapter we will focus on Interleukin 5 (IL5).

IL5 is an important immune cytokine that is released from T-cells and induces activated B-cells into antibody producing cells. In addition, IL5 also acts as a differentiating factor for eosinophils. From a clinical perspective, the role of IL5 is important for immune diseases that involve hyperproliferation and invasion of eosinophils, such as in asthma (7). IL5 works as a homodimer that binds specifically to its membrane-bound receptor (IL5R). In the case of IL5, chemical digestion of either the N-linked or O-linked sugar residues on recombinant hIL5 had profound effects on the biological activity of the cytokine in terms of its ability to stimulate release of IgM from BCL_1 cells (8). Removal of the N-linked glycosylation on IL5 improved potency of the cytokine by approximately 3 fold. Interestingly, removal of the O-linked sugars led to an approximate 10 fold improvement in potency of IL5 which was equivalent to fully deglycosylated IL5 (8). In this same study, the authors also demonstrate that the N-linked glycosylation but not the O-linked glycosylation significantly improved the thermostability of IL5 *in vitro*. These data suggest that both N-linked and O-linked glycosylation play important roles in the biological activity of IL5.

The IL5R is composed of two subunits, the IL5Rα and βc subunits. Mechanistic studies have revealed that IL5 induces biological activity through a two step process in which IL5 binds to the IL5α subunit leading to interaction with the preformed βc subunit (9, 10). The βc then induces the signaling cascade within the target cell through activation of a kinase cascade by way of associated JAK kinases. The IL5α subunit is highly glycosylated having 4 N-glycosylation sites (Asn^{15}, Asn^{111}, Asn^{196} and Asn^{224}) in the extracellular region (11, 12). Complete removal of the glycosylation of IL5α leads to a loss of ligand binding. More detailed studies of the contributions of the N-glycosylation sites on IL5α revealed that Asn^{196} is required for ligand binding (9). Loss of the other three sites by mutation had no effect on IL5 affinity and biological activity (B-cell proliferation assay). Interestingly, mutation of Asn^{196} led to a complete loss of binding and biological activity, suggesting that glycosylation of that residue is absolutely required for IL5 recognition (9).

IL5R βc subunit is also glycosylated. This protein interacts with a number of cytokine receptors including IL5α, interleukin 3 α (IL3 α) and granulocyte-macrophage colony-stimulating factor α (GM-CSF α). Therefore, the βc protein is a common signal transducing partner to many cytokine receptors. The βc protein contains an N-linked glycosylation site at Asn^{328}. Conflicting reports have been published suggesting that glycosylation of Asn^{328} is either required for signaling activity of the βc subunit or not required. However, a recent publication by Murphy et al, suggests strongly that glycosylation at Asn^{328} does not play a role in either ligand binding or receptor activation.

3.3. Glycoprotein hormone family

The reproductive hormones called gonadotropins (luteininzing hormone (LH) and follicle-stimulating hormone (FSH)) are important to proper regulation of reproduction. These proteins are found in the circulation as alpha subunit and beta subunit heterodimers that contain multiple glycosylation sites on both subunits (13, 14). Along with the thyroid-stimulating hormone (TSH), they comprise the glycoprotein hormone family. Interestingly, the degree of glycosylation added to these hormones varies depending upon the physiological state and therefore, they are found in the plasma as a series of isoforms that vary in glycosylation complexity.

Historically, it has been accepted that glycosylation complexity has an impact on the overall acidity of each isoform with more complex variants (higher degree of terminal sialylation and sulfation) possessing more acidic isoelectric points (pI) (15, 16) with less terminally sialylated /sulfonated isoforms more basic in pI. Chromoatofocusing has been used as a way of purifying these differently glycosylated isoforms. Recently though, Bousfield has generated data demonstrating that chromatofocusing does not separate isoforms on the basis of glycan structure suggesting that the isoelectric point of gonadotropins is not completely determined by the glycosylation structure (17). Nevertheless, the physiological significance of these glycosylated variants is suggested by data demonstrating that the degree of glycosylation that occurs within the anterior pituitary synthetic cells is regulated by exogenous factors including ovarian steroids (18, 19). Tight regulation of this secondary protein processing of glycoprotein hormones suggests an important physiological role for the presence of glycosylated variants of TSH, LH and FSH. Numerous reports have detailed the effects of partial or complete deglycosylation on action of these hormones (14, 20, 21). The data in this area are varying with some noting no effects on binding (14, 22-24) and others noting increased binding affinity (21, 25). However, recent work in this area using more sophisticated separation techniques strongly suggest that hormone glycosylation does play a significant role in receptor binding (25). There is a significant effect on signaling. Indeed, using a baculovirus expression system to create partially glycosylated isoforms of FSH has shown that glycosylation can change the pharmacological properties of the hormone *via* alterations in interaction of the FSH receptor with the G protein signaling machinery (13, 24). In addition, numerous reports have found a distinct difference in the observed bioactivity of basic isoforms of glycoprotein hormones as compared to acidic isoforms (13, 14).

The glycoprotein hormone receptors are G protein-coupled receptors (GPCR) (26). These receptors are characterized by long amino-terminal extracellular domains (>300 aa) that are required for binding of ligand, seven lipophilic, membrane-spanning domains and relatively short, cytoplasmic carboxy-terminal tails (27, 28). The extracellular domains of the glycoprotein hormones are characterized by numerous leucine rich repeats (LRR) that have been shown to be important to binding of the receptors to their respective ligands (27). Similar to their hormone ligands, glycoprotein hormone receptors also contain sugar residues on their extracellular domains. Studies of the contribution of this glycosylation *via* mutation of the Asp residues acting as glycosylation sites has found that the glycosylation is responsible for proper folding of the receptor during protein synthesis (29, 30). This is more so for TSHR and FSHR than for LH/hCGR (31). Overall, the glycosylation state of the receptor does not seem to have an effect on ligand binding affinity or activation of signal transduction pathways (32).

4. Glycosylation of signaling proteins

4.1. Adenylate cyclase

In addition, to determining ligand-receptor interactions in some systems, glycosylation can also play a role in regulating intracellular signaling proteins. Adenylate cyclase is a key enzyme that produces the second messenger cAMP from ATP. It is best described for its ability to be stimulated or inhibited by activation of heterotrimeric GTP binding proteins (G-proteins) following G-protein-coupled receptor (GPCR) binding to agonist. There are nine recognized adenylate cyclase isoforms and three general classes of membrane bound adenylate cyclases: the calcium activated family (AC1, AC3 and AC8) , the calcium-inhibited family (AC5 and AC6) and the G-protein activated family (AC2, AC4 and AC7) (33). All nine adenylate cyclases are regulated by Gαs and magnesium. The calcium activated family members respond to calcium in a calmodulin-dependent manner (33). The calcium-inhibited adenylate cyclases are inhibited by calcium at less than micromolar concentrations. The G-protein activated family responds to activation by G $\beta\gamma$ subunits following GPCR agonist binding (33). In addition, splice variants of some forms of adenylate cyclases have been noted (34). Generally speaking, adenylate cyclases tend to be clustered together within cell membranes with other signaling components such as receptors and G-proteins in lipid rich domains such as those formed by the scaffolding protein caveolin-1 (35).

Several types of post-translational modifications to adenylate cyclases have been found to affect their activity; including nitrosylation, phosphorylation and glycosylation (36). In terms of glycosylation, regulation of several adenylate cyclase family members has been reported. N-linked glycosylation has been observed on the extracellular domains of some adenylate cyclases and for this reason there had been some controversy concerning the functional role it played in enzyme activity. This was mainly due to the fact that adenylate cyclase interacts with its protein partners within domains found on the cytoplasmic side of the enzyme. However, deglycosylation using metabolic inhibitors or site-directed mutants have revealed a critical role for glycosylation in adenylate cyclase activity.

Glycosylation of Type 8 adenylate cyclase (AC8) has been found on two of its three isoforms (34). These glycosylation sites are found the extracellular surface of two of these isoforms (AC8-A and AC8-C) but not on the AC8-B isoform. This is presumably due the excision of a portion of the extracellular domain between transmembrane spans 9 and 10 in this isoform that contains the N-linked glycosylation site. Recent studies of the role of these glycosylation sites in AC8 have shown that they are critical to localization of the enzyme to the lipid rich rafts in membranes (37). Thus, potentially determining function of AC8 in cells where it is expressed. This would imply a differential localization and functional role for AC8-B.

Glycosylation of AC6 is required for response to several stimulators of AC activity since mutagenesis or glycosylation inhibitors affect AC6 response to forskolin and G-proteins (38). The other member of the calcium-inhibited ACs, AC5, has two putative glycosylation sites but it is still unclear as to whether these sites are glycosylated (36).

The other member of the adenylate cyclase family of enzymes is adenylate cyclase 9 (AC9). This particular adenylate cyclase is the most divergent in terms of sequence from the other known ACs. AC9 activity is regulated by G-proteins and by protein kinase C. Gs stimulates activation of AC9, while Gi and PKC have been shown to negatively regulate AC9. AC9 is glycosylated on two sites; removal of the N-linked glycosylation on these sites by site directed mutagenesis did not affect the stimulation of AC9 by forskolin (39). However, removal of the glycosylation sites on AC9 did affect Gs -mediated stimulation of AC9 in HEK cells (39).

Taken together, these data reveal an integral role for glycosylation in determining and modulating adenylate cyclase localization, protein-protein interactions and function.

4.2. Insulin receptor signaling

The insulin receptor is a hetero-tetrameric receptor tyrosine kinase that is well known for its regulation of glucose metabolism. The insulin receptor contains numerous glycosylation sites that include both O- and N-linked glycosylation. The glycosylation of the insulin receptor is metabolically regulated since glucose deprivation has been shown to preferentially affect O-linked but not N-linked glycosylation of the receptor. Since insulin is a master metabolic regulator, this suggests that glycosylation plays a significant physiological/pathophysiological role in insulin action (40). Indeed, mutational analysis of the potential O-glycosylation sites on the insulin receptor has revealed significant effects on functioning of the receptor. This may be due to the fact that these sites tend to be near phosphorylation sites on the receptor important to regulation of the receptor activity (41). Removal of glycosylation on the receptor does not affect receptor binding but partial loss of glycosylation leads to a constitutively active kinase activity of the receptor. In pancreatic β-cells, an increase in O-linked glycosylation results in an increased β-cell apoptosis (42). Downstream of the insulin receptor, an increase in O-linked glycosylation leads to decreased phosphorylation of key insulin signaling molecules, insulin receptor substrate 1 (IRS1) and 2 (IRS2), Akt and FOXO1a (42). These data demonstrate that glycosylation is a key regulator of insulin receptor function. Taken together with the observation that the glycemic state of the cell can modulate the pattern of glycosylation of the receptor (40, 43), these data suggest that insulin receptor activity is dynamically regulated within insulin target cells and is sensitive to the metabolic state of the cell.

In addition to the insulin receptor, insulin signaling molecules have been found to be regulated by glycosylation. Specifically, IRS-1 and β-catenin, two important downstream effectors of insulin receptor activation, are known to be glycosylated. Furthermore, it is thought that shunting of glucose metabolism through the hexosamine biosynthetic pathway leads to a general increase in O-linked glycosylation of nuclear and cytoplasmic proteins through increased substrate for O-linked N-acetylglucosamine transferase (41). Increased glycosylation of IRS and β-catenin and insulin receptor have been linked with decreased phosphorylation and activity of these key metabolic enzymes. The end result of this process is loss of cellular insulin sensitivity (41).

5. Glycosylation in receptor pharmacology

5.1. Gonadotropins

It is now well documented that most GPCRs have the capability of signaling *via* multiple pathways in a given cell type. For many years, this hypothesis was poorly understood and was thought to be an artifact of recombinant cell systems. With the recent development of allosteric agonists, antagonists and modulators to GPCRs, more light has been shed on this concept; e.g., direct targeting of specific signaling pathways has been demonstrated for various ligands (4, 44). This phenomenon has most recently been termed, biased signaling. Simply put, the concept of biased signaling describes the ability of ligands to direct specific and distinct biological responses *via* activation of select signaling pathways in a ligand-specific manner (4). There are many receptors that are known to associate with multiple naturally occurring ligands. The gonadotropins, LH and FSH are excellent examples of this phenomenon. Since varying glycosylation of gonadotropin isoforms is known to alter their physicochemical properties, one can consider gonadotropin isoforms as different ligands with potentially subtle, but unique association with their cognate receptors (4, 24). In addition, interaction of these diverse ligands with the receptor would result in multiple ligand-receptor conformations, which in turn lead to the observed activation of differing biological signaling pathways for LH, hCG and FSH (13, 45, 46). Thus, it has been suggested that gonadotropin isforms are naturally occurring biased agonists for their receptors (4, 24).

The molecular basis for biased agonism lies in the stabilization of conformation(s) of the receptor which increases the affinity of the biased agonist-receptor complex for a distinct and specific signaling pathway over another (44). Since GPCRs primarily utilize G proteins as signal transducers, biased agonism would imply ligand-dependent preference of the ligand-receptor complex for a specific G-protein over another. Since GPCR signaling is not exclusive *via* G-proteins, biased agonism is not restricted to G-protein signaling, and recent descriptions of biased ligand-mediated activation of non-G-protein-dependent signaling of GPCRs have appeared, such as is the case with β-arrestin signaling.

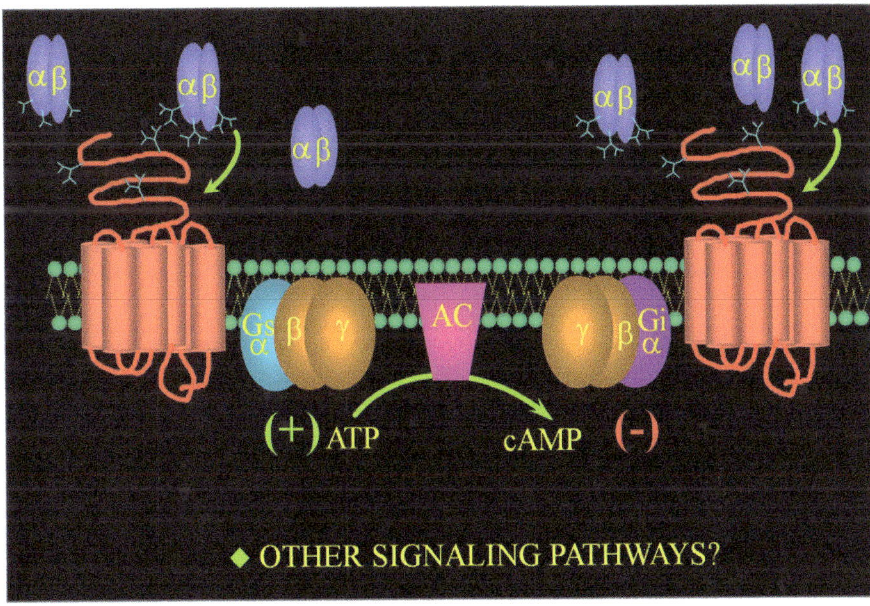

Figure 2. Glycosylated variants of gonadotropins are biased agonists at their receptors. Gonadotropins are secreted into the blood as isoforms that vary in the degree of complexity of glycosylation. Variations in glycosylation has been shown to be important to plasma half-life and receptor binding. It is now appreciated that these isoforms also impact the biological activity of the gonadotropins (4). In the case of FSH (depicted), glycosylated variants stabilize different conformations of the receptor-ligand complex. This leads to different affinities of the receptor-ligand complex for association with G proteins. Highly complex, terminally sialylated FSH induces activation of Gs signaling pathways. Less complex glycosylation associated with terminal mannose leads to activation of both Gs and Gi signaling pathways. Deglycosylated FSH activates the Gi signaling pathway. Since the FSHR is also know to affect other signaling pathways, such as Gq, these isoforms may also differentially activate Gq signaling as well.

For many years, FSH has been used as a model to understand the role of glycosylation in determining glycoprotein hormone function. Several years ago, we noted that differently glycosylated variants of hFSH could induce activation of both the Gs and Gi signaling pathways (24, 47). The phenomenon appeared as a bell-shaped concentration-response curve in *in vitro* assay systems for less glycosylated insect cell expressed hFSH (BV-hFSH). Pertussis toxin was found to block the down-turn in the dose-response relationship, indicating that the descending phase of the curve for the BV-hFSH was due to activation of Gi at higher concentrations of the hormone. These pharmacological relationships had been described previously for other receptors such as the catacholamines and adenosine receptors (48, 49). In the case of the insect cell expressed hFSH (BV-hFSH), glycosylation was terminated at short branched mannose residues, and the protein displayed a more basic migration pattern in chromatofocusing (Arey, unpublished data). Subsequent experiments using an ADP-ribosylation assay, along with immunoprecipitation and Western blotting of specific G-proteins, revealed that these pharmacological responses were definitively associated with activation of specific G-proteins (50). Taken together with *in vivo* effects of the different FSH preparations (4), these data demonstrate that the activities observed in signaling are directly translated into organ growth responses and illustrate the ability of the biased ligand (e.g., BV-hFSH) to elicit a different response pattern than that of the native ligand (phFSH) (Figure 2). Interestingly, similar glycosylation-dependent signal biasing has been noted for other secreted glycoproteins [e.g. IL22 and BMP6, (51, 52)] including LH/hCG (53, 54). Ligand glycosylation has also been suggested to be required for LH/hCG receptor dimerization (53, 55). In the case of BMP6, detailed mutagenesis around key asparagine residues has revealed the importance of glycosylation in interactions with its receptor (52). In the case of IL22, a single fucose residue on Asn54 was shown to be required for full efficacy of the cytokine at its receptor. It is worth mentioning that the binding kinetics of the receptor were altered by more complex glycosylation at this site (51). Similar but more dramatic effects of glycosylation on binding kinetics have also been noted for erythropoeitin (56). These examples lay the foundation for the concept that a variety of related natural ligands talk to the receptor by inducing specific receptor conformations and that glycosylation plays a role in aiding in this stabilization, thus transducing specific signals that are unique to a given physiological state. Similar activities were ultimately discovered for other GPCRs (57). Therefore, from a mechanistic viewpoint, there is strong support for the notion that alteration of the glycosylation pattern on glycoprotein hormones leads to biased ligands that direct activation of one signaling pathway over another. The data support the notion that more basic isoforms of the gonadotropins bind with a higher affinity but are "less" bioactive. Therefore, different glycosylated variants may interact with the receptor in subtle, but unique ways to result in different signaling and biological responses . Overall, the data indicate that receptor signaling has evolved to convey complex regulatory signals in response to varying ligands which are dynamically adjusted to accommodate to external/internal influences and ultimately maintain homeostasis.

Author details

Brian J. Arey
Bristol-Myers Squibb Co, USA

Acknowledgement

This chapter is dedicated to my father, James, my first and most important mentor.

6. References

[1] Ruddon RW, Bedows E 1997 Assisted protein folding. *J Biol Chem* 272:3125-3128

[2] Shental-Bechor D, Levy Y 2009 Folding of glycoproteins: toward understanding the biophysics of the glycosylation code. *Curr Opin Struct Biol* 19:524-533

[3] Shental-Bechor D, Levy Y 2008 Effect of glycosylation on protein folding: A close look at thermodynamic stabilization. *PNAS* 105:8256-8261

[4] Arey BJ, Lopez FJ 2011 Are circulating gonadotropin isoforms naturally occurring biased agonists? Basic and therapeutic implications. Rev Endocr Metab Disorders

[5] Meyer B, Moller H 2007 Conformation of glycopeptides and glycoproteins. *Topics Curr Chem* 267:187-251

[6] Chamorey A-L, Magne N, Pivot X, Milano G 2002 Impact of glycosylation on the effect of cytokines. A special focus on oncology. *Eur Cytokine Net* 13:154-160

[7] Foster P, Hogan S, Ramsay A, Matthaei K, Young I 1996 Interleukin-5 deficiency abolishes eosiniphilia, airways hyperreactivity and lung damage in a mouse asthma model. *J Exp Med* 183:195-201

[8] Kodama S, Tsulimoto M, Tsuruoka N, Suko T, Edno T, Kobata A 1993 Role of sugar chains in the in vitro activity of recombinant interleukin 5. *Eur J Biochem* 211:903-908

[9] Ishino T, Economou NJ, McFadden K, Zaks-Zilberman M, Jost M, Baxter S, Contarino MR, Harrington AE, Loll PJ, Pasut G, Lievens S, Tavernier J, Chaiken I 2011 A protein engineering approach differentiates the functional importance of carbohydrate moieties of interleukin-5 receptor. *Biochemistry* 50:7546-7556

[10] Ishino T, Harrington AE, Zaks-Zilberman M, Scibek JJ, Chaiken I 2008 Slow dissociation effect of common signaling subunit bc on IL5 and GM-CSF receptor assembly. *Cytokine* 42:179-190

[11] Cornelius S, Plaetinck G, Devos R, Van der Heyden J, Tavernier J, Sanderson C, Guisez Y, Fiers W 1995 Detailed analysis of the IL-5-IL5Ra interaction: Characterization of crucial residues on the ligand and the receptor. *EMBO J* 14:3395-3402

[12] Ishino T, Pasut G, Scibek J, Chaiken I 2004 Kinetic interaction analysis of human interleukin 5 receptor mutants reveals a unique binding topology and charge distribution for cytokine recognition. *J Biol Chem* 279:9547-9556

[13] Ulloa-Aguirre A, Timossi C, Barrios-de-Tomasi J, Maldonado A, Nayudu P 2003 Impact of carbohydrate heterogeneity in function of follicle-stimulating hormone: studies derived from in vitro and in vivo models. *Biol Reprod* 69:379-389

[14] Ulloa-Aguirre A, Midgley AR, Jr., Beitins IZ, Padmanabhan V 1995 Follicle-stimulating isohormones: characterization and physiological relevance. *Endocr Rev* 16:765-787

[15] Chappel SC, Ulloa-Aguirre A, Ramaley JA 1983 Sexual maturation in female rats: time-related changes in the isoelectric focusing pattern of anterior pituitary follicle-stimulating hormone. *Biol Reprod* 28:196-205

[16] Ulloa-Aguirre A, Coutifaris C, SC. C 1983 Multiple species of FSH are present within hamster anterior pituitary cells cultured in vitro. *Acta Endocrinol Copen* 102:343-350

[17] Bousfield GR, Butnev VY, Bidart JM, Dalpathado D, Irungu J, Desaire H 2008 Chromatofocusing fails To separate hFSH isoforms on the basis of glycan structure. *Biochemistry* 47:1708-1720

[18] Chappell SC, Bethea CL, Spies HG 1984 Existence of multiple forms of follicle-stimulating hormone within anterior pituitaries of cynomolgus monkeys. *J Primatol* 14:177-194

[19] Ulloa-Aguirre A, Espinoza R, Damian-Matsumura P, Larrea F, Flores A, Morales L, Dominguez R 1988 Studies on the microheterogeneity of anterior pituitary follicle-stimulating hormone in the female rat: isoelectric focusing throughout the estrous cycle. *Biol Reprod* 38:70-78

[20] Smith PL, Kaetzel D, Nilson J, Baenziger JU 1990 The sialylated oligosaccharides of recombinant bovine lutropin modulate hormone bioactivity. *J Biol Chem* 265:874-881

[21] Grossmann M, Szkudlinski MW, Tropea JE, Bishop LA, Thotakura NR, Schofield PR, Weintraub BD 1995 Expression of human thyrotropin in cell lines with different glycosylation patterns combined with mutagenesis of specific glycosylation sites. *J Biol Chem* 270:29378-29385

[22] Matzuk MM, Keene JL, Boime I 1989 Site specificity of the chorionic gonadotropin N-linked oligosaccharides in signal transduction. *J Biol Chem* 264:2409-2414

[23] Bishop LA, Robertson DM, Cahir N, Schofield PR 1994 Specific roles for the asparagine-linked carbohydrate residues of recombinant human follicle stimulating hormone in receptor binding and signal transduction. *Mol Endocrinol* 8:722-731

[24] Arey BJ, Stevis PE, Lopez FJ 1997 Induction of promiscuous G protein coupling of the follicle-stimulating hormone (FSH) receptor: a novel mechanism for transducing pleiotropic actions of FSH isoforms. *Mol Endocrinol* 11:517

[25] Bousfield GR, Butnev VY, Butnev VY, Nguyen VT, Gray CM, Dias JA, MacColl R, Eisele L, Harvey DJ 2004 Differential effects of subunit asparagine56 oligosaccharide structure on equine lutropin and follitropin hybrid conformation and receptor-binding activity. *Biochemistry* 43:10817-10833

[26] Foord SM, Bonner TI, Neubig RR, Rosser EM, Pin J-P, Davenport AP, Spedding M, Harmar AJ 2005 International Union of Pharmacology. XLVI. G Protein-Coupled Receptor List. *Pharmacol Rev* 57:279-288

[27] Heckert LL, Daley IJ, Griswold MD 1992 Structural organization of the follicle-stimulating hormone receptor gene. *Mol Endocrinol* 6:70-80

[28] Sprengel R, Braun T, Nikolics K, Segaloff DL, Seeburg PH 1990 The testicular receptor for follicle stimulating hormone: structure and functional expression of cloned cDNA. *Mol Endocrinol* 4:525-530

[29] Davis D, Liu X, Segaloff DL 1995 Identification of the sites of N-linked glycosylation on the follicle- stimulating hormone (FSH) receptor and assessment of their role in FSH receptor function. *Mol Endocrinol* 9:159-170

[30] Davis DP, Rozell TG, Liu X, Segaloff DL 1997 The six N-linked carbohydrates of the lutropin/choriogonadotropin receptor are not absolutely required for correct folding, cell surface expression, hormone binding, or signal transduction. *Mol Endocrinol* 11:550-562

[31] Davis DP, Segaloff DL, Ravi Iyengar and John DH 2002 N-linked carbohydrates on G protein-coupled receptors: mapping sites of attachment and determining functional roles. In: Methods in Enzymol Academic Press; 200-212

[32] Ascoli M, Fanelli F, Segaloff DL 2002 The lutropin/choriogonadotropin receptor, a 2002 perspective. *Endocr Rev* 23:141-174
[33] Hanoune J, Defer N 2001 Regulation and role of adenylyl cyclase isoforms. *Ann Rev of Pharmacol Toxicol* 41:145-174
[34] Cali JJ, Parekh RS, Krupinski J 1996 Splice variants of type VIII adenylyl cyclase. *J Biol Chem* 271:1089-1095
[35] Insel PA, Head BP, Ostrom RS, Patel HH, Swaney JS, Tang C-M, Roth DM 2005 Caveolae and lipid rafts: G protein-coupled receptor signaling microdomains in cardiac myocytes. *Annals New York Acad Sci* 1047:166-172
[36] Beazely MA, Watts VJ 2006 Regulatory properties of adenylate cyclases type 5 and 6: a progress report. *Eur J Pharmacol* 535:1-12
[37] Pagano M, Clynes MA, Masada N, Cirulea A, Ayling L-J, Wachten S, Cooper DMF 2009 Insights into the residence in lipid rafts of adenylyl cyclase AC8 and its regulation by capacitative calcium entry. *Am J Physiol* 296:C607-C619
[38] Wu G-C, Lai H-L, Lin Y-W, Chu Y-T, Chern Y 2001 N-Glycosylation and residues Asn805 and Asn890 are involved in the functional properties of type VI adenylyl cyclase. *J Biol Chem* 276:35450-35457
[39] Cumbay MG, Watts VJ 2004 Novel regulatory properties of human type 9 adenylate cyclase. *J Pharmacol Exp Ther* 310:108-115
[40] Ohtsubo K, Chen MZ, Olefsky JM, Marth JD Pathway to diabetes through attenuation of pancreatic beta cell glycosylation and glucose transport. *Nature Med* 17:1067-1075
[41] Wells L, Vosseller K, Hart GW 2001 Glycosylation of nucleocytoplasmic proteins: signal transduction and O-GlcNAc. *Science* 291:2376-2378
[42] D'Alessandris C, Andreozzi F, Federici M, Cardellini M, Brunetti A, Ranalli M, Del Guerra S, Lauro D, Del Prato S, Marchetti P, Lauro R, Sesti G 2004 Increased O-glycosylation of insulin signaling proteins results in their impaired activation and enhanced susceptibility to apoptosis in pancreatic Î²-cells. *FASEB J*
[43] Maggi D, Andraghetti G, Carpentier J-L, Renzo C 1998 Cys^{860} in the extracellular domain of insulin receptor b-subunit is critical for internalization and signal transduction. *Endocrinology* 139:496-504
[44] Kenakin T 2007 Functional selectivity through protean and biased agonism: who steers the ship? *Mol Pharmacol* 72:1393-1401
[45] Grewal N, Nagpal S, Chavali GB, Majumdar SS, Pal R, Salunke DM 1997 Ligand-induced receptor dimerization may be critical for signal transduction by choriogonadotropin. *Biophys J* 73:1190-1197
[46] Bousfield GR, Butnev VY, Butnev VY, Nguyen VT, Gray CM, Dias JA, MacColl R, Eisele L, Harvey DJ 2004 Differential effects of a subunit asparagine56 oligosaccharide structure on equine lutropin and follitropin hybrid conformation and receptor-binding activity. *Biochemistry* 43:10817-10833
[47] Arey B 2008 Allosteric modulators of glycoprotein hormone receptors: discovery and therapeutic potential. *Endocrine* 34:1-10
[48] Munshi R, Linden J 1989 Co-purification of A1 adenosine receptors and guanine nucleotide-binding proteins from bovine brain. *J Biol Chem* 264:14853-14859

[49] Kimura K, White BH, Sidhu A 1995 Coupling of human D-1 dopamine receptors to different guanine nucleotide binding proteins: evidence that D-1 dopamine receptors can couple to both Gs and G(o). *J Biol Chem* 270:14672-14678

[50] Arey BJ, Yanofsky SD, Claudia Perez M, Holmes CP, Wrobel J, Gopalsamy A, Stevis PE, Lopez FJ, Winneker RC 2008 Differing pharmacological activities of thiazolidinone analogs at the FSH receptor. *Biochem Biophys Res Comm* 368:723-728

[51] Logsdon NJ, Jones BC, Allman JC, Izotova L, Schwartz B, Pestka S, Walter MR 2004 The IL-10R2 binding hot spot on IL-22 is located on the N-terminal helix and is dependent on N-linked glycosylation. *J Molec Biol* 342:503-514

[52] Saremba S, Nickel J, Seher A, Kotzsch A, Sebald W, Mueller TD 2008 Type I receptor binding of bone morphogenetic protein 6 is dependent on N-glycosylation of the ligand. *FEBS J* 275:172-183

[53] Nguyen VT, Singh V, Butnev VY, Gray CM, Westfall S, Davis JS, Dias JA, Bousfield GR 2003 Inositol phosphate stimulation by LH requires the entire a Asn56 oligosaccharide. *Mol Cell Endocrinol* 199:73-86

[54] Matzuk MM, Keene JL, Boime I 1989 Site specificity of the chorionic gonadotropin N-linked oligosaccharides in signal transduction. *J Biol Chem* 264:2409-2414

[55] Roess DA, Horvat RD, Munnelly H, Barisas BG 2000 Luteinizing Hormone Receptors Are Self-Associated in the Plasma Membrane. *Endocrinology* 141:4518-4523

[56] Darling RJ, Kuchibhotla U, Glaesner W, Micanovic R, Witcher DR, Beals JM 2002 Glycosylation of Erythropoietin Affects Receptor Binding Kinetics:â€‰ Role of Electrostatic Interactions. *Biochemistry* 41:14524-14531

[57] Kenakin T, Miller LJ 2010 Seven transmembrane receptors as shapeshifting proteins: the impact of allosteric modulation and functional selectivity on new drug discovery *Pharmacol Rev* 62:265-304

The Role of Glycosylation in the Control of Processing and Cellular Transport of the Functional Amyloid PMEL17

Julio C. Valencia and Vincent J. Hearing

Additional information is available at the end of the chapter

1. Introduction

In this chapter, we focus on the effects of glycosylation in the cell biology of an amyloid, specifically in melanocytic cells. The actual operational definition of amyloid is that amyloid fibers display the cross-β fiber diffraction pattern, which is the basis of fiber formation, that leads to aggregation of misfolded proteins [1]. Under this concept, classical disease-associated amyloids such as amyloid β-peptide (Ab) with Alzheimer's disease join other non-disease associated or functional amyloids such as PMEL/Pmel17/gp100/silver (herein referred to as PMEL), which is involved in the deposition of melanin [2-4]. Early reports suggested that two domains inside PMEL are contributors to fibril formation [5]. Today, unequivocal evidence recognizes that the so called repeat (RPT) domain is the only region capable to form amyloid fibrils *in vivo* and *in vitro* [6-8]. Unlike disease-associated amyloids, functional amyloids are the product of coordinated and regular cellular processes that ensure that amyloidogenesis does not result in cell damage or death [2;9]. Several reports have indicated the role of post-translational modifications, specifically glycosylation, in the protein aggregation that leads either to neurodegeneration [10;11] or physiological amyloid deposition [12]. In Alzheimer's disease, a cytosolic phosphoprotein called Tau, is abnormally phosphorylated [13]. The reason for this was attributed to an overall decrease in O-GlcNAcylation, a novel type of O-glycosylation by which the monosaccharide β-N-acetylglucosamine (GlcNAc) attaches to serine /threonine residues via an O-linked glycosidic bond. O-glycosylated tau protein aggregates which leads to the formation of neurofibrillary tangles leading to toxicity and neuron death [13;14]. The aberrant deposition of proteins as cellular inclusions or plaques in the form of amyloid fibrils is a characteristic hallmark of all amyloid diseases (or amyloidosis) and of the so-called conformational diseases. Additionally, the β-amyloid precursor protein (βAPP) has also been found to be modified in its cytoplasmic domain by both phosphorylation and glycosylation [15]. Large glycosylated secreted (βAPP) derivates are present in human cerebrospinal fluid, in brain, and in conditioned cell culture medium.

Despite the vast number of reports and clinical studies conducted to date, neurodegeneration and amyloid formation are still one of the least understood processes, which is mostly due to the complex nature of molecular interactions as well as the challenging properties of naturally aggregating systems. In particular, the low solubility of the proteins involved, their tendency to aggregate and their difficult synthesis/isolation in vivo. In this chapter, we detail the influence of glycosylation in the formation of functional amyloid, specifically for PMEL. The study of this type of amyloid raises the question on why these amyloids are non-toxic and whether there are lessons to be learned related with the cellular management of these amyloids that could be used to understand and possibly alter the detrimental effects of disease-associated amyloids.

2. Melanosomes and the functional amyloid PMEL

The ability to form pigment or melanin requires that the melanocyte handle the biochemical steps it takes in a safe and controlled environment. In many organisms, melanocytes build a specialized organelle called the melanosome. Melanosomes are membrane-bound organelles, specialized in the production and distribution of melanin pigment, that are conserved in structure from primitive organisms to mammals. However, the nature of this organelle has recently been studied in detail using novel techniques for isolation and global proteome analysis [16-18]. Those studies shed light on the complex nature of this organelle. As shown in Figure 1, melanosomes mature from undifferentiated vesicles (stage I) to elongated vesicles containing internal fibrils (stage II). In the presence of specific enzymes such as tyrosinase, melanin is synthesized and deposited on the internal fibrils (stage III) of melanosome until they become uniformly dense (stage IV) in pigmented cells. Previously, we detailed that this maturation process involves the acquisition of several specific proteins that we will call the maturation package that stabilizes the organelle for the subsequent maturation steps. As any structure, melanosomes build an internal matrix to store melanin in an orderly and efficient manner. The internal fibrils identified in stage II melanosomes have been unequivocally identified as amyloid aggregates [5;7]. Amyloid fibers derive from a class I transmembrane glycoprotein called PMEL that undergoes a series of post-translational modifications that involve processing by pro-protein convertases (PC) [5;19;19;20] and glycosylation [12;21] (Figure 2). These steps are highly regulated and follow an exact succession of processing steps which prevents the formation of amyloids during assembly or intracellular trafficking of PMEL.

Despite evidence that PMEL is a good substrate for PC in melanoma cells, Leonhard and coworkers confirmed that PC does not trigger immediate amyloid formation [19]. Our evidence suggests that glycosylation is responsible for such protection [12]. Glycosylation of proteins involves the addition of N- and/or O-linked oligosaccharide chains to acceptor asparagine residues or serine/threonine residues, respectively. N-linked core glycosylation begins in the ER, while O-linked glycosylation is a post-translational event which starts in the *cis*-Golgi [22]. PMEL has 5 potential N-glycosylation sites distributed at the N-terminal

(3), RPT (1) and C-terminal (1) domains [12;23]; while O-glycosylation sites have been predicted at the repeat sequences of proline, serine, and threonine in the RPT domain [24].

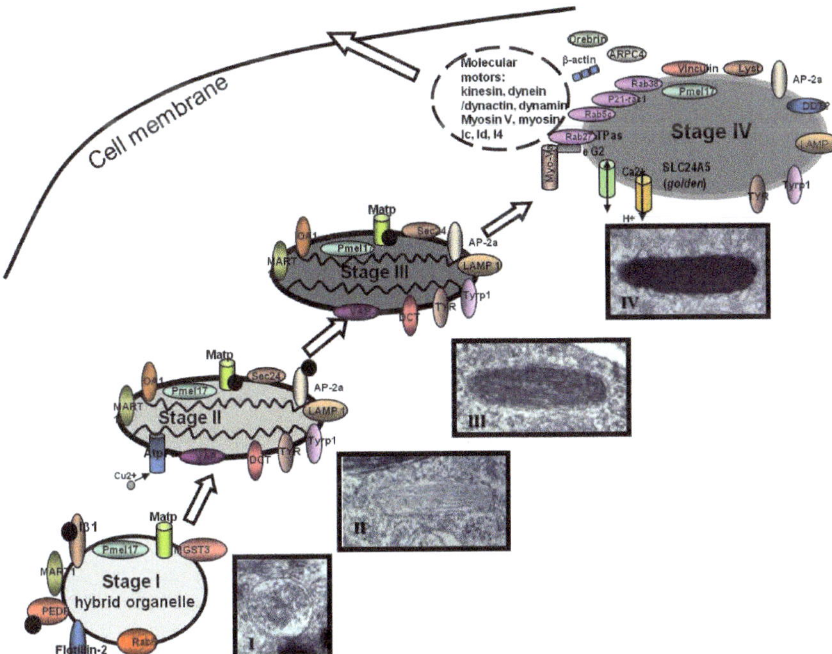

Figure 1. Melanosome maturation at molecular and morphological levels. The scheme details the process of melanosome maturation including variations at the molecular level (drawings) and morphological level (images). Note that every stage is characterized by a group of melanosome-specific and non-specific proteins. Electron microscopy images show the transformation of this organelle from an amorphous, non-pigmented and non-differentiated organelle (stage I) to a well-defined, differentiated and pigmented organelle (stage IV). Scheme adapted from [16].

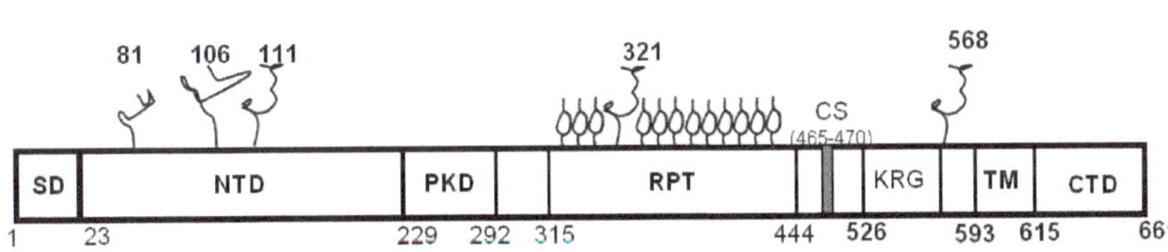

Figure 2. Scheme of PMEL domains and modification sites. PMEL has only one conserved domain called polycystic Kidney disease (PKD) that is located near the center of the protein. During processing, the signal domain (SD) is removed early and then PMEL is cleaved (CS) by PC. The N-terminal (NTD) and C-terminal (CTD) domains undergo N-glycosylation, while the repeat domain (RPT) undergoes O-glycosylation. Numbers above the structures indicate the predicted aminoacid position for the N-glycan sites; while numbers below the protein graph indicate the aminoacids where each domain approximately starts and ends. Scheme adapted from [12].

a. N-Glycosylation in PMEL

The β-glycosylamine linkage of GlcNAc to Asn represents the most widely distributed carbohydrate-peptide bond and is the site of attachment for a large variety of complex and poly-mannose oligosaccharides in proteins with demonstrated biological importance (See [25;26]). A summary of this process is presented in Figure 3.

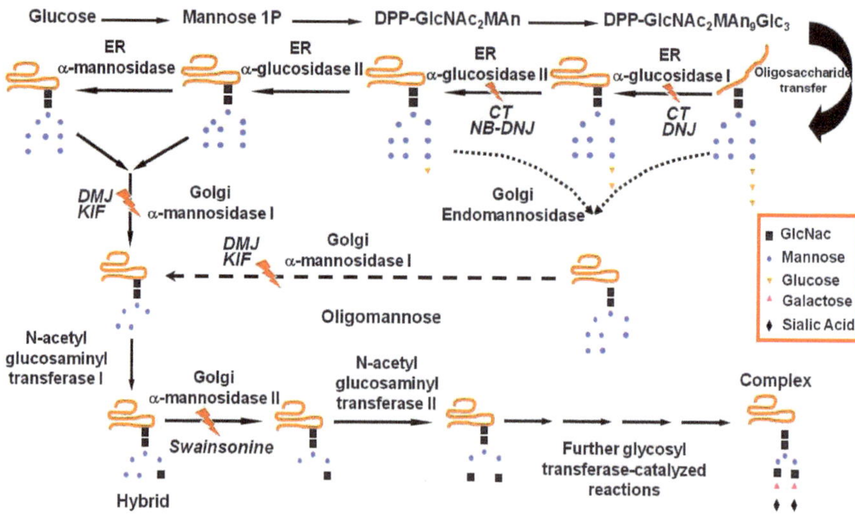

Figure 3. N-glycosylation processing in mammalian cells. The scheme details the place and different enzymes involved in the post-translational modification of glycoproteins with N-linked oligosaccharides. It also includes the different enzymes or steps targeted by different glucosidase inhibitors such as castanospermine (CT), deoxynojirimycin (DNJ), deoxymannojirimycin (DMJ), and kifunesine (KIF). Note that a series of sialyltransferases (not shown) are in charge of the addition of sialic acid at the end of the glycan chains late in the process. Peptide sequence is represented as an orange line. Scheme has been modified and adapted from [27].

In the ER, N-glycosylation starts with the attachment of a core unit of 14 monosaccharide residues ($Glc_3Man_9GlcNAc_2$), which is then processed by the removal of the 3 glucose residues attached to the terminal mannose D1 on the α 1-3 arm of the oligosaccharide by α-glucosidases I and II [27]. In the Golgi, N-glycan processing starts when $Man_6GlcNAc_2$ is trimmed to $Man_5GlcNAc_2$ by Golgi mannosidase I (Figure 3). The extent of N-glycan processing towards triantenary and tetraantenary structures is limited by glycosidase availability and their accessibility to the processing sites. Noticeably, the enzyme endo-α-D-mannosidase provides an alternate route for de-glucosylation during a glucosidase blockade [27]. To determine the role of N-glycosylation in the processing and the trafficking of PMEL, we used several inhibitors that block the modification reactions either in the ER or Golgi followed by metabolic labeling with [^{35}S] Met/Cys in MNT-1 melanoma cells. We first targeted ER glucosidases with either Castanospermine (CT), a competitive inhibitor of α- and β-glucosidases, or N-butyl-deoxynojirimycin (NBDNJ), an inhibitor of both glucosidases with preference towards glucosidase II[28]. These inhibitors prevent the processing of complex glycoproteins and cause the production of immature N-linked glycoproteins with $Glc_3Man_{7-9}GlcNAc_2$ structure. As detected by αPEP13h (an antibody targeting the C-terminal domain of PMEL), treatment with either CT or NBDNJ did not affect processing or stability of PMEL. Similar results have been obtained for tyrosinase, a key melanin producing enzyme, although it was inactive [28;29]. To confirm the effect of this intervention, cells were treated with Endo-H, which removes high mannose/hybrid N-glycans. Endo H treatment revealed that all forms of PMEL contained immature N-glycans sensitive to Endo H (Figure 4A). These results indicate that PMEL, similar to tyrosinase [27;30], is not a substrate for Golgi endo-α-mannosidase which is capable of by passing ER glucosidases that remove end glucose residues and proceed with the glycosylation process in the Golgi. Confocal microscopy analysis revealed that PMEL (red) reaches intracellular targets but does not localize to the plasma membrane (Figure 4B). Thus, these results suggest that N-glycans may have a limited impact on PMEL processing but may play a role in its trafficking to the plasma membrane, possibly for secretion.

To further verify our results, we blocked the synthesis of complex type N-glycans using a fairly potent and specific inhibitors of mannosidase I and α-mannosidase, called deoxymannojirimycin (DMJ) and Mannostatin A (ManA), respectively. These inhibitors have been extensively studied and have proven to be useful for studying the processing pathway and for comparing processing enzymes (For review see [28]). Pre-treatment of melanoma cells was followed by metabolic labeling with [^{35}S] Met/Cys for at 3 hrs at 37°C . Consistently these inhibitors affected the formation of the mature form of PMEL without affecting cleavage of the C-terminal domain. As expected, digestion with Endo H revealed that the resulting "mature" PMEL contains non-complex glycans and as such that band was sensitive to this enzyme and suggest that this glycoprotein is getting additional modifications independent of N-glycosylation. These results further suggest that the mature PMEL could be distinguished by SDS-PAGE despite it has non-complex modified N-glycans (Figure 5, A and B). We then assessed whether PMEL intracellular localization was altered after these conditions in the same melanoma cells. To explore this possibility, cells were cultured in the presence of CT and ManA for 3 hrs previous to fixation. Confocal microscopy analysis revealed that PMEL (red) had a limited distribution in the cytoplasm, mostly perinuclear, compared to HMB-45 (green) which is already processed and localized in stage II melanosomes (Figure 5C). Superimposed confocal images with DIC showed that treatment did not compromised cell morphology or adhesion. Noticeably, PMEL staining was absent nearby the plasma membrane further supporting that N-glycosylation plays a role in the distribution of PMEL at the plasma membrane. Toxicity and viability assays were performed using different combinations of inhibitors and incubation times to determine optical treatment conditions in culture for pigmented and non-pigmented melanoma cells (data not shown).

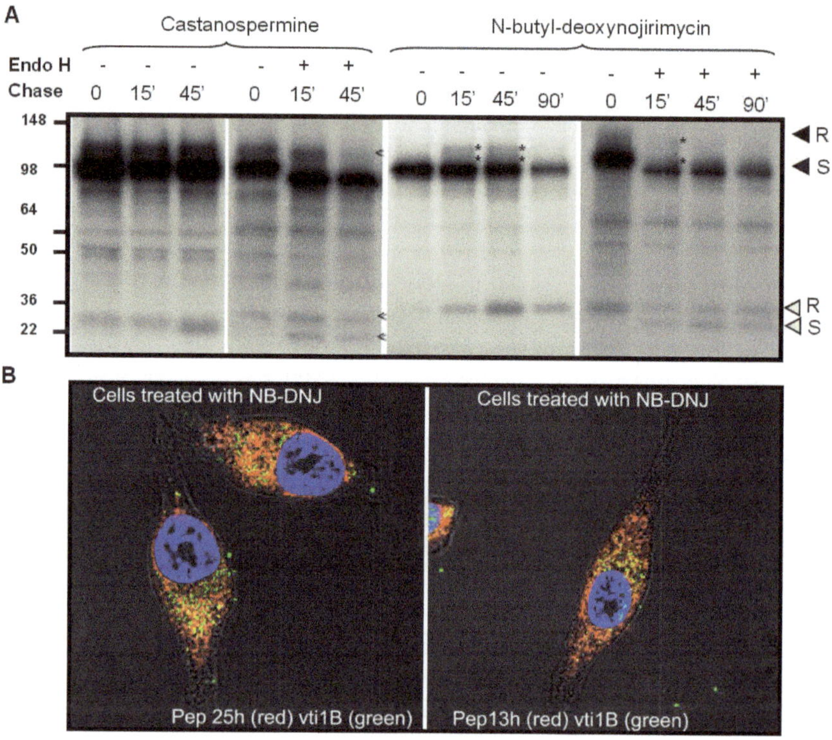

Figure 4. Inhibition of N-Glycan maturation does not alter PMEL processing. A. MNT-1 melanoma cells were treated for 3hrs with 5 μg/ml of CT, which is known to produce minimal toxicity as reported by Takahashi et al.[31], were [^{35}S]-labeled for 30 minutes and chased for the indicated times. Collected samples were left undigested or digested with Endo H where noted for 3h at 37 C, after which samples were immunoprecipitated with the αPEP13h antibody, which recognize the C-terminal domain of PMEL. Immunoprecipitated bands were visualized by autoradiography. Faint bands are indicated with either an open arrow (<) or an asterix (*). B. MNT-1 cells were cultured in the presence of 5 mM of NBDNJ for 3 hrs prior to fixation with 4% p-formaldehyde. Cells were dual stained either with αPEP13h or αPEP25h (red), anti-PMEL antibodies, and vti1b, a Golgi marker (green). Images are representative of the staining pattern after three repeated experiments in melanoma cells. Note the granular distribution patterns of PMEL. Images are merged with DIC to visualize the unstained parts of the melanoma cells. Colocalization of the red and green signals is shown in yellow. Nuclear counterstaining was done with DAPI (blue).

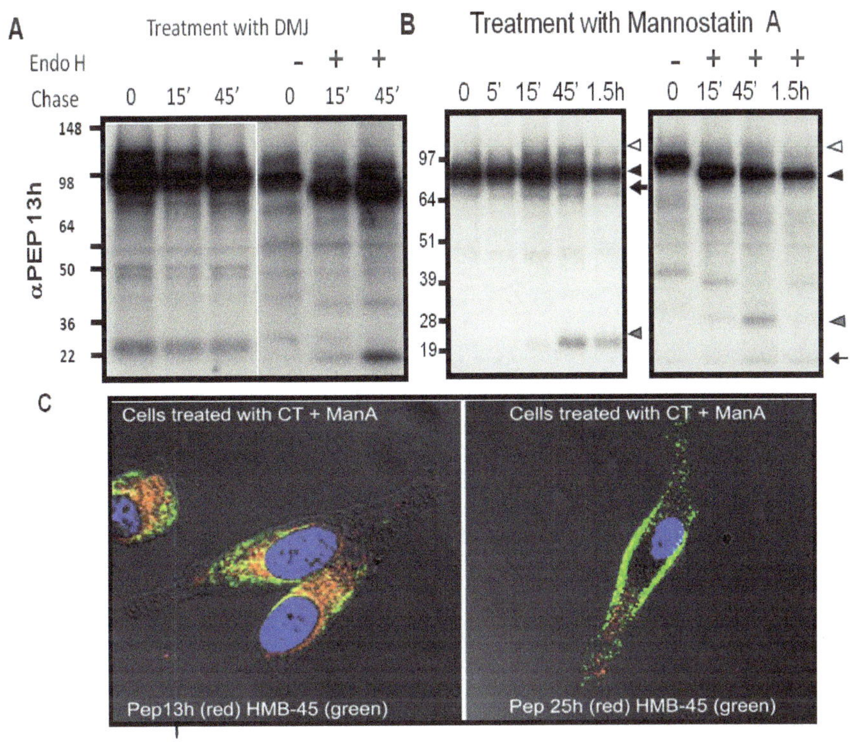

Figure 5. Complete inhibition of N-Glycan maturation alters the intracellular localization of PMEL. A and B. MNT-1 melanoma cells were treated for 3hrs with 5 μg/ml of DMJ or overnight with 50 μg/ml of mannostatin A (ManA). Samples were [^{35}S]-labeled for 30 minutes and were chased for the indicated times. Collected samples were left undigested or were digested with Endo H where noted for 3h at 37 °C, after which samples were immunoprecipitated with the αPEP13h antibody. Immunoprecipitated bands were visualized by autoradiography. C. MNT-1 cells were cultured in the presence of both CT and ManA for 3 hrs previous to fixation with 4% p-formaldehyde. Cells were dual stained either with αPEP13h or αPEP25h (red) and HMB-45 (green). Images are representative of the staining pattern after three repeated experiments in melanoma cells. Note limited granular distribution patterns of PMEL antibodies, while HMB-45 reactivity seems to be conserved. Images are merged with DIC to visualize the unstained parts of the melanoma cells. Colocalization of the red and green signals is shown in yellow. Nuclear counterstaining was done with DAPI (blue).

To establish a specific role for individual N-glycan sites in PMEL, Hoashi et al., performed mutational analysis using several PMEL mutants at the predicted N-glycosylation sites [23]. In that study, it was determined that 2 out of 4 N-glycosylation sites were not fully glycosylated in Hela cells, which were derived from an adenocarcinoma of the cervix. Consistent with our data, the authors concluded that this does not affect PMEL processing but pointed that the mutation of the site at S570A altered secretion of PMEL. It is worthwhile to note that a glycoform of PMEL is known to be secreted to the extracellular space and in the medium of cells in culture [23;32]. However, there is no evidence of amyloid accumulation either in skin samples or in skin keratinocytes when co-cultured with normal human melanocytes (JVC, unpublished data). One possible reason for this is that PMEL amyloid formation is sensitive to pH [7]. Thus, the neutral pH in the extracellular compartment most likely undermines PMEL amyloid fiber ability or if any amyloid is formed it is dissolved; while the intra organelle acidic pH (5.0) will favor amyloid formation. Interestingly, the secretion of amyloid generating proteins has been reported previously for other functional (hormones) and disease-associated amyloids [9]. In the case of secreted PMEL, the effects of this activity are still unknown both in the context of human skin physiology and melanoma disease progression. Taken all together, we conclude that the N-glycosylation of PMEL might influence trafficking to and from the plasma membrane while playing a limited role in the protein shedding process and direct trafficking to melanosomes in melanocytic cells.

b. O-glycosylation in PMEL maturation

Prediction of O-glycosylation sites is not standardized but is known in several proteins as linkages in which the sugar is attached to an aminoacid containing a hydroxyl group. However, every aminoacid with a hydroxyl functional group (i.e. Serine (Ser), Threonine (Thr), Tyrosine (tyr), Hydroxyproline, and Hydroxylysine) has been implicated in that process (See details in [25;26]). As summarized in Figure 6, O-glycan structures are generated following the action of polypeptide N-acetylgalactosaminyltransferase (GalNAcT) enzyme and include 4 common subtypes based on differential monosaccharide linkage reactions to the unsubstituted GalNAc (GalNAcα-Ser/Thr). All O-glycans may be modified by one of several enzymes to generate different core structures known as core 1, core 2, core 3, etc. The core 1 O-glycan is generated by the core 1 β3 galactosyltransferase (Gal-T), which adds galactose to generate Galβ1-3GalNAcα1-Ser/Thr. This is a key precursor for all core 1 and 2 mucin-type O-glycans in vertebrates and invertebrates. The biosynthesis of O-glycans can be modified and terminated with the addition of sialic acid residues relatively early in biosynthesis. For example, certain sialyltransferase enzymes are capable of acting on GalNAcα-Ser/Thr, or early O-glycan core subtypes after Core-1-GalT action. These sialic acid additions give rise to a series of O-glycan structures that generally restrict further biosynthetic steps and have been commonly referred to as tumor-associated (T) antigens.

In PMEL, the RPT domain has 10 imperfect repeats rich in proline, serine, threonine and glutamic acid aminoacids [5;33]. Recognition of this domain in amyloid fibers inside stage II melanosomes by the HMB-45 antibody, which recognizes an unknown sialic acid structure, suggests that glycan structures in the RPT domain are modified with sialic acid.

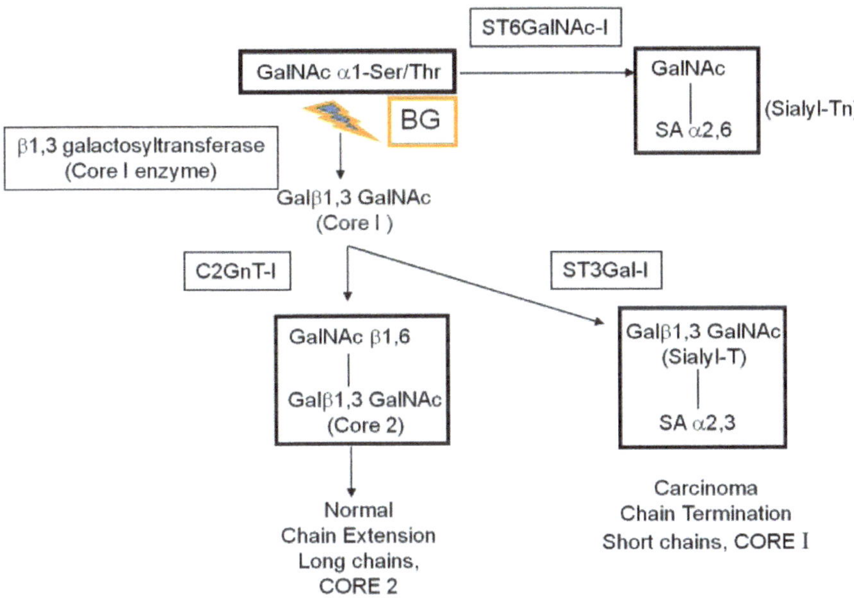

Figure 6. O-glycosylation process in mammalian cells. The scheme details the different enzymes involved in the post-translational modification of glycoproteins with O-linked oligosaccharides. It includes the step targeted by the inhibitor benzyl-N-acetyl-α-D-galactosaminide (BG). Note that the conformation of O-glycan chains varies in cancer cells compared to normal cells due an early addition of sialic acid. Adapted from [37].

Previously, we have confirmed that PMEL is modified with sialylated core-1 type O-glycans in MNT-1 melanoma cells [12]. To assess the role of O-glycosylation in PMEL function, we used the sugar analog benzyl-N-acetyl-α-D-galactosaminide (BG), known to inhibit O-glycosidation [34-37]. MNT-1 melanoma cells were cultured overnight in the presence of 1 to 4nM BG. Immunoblotting analysis revealed that maturation of PMEL (upper band) was completely inhibited in a dose dependent manner reaching a maximum inhibition at 4 nM (Figure 7A). As expected, HMB-45 staining decreased substantially at the doses of 2 nM and 4 nM of BG (Figure XB). Because HMB-45 reactivity depends on sialic acid, we then infer that O-glycan structures in PMEL were modified with sialic acid. To further confirm this and assess the impact on PMEL stability, MNT-1 cells were metabolically labeled with [^{35}S] met/cys and chased at the indicated times. As shown in Figure 7B, the mature form of PMEL (white arrow) was inhibited as early as 45 min and the stability of PMEL at 90 min. The observed shift in electrophoretic mobility following synthesis reflects mainly the loss of negatively charged sialic acid residues to O-linked sugars, with the maturation of N-linked chains to hybrid or complex forms contributing only slightly to the change in molecular weight. Noticeably, the observation of the cleaved 26 kDa band at all time points further suggest that PMEL processing by PC is also independent of O-glycan modification. At this point, there is enough evidence to conclude that protein cleavage is independent of post-translational glycosylation.

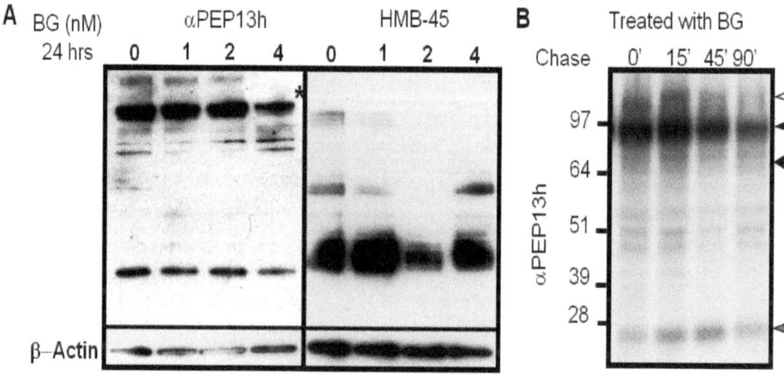

Figure 7. O-glycosylation is important for proper PMEL maturation. A. MNT-1 melanoma cells were cultured in the absence or in the presence of 4 nM BG overnight. Cells were then harvested and cell lysates were analyzed using SDS-PAGE. Detection of PMEL was done with αPEP13h or HMB-45 as indicated. Actin was used as a loading control. B. After treatment with BG overnight, samples were [^{35}S]-labeled for 30 min and then chased for the indicated times. Collected samples were immunoprecipitated with the αPEP13h antibody. Immunoprecipitated bands were visualized by autoradiography. Arrows next to PMEL bands represent: mature PMEL (white), cleaved (grey), intermediate (black).

c. Sialylation in PMEL

The negatively charged residues of sialic acid exert a particular influence on the 3 dimensional structure of the protein backbone. Sialic acid can be added to complex N- and O- linked glycans preferentially in α2,3 and α2,6 linkages to a penultimate galactose residue [38;39]. The transfer of these residues is catalyzed by several enzymes known as sialyltransferases (ST). To examine the impact of sialic acid addition to N- and O-glycan chains in PMEL, we transfected full length PMEL into wild-type and mutant Chinese hamster ovary (CHO) cells that exhibit a specific glycosylation defect. These cells provide an excellent tool to study the role of individual sugar residues in protein processing. Thus, the outcome of the glycoproteins could be predicted and it is summarized in Figure 8.

To test whether sialic acid addition affects the PMEL amyloid formation ability, we transfected full length PMEL into wild type- and mutant- (Lec1, Lec2 and Lec8) CHO cells. The Lec1 mutant is a leuco-phytohemagglutinin resistant cell line unable to synthesize complex and hybrid *N*-glycans due to the lack of *N*-acetylglucosaminyltransferase I (GnTI) activity [40]. The Lec2 mutant has a deletion mutation in the CMP-sialic acid transporter resulting in N- and O-glycans with a greater than 90% decrease in sialic acid content [41]. The Lec8 mutant has a deletion mutation in the UDP-galactose transporter resulting in a truncated protein with a greatly reduced ability to translocate UDP-galactose (UDP-Gal) inside the Golgi [42]. Thus, Lec8 cells generate non-galactosylated and non-sialylated N- and O-glycans. Immunoblotting analysis showed PMEL was successfully expressed in these cell lines. Noticeably, PMEL was correctly processed in all mutant cells showing a band pattern that was no different than those in melanoma cells, except for minor differences in the molecular weight of the mature PMEL (white arrow) which reflects the different glycosylation background of the host cells (Figure 9). We noted that the mature PMEL (white arrow) band intensity is stronger in wild-type CHO cells than that observed in melanoma cells and in melanocytes. On the contrary, this same band in the absence of sialic acid modification is undefined despite correct modification with core 1-O-glycans in Lec2 cells.

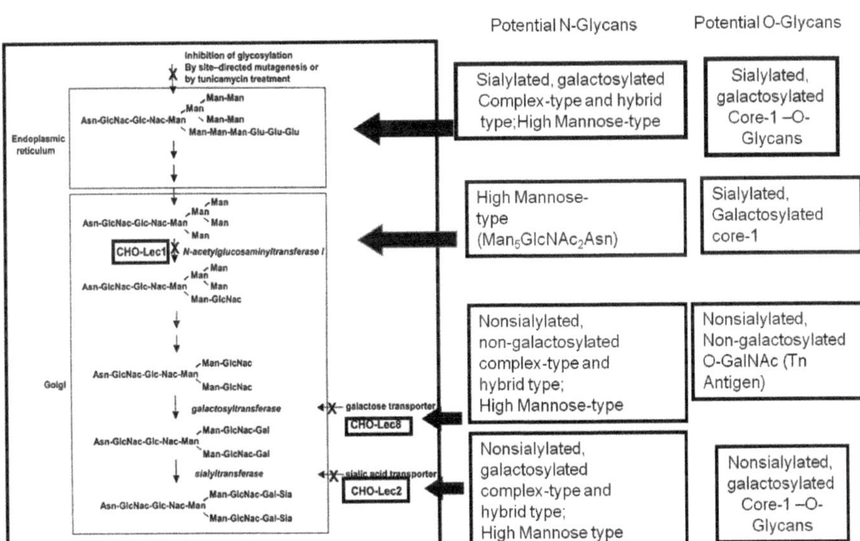

Figure 8. Scheme detailing the specific mutation and the affected step in glycosylation processing. Summary of the outcomes for N- and O-glycans expected from the different CHO cell mutants (Lec1, Lec2 and Lec8) Adapted and modified from [41].

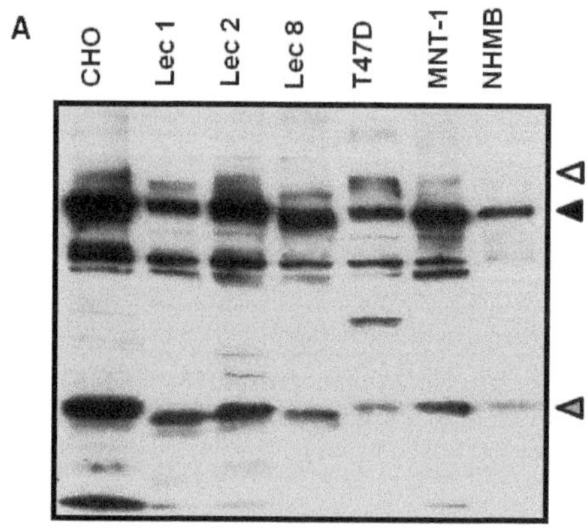

Figure 9. PMEL expression in CHO cells results in different glycoforms. A V5-His 6 tag full length PMEL was transiently transfected into the indicated cells lines. At 48 hr post-transfection, cells were harvested and lysates were analyzed by immunoblotting. PMEL was detected using the αPEP13h antibody. Lysates from normal human melanocytes from black donors (NHMB) and from MNT-1 melanoma cells were used as controls for the PMEL band pattern. Arrows next to PMEL bands represent: mature PMEL (white), cleaved (grey), intermediate (black).

These results suggest that the band shifts in the mature PMEL band mainly reflect either the addition or loss of sialic acid residues. Previous reports have focused on the band size and location to infer about the maturation state of PMEL [20;23]. However, our results clearly indicate that PMEL shares a similar band pattern in all transfected cell lines despite the different N- and O-glycan composition. In other words these band patterns represent glycoforms of PMEL. Glycoform is a term that indicates differences in the composition of the glycan chains derived from an identical backbone protein [25;26;43]. This phenomenon is also known as microheterogeneity. From the practical point of view, microheterogeneity explains the anomalous behavior of glycoproteins in various forms of chromatography (such as the diffuse bands observed on SDS-PAGE gels) and makes the complete structural analysis of most 'glycoproteins a difficult task. From a functional point of view, the meaning of this heterogeneity remains unclear. It is possible that this is a type of "diversity generator" intended either to diversify endogenous recognition functions and/or to evade detection after malignant transformation. Interestingly, we reported that the amyloid formation abilities of PMEL were altered *in vivo* at different levels depending on the glycosylation defect [12]. Thus, lack of sialic acid resulted in the accumulation of thick amyloid fibers (Lec 2 cells), while the lack of the end terminal galactose modified with sialic acid at core-1-O-glycans resulted in complete loss of amyloid fibrils at target organelles (Lec 8 cells). Thus, these results indicate that mature PMEL is mainly formed by O-glycan derived-structures and the loss of this modification decreased PMEL stability in melanoma cells.

There have been several reports indicating that changes in sialylation patterns correlate with changes in cell motility and invasiveness [44-47]. In melanoma, it has been shown that adhesion molecules such as N-cadherin and integrins undergo altered glycosylation and sialic acid modification [48-51]. Thus, we hypothesize that melanoma cells may exhibit altered sialyltransferase (ST) activities compared to either melanocytes or other cell lines that affects the glycosylation pattern of most proteins, including PMEL. Analysis of sialyltransferase activity in MNT-1 melanoma cells compared to normal human melanocytes (NHM) using different glycoproteins as acceptor substrates confirmed differences between these cell lines. As shown in Figure 10, both cell lines exhibited sialyltransferase activity towards asialofetuin, which bear unsialylated N-glycans and unsialylated core 1 O-glycans, although MNT-1 cells were less active than NHM. In contrast, very high levels of sialyltransferase activities were detected using fully sialylated fetuin as acceptor (Figure 10A), showing that these cells can substitute sialic acid on already sialylated carbohydrates an ability known as oligo- or poly-sialylation [52]. Thus, NHM is 5 times more active than MNT-1 and 25 times more active than the breast cancer cell line T47-D, used as a control, in the addition of sialic acid to glycan chains. Using real-time PCR, we confirmed that the mRNA levels for α2,3-sialyltransferase (ST3 GalNacI) and α2,6-sialyltransferase (ST6 GalNac II) were higher in melanocytes, which correlated with the increased ST activity detected in these cells. Furthermore, we reported that melanocytes preferentially add sialic acid to O-glycans while melanoma cells do it without specific preference, although in a less active fashion [12]. Thus, our results prove that there are cell-specific differences related to addition of sialic acid to the ends of both N- and O-glycans.

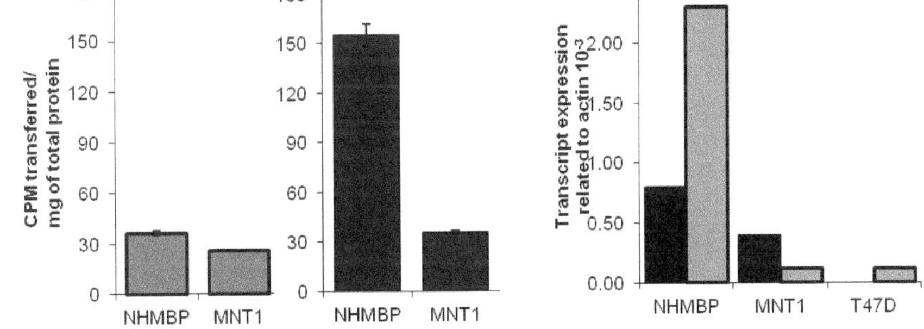

Figure 10. Melanocytes exhibit increased ability to transfer sialic acid to glycans. A. Sialyltransferase activity against the acceptor substrates asialofetuin or fetuin measured in lysates of normal human melanocytes from black donors (NHMB) or MNT-1 melanoma cells; values are expressed as total counts/min of ^{14}C against µg of protein. B. Real-time PCR mRNA expression of ST3 GalNac I and ST6 GalNac II enzymes in NHMB and the human cancer cell lines, MNT1 (melanoma) and T47D (breast). Values are expressed in transcript units related to actin. PCR primer sequences have been reported in Valencia et al., [12].

Lectins are sugar-binding proteins (not to be confused with glycoproteins, which are proteins containing sugar chains or residues) that are highly specific for their sugar moieties. They play a role in biological recognition phenomena involving cells and proteins. Lectins have a multimeric structure, which is responsible for the ability to agglutinate cells or form precipitates glycoconjugated in a manner similar to antigen-antibody interactions. During the past few years, lectins that discriminate between various types of sialylated sequences have been reported [53]. The lectin from elderberry (Sambucus nigra) bark (SNA I) has been shown to bind with high affinity to glycoconjugates containing the terminal sequence sialic acid α2,6-galactose/N-acetylgalactosamine. Similarly, the leukoagglutinin from Maackia amurensis (MAL) binds with high affinity to sialic acid α2,3 galactose β1,4 N-acetylglucosamine but not to 2,6-linked isomers and, therefore, is a most interesting complementary probe to the SNA [53]. To more specifically determine the most frequent sialic addition related with PMEL, we performed a detail cytochemical analysis of the distribution of α2,6- and α2,3- linked sialic acid residues using MAL and SNA I in MNT-1 melanoma cells (Figure 12). Confocal microscopy dual immunofluorescence revealed that PMEL (red), as detected by αPEP13h or HMB-45 antibodies, colocalized more frequently with MAL (green) in the cytoplasm of melanoma cells (Figure 12,A-C). Specifically, the staining pattern of HMB-45 antibody colocalized almost exclusively with MAL (Figure 12, D-F) suggesting that the α2,3 addition is more likely recognized by this antibody. These result also suggest that α2,3 is the most frequent modification in the glycan chains of PMEL. On the other hand, SNA staining was limited to the plasma membrane where it colocalized with PMEL (Figure 12, G-I) suggesting that plasma membrane forms of PMEL, including the secreted PMEL glycoform, are more likely to be modified with α2,6 instead of α2,3 terminal sialic acids ends.

In melanoma cells, immunopurified forms of PMEL show that N-glycosylation sites located at the N- and C-terminal domains were modified with sialic acid [12]. Detection of unsialylated N-glycan chains is an indication of the predominantly hybrid nature of the structures in PMEL. This type of variability opens the possibility that at least two glycoforms of PMEL are produced in one cell type or in cell lines from the same lineage. To effectively prove that PMEL glycan chains are differentially managed in a cell-specific manner, lysates from different melanoma cell lines and NHM were digested with a combination of the following enzymes: α2,3,6,8,9 neuroaminidase, β1,4 galactosidase, β-N-acetylglucosaminidase known to remove complex and hybrid structures (Figure 13). As detected by the anti-PMEL antibody αPEP13h, enzyme digestion rendered new bands that migrated different between cell lines, specially between melanoma cells and melanocytes. Thus, all cells exhibited bands of around 85 kDa and 20 kDa after the combined enzyme digestion (red arrows). These new bands are composed by non-complex unsialylated N- and O-glycans. Interestingly, the band patterns of pigmented cells (NHM, MNT-1 and UACC-257) were somewhat different than those in non-pigmented cells (SK-Mel-28 and M14). The only consistency was the minor molecular shift of the cleaved band at 26 kDa suggesting that post-translational modification of this domain in PMEL was similarly modified independent of the cell type. These results confirm that the majority of N-glycans on this band are hybrid not complex type.

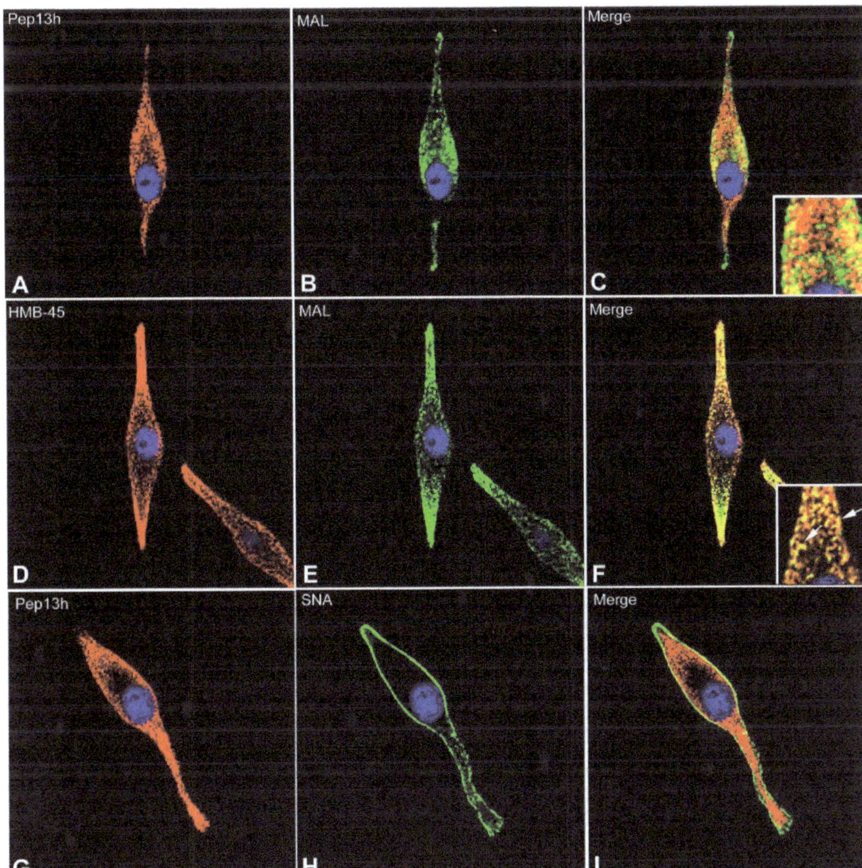

Figure 11. Immunohistochemical distribution of α2,3 and α2,6 sialic acid additions in MNT-1 melanoma cells. MNT-1 cells were grown in Lab TeK II chamber slides with cover at a density of 1 X10^4 cells per chamber. After 48hrs, cells were fixed with 4% p-formaldehyde and dual stained either with αPEP13h or HMB-45 (red) and the lectins MAL or SNA (green). Images are representative of the staining pattern after three repeated experiments in melanoma cells. Note the granular distribution patterns of PMEL antibodies and MAL that differ from the predominantly plasma membrane distribution of the lectin SNA. Colocalization of the red and green signals is shown in yellow. Nuclear counterstaining was done with DAPI (blue).

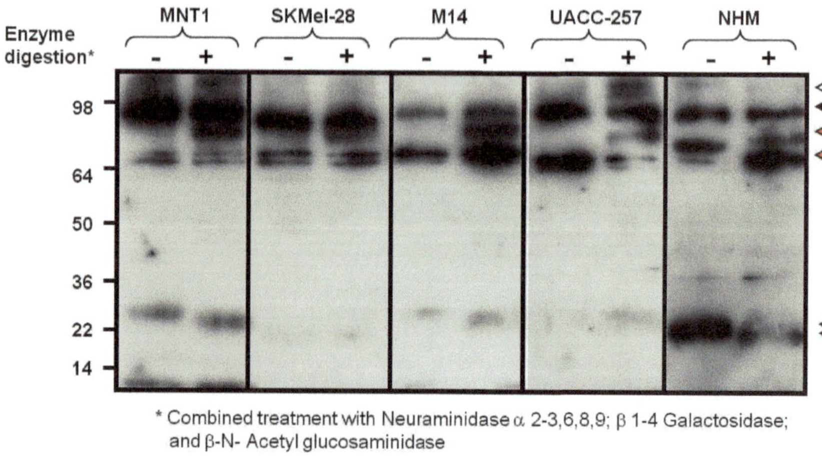

Figure 12. PMEL is differently glycosylated in human melanoma cells compared to melanocytes. Lysates from indicated cell lines were digested in the presence or absence of the enzymes noted and were analyzed by SDS-PAGE. Samples were separated on a 4-20% Tris glycine gels for 2 hr. Arrows next to PMEL bands represent: mature PMEL (white), cleaved (grey), intermediate (black), digested bands (red).

d. Glycosylation enzymes in melanosomes

Previously, we made possible global melanosome proteome characterization by using LC/MS to analyze both non-pigmented (pre-melanosomes) and pigmented melanosomes in solution digest after removal of melanin by immobilized metal affinity chromatograpy (IMAC) [16;17]. Briefly, we took advantage of the heavy metal ion sequestering property of melanin by loading the resulting peptides onto an IMAC column activated with an excessive volume of FeCl3, assembled back-to-back with a reverse-phase (RP) precolumn. The melanin was retained on the IMAC column due to the high affinity of Fe (III) for the o-diOH groups of melanin, while the peptides passed through and were subsequently caught on the C18 RP-precolumn. This assembly allowed the efficient and concurrent removal of melanin and the loading of the sample onto the column in a single step prior to the LC/MS analysis. Thus, proteomic analysis of melanosomes required an effective approach for purifying and solubilizing them, and removing the melanin.

Arrival of PMEL to melanosomes initiates a well coordinated sequence of events to incorporate and shred the protein into smaller pieces to expose the RPT domain and start the formation of amyloid fibers. Several reports agree that this process is initiated by a series of PC [19;54]. After that process, we hypothesized that the sialylated O-glycans covering the RPT domain of PMEL must be removed before amyloid formation. This is what we propose as the protective role these glycans exert in the amyloid processing. To perform such task, melanosomes will have to be equipped with enzymes specialized in the removal of carbohydrate modifications, such as sialyltransferases or hexosaminidases. To evaluate whether or not this is possible, we analyzed our proteome database for the presence of enzymes involved in the modification of glycan chains. As shown in Table 1, there were several enzymes involved in either the transfer of sialic acid (like α2,6 ST), removal of fucose (another terminal modification of N- and O-glycan products), or removal of carbohydrates such as glucose (α-glucosidase II, glucosidase I), mannose (mannosidase) and galactose (GalNAc-T2). Many of these enzymes have been identified to localize throughout the Golgi cisternae like GalNac-T2 or T3. Note that most of these enzymes were identified in early maturation stages of melanosomes (stage I and II), not the fully mature late stage, of the pigmented MNT-1 and non-pigmented SK-Mel-28 cells. Interestingly, there is evidence that the localization of these enzymes is highly variable not only with respect to their normal localization inside the Golgi stack [55;56], but also in ectopic Golgi sites including the plasma membrane (see review [57]).

Protein AC	Protein ID	Group	Protein Name
Q9H553	ALG2	MNT1_stage2 MNT1_stage2	Alpha-1, 3-mannosyltransferase ALG2 (EC 2.4.1.-) (GDP- Man:Man(1)GlcNAc(2)-PP-dolichol mannosyltransferase
Q9BTY2	FUCO2	MNT1_stage1	alpha-L-fucosidase precursor (EC 3.2.1.51) (Alpha-L-fucosidase 2) (Alpha-L-fucoside fucohydrolase 2)
Q96L58	B3GT6	Skmel28_stage1	Beta-1, 3-galactosyltransferase 6 (EC 2.4.1.134) (beta 3GalT6) (Galactosylxylosylprotein 3-beta-galactosyltransferase) (Galactosyltransferase II) (GAG GalTII)
Q92685	ALG3	Skmel28_stage1; Skmel28_stage2; uniqueSkmel28	Dolichyl-P-Man:Man(5)GlcNAc(2)-PP-dolichyl mannosyltransferase (EC 2.4.1.-) (Dol-P-Man-dependent alpha(1-3)-mannosyltransferase)
Q16739	CEGT	MNT1_stage 1 and 2; Skmel28_stage1; Skmel28_stage2; common_earlystage	Ceramide glucosyltransferase (EC 2.4.1.80) (Glucosylceramide synthase) (GCS) (UDP-glucose:N-acylsphingosine D-glucosyltransferase) (UDP- glucose ceramide glucosyltransferase) (GLCT-1)
Q14697	GANAB	MNT1_stage1,2, and 4; Skmel28_ stage1; Skmel28_stage2; common_early_ stage	Neutral alpha-glucosidase AB precursor (EC 3.2.1.84) (Glucosidase II subunit alpha)
Q14435	GALT3	Skmel28_stage1	Polypeptide N-acetylgalactosaminyltransferase 3 (EC 2.4.1.41) (Protein-UDP acetylgalactosaminyltransferase 3) (UDP-GalNAc:polypeptide N-acetylgalactosaminyltransferase 3) (Polypeptide GalNAc transferase 3) (GalNAc-T3) (pp-GaNTase 3)

Protein AC	Protein ID	Group	Protein Name
Q13724	GCS1	MNT1_stage1; Skmel28_stage2	Mannosyl-oligosaccharide glucosidase (EC 3.2.1.106) (Processing A- glucosidase I)
Q10471	GALT2	Skmel28_stage1	Polypeptide N-acetylgalactosaminyltransferase 2 (EC 2.4.1.41) (Protein-UDP acetylgalactosaminyltransferase 2) (UDP-GalNAc:polypeptide N-acetylgalactosaminyltransferase 2) (Polypeptide GalNAc transferase 2) (GalNAc-T2) (pp-GaNTase 2)
P34059	GA6S	NG; Skmel28_stage1	N-acetylgalactosamine-6-sulfatase precursor (EC 3.1.6.4) (N- acetylgalactosamine-6-sulfate sulfatase) (Galactose-6-sulfate sulfatase) (GalNAc6S sulfatase) (Chondroitinsulfatase) (Chondroitinase)
P15907	SIAT1	Skmel28_stage1	CMP-N-acetylneuraminate-beta-galactosamide-alpha-2,6-sialyltransferase (EC 2.4.99.1) (Beta-galactoside alpha-2,6-sialyltransferase) (Alpha 2,6-ST) (Sialyltransferase 1) (ST6Gal I)
P15586	GL6S	MNT1_stage 1,2,4; commonMNT1; unique MNT1	N-acetylglucosamine-6-sulfatase precursor (EC 3.1.6.14) (G6S) (Glucosamine-6-sulfatase)
P07686	HEXB	MNT1_stage4; NG; Skmel28_stage1; Skmel28_stage2;	Beta-hexosaminidase beta chain precursor (EC 3.2.1.52) (N-acetyl-beta- glucosaminidase) (Beta-N-acetylhexosaminidase) (Hexosaminidase B) (Cervical cancer proto-oncogene 7) (HCC-7)
P06865	HEXA	MNT1_stage1; Skmel28_stage1;	Beta-hexosaminidase alpha chain precursor (EC 3.2.1.52) (N-acetyl- beta-glucosaminidase) (Hexosaminidase A)
O94766	B3GA3	Skmel28_stage1	Galactosylgalactosylxylosylprotein 3-beta-glucuronosyltransferase 3 (EC 2.4.1.135) (Beta-1,3-glucuronyltransferase 3) (Glucuronosyltransferase-I) (GlcAT-I) (UDP-GlcUA:Gal beta-1,3-Gal-R glucuronyltransferase) (GlcUAT-I)
O60701	UGDH	Skmel28_stage1 Skmel28_stage1	UDP-glucose 6-dehydrogenase (EC 1.1.1.22) (UDP-Glc dehydrogenase) (UDP-GlcDH) (UDPGDH)
O00754	MA2B1	MNT1_stage1; MNT1_stage4; Skmel28_stage1;	Lysosomal alpha-mannosidase precursor (EC 3.2.1.24) (Mannosidase, alpha B) (Lysosomal acid alpha-mannosidase) (Laman) (Mannosidase alpha class 2B member 1

Source: Full database access could be found at http://pir.georgetown.edu/iproxpress/melanosome/suppl_data/. For details of the database features please refer to [16;17]

Table 1. List of glycosylation enzymes identified in the proteome of melanosomes

Despite these initial data, we should consider that the identification of these enzymes in melanosomes might be due to contamination of the purified melanosome sample. A crucial issue in proteomic analysis of subcellular proteomes is the ability to distinguish between true components and contaminants coming from other cell compartments, especially when working with low abundance proteins. Interestingly, except for known melanosome proteins that give the organelle its unique structure and functions, a majority of proteins detected in the melanosome proteome are not organelle-specific. Some, such as ribosomal protein complexes, are obvious minor contaminants that were co-purified during sucrose gradient fractionation. Even though extra precaution was taken, sensitive mass spectrometers can always detect trace amounts of peptides that originate either from resident low abundance proteins or from low-level contaminants. By searching our data against known human mitochondrial proteins annotated in UniProtKB, we estimate that the melanosome fractions at various stages are of high purity (94 % at early stage, and 98-99% at late stage). On the other hand, many proteins identified in our study demonstrate that melanosomes are highly dynamic. They may be viewed as a microcosm of organelles, representing a dynamic balance of proteins as well as small molecules being transported in and out. Many of the "nonspecific" proteins might be associated with melanosomes only for a short period of time, or they may be proteins that reside in other subcellular compartments. In that sense, we considered the possibility that true permanent "resident" molecules for organelles may not exist. Because of the uniqueness of melanosomes our study has become the gold-standard for future analysis. Taken all together, it is feasible that certain enzymes relocate to early melanosomes and become involved in the removal of glycan chains from PMEL to start the amyloid formation process. Further studies will be necessary to convincingly answer this question.

3. Conclusions and future directions

We suggest that the dynamics of amyloid formation could be managed or restored by jointly manipulating protein synthesis, *N*- and O-glycosylation, and Golgi remodeling. In the case of PMEL, we believe that the main role of glycosidation is to avoid the early start of the amyloid formation process in non-suitable intra or extra cellular locations. Thus, the sialic acid modified core-1-Oglycans in the RTP domain protect this domain from early protein shredding. The influence of glycosylation can be attributed to both the steric demand of the sugar moiety and hydrogen bonding, respectively. First, the incorporation of a single or a series of sugars will increase the serine side-chain volume dramatically, which results in a tight packing interaction which disrupts protein self-assembly into fibrils and prevents the exposure of this domain. The reported finding that the N-glycan site at aminoacid 321, located in the RPT domain of PMEL, is minimally modified indirectly supports this notion [12;33]. However, carbohydrate hydroxyl groups can still be involved in hydrogen bonding and sugar competition for intermolecular hydrogen bonds and could have an effect on the overall integrity of the structure. To avoid this situation, sialic acids are added at the non-reducing ends of glycans, conferring strong negative charges on the protein. We predicted that sialic acid when added to N-glycans makes then hydrophilic and leads to exposure of hydrophobic protein domains, as occurs during cleavage of PMEL, favoring the formation of polymers through hydrophobic interactions. In contrast, sialic acid added to O-glycans make them hydrophobic and protects them from early cleavage and degradation, a function that has been reported previously in other cells [12;33;58;58]. Thus, the extent and type of O-glycans modified with sialic acid in the RPT domain of PMEL play key roles to determine the proper PMEL glycoform destined to form fibrils [59], or to be targeted to the plasma membrane [60; 61]

In light of the presented evidence, it is critical to understand that cancer cells present altered glycosylation patterns compared to their normal counterparts and that it will impact all glycoproteins. This is the case in murine melanoma cells, the interaction of cell surface lectins (such as calreticulin [62]) that bind proteins modified with mannose type glycans are necessary to initiate the increases of murine melanoma [63]. Similarly, the expression of high-mannose type glycans increases the development of hepatic metastasis in B16 melanoma cells [64] or that high-mannose type glycans are also involved in the functionality and cell surface expression of the "mature" human transferrin receptor [65;66]. Furthermore, differences in the patterns of expression of sialyltransferases, such as ST3 Gal I, between normal and malignant cells had already been associated with a cancer-specific regulation of glyco-epitopes [67]. Therefore, it is not unusual that "mature" proteins contain unmodified mannose-rich/ hybrid type N-glycans and that such conformation plays an active role in cellular processes. On the contrary, these findings reveal a new factor to consider in the processing of cancer-specific proteins, especially in Melanomas. Future research should be aimed to understand in depth and consider these alternatives.

Author details

Julio C. Valencia* and Vincent J. Hearing
Pigment Cell Biology Section, Laboratory of Cell Biology, National Cancer Institute, Bethesda, USA

Acknowledgements

This research was supported in part by the Intramural Research Program of the NIH, National Cancer Institute, Center for Cancer Research.

4. References

[1] Eisenberg D, Jucker M. The amyloid state of proteins in human diseases. Cell 2012;148: 1188-203.

[2] Fowler DM, Koulov AV, Balch WE, Kelly JW. Functional amyloid--from bacteria to humans. Trends Biochem Sci 2007;32: 217-24.

[3] Chapman MR, Robinson LS, Pinkner JS, et al. Role of Escherichia coli curli operons in directing amyloid fiber formation. Science 2002;295: 851-5.

---

* Corresponding Author

[4] Maji SK, Perrin MH, Sawaya MR, et al. Functional amyloids as natural storage of peptide hormones in pituitary secretory granules. Science 2009;325: 328-32.
[5] Fowler DM, Koulov AV, Alory-Jost C, et al. Functional amyloid formation within mammalian tissue. PLoS Biol 2006;4: e6.
[6] McGlinchey RP, Shewmaker F, McPhie P, et al. The repeat domain of the melanosome fibril protein Pmel17 forms the amyloid core promoting melanin synthesis. Proc Natl Acad Sci U S A 2009;106: 13731-6.
[7] McGlinchey RP, Shewmaker F, Hu KN, et al. Repeat domains of melanosome matrix protein Pmel17 orthologs form amyloid fibrils at the acidic melanosomal pH. J Biol Chem 2011;286: 8385-93.
[8] Pfefferkorn CM, McGlinchey RP, Lee JC. Effects of pH on aggregation kinetics of the repeat domain of a functional amyloid, Pmel17. Proc Natl Acad Sci U S A 2010;107: 21447-52.
[9] Badtke MP, Hammer ND, Chapman MR. Functional amyloids signal their arrival. Sci Signal 2009;2: e43.
[10] Lefebvre T, Guinez C, Dehennaut V, et al. Does O-GlcNAc play a role in neurodegenerative diseases? Expert Rev Proteomics 2005;2: 265-75.
[11] Diaz-Nido J, Wandosell F, Avila J. Glycosaminoglycans and beta-amyloid, prion and tau peptides in neurodegenerative diseases. Peptides 2002;23: 1323-32.
[12] Valencia JC, Rouzaud F, Julien S, et al. Sialylated core 1 O-glycans influence the sorting of Pmel17/gp100 and determine its capacity to form fibrils. J Biol Chem 2007;282: 11266-80.
[13] Liu F, Iqbal K, Grundke-Iqbal I, Hart GW, Gong CX. O-GlcNAcylation regulates phosphorylation of tau: a mechanism involved in Alzheimer's disease. Proc Natl Acad Sci U S A 2004;101: 10804-9.
[14] Broncel M, Wagner SC, Hackenberger CP, Koksch B. Enzymatically triggered amyloid formation: an approach for studying peptide aggregation. Chem Commun (Camb) 2010;46: 3080-2.
[15] Griffith LS, Mathes M, Schmitz B. Beta-amyloid precursor protein is modified with O-linked N-acetylglucosamine. J Neurosci Res 1995;41: 270-8.
[16] Chi A, Valencia JC, Hu ZZ, et al. Proteomic and bioinformatic characterization of the biogenesis and function of melanosomes. J Proteome Res 2006;5: 3135-44.
[17] Hu ZZ, Valencia JC, Huang H, et al. Comparative Bioinformatics Analyses and Profiling of Lysosome-Related Organelle Proteomes. Int J Mass Spectrom 2007;259: 147-60.
[18] Watabe H, Kushimoto T, Valencia JC, Hearing VJ. Isolation of melanosomes. Curr Protoc Cell Biol 2005;Chapter 3: Unit.
[19] Leonhardt RM, Vigneron N, Rahner C, Cresswell P. Proprotein convertases process Pmel17 during secretion. J Biol Chem 2011;286: 9321-37.
[20] Berson JF, Harper DC, Tenza D, Raposo G, Marks MS. Pmel17 initiates premelanosome morphogenesis within multivesicular bodies. Mol Biol Cell 2001;12: 3451-64.

[21] Kobayashi T, Urabe K, Orlow SJ, et al. The Pmel 17/silver locus protein. Characterization and investigation of its melanogenic function. J Biol Chem 1994;269: 29198-205.
[22] Naim HY, Joberty G, Alfalah M, Jacob R. Temporal association of the N- and O-linked glycosylation events and their implication in the polarized sorting of intestinal brush border sucrase-isomaltase, aminopeptidase N, and dipeptidyl peptidase IV. J Biol Chem 1999;274: 17961-7.
[23] Hoashi T, Tamaki K, Hearing VJ. The secreted form of a melanocyte membrane-bound glycoprotein (Pmel17/gp100) is released by ectodomain shedding. FASEB J 2010;24: 916-30.
[24] Kwon BS, Chintamaneni C, Kozak CA, et al. A melanocyte-specific gene, Pmel 17, maps near the silver coat color locus on mouse chromosome 10 and is in a syntenic region on human chromosome 12. Proc Natl Acad Sci U S A 1991;88: 9228-32.
[25] Essentials of Glycobiology. Woodbury: Cold Spring Harbor Laboratory Press, 2002.
[26] Varki AP, Cummings R, Esko J, et al. Essentials of Glycobiology. 1st ed. Plainview, NY: Cold Spring Harbor Laboratory Press, 1999.
[27] Branza-Nichita N, Petrescu AJ, Negroiu G, Dwek RA, Petrescu SM. N-glycosylation processing and glycoprotein folding-lessons from the tyrosinase-related proteins. Chem Rev 2000;100: 4697-712.
[28] Elbein AD. Glycosidase inhibitors: inhibitors of N-linked oligosaccharide processing. FASEB J 1991;5: 3055-63.
[29] Petrescu SM, Branza-Nichita N, Negroiu G, Petrescu AJ, Dwek RA. Tyrosinase and glycoprotein folding: roles of chaperones that recognize glycans. Biochemistry 2000;39: 5229-37.
[30] Branza-Nichita N, Negroiu G, Petrescu AJ, et al. Mutations at critical N-glycosylation sites reduce tyrosinase activity by altering folding and quality control. J Biol Chem 2000;275: 8169-75.
[31] Takahashi H, Parsons PG. Rapid and reversible inhibition of tyrosinase activity by glucosidase inhibitors in human melanoma cells. J Invest Dermatol 1992;98: 481-7.
[32] Valencia JC, Watabe H, Chi A, et al. Sorting of Pmel17 to melanosomes through the plasma membrane by AP1 and AP2: evidence for the polarized nature of melanocytes. J Cell Sci 2006;119: 1080-91.
[33] Hoashi T, Muller J, Vieira WD, et al. The repeat domain of the melanosomal matrix protein PMEL17/GP100 is required for the formation of organellar fibers. J Biol Chem 2006;281: 21198-208.
[34] Ulloa F, Real FX. Benzyl-N-acetyl-alpha-D-galactosaminide induces a storage disease-like phenotype by perturbing the endocytic pathway. J Biol Chem 2003;278: 12374-83.
[35] Kuan SF, Byrd JC, Basbaum C, Kim YS. Inhibition of mucin glycosylation by aryl-N-acetyl-alpha-galactosaminides in human colon cancer cells. J Biol Chem 1989;264: 19271-7.
[36] Delacour D, Gouyer V, Leteurtre E, et al. 1-benzyl-2-acetamido-2-deoxy-alpha-D-galactopyranoside blocks the apical biosynthetic pathway in polarized HT-29 cells. J Biol Chem 2003;278: 37799-809.

[37] Gouyer V, Leteurtre E, Zanetta JP, et al. Inhibition of the glycosylation and alteration in the intracellular trafficking of mucins and other glycoproteins by GalNAcalpha-O-bn in mucosal cell lines: an effect mediated through the intracellular synthesis of complex GalNAcalpha-O-bn oligosaccharides. Front Biosci 2001;6: D1235-D1244.

[38] Breen KC. The role of protein glycosylation in the control of cellular N-sialyltransferase activity. FEBS Lett 2002;517: 215-8.

[39] Sewell R, Backstrom M, Dalziel M, et al. The ST6GalNAc-I sialyltransferase localizes throughout the Golgi and is responsible for the synthesis of the tumor-associated sialyl-Tn O-glycan in human breast cancer. J Biol Chem 2006;281: 3586-94.

[40] Puthalakath H, Burke J, Gleeson PA. Glycosylation defect in Lecl Chinese hamster ovary mutant is due to a point mutation in N-acetylglucosaminyltransferase I gene. J Biol Chem 1996;271: 27818-22.

[41] Deutscher SL, Hirschberg CB. Mechanism of galactosylation in the Golgi apparatus. A Chinese hamster ovary cell mutant deficient in translocation of UDP-galactose across Golgi vesicle membranes. J Biol Chem 1986;261: 96-100.

[42] Oelmann S, Stanley P, Gerardy-Schahn R. Point mutations identified in Lec8 Chinese hamster ovary glycosylation mutants that inactivate both the UDP-galactose and CMP-sialic acid transporters. J Biol Chem 2001;276: 26291-300.

[43] Cheresh DA, Varki AP, Varki NM, et al. A monoclonal antibody recognizes an O-acylated sialic acid in a human melanoma-associated ganglioside. J Biol Chem 1984;259: 7453-9.

[44] Bellis SL, Newman E, Friedman EA. Steps in integrin beta1-chain glycosylation mediated by TGFbeta1 signaling through Ras. J Cell Physiol 1999;181: 33-44.

[45] Bellis SL. Variant glycosylation: an underappreciated regulatory mechanism for beta1 integrins. Biochim Biophys Acta 2004;1663: 52-60.

[46] Christie DR, Shaikh FM, Lucas JA, Lucas JA, III, Bellis SL. ST6Gal-I expression in ovarian cancer cells promotes an invasive phenotype by altering integrin glycosylation and function. J Ovarian Res 2008;1: 3.

[47] Liu Y, Pan D, Bellis SL, Song Y. Effect of altered glycosylation on the structure of the I-like domain of beta1 integrin: a molecular dynamics study. Proteins 2008;73: 989-1000.

[48] Ciolczyk-Wierzbicka D, Amoresano A, Casbarra A, et al. The structure of the oligosaccharides of N-cadherin from human melanoma cell lines. Glycoconj J 2004;20: 483-92.

[49] Ciolczyk-Wierzbicka D, Bodzioch M, Gil D, et al. Expression of fucosyltransferases contributes to melanoma invasive phenotype. Med Chem 2007;3: 418-24.

[50] Ciolczyk-Wierzbicka D, Gil D, Laidler P. The inhibition of cell proliferation using silencing of N-cadherin gene by siRNA process in human melanoma cell lines. Curr Med Chem 2012;19: 145-51.

[51] Laidler P, Litynska A, Hoja-Lukowicz D, et al. Characterization of glycosylation and adherent properties of melanoma cell lines. Cancer Immunol Immunother 2006;55: 112-8.

[52] Angata K, Fukuda M. Polysialyltransferases: major players in polysialic acid synthesis on the neural cell adhesion molecule. Biochimie 2003;85: 195-206.

[53] Sata T, Roth J, Zuber C, Stamm B, Heitz PU. Expression of alpha 2,6-linked sialic acid residues in neoplastic but not in normal human colonic mucosa. A lectin-gold cytochemical study with Sambucus nigra and Maackia amurensis lectins. Am J Pathol 1991;139: 1435-48.
[54] Leonhardt RM, Vigneron N, Rahner C, Van den Eynde BJ, Cresswell P. Endoplasmic reticulum export, subcellular distribution, and fibril formation by Pmel17 require an intact N-terminal domain junction. J Biol Chem 2010;285: 16166-83.
[55] Axelsson MA, Karlsson NG, Steel DM, et al. Neutralization of pH in the Golgi apparatus causes redistribution of glycosyltransferases and changes in the O-glycosylation of mucins. Glycobiology 2001;11: 633-44.
[56] Rottger S, White J, Wandall HH, et al. Localization of three human polypeptide GalNAc-transferases in HeLa cells suggests initiation of O-linked glycosylation throughout the Golgi apparatus. J Cell Sci 1998;111 (Pt 1): 45-60.
[57] Berger EG. Ectopic localizations of Golgi glycosyltransferases. Glycobiology 2002;12: 29R-36R.
[58] Azuma Y, Taniguchi A, Matsumoto K. Decrease in cell surface sialic acid in etoposide-treated Jurkat cells and the role of cell surface sialidase. Glycoconj J 2000;17: 301-6.
[59] Hoashi T, Muller J, Vieira WD, et al. The repeat domain of the melanosomal matrix protein PMEL17/GP100 is required for the formation of organellar fibers. J Biol Chem 2006;281: 21198-208.
[60] Ikonen E, Simons K. Protein and lipid sorting from the trans-Golgi network to the plasma membrane in polarized cells. Semin Cell Dev Biol 1998;9: 503-9.
[61] Gasbarri A, Del Prete F, Girnita L, et al. CD44s adhesive function spontaneous and PMA-inducible CD44 cleavage are regulated at post-translational level in cells of melanocytic lineage. Melanoma Res 2003;13: 325-37.
[62] White TK, Zhu Q, Tanzer ML. Cell surface calreticulin is a putative mannoside lectin which triggers mouse melanoma cell spreading. J Biol Chem 1995;270: 15926-9.
[63] Chandrasekaran S, Tanzer ML, Giniger MS. Oligomannosides initiate cell spreading of laminin-adherent murine melanoma cells. J Biol Chem 1994;269: 3356-66.
[64] Mendoza L, Olaso E, Anasagasti MJ, Fuentes AM, Vidal-Vanaclocha F. Mannose receptor-mediated endothelial cell activation contributes to B16 melanoma cell adhesion and metastasis in liver. J Cell Physiol 1998;174: 322-30.
[65] Williams AM, Enns CA. A region of the C-terminal portion of the human transferrin receptor contains an asparagine-linked glycosylation site critical for receptor structure and function. J Biol Chem 1993;268: 12780-6.
[66] Hayes GR, Williams AM, Lucas JJ, Enns CA. Structure of human transferrin receptor oligosaccharides: conservation of site-specific processing. Biochemistry 1997;36: 5276-84.
[67] Dalziel M, Whitehouse C, McFarlane I, et al. The relative activities of the C2GnT1 and ST3Gal-I glycosyltransferases determine O-glycan structure and expression of a tumor-associated epitope on MUC1. J Biol Chem 2001;276: 11007-15.

Section 5

Glycoengineering and Therapy

Chapter 16

Production of Highly Sialylated Monoclonal Antibodies

Céline Raymond, Anna Robotham, John Kelly, Erika Lattová, Hélène Perreault and Yves Durocher

Additional information is available at the end of the chapter

1. Introduction

The first monoclonal antibody (Mab), developed against kidney transplant rejection, was accepted by the FDA in 1986 [1]. Today, Mabs are leading the biotherapeutics market as 28 have been approved in Europe and the USA, and hundreds are in clinical trials [2-4]. Most of them are of IgG1 subtype, developed for cancer and immune disease treatments. Mabs clinical efficacy not only relies on specific target binding provided by their variable region, but also on their ability to trigger defense mechanisms such as antibody-dependent cellular cytotoxicity (ADCC) and complement dependent cytotoxicity (CDC). These effector functions are mediated by the interaction between the antibody Fc fragment and the Fcγ-receptors expressed on immune cell surfaces or the molecules of the complement involved in ADCC and CDC respectively. In the last decade, these interactions were found to be highly dependent on on the presence and structure of the N-glycan linked to the Fc fragment [5, 6].

Fc fragments possess two conserved N-glycosylation sites on asparagine 297 in the CH2 domain of each heavy chain. Mabs produced in mammalian cells possess a wide variety of glycoforms, as the attached glycans are modified to different extents with core-fucosylation, bisecting N-acetylglucosamine addition, galactosylation and sialylation. The glycan composition is crucial, as the presence or absence of a single monosaccharide residue can remarkably affect the affinity of the Mab for the different Fcγ-receptors. Among the variety of monosaccharides present on Fc glycans, terminal sialic acids are particularly interesting, as their role in Mab functions is both positive and negative. Sialylation of the Fc glycan dramatically decreases Mab affinity for the canonical Fc receptors, thereby inhibiting ADCC. However, recent studies on the anti-inflammatory properties of intravenous immunoglobulins (IVIg) suggest that this biological activity could be conferred by the presence of α2,6-sialic acid residues on the Fc glycans. Although this hypothesis is still controversial, the study has raised a new interest in α2,6-sialylated IgGs. While glycosylation of therapeutic Mabs significantly impact their biological activity, the production of Mabs with a specific homogenous glycoform profile is in general beyond the reach of manufacturing bioprocesses. In this chapter, we describe and compare two large-scale transient expression platforms using chinese hamster ovary (CHO) and human embryonic kidney 293 (HEK293) cells for the production of highly sialylated monoclonal antibodies.

2. IgG N-glycans and their interactions with the Fc

Glycosylation is a complex process that involves several glycosyltransferases and glycosidases. Most glycosylation sites are located on the glycoprotein surface, whereas IgGs' N-glycans are embedded within the Fc fragment. This particular location restricts the access of glycosyltransferases to their substrates, thereby reducing glycan complexity. Subsequently, while tri- and tetra-antennary glycans can be found on many glycoproteins such as EPO or IgG Fab fragments, Fc N-glycans are of the complex biantennary type [7] that consists of a heptasaccharide core structure comprising four N-acetylgalactosamines (GlcNAc) and three mannoses (Figure 1). The α1,3 and α1,6 arms can be further elongated with galactose and sialic acid. Fucose and bisecting-GlcNAc can be found on the core GlcNAc and on the central mannose respectively. This glycan is rarely fully processed; the predominant glycoform found on antibodies produced in CHO and 293 cell cultures is the fucosylated core-structure.

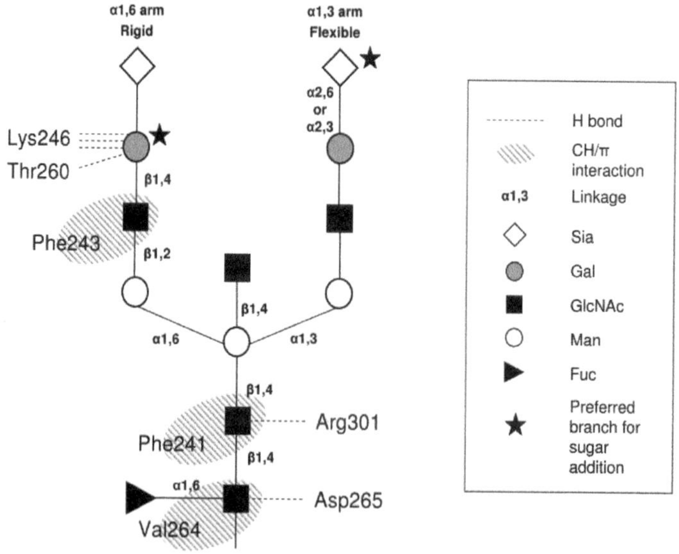

Figure 1. IgG1 glycan bi-antennary complex structure. Glycan interactions with CH2 amino acids. Sia: sialic acid; Gal: galactose; GlcNAc: N-acetylglucosamine; Man: mannose; Fuc: fucose.

The C_H2 amino acid sequence around Asn297 is very well conserved amongst IgG subtypes. Several amino acids have been shown to interact with the glycan located on the same heavy chain (HC), whereas no interaction is likely to happen with the other heavy chain. Amino acid-glycan interactions determine the glycan position within the Fc pocket and its availability for glycosyltransferases. The galactose (Gal) on the α1,6 arm is the main residue retaining the glycan on the protein surface [6, 8, 9] by generating H-bonds with Lys246 and Thr260 [6]. In IgG1 and IgG4 subtypes, galactosylation occurs preferentially on α1,6 branch [8, 10]. The inner saccharides have less interaction with the protein. The first and second GlcNAc residues generate H-bonds with Asp265 and Arg301 respectively [6]. They also form a CH/π interaction with the non-polar moieties of Val264 and Phe241 [6]. Similarly, the GlcNAc residue on the α1,6 branch also forms a CH/π interaction with Phe243 [11]. In contrast to galactosylation, sialylation preferentially occurs on the α1,3 arm. It has been suggested that the galactose on the α1,6 arm may limit further elongation by maintaining the branch close to the C_H2 protein surface [8]. However, even if steric hindrance may play a role, branch α2,6-sialylation specificity was shown to be a consequence of ST6Gal-I α1,3 arm preference [12], a phenomenon that is protein independent. The presence or absence of a sugar residue on the Fc glycan thus affects the conformation of the Fc, thereby affecting the Fc-mediated effector functions. These effects are summarized in Table 1.

	Mab-FcγR Affinity	Mab-C1q Affinity	Impacted effector function	Reference
b-GlcNAc	+		ADCC	Umaña 1999
Fucose	-		ADCC	Shields 2002, Shinkawa 2003
Galactose		+	CDC	Hodoniczky 2005
Sialic acid	-		ADCC	Scallon 2007

Table 1. Impact of the presence of extra-core monosaccharides on IgG ADCC and CDC effector functions.

3. Sialic acids in Fc N-glycans

The half-life of a number of glycoproteins can be enhanced by sialylation, as sialic acid acts as a cap that hides the penultimate galactose residue recognized by the hepatic asialoglycoprotein receptor (ASGPR, or Ashwell-Morell receptor) [13]. Sialic acid can be linked to the galactose either with α2,3 or α2,6 linkage. Recent studies showed that α2,3 sialylation provides a better protection to the protein than α2,6 sialylation, as ASGPR recognizes Siaα2,6Gal and Siaα2,6GalNAc moieties in addition to the well-known Gal and GalNAc residues [14, 15]. However, Fc-glycans have no apparent impact on IgG half-life [16]. The benefits of IgG sialylation on *in vivo* properties of IgGs are still to be understood. Recent studies suggest that sialylation provides anti-inflammatory properties to IgGs. It was observed that IVIg injected at very high doses have a therapeutic effect in several auto-immune and inflammatory diseases such as immune thrombocytopenic purpura

(ITP) and rheumatoid arthritis (RA). Kaneko et al. demonstrated that IgGs bearing sialylated Fc-glycans have anti-inflammatory properties in a RA murine model [17]. The inhibitory Fcγ-receptor FcγRIIb as well as Dendritic Cell-Specific Intercellular adhesion molecule-3-Grabbing Non-integrin (DC-SIGN) receptor were shown to be involved [18-21], but the exact mechanism has not been elucidated. In parallel, Van de Geijn et al. reported that the increased levels of IgG1 galactosylation and sialylation during pregnancy may be responsible for the improved condition of RA-affected pregnant women [22].

IVIg are IgGs pooled and purified from sera of 3000 to 10000 donors. The sialylated fraction represents around 10% of the total IgGs present in the pool. To be therapeutically effective, IVIg used as an anti-inflammatory drug require repeated injections of very high doses (1-3g/kg). As IVIg are successfully used in an increasing number of applications, a lack of donors is expected in the near future [23, 24]. To increase the therapeutic efficacy of anti-inflammatory IgGs and avoid IVIg shortage, one strategy may reside in the production of recombinant sialylated IgGs [25].

4. Production of sialylated recombinant antibody

The CHO cell line is the most widely accepted production cell line in the industry for therapeutic manufacturing, including monoclonal antibodies. CHO cells have proven to be a safe expression system and provide high production yields. IgG and Fc fragment produced in CHO cells exhibit a very low level of sialylation (<2%). However, it was shown that a single mutation in the CH2 domain, the replacement of Phe243 by an alanine, enhances the sialylation level of an IgG3 [26]. CHO glycosylation machinery is very similar to that found in human cells but with two major differences: they lack a functional N-acetylglucosaminyltransferase-III (GnTIII) for the addition of bisecting-GlcNAc and, more importantly, they lack the alpha-2,6-sialyltransferase-I activity (ST6Gal-I or SIAT1) responsible for the addition of sialic acids on galactose residues with α2,6 an linkage [27]. The expression of a recombinant ST6Gal-I is therefore necessary to achieve the production of α2,6 sialylated IgGs in CHO cells.

Large-scale transfection strategies allow for the rapid expression of recombinant glycoproteins. The HEK293 cell line is easily transfectable with a variety of gene transfer reagents, and is probably the most utilized cell line for large-scale transient gene expression [28, 29]. More recently, approaches using CHO cells have also been developed that provide tens to hundreds of milligrams of protein per litre [30]. In order to generate high titres of antibodies with enhanced α2,6-sialylation, we optimized the conditions for the transient co-expression of human ST6Gal-I together with IgG light and heavy chains in CHO and HEK293 cells. Our model antibody, Herceptin (or trastuzumab, herein abbreviated TZM), is a humanized mouse IgG1 used in the treatment of HER2-positive breast cancer. TZM has no glycosylation site on the Fab fragment but only the N297 sites on each CH2 domains of the Fc. We compared sialylation of the wild-type antibody to the F246A mutant (TZMm), which is equivalent to the IgG3 F243A mutant described previously [26].

5. Materials and methods

Mammalian cell culture

The human embryonic kidney 293-6E cell line stably expressing truncated Epstein-Barr virus Nuclear Antigen-1 (EBNA1) and the Chinese hamster ovary cell line also expressing a truncated EBNA1 protein (clone 3E7, or CHO-3E7) were grown in suspension culture in serum-free F17 medium (Invitrogen, Carlsbad, CA) supplemented with 0.1% Pluronic-F68, 25 μg/mL geneticin G418 (for 293-6E cells only) and 4 mM glutamine [31]. Cultures were maintained between 0.1 and 2.0 x 10^6 cells/mL in 125 mL ventilated Erlenmeyer flasks shaken at 120 rpm in a humidified incubator at 37°C with 5% CO_2.

Plasmids

The light (LC), heavy (HC), and F246A mutated heavy (HCF246A) chains were cloned between the EcoRI and BamHI sites of the pTT5 vector [32, 33]. The human ST6Gal1 gene was cloned between the HindIII and BamHI sites of the pYD7 vector [34]. Green fluorescent protein (GFP) cloned into the pTT vector was used as a reporter gene to evaluate transfection efficiency [35]. Plasmids were amplified in Escherichia coli (DH5α) grown overnight in CircleGrow medium (MP Biomedical, Solon, OH) supplemented with 100ug/mL ampicillin and purified using MAXIprep or QIAprep spin Miniprep columns (Qiagen, Mississauga, ON).

Transfection

Linear 25 kDa polyethylenimine[1] (LPEI), and linear deacylated polyethylenimine (PEI max) were obtained from Polysciences (Warrington, PA). Stock solutions (1 mg/mL and 3 mg/mL for LPEI and PEImax respectively) were prepared in ultrapure water, sterilized by filtration (0.2 μm), aliquoted and stored at 4°C. Cells were diluted 2 days before transfection in fresh medium at 0.5 and 0.2 x 10^6 cells/mL for 293-6E and CHO 3E7 cells respectively. Cells were transfected at densities between 1.5 and 2 x 10^6 cells/mL. DNA and PEI were separately diluted in complete serum-free F17 medium in sterile tubes. Transfection reagents volume was 10% of the final culture volume. The final DNA concentration was 1 μg per mL of 293-6E culture, and 1.5 μg per mL of 3E7 culture. PEI was used at the final concentration of 3 μg per mL in 293-6E culture and 7 μg per mL in 3E7 culture. Plasmid DNA mix was directly added to the cells and the suspension was allowed to incubate under agitation for 5 min at 37°C before the addition of PEI according to the direct transfection protocol [46]. Cells were fed 24 hours post-transfection (hpt) with TN1 peptone to a final concentration of 0.5% (w/w) to enhance productivity [28]. Transfection efficiency was assessed 48 hpt by determining the percentage of GFP positive cells and GFP fluorescence intensity by flow cytometry with a BD LSRII cytometer (BD Biosciences, Mississauga, ON). Only viable single cells were taken in account. Cell density and viability were determined using the Cedex Innovatis

[1] "Use of PEI for transfection may be covered by existing intellectual property rights, including US Patent 6,013,240, European Patent 0,770,140, and foreign equivalents for which further information may be obtained by contacting licensing@polyplus-transfection.com"

automated cell counter Cedex Analyzer (Roche, Laval, QC) based on the trypan blue exclusion method.

Purification of Mabs from cell culture supernatants

Cell cultures were centrifuged 20 min at 3000g. The supernatant was collected and loaded on a 4mL MabSelect SuRe column (GE Healthcare, Mississauga, ON) equilibrated in PBS. The column was washed with PBS and Mabs were eluted with 100mM citrate buffer pH 3.0. The fractions containing Mabs were pooled and the citrate buffer was exchanged against water on Econo-Pac® 10DG columns (Bio-Rad, Mississauga, ON). Purified Mabs were sterilized by passing through 0.2 μm filters, aliquoted, and stored at -80°C.

Quantification of Mabs

Concentration of TZM and TZMm in culture supernatants were determined by protein-A HPLC using a 800 μL POROS® 20 micron Protein A ID Cartridge (Applied BioSystems, Foster City, CA) according to the manufacturer's recommendations. The antibody present in the culture medium was also visualized following reducing and non-reducing SDS-PAGE stained by Bio-Safe Coomassie Stain (Bio-Rad, Mississauga, ON). Purified Mab were quantified by absorbance at 280 nm using a Nanodrop™ spectrophotometer (ThermoScientific).

Lectin-Blot for α2,6-sialylation evaluation

α2,6 sialylation was assessed by lectin-blotting on denatured and reduced antibody to separate HC from LC. After protein transfer, nitrocellulose membrane was incubated 3 hours with biotinylated Sambucus Nigra (SNA) lectin (Vector Laboratories, Burlingame, CA), then incubated with Streptavidin-Peroxidase Polymer (Sigma, Saint Louis, MO) for one hour. Signal was revealed with BM Chemiluminescence Blotting Substrate (POD) (Roche Applied Science, Indianapolis, IN).

Isoelectric focusing (IEF)

Purified Mabs were analysed on PhastGel™ 3-10, run on the PhastSystem™ (Amersham Biosciences, Baie d'Urfe, QC) according to the manufacturer's recommendations. Gels were fixed in trichloroacetic acid (TCA) 5% in water (w/w) and stained with Coomassie blue 0.02%.

Enzymatic removal of sialic acids

α2,3 linked sialic acids were enzymatically removed from the purified antibody by *Streptococcus pneumoniae* Glyko Sialidase S (PROzyme, Hayward, CA) after one hour incubation at 37°C. Total sialylation (α2,6 and α2,3 sialic acids) was removed after incubation overnight at 37°C with *Arthrobacter ureafaciens* Neuraminidase (MP Biomedicals, Solon, OH).

Mass spectrometry

Peptide-N-glycosidase F (PNGaseF) was purchased from Roche (Mannheim, Germany). Phenylhydrazine (PHN), phenylhydrazine hydrochloride (PHN.HCl), iodomethane (MeI), dimethylsulfoxide (DMSO), 2,5-dihydroxybenzoic acid (DHB) and 2-aza-2-thiothymine (ATT) were obtained from Sigma (St. Louis, MO, USA). Carbon and STRATA-X-C cartridges were purchased from Phenomenex (Torrance, CA). Solid sodium hydroxide (NaOH) and HPLC-grade solvents (acetonitrile (ACN), chloroform, methanol) were obtained from Fisher Scientific (Fair Lawn, NJ). HPLC-grade water was obtained with a Milli-Q® plus TOC water purification system (Millipore, Bedford, MA).

Release of N-linked oligosaccharides from samples and MALDI-MS analysis

To the sample solution (100 μL; 50μg glycoprotein) 2 μL of PNGaseF were added and incubated at 37°C for 5-18 h. One μL of digested mixture was loaded onto a spot with freshly prepared matrix solution (0.8 μL; 5 mg of ATT and 2.5 mg of PHN.HCl dissolved in 350 μL 50% ACN in water) predeposited on the surface of a MALDI target. When the mixture was partially dried, 0.4 μL of PHN-labelling solution (phenylhydrazine:deionized water:ACN/1:4:1) was added to the spot with sample-matrix and left to air air-dry. The samples were analysed by MALDI-TOF/TOF-MS (UltrafleXtremeTM, Bruker) in both positive and negative ion modes. For the structural analysis individual parent ions were manually selected. MS/MS spectra of oligosaccharides were interpreted manually. The structures of N-glycans were derived on the basis of fragmentation patterns which were produced under MS/MS conditions for single precursor ions. For a discrimination of isomeric structures, general rules described in previous study have been applied [36].

To verify glycan profiles obtained by PHN-target derivatization immediately after PNGaseF treatment, all digested samples were also purified on STRATA-XC cartridges combined with carbon cartridges. After washing carbon cartridge with 4x1000μL of deionized water, glycans were eluted with 40% ACN (1000μL). After total evaporation the glycans were labelled with PHN as described above or permethylated according to the procedure described by Ciucanu and Kerek [47]. Briefly, dried fractions containing oligosaccharides were dissolved in DMSO (40 μL), to which NaOH (2 mg) and methyl iodide (8 μL) were added. The mixture was vortexed vigorously (30 min) at room temperature. Then, the reaction was stopped by adding ice-cold water (500 μL) followed by chloroform (200 μL). After mixing and centrifugation, the upper aqueous layer was discarded and the chloroform portion was again washed with distilled water (4 X 500 μL each). Chloroform was evaporated and the sample was reconstituted in 70% aqueous methanol (5 μL).One μL of permethylated sample solution was spotted into DHB matrix and analysed by MS.

Intact mass analysis of Mabs

The antibody isolates were analysed by Trap-LC-MS using an Agilent 1100 series HPLC system coupled to a Q-TOF2 hybrid quadrupole time-of-flight mass spectrometer (Waters) equipped with an electrospray interface. Approximately 3 μg of antibody was injected onto a Protein MicroTrap (Michrom BioResources/Dionex). The trap was washed with deionized water (~500 μL) to remove buffers prior to being switched on-line with the LC-MS system. The antibody was eluted using a linear gradient from 1% to 85% solvent B in 27min (solvent A: 0.1% formic acid, solvent B: 0.1% formic acid in acetonitrile, flow rate: 0.45mL/min). The mobile phase was split before the trap to 50μL/min. The electrospray voltage was 4200V and cone voltage was 80V. The Q-TOF2 mass spectrometer was calibrated for high mass range using cesium iodide. Mass spectra were acquired every 2 seconds during the analyses over the full mass range (m/z 50-4000). The mass spectra acquired across the protein peak were combined and smoothed (smooth window: ± 3, number of smooths: 8, smoothing method: Savitsky-Golay) and deconvoluted using the MassLynx MaxEnt 1 software from Waters: spectral window: m/z 2500-3200, molecular weight range: 140,000-160,000 Da, resolution: 1.00Da/channel, damage model: simulated isotope pattern spectrometer blur width of 0.500Da, minimum intensity ratios: 60% left, 60% right, number of iterations: 15). Peaks in the molecular weight profile were integrated using the MassLynx software.

6. Results

6.1. Transfection conditions optimization

DNA ratio parameters

Transfection efficacy relies on a variety of parameters that are, amongst others, cell line- protein- and plasmid-dependent. We expressed our model Mab together with the human α2,6 sialyltransferase (ST6Gal-I) by performing co-transfection of the Mab heavy chain (HC), light chain (LC) and ST6Gal-I expressing plasmids. The optimal proportions of each plasmid were determined to ensure 1) the proper assembly of the Mab, 2) high sialylation levels and 3) high Mab yields. The LC:HC and (LC:HC):ST6Gal-I ratios were then assessed in CHO and 293 cells in suspension in 6-well plates, after 5 days at 37°C. The supernatants were collected and analysed by protein-A HPLC (Fig. 2A) and by SDS-PAGE followed by Coomassie staining (Fig. 2B) to evaluate the expression yield and proper assembly of antibodies. Sialylation was assessed by SNA-blotting (Fig. 2C). For TZMm, a slight excess of LC plasmid (LC:HC of 6:4) provided a better expression yield in both cell lines. For wild-type TZM, this ratio was of 5:5 in 293 and 6:4 in CHO cells. A ratio 6:4 was chosen for the expression of TZM and TZMm in both cell lines. As for sialylation, 20% of ST6Gal-I plasmid was found to allow maximum signal in SNA blots without significantly affecting TZMm expression yields in both cell lines. It is worth noticing that the sialylation of the wild-type antibody remained very low for all ST6Gal-I plasmid ratios, suggesting the inability of the enzyme to sialylate TZM. Mab yields were increased in 293 cells with the incorporation of 30% of non-coding DNA (salmon sperm DNA) in the final DNA mix, while this was unnecessary in CHO cells (data not shown).

Transfection protocol

The usual polyethylenimine (PEI) mediated transfection protocol involves the formation of DNA-PEI complexes (polyplexes) prior to their addition to the cells. Another approach, the direct transfection, consists of the successive addition of DNA and PEI to the cell suspension, and was validated with 293 cells to facilitate large-scale transfection handling [46, 48]. This strategy was adapted here for CHO cells. 1.5 μg of DNA and 7 μg of PEI per mL of culture allowed the expression of amounts of IgGs equal to the classic indirect protocol (Fig. 3). Transfection parameters for CHO and 293 cells are summarized in Table 2.

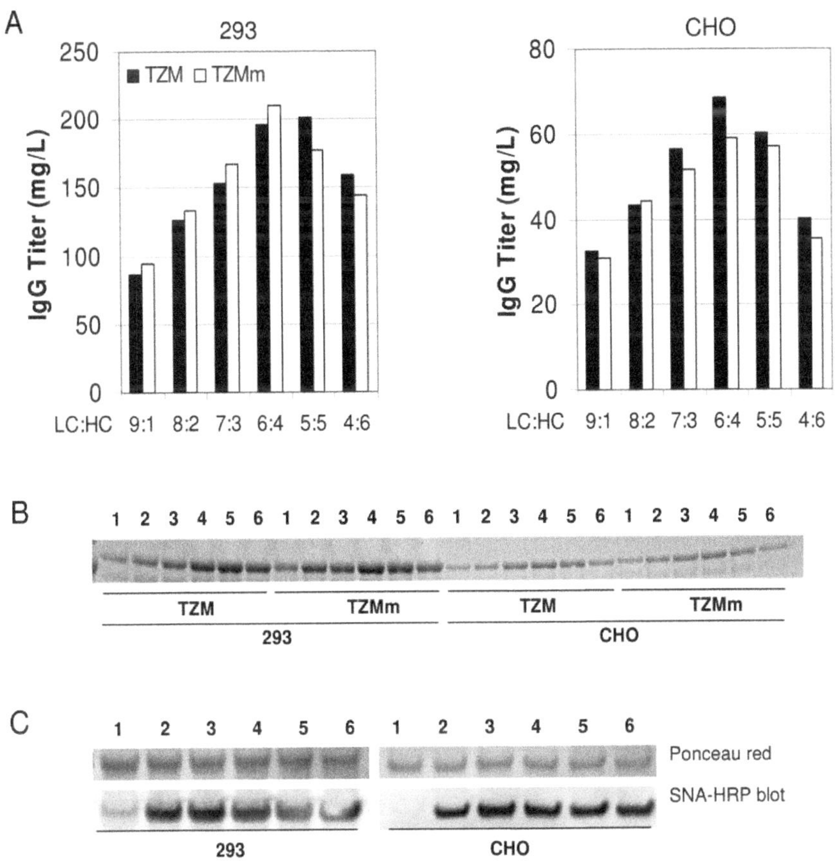

Figure 2. Optimization of LC:HC and ST6Gal-I ratio. Panel A – Mabs were quantified by protein-A HPLC in 293 and CHO supernatants 5 days post-transfection. Different LC:HC and LC:HC F246A DNA ratios were tested for proper expression of TZM and TZMm respectively. Panel B – The supernatants were loaded on SDS-PAGE (non-reducing) and stained with Coomassie blue. LC:HC ratios are: Lane 1: 9:1; Lane 2: 8:2; Lane 3: 7:3; Lane 4: 6:4; Lane 5: 5:5; Lane 6: 4:6. Panel C – Determination of the optimal ST6Gal-I ratio. Supernatants from cells transfected with TZMm HC and LC and various ST6Gal-I ratios were harvested 5 days post-transfection, separated by SDS-PAGE (reducing conditions) and transferred on nitrocellulose membrane for the determination of the sialylation level by blotting with SNA-HRP. Nitrocellulose membrane was stained with Ponceau red for protein load control, or incubated with SNA-HRP. Proportion of ST6Gal-I plasmid in transfection mixture was: Lane 1: 0%; Lane 2: 10%; Lane 3: 20%; Lane 4: 30%; Lane 5: 40%; Lane 6: 50%.

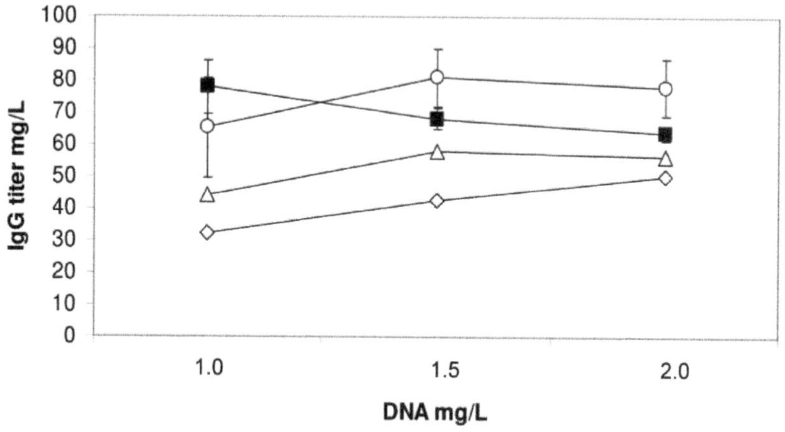

Figure 3. Determination of the optimal transfection parameters. Final concentrations of DNA and PEI per mL of culture for direct transfection of CHO cells were determined for TZM. Closed squares represent the TZM titres obtained using the traditional indirect protocol, i.e. with DNA-PEI complexation prior to the addition to the cell culture (PEI 5μg/mL). Open symbols represent the results for 3 different PEI concentrations with direct protocol: open circle: 7 μg/mL; open triangle: 6 μg/mL; open diamond 5 μg/mL. Results represent the average of three independent experiments ± SD.

	293	CHO
LC:HC	6:4	6:4
(LC:HC):ST6Gal-I	(80):20	(80):20
((LC:HC):ST6Gal-I):ssDNA	(70):30	(100):0
DNA:PEI for Direct protocol	1:3 (LPEI)	1.5:7 (PEI max)

Table 2. Optimal DNA and PEI ratios determined for the expression of TZM and TZM F246A mutant (TZMm) in 293-6E and CHO cells using the direct transfection protocol.

6.2. Mab expression and sialylation kinetics

Sialylation is likely to be limited after a certain time in batch culture due to a combination of adverse factors such as, but not limited to, ammonia accumulation [37, 38], ST6Gal-I plasmid loss upon successive cell division and sialidases release in medium [39]. As Mab titre increases over time in the culture medium, the harvest time point selected must provide an acceptable compromise between high Mab yield and high sialylation. In this study, we gave priority to sialylation over Mab expression yield.

Sialylation kinetics was studied on the mutant antibody as sialylation of TZM was barely detectable. TZMm co-expression with ST6Gal-I was carried out for 7 days in 293 and CHO cells, until viability dropped to approximately 60% (Fig. 4A). A fraction of each culture was harvested daily from day 3 to day 7 and purified by protein-A affinity chromatography. Sialylation level of the purified TZMm fractions were assessed by SNA-blotting and IEF analyses.

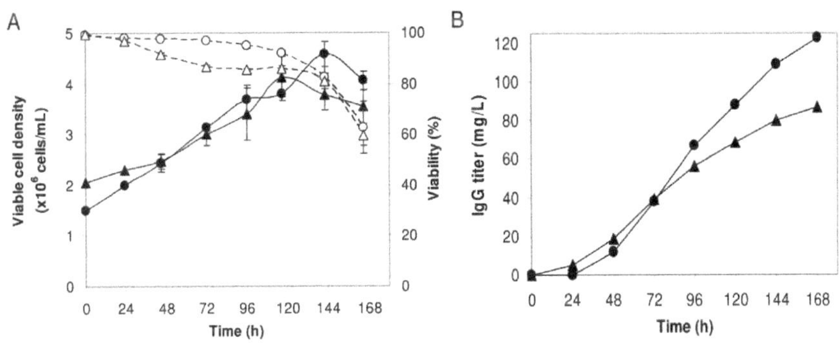

Figure 4. Production kinetic of TZMm in 293 and CHO cells. Panel A - Cell growth and viability for TZMm co-expression with ST6Gal-I. Cells were transfected at t=0h and fed at 24h post-transfection with TN1 peptone (0.5% w/v final). Open and closed circles: viability and viable cell density in 293 cell cultures. Open and closed triangles: viability and viable cell density in CHO cell cultures. Panel B - TZMm yield over time as determined by protein-A HPLC quantification of each fraction. Closed circles: titres in 293 cells; closed triangles: titres in CHO cells.

In 293 cells, SNA-blot showed a maximum α2,6-sialylation on day 3, which then decreased over time (Fig. 5A). In CHO cells, no significant loss of intensity was observed, suggesting that α2,6-sialylation of the heavy chain was more stable over time. These patterns were supported by IEF analyses. Due to the negative charge of sialic acid, mono, di, tri and tetra-sialylated Mabs are expected to have lower pI than asialylated glycoforms. Even though these variations in pI are small, they could be easily distinguished on IEF gels. The upper band intensity showed a higher amount of asialylated glycoforms in 293 than in CHO as seen on the blot (Fig. 5B). The decline in the sialylation level was less obvious by IEF vs SNA-blot analysis for the 293 cells, which could be explained by a change in the α2,3 / α2,6 sialylation ratio over time. For CHO cells, the IEF profiles mirrored the SNA-blot results as there was no significant band intensity variation among the samples from day 3 to day 7. TZM and TZMm treated overnight with α2,3-6 neuraminidase (*Arthrobacter ureafaciens*) at 37°C exhibited the same pattern than aglycosylated TZM mutant N297A (Fig. 5C), confirming that the lower bands correspond to sialylated states of the Mabs.

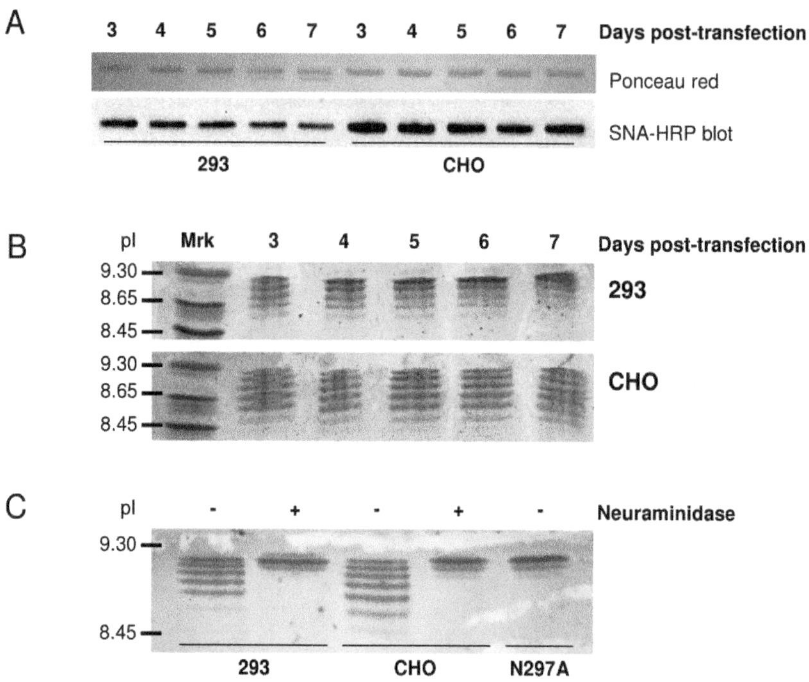

Figure 5. Sialylation kinetic of TZMm in 293 and CHO cells. Panel A –TZMm was co-expressed with ST6Gal-I, harvested from day 3 to day 7 and purified by protein-A chromatography. 300ng were resolved by SDS-PAGE (reducing condition), transferred to nitrocellulose membrane and probed with SNA-HRP. Only the heavy chain is shown. The membrane was stained with Ponceau red to monitor protein loads. Panel B – IEF profiles of purified TZMm co-expressed with ST6Gal-I and harvested between day 3 and 7 post-transfection. Mrk: IEF calibration markers. Panel C – IEF profiles of purified TZMm from 293 and CHO cells digested or not with *A. ureafaciens* neuraminidase and of aglycosylated TZM (N297A mutant produced in 293 cells).

6.3. Impact of ST6Gal-I expression and F246A mutation on sialylation patterns

Impact of ST6Gal-I overexpression and F246A mutation on TZM sialylation

To evaluate the impact of ST6Gal-I expression on overall sialylation levels of TZM and TZMm, purified Mabs were submitted to SNA blotting and IEF analyses (Figure 6). While SNA blot (panel A) shows a significant increase in α2,6-sialylation levels of TZMm, no increase in sialylation of the wild-type antibody could be observed in 293 cells while only a faint band could be seen in CHO cells. In 293 cells, only a marginal signal with SNA lectin could be seen when TZMm was expressed alone, suggesting that endogenous levels of ST6Gal-I activity are low in this cell line. The main consequence of ST6Gal-I co-expression was the significant preference for α2,6-sialylation over α2,3-sialylation of TZMm in CHO cells. While all of the acidic TZMm species observed when the Mab was expressed alone disappeared after treatment with Sialidase S (an α2,3-specific neuraminidase), these acidic species became resistant to Sialidase S when ST6Gal-I was co-expressed. This suggests that sialylation of TZMm in the presence of ST6Gal-I was mostly due to the preferential addition of α2,6- over α2,3-sialic acid residues.

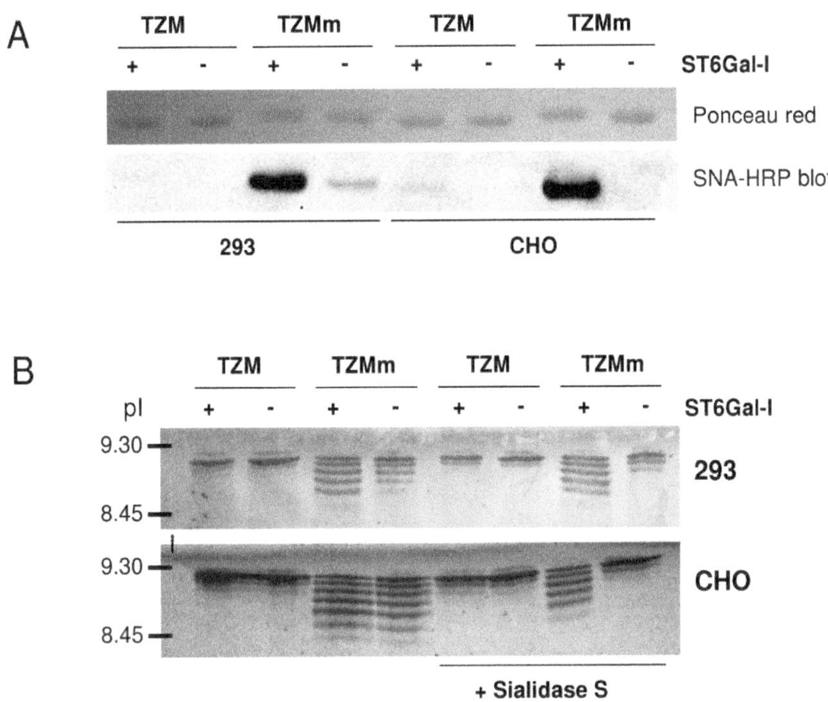

Figure 6. Impact of ST6Gal-I expression and F246A mutation on TZM sialylation. Transfected CHO and 293 cultures (with or without ST6Gal-I) were harvested on day 3 and day 4, respectively and Mabs were purified by protein A chromatography. Panel A – Mabs (300ng per lane) were resolved by SDS-PAGE (reducing conditions) and submitted to SNA-blotting (only the heavy chains are shown). Panel B – IEF analyses of Mabs from 293 and CHO cells treated or not with Sialidase S (α2,3-neuraminidase).

Glycan analysis by MALDI-MS spectrometry

After PNGaseF digestion of the purified Mabs, oligosaccharide detection was firstly achieved by MALDI-MS directly after non-reductive on-target derivatization with phenylhydrazine (PHN) without using any purification procedure. Despite a presence of side products and salts coming from original buffer, which generally have a tendency to suppress signal peaks corresponding to glycans, all samples provided good evidence for the presence of N-glycans.

These N-glycans were known to be located on the Fc fragment as Asparagine 297 is the only glycosylation site in our model antibody. The profile differences between the samples in CHO and 293 were consistent with what was observed by SNA-blot and IEF. The dominant oligosaccharides derived from all Mab samples were observed in positive mode in PHN-derivatized samples at *m/z* 1575, 1737 and were consistent with biantennary core-fucosylated structures with zero and one galactose. Glycan having both antennae occupied with galactoses (*m/z* 1899) was observed with higher intensity only in TZMm samples without ST6Gal-I.

The spectra recorded in the negative mode from TZM samples co-expressed with ST6Gal-I and derivatized with PHN, provided evidence for the presence of small amount of monosialylated glycans structure - $NeuAcGal_{1\text{-}2}GlcNAc_2Man_3GlcNAc_2Fuc$ (Figure 7A & 7E).

Sialylated glycans were clearly detected at higher levels in TZMm samples. TZMm samples co-expressed with ST6Gal-I in both cell lines (Fig. 7C & 7G) indicated higher sialylation degree than that obtained under control conditions (Fig. 7D & 7H).

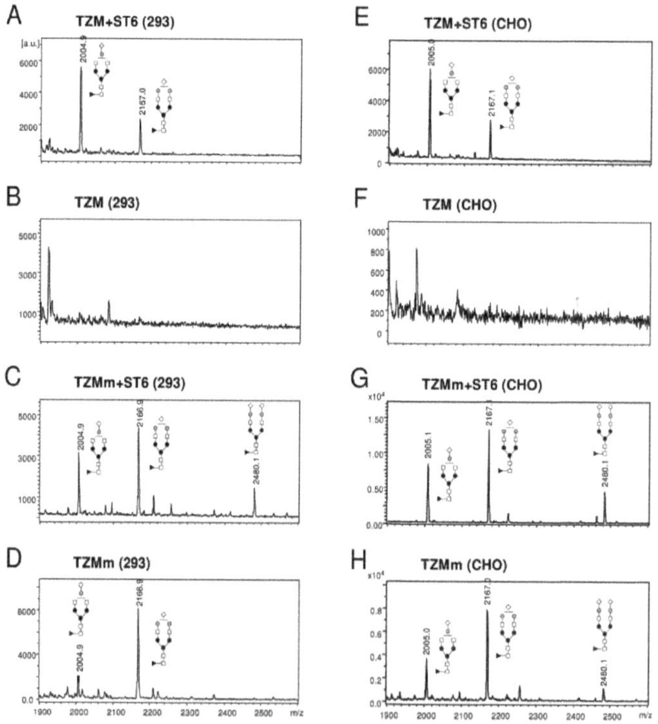

Figure 7. Glycan analysis by MALDI-MS in negative mode. Negative MALDI-MS spectra of N-Glycans obtained after PNGase F digestion and on-target derivatization with PHN (+90 Da) of MAbs from 293 cells (panels A to D) and CHO cells (panels E to H). Monosialylated glycans (m/z at 2005 and 2167) are detected as $[M-H]^-$ ions.

The high level of sialylated glycans was even better observed in the fractions obtained after purification prior to derivatization, as shown in Figure 8 which represents MS spectra recorded from TZMm samples produced in 293 cells. Interestingly, this analysis revealed traces of triantennary glycans in TZMm that cannot naturally occur in wild-type IgG1, supporting the vision of an open Fc conformation presenting reduced space constraint to the glycosyltransferases.

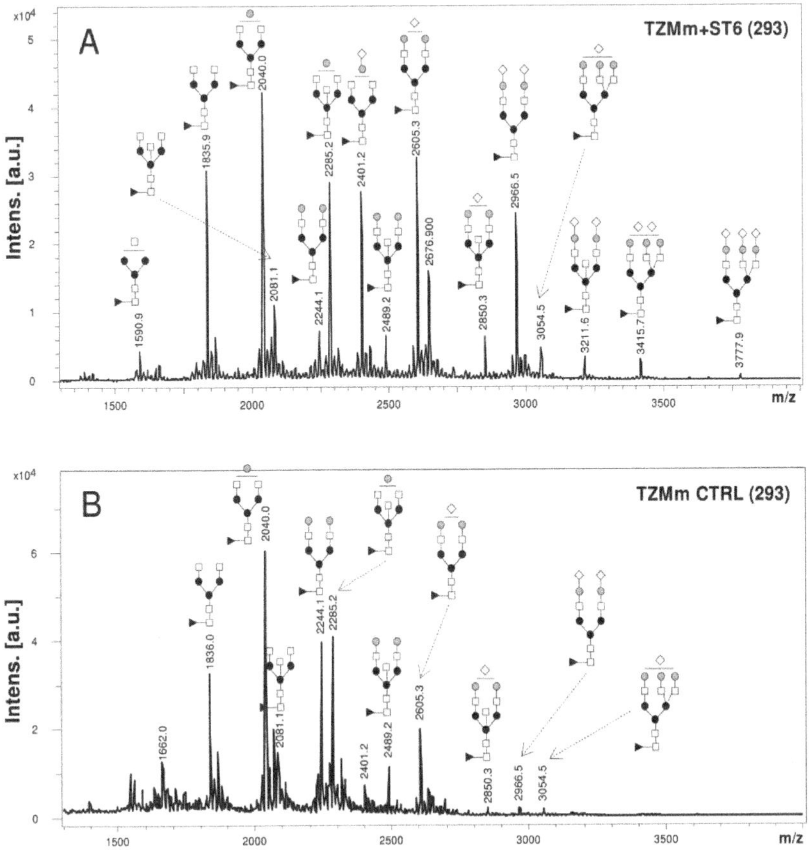

Figure 8. Glycan analysis by MALDI-MS in positive mode. Positive MALDI-MS spectra of N-glycans recorded after glycans purification and permethylation from PNGase digested TZMm produced in 293 cells. All glycans are as $[M+Na]^+$.

Intact mass analysis of antibodies by LC-ESI-MS using a Q-TOF2 spectrometer

The intact mass analysis approach allowed us to see the outcome of the pairing of the glycans on the overall glycosylation of the antibodies, and to distinguish with a high resolution between galactosylated and sialylated Mabs glycoforms. The prevalence of agalactosylated and monogalactosylated glycoforms in TZM+/-ST6 samples in both cell lines was consistent with the glycan analysis by MALDI-MS (Fig. 9A). No trace of sialylated

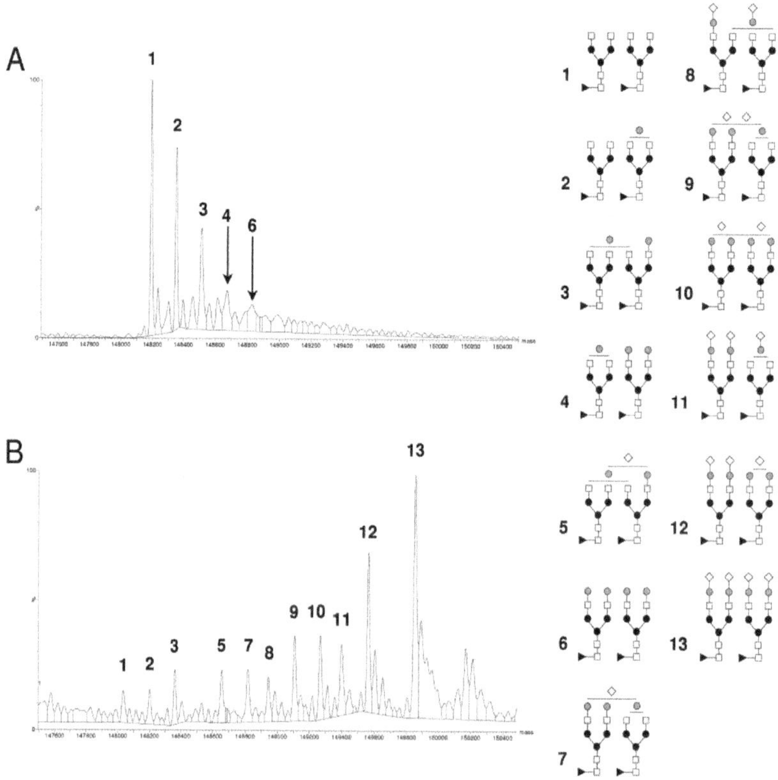

Figure 9. Intact antibody analysis of Mabs by by LC-ESI-MS. Purified Mabs produced in CHO cells were analyzed by LC-ESI-MS on a Q-TOF2 spectrometer as described in Materials and Methods. Panel A: TZM+ST6; panel B: TZMm+ST6.

antibodies was detected. The close resemblance of the profiles of TZM produced in the presence or absence of ST6Gal-I confirmed the poor sialylation efficacy of the wild-type antibody. In contrast, the association of the Fc F246A mutation with the expression of ST6Gal-I in CHO cells resulted in a massive conversion of the neutral Mabs into a variety of mono-, di-, tri- and tetrasialylated antibodies, where the tetrasialylated version was predominant (Fig. 9B). The combination of the two approaches in 293 cells, as well as the Fc mutation alone in both cell lines provided intermediate results (not shown).

7. Discussion

IVIg are successfully used for the treatment of an increasing number of autoimmune and inflammatory disorders. Their efficacy results from several distinct mechanisms working together in a complex fashion that is not yet understood. Anti-inflammatory mechanisms of IVIg may involve distinct pathways via FcRn or FcγRIIb receptors. Kaneko *et al.* highlighted the implication of the Fc α2,6-sialylated fraction of IVIg in a murine model of rheumatoid arthritis [17]. The production of sialylated IgGs with potentially enhanced specificity and efficacy may provide a safe alternative to IVIg and could also compensate for possible IVIg shortage. The ability to generate highly sialylated IgGs could also help in gaining a better understanding of their mechanism of action.

Several approaches are used to enhance the sialylation level of recombinant proteins produced in mammalian cell cultures, such as cell line engineering, bioprocess control, post-production enrichment (such as lectin-affinity purification) and *in vitro* glycan remodelling [27, 40]. In the field of mammalian cell engineering, a variety of approaches were developed. Antisense RNA and short interfering RNA were used in CHO cell lines to knock-down genes encoding for sialidases, leading to a significant increase in total amount of sialic acid in recombinant proteins [39, 41]. Smaller but still significant improvement was achieved by the overexpression of CMP-sialic acid transporter, either alone [42] or in combination with CMP-sialic acid substrate [43]. Among the variety of approaches, the expression of the ST6Gal-I gene remains the only way to introduce α2,6-linked sialic acids on proteins in CHO cells.

The results obtained in this study were very similar to those of Jassal et al. on a monoclonal antibody of the IgG3 subtype [26]. The replacement of phenylalanine 246, equivalent to F243 in IgG3, by an alanine residue was necessary to reach significant sialylation levels, and the impact of this mutation was much stronger than that of human ST6Gal-I over-expression in both 293 and CHO cell lines. This confirms that IgG1 Fc structure strongly limits glycan accessibility to the sialyltransferases. The intact mass analysis profile of the mutant co-expressed with ST6Gal-I in CHO was dominated by the fully sialylated Mab, followed by tri-, di- and monosialylated forms.

Replacing the phenylalanine with an alanine residue decreases the hydrophobic environment that maintains the glycan buried in the Fc. Multiple mutation strategies are a powerful mean to promote glycosylation enhancement and Fc affinity for FcγRs. Computational modeling allows the design of antibodies with amino acid mutations that are not even exposed on the surface of the glycan pocket. However, the accumulation of mutations in a therapeutic Mab can have unpredictable side effects [44]. In the first place, mutation may compromise the affinity of the Mab for potential receptors involved in the anti-inflammatory mechanism. Also, non-natural Mab glycan structures (like sialylated tri-antennary glycans) may reduce their biological activity or even render them immunogenic. In a recent study, the CHO ST6Gal-I gene, present but not expressed, was cloned and stably expressed in these cells [45]. As a result, 70% of the glycans of a stably expressed IgG1 and released by PNGaseF were sialylated. This striking result was thought to be conferred by a higher propensity of the CHO ST6Gal-I gene product for Fc glycan compared to its human or rat homologues, possibly due to some mutations in the substrate binding region. The use of this particular CHO ST6Gal-I gene may provide an interesting alternative strategy independent any IgG mutations to enhance sialylation of Mabs.

The production of highly sialylated IgGs in large quantities will probably be achieved by the simultaneous use of cell line engineering, Fc engineering, bioprocess control and downstream processing. Alternatively, optimizing *in vitro* glycan remodeling using ST6Gal-I and CMP-sialic acid may represent another cost-effective way to produce industrial amounts of sialylated IgGs [12]. From the perspective of obtaining a safe and reliable alternative to IVIg for anti-inflammatory applications, the production of sialylated recombinant Fc could be an attractive approach. However, developing these molecules as therapeutics awaits a better understanding of their mode of action.

Author details

Céline Raymond and Yves Durocher
Département de biochimie, Université de Montréal, Montréal, Canada

Céline Raymond, Erika Lattová, Hélène Perreault and Yves Durocher
MabNet NSERC Network for the Manufacturing of Single-Type Glycoform Monoclonal Antibody

Anna Robotham, John Kelly and Yves Durocher
Human Health Therapeutic Portfolio of the National Research Council Canada, Montreal and Ottawa, Canada

Erika Lattová and Hélène Perreault
Chemistry Department, University of Manitoba, Winnipeg, Canada

Acknowledgement

We thank Louis Bisson for protein-A HPLC quantifications.

8. References

[1] Kozlowski, S. and P. Swann, Current and future issues in the manufacturing and development of monoclonal antibodies. Advanced drug delivery reviews, 2006. 58: p. 707-22.

[2] Munro, T.P., et al., Bridging the gap: facilities and technologies for development of early stage therapeutic mAb candidates. mAbs, 2011. 3: p. 440-52.

[3] Carter, P.J., Introduction to current and future protein therapeutics: a protein engineering perspective. Experimental cell research, 2011. 317: p. 1261-9.

[4] Reichert, J., M, Marketed therapeutic antibodies compendium. Mabs, 2012. 4(3): p. 413-415.

[5] Lund, J., et al., Oligosaccharide-protein interactions in IgG can modulate recognition by Fc gamma receptors. FASEB J, 1995. 9(1): p. 115-9.

[6] Raju, T.S., Terminal sugars of Fc glycans influence antibody effector functions of IgGs. Curr Opin Immunol, 2008. 20(4): p. 471-8.

[7] Mizuochi, T., et al., Structural and numerical variations of the carbohydrate moiety of immunoglobulin G. The Journal of Immunology, 1982. 129: p. 2016-2020.

[8] Wormald, M.R., et al., Variations in oligosaccharide-protein interactions in immunoglobulin G determine the site-specific glycosylation profiles and modulate the dynamic motion of the Fc oligosaccharides. Biochemistry, 1997. 36: p. 1370-80.

[9] Deisenhofer, J., Crystallographic refinement and atomic models of a human Fc fragment and its complex with fragment B of protein A from Staphylococcus aureus at 2.9- and 2.8-A resolution. Biochemistry, 1981. 20: p. 2361-2370.

[10] Jefferis, R., et al., A comparative study of the N-linked oligosaccharide structures of human IgG subclass proteins. 1990. 268: p. 529-537.

[11] Krapp, S., et al., Structural analysis of human IgG-Fc glycoforms reveals a correlation between glycosylation and structural integrity. J Mol Biol, 2003. 325(5): p. 979-989.

[12] Barb, A.W., E.K. Brady, and J.H. Prestegard, Branch-specific sialylation of IgG-Fc glycans by ST6Gal-I. Biochemistry, 2009. 48(41): p. 9705-7.

[13] Morell, G., G. Gregoriadis, and I.H. Scheinberg, The Role of Sialic Acid in Determining Glycoproteins in the Circulation. The Journal of biological chemistry, 1971. 246: p. 1461-1467.

[14] Park, E.I., et al., The asialoglycoprotein receptor clears glycoconjugates terminating with sialic acid alpha 2,6GalNAc. Proc Natl Acad Sci U S A, 2005. 102(47): p. 17125-9.

[15] Unverzagt, C., et al., Structure-activity profiles of complex biantennary glycans with core fucosylation and with/without additional alpha 2,3/alpha 2,6 sialylation: synthesis of neoglycoproteins and their properties in lectin assays, cell binding, and organ uptake. J Med Chem, 2002. 45(2): p. 478-91.

[16] Millward, T., et al., Effect of constant and variable domain glycosylation on pharmacokinetics of therapeutic antibodies in mice. Biologicals : journal of the International Association of Biological Standardization, 2008. 36: p. 41-7.

[17] Kaneko, Y., F. Nimmerjahn, and J.V. Ravetch, Anti-inflammatory activity of immunoglobulin G resulting from Fc sialylation. Science, 2006. 313(5787): p. 670-673.

[18] Samuelsson, a., T.L. Towers, and J.V. Ravetch, Anti-inflammatory activity of IVIG mediated through the inhibitory Fc receptor. Science (New York, N.Y.), 2001. 291: p. 484-6.

[19] Anthony, R.M., et al., Identification of a receptor required for the anti-inflammatory activity of IVIG. Proc Natl Acad Sci U S A, 2008. 105(50): p. 19571-8.

[20] Siragam, V., et al., Intravenous immunoglobulin ameliorates ITP via activating Fc gamma receptors on dendritic cells. Nature Medicine, 2006. 12: p. 688-692.

[21] Crow, A.R., D. Brinc, and A.H. Lazarus, New insight into the mechanism of action of IVIg: the role of dendritic cells. Journal of thrombosis and haemostasis, 2009. 7 Suppl 1: p. 245-248.

[22] van de Geijn, F.E., et al., Immunoglobulin G galactosylation and sialylation are associated with pregnancy-induced improvement of rheumatoid arthritis and the postpartum flare: results from a large prospective cohort study. Arthritis research & therapy, 2009. 11: p. R193.

[23] Patel, D.a., et al., Neonatal Fc receptor blockade by Fc engineering ameliorates arthritis in a murine model. Journal of immunology, 2011. 187: p. 1015-22.

[24] Ishii, N., et al., High-dose intravenous immunoglobulin (IVIG) therapy in autoimmune skin blistering diseases. Clinical reviews in allergy & immunology, 2010. 38(2-3): p. 186-95.

[25] Siberil, S., et al., Intravenous immunoglobulins in autoimmune and inflammatory diseases: a mechanistic perspective. Annals of the New York Academy of Sciences, 2007. 1110: p. 497-506.

[26] Jassal, R., et al., Sialylation of human IgG-Fc carbohydrate by transfected rat alpha2,6-sialyltransferase. Biochem Biophys Res Commun, 2001. 286(2): p. 243-9.

[27] Durocher, Y. and M. Butler, Expression systems for therapeutic glycoprotein production. Curr Opin Biotechnol, 2009. 20(6): p. 700-7.

[28] Pham, P.L., A. Kamen, and Y. Durocher, Large-scale transfection of mammalian cells for the fast production of recombinant protein. Mol Biotechnol, 2006. 34(2): p. 225-37.

[29] Geisse, S. and C. Fux, Recombinant protein production by transient gene transfer into Mammalian cells. Methods in enzymology, 2009. 463: p. 223-38.

[30] Rajendra, Y., et al., A simple high-yielding process for transient gene expression in CHO cells. Journal of biotechnology, 2011. 153(1-2): p. 22-6.

[31] Tom, R., L. Bisson, and Y. Durocher, Culture of HEK293-EBNA1 Cells for Production of Recombinant Proteins. Cold Spring Harbor Protocols, 2008. 3(4): p. pdb.prot4976.

[32] Shi, C., et al., Purification and Characterization of a Recombinant G-Protein-Coupled Receptor , Saccharomyces cerevisiae Step 2p, Transiently Expressed in HEK293 EBNA1 Cells. Biochemistry, 2005. 44: p. 15705-15714.

[33] Zhang, J., et al., Transient expression and purification of chimeric heavy chain antibodies. Protein Expr Purif, 2009. 65(1): p. 77-82.

[34] Loignon, M., et al., Stable high volumetric production of glycosylated human recombinant IFNalpha2b in HEK293 cells. BMC Biotechnol, 2008. 8: p. 65.

[35] Durocher, Y., S. Perret, and A. Kamen, High-level and high-throughput recombinant protein production by transient transfection of suspension-growing human 293-EBNA1 cells. Nucleic Acids Res., 2002. 30(2): p. E9.

[36] Lattová, E., O. Krokhin, and H. Perreault, Matrix-assisted laser desorption/ionization tandem mass spectrometry and post-source decay fragmentation study of phenylhydrazones of N-linked oligosaccharides from ovalbumin. J. Am. Chem. Soc. Mass Spectrom., 2004. 15: p.725-735.

[37] Yang, M. and M. Butler, Effects of ammonia and glucosamine on the heterogeneity of erythropoietin glycoforms. Biotechnol.Prog., 2002. 18(1): p. 129-138.

[38] Gawlitzek, M., et al., Ammonium alters N-glycan structures of recombinant TNFR-IgG: degradative versus biosynthetic mechanisms. Biotechnol.Bioeng., 2000. 68(6): p. 637-646.

[39] Ferrari, J., et al., Chinese hamster ovary cells with constitutively expressed sialidase antisense RNA produce recombinant DNase in batch culture with increased sialic acid. Biotechnology and Bioengineering, 1998. 60: p. 589-595.

[40] Raju, T.S., et al., Glycoengineering of therapeutic glycoproteins: in vitro galactosylation and sialylation of glycoproteins with terminal N-acetylglucosamine and galactose residues. Biochemistry, 2001. 40(30): p. 8868-76.

[41] Ngantung, F.A., et al., RNA interference of sialidase improves glycoprotein sialic acid content consistency. Biotechnology and Bioengineering, 2006. 95(1): p. 106-19.

[42] Wong, N.S., M.G. Yap, and D.I. Wang, Enhancing recombinant glycoprotein sialylation through CMP-sialic acid transporter over expression in Chinese hamster ovary cells. Biotechnol.Bioeng., 2006. 93(5): p. 1005-1016.

[43] Jeong, Y.T., et al., Enhanced sialylation of recombinant erythropoietin in genetically engineered Chinese-hamster ovary cells. Biotechnol Appl Biochem, 2009. 52(Pt 4): p. 283-91.

[44] Kaneko, E. and R. Niwa, Optimizing therapeutic antibody function: progress with Fc domain engineering. BioDrugs : clinical immunotherapeutics, biopharmaceuticals and gene therapy, 2011. 25: p. 1-11.

[45] Onitsuka, M., et al., Enhancement of sialylation on humanized IgG-like bispecific antibody by overexpression of α2,6-sialyltransferase derived from Chinese hamster ovary cells. Applied microbiology and biotechnology, 2012. 94: p. 69-80.

[46] Raymond, C., et al., A simplified polyethylenimine-mediated transfection process for large-scale and high-throughput applications. Methods, 2011. 55(1): p.44-51.

[47] Ciucanu, I., and F. Kerek, A simple method for the permethylation of carbohydrates. Carbohydr. Res. 1984; 131: p. 209-217.
[48] Schlaeger, E., and K. Christensen, Transient gene expression in mammalian cells grown in serum-free suspension culture. Cytotechnology, 1999, 30(1-3): p.71-83.

Chapter 17

Glycoengineered Yeast as an Alternative Monoclonal Antibody Discovery and Production Platform

Dongxing Zha

Additional information is available at the end of the chapter

1. Introduction

Biologic pharmaceuticals are gaining in both market share and clinical utility compared with small molecule therapeutics (Projan et al., 2004). Global biologic drug sales reached $93 billion in 2009 and the sales are expected to grow at least twice as fast as those of small molecules (McCamish and Woollett, 2011). The rapid market growth and the promise of successful rate of developing biologic drugs has drawn the attention of traditional small molecule pharma into the biotechnology business. Today, more than ever, large pharmaceutical companies are venturing into the biotechnology arena with the hope that novel therapeutic proteins will augment the traditional pharmaceutical business enough to fundamentally reshape the market landscape. The acquisition/merging of pipeline and research capacity of biotech companies by big pharma is a great example of this trend. Several companies are even projecting optimistically that, within the decade, therapeutic biologics will comprise a majority of their commercial portfolios (Zhou, 2007). Among highly successful biologic products, monoclonal antibodies (mAbs) are the largest and fastest growing segment. MAbs are established as a key therapeutic modality for a range of diseases most notably rheumatoid diseases and various cancers. Due to the high degree of selectivity of these agents, in particular for cancer targets, they can be designed to selectively target tumor cells and elicit a variety of responses. These agents can kill cells directly by carrying a toxic payload to the target or can orchestrate the destruction of cells by activating immune system components, blocking receptors or sequestering growth factors (Nicolaides et al., 2010).

It is well known that glycosylation can impact the pharmacokinetics, efficacy and tissue targeting of therapeutic proteins (Li and d'Anjou, 2009). N-glycosylation of immunoglobulin

G (IgG) at asparagine residue 297 plays a role in antibody stability and is required for immune cell-mediated Fc effector function (Correia, 2010; Kayser et al., 2011; Mimura et al., 2000). Preclinical and clinical studies have demonstrated that modulating the glycosylation of antibody and non-antibody therapeutics can be an effective means to improve the properties of biologic medicines leading to a class of drugs termed "biobetters"(Jefferis, 2009; Walsh, 2010). The strategy of engineering expression hosts to express glycoproteins with "optimized" glycosylation has been applied in both prokaryotic and eukaryotic cells (Beck et al., 2010; Hamilton et al., 2006; Hamilton and Gerngross, 2007; Jacobs and Callewaert, 2009b; Pandhal and Wright, 2010; Tomiya, 2009). Chinese Hamster Ovary (CHO) cells, a widely used host for producing therapeutic antibodies with similar human glycosylation, have been engineered in multiple ways to eliminate the core fucose on secreted mAbs. Recent preclinical and clinical studies have reported superior efficacy utilizing afucosylated mAbs(Herbst et al., 2011; Junttila et al., 2010; von Horsten et al., 2010; Ward et al., 2011; Wong et al., 2010). Non-mammalian expression hosts have also been utilized for producing afucosylated mAbs including insect cells, plants and yeast (Gasdaska et al., 2007; Barbin et al., 2006; Zhang et al., 2011).

Following marketing of the first therapeutic antibody, Muromonab-CD3 in 1986, mAbs used in the clinic have evolved over the years from entirely murine to mouse-human chimeric, and then to humanized and finally fully human antibodies in order to minimize anti-drug related immunogenicity in patients and maintain maximum potency (Li and Zhu, 2010). Several different antibody discovery technologies co-exist today in therapeutic antibody development ranging from isolating antibodies from immunized mice or engineered mice carrying human Ig-repertoire genes to flow-cytometric isolation of human antibodies from non-immune yeast display or panning a large non-immunized phage display libraries (Feldhaus et al., 2003; Vaughan et al., 1996; Weaver-Feldhaus et al., 2004). Recently *de novo* computer designed and synthetic antibody libraries based on a human Ig-repertoire and coupled with yeast or phage display have attracted considerable attention (Knappik et al., 2000). However, one key issue that is not solved by current antibody discovery technologies, is the need to switch expression hosts from antibody discovery (E. coli, yeast) to production host (often CHO or other mammalian cell lines). Moreover, for many current mAb discovery technologies reformatting steps from scFv/Fab to full-length antibody are required. These additional steps increase cycle times and often decrease the probabilities of success both of which have negative impacts on development timeline (Lin et al., 2012).

Yeast has been widely used for expressing proteins in research and development. There are multiple advantages to using yeast as an expression system for therapeutic glycoprotein production, including ease of genetic manipulation, stable expression, rapid cell growth, high yield of secreted protein, low-cost scalable fermentation processes and no risk of human pathogenic virus contamination. However, glycoproteins expressed in wild type yeast generally cannot be used for therapeutic applications due to fungal type high-mannose glycans which can result in immunogenicity and poor PK *in vivo* (Zhang et al., 2011). Humanization of the N-glycosylation pathways of the yeast *Pichia pastoris* has been achieved by eliminating fungal genes responsible for adding high-mannose and concomitantly reconstituting the canonical human pathway (Hopkins et al., 2011; Jiang et al., 2011; Li et al., 2006; Liu et al., 2011; Potgieter et al., 2010; Ye et al., 2011b). This result of the engineering is a platform that allows not only yeast-based production of therapeutic glycoproteins with human glycosylation, but also provides a versatile tool for glycosylation-based structure-activity-relationship (SAR) studies to optimize therapeutic proteins for better efficacy, PK and drugability (Li et al., 2006; Nett et al., 2012; Zhang et al., 2011). In addition, development of antibody surface display on glycoengineered yeast strain as an antibody discovery tool facilitates the earliest stage of antibody discovery and development in the same expression host which is expected to have a positive impact on cycle times and probabilities of success (Lin et al., 2012).

2. Glycosylation in therapeutic proteins is important for its efficacy, PK, tissue targeting

2.1. Glycosylation on Erythropoietin affects its potency

Recombinant human Erythropoietin (rHuEPO) is a 30.4 kDa glycoprotein hormone containing multiple N-linked glycosylation sites and currently is used to treat patients with anemia. The marketed forms of recombinant erythropoietin include Epogen with three native N-glycan structures and Aranesp® (darbepoetin), an epoetin engineered to contain two additional N-glycosylation sites, conferring greater metabolic stability *in vivo*. Typical mammalian CHO cell-produced epoetin are secreted with a heterogeneous mixture of sialylated N-glycan structures. Usually the manufacturing process is controlled to enrich for the tetra-antennary sialylated glycoforms, which along with tri-antennary forms, are required for maximum *in vivo* efficacy (Egrie and Browne, 2001). However, in cell-based and receptor binding assays, tri- and tetra-sialylated erythropoietin and darbepoetin exhibit decreased potency relative to bi-antennary sialylated erythropoietin (Takeuchi et al., 1989). This paradox can be explained by the extended serum half-lives of the tetra-antennary sialylated erythropoietin and darbepoetin compared with the faster clearance rate of bi-antennary sialylated erythropoietin (Misaizu et al., 1995). Recently, glycoengineered *Pichia pastoris* has been used to produce rHuEPO with highly homogeneous bi-antennary structures. As expected, bi-antennary EPO produced by glycoengineered *Pichia* showed significantly more potent *in vitro* activity compared to predominantly tri- and tetra-antennary EPO, such as Aranesp®. The faster clearance rate of bi-antennary rHuEPO *in vivo* was compensated with Peglylation to elongate its half life in this study. The authors concluded overall *in vivo* activity of this novel glycoengineered molecule is comparable *in vitro* and *in vivo* characteristics to its counterparts produced in mammalian systems (Nett et al., 2012).

2.2. Glycosylation on recombinant human glucocerebrosidase is critical for its targeting in enzyme replacement therapy

Gaucher's disease is a lysosomal storage disorder caused by mutations of glucocerebrosidase (GCD), and an enzyme replacement treatment has been developed using recombinant GCD. GCD is glycoprotein and its glycosylation plays an important role in its targeting in therapeutic setting. GCD produced in Chinese Hamster ovary (CHO) cells contains major complex glycan but it has failed to provide clinical benefit in direct infusion due to the preferential uptake of enzyme by hepatocytes rather than Kupffer cells. An *in vitro* glycan modification is required in order to expose the mannose residues on the glycans of Cerezyme (commercialized therapeutic GCD)(Bergh et al., 1990;Brumshtein et al., 2010;Pastores, 2010). Cerezyme production involves sequential *in vitro* deglycosylation, using α-neuraminidase, β-galactosidase and β-N-acetylglucosaminidase, to expose terminal mannose residues, a procedure which dramatically improves targeting and internalization. Recently the recombinant plant-derived GCD (prGCD) is targeted to the storage vacuoles, using a plant-specific C-terminal sorting signal. Notably, the recombinant human GCD expressed in the carrot cells naturally contains terminal mannose residues on its glycans, apparently as a result of the activity of a special vacuolar enzyme that modifies complex glycans(Shaaltiel et al., 2007). Hence, the plant-produced recombinant human GCD does not require exposure of mannose residues through *in vitro* enzymatic modification. Interestingly, the N-glycosylation pathway of H. polymorpha has been remodeled by deleting the HpALG3 gene in the Hpoch1 null mutant strain and blocked in the yeast-specific outer mannose chain synthesis and by introducing an ER-targeted Aspergillus saitoi α-1,2-mannosidase gene. This glycoengineered H. polymorpha strain produced glycoproteins mainly containing trimannosyl core N-glycan ($Man_3GlcNAc_2$), which is the common core backbone of various human-type N-glycans and the glycoform is the same as that of Cerezyme which was achieved through *in vitro* sequential deglycosylation of complex glycans (Oh et al., 2008). Similar glycoengineering effort was published in a glycoengineered *Pichia* with the ability of producing glycoprotein carrying similar mannose type of glycan (Davidson et al., 2004).

2.3. Afucosylated antibody has enhanced ADCC and can translate into better efficacy

Antibody Dependent Cell-Mediated Cytotoxicity (ADCC) is a mechanism of cell mediated immune defense whereby an effector cell of the immune system actively lyses a target cell that has been bound by specific antibodies. The typical ADCC involves activation of effector cells such as Natural Killer (NK) cells by antibodies. An NK cell expresses Fc receptor IIIa or CD16a (Nimmerjahn and Ravetch, 2007; Nimmerjahn and Ravetch, 2008), and this receptor recognizes and binds to the Fc portion of an antibody which it has already bound to the surface of a pathogen-infected target cell or tumor cell. The NK cell releases cytokines such as IFN-γ, and cytotoxic granules containing perforin and granzymes that enter the target cell and promote cell death by triggering apoptosis. Monoclonal antibody IgG1 Fc has single N-linked site at N-297 and its glycosylation is involved with antibody and Fc γ receptor III binding. Antibodies lacking core fucosylation show a large increase in affinity for Fc γ RIIIa leading to an improved receptor-mediated effector function. The structure study reveals that a unique type of interface consisting of carbohydrate-carbohydrate interactions between glycans of the receptor and the afucosylated Fc. In contrast, in the complex structure with fucosylated Fc, these contacts are weakened or nonexistent (Ferrara et al., 2011). Although afucosylated IgGs exist naturally, a next generation of recombinant therapeutic, glycoenginereed antibodies with enhanced ADCC activity is currently being developed for better efficacy(Mori et al., 2007; Ysebaert et al., 2010; Robak and Robak, 2011).

2.4. Glycans at Fc region impact on its pharmacokinetics

Glycosylation at mAb Fc regions not only plays a role in its effector functions, but it also can impact its pharmacokinetics. The neonatal Fc receptor (FcRn) has a major role in prolonging the exposure of therapeutic mAbs. MAbs internalized by fluid-phase or receptor-mediated endocytosis can be redirected to the cell surface by FcRn mediated recycling and released into plasma or interstitial fluids thus preventing lysosomal degradation(Chowdhury and Wu, 2005; Roopenian and Akilesh, 2007). This recycling dramatically increases systemic exposure to therapeutic mAbs by studies comparing in wild type and FcRn-deficient mice which demonstrated over an order of magnitude increase in IgG half-life as a consequence of the FcRn salvage pathway. Another mechanism for mAb clearance may be through Fc γ receptor (FcγR) binding on some immune effector cells. Interactions with Fc γ receptors can be influenced by the glycosylation pattern. To produce mAbs in lower eukaryotic hosts such as yeast have been reported, however, wild type yeast host expressed antibody secretes full length mAbs with hyper-mannose type glycans. MAb produced in wild type yeast exhibited 2 to 3-fold faster clearance, which might be due to the high mannose content interacting with mannose receptors. On the other hand, *in vitro* binding affinity to human FcRn or mouse FcRn was similar between the mAb produced in CHO cell line and mAbs produced in glycoengineered yeast, and the glycovariants produced in glycoengineered yeast exhibited similar PK patterns in human FcRn transgenic mice and in wild type mice (Liu et al., 2011).

3. Glycoengineered *Pichia* expresses protein with human glycosylation

3.1. Difference of glycosylation pathway between human and yeast

Yeast N-glycosylation is of the high-mannose type, which differs from human complex glycans. Fungal type of glycans confers a short half-life *in vivo* and thereby compromises efficacy of most therapeutic glycoproteins. In addition non-human types of glycans can cause immunogenicity and therefore inactive the drug substance and even more it may cast safety concerns (Jacobs and Callewaert, 2009a; Sinclair and Elliott, 2005; Sola and Griebenow, 2009). However, the initiation of yeast N-glycosylation is similar to that of human. In ER, a core oligosaccharide (Glc3Man9GlcNAc2) is transferred onto the nascent polypeptide. Following the transfer of the core oligosaccharide to the asparagine residue within the Asn-X-Ser/Thr motif, three glucose moieties and one terminal α-1,2-mannose moiety are removed by glucosidase I and glucosidase II, and an ER-residing α-1,2-mannosidase, respectively. The resulting Man8GlcNAc2-containing glycoprotein is then transported to the Golgi apparatus where N-glycan processing differs markedly between yeast and human. In human and mammals, early Golgi N-glycan processing involves the trimming of Man8GlcNAc2 to Man5GlcNAc2 by α-1,2-mannosidase(s) (MNS1), a process that generates the substrate for *N*-acetylglucosaminyl transferase I (GNT1), which transfers a single *N*-acetylglucosamine (GlcNAc) sugar onto the terminal 1,3-mannose of the tri-mannose core. Following this transfer, mannosidase II (MNS II) removes the two remaining α-1,3- and α-1,6 terminal mannose sugars to produce GlcNAcMan3GlcNAc2.This is the substrate for N-acetylglucosaminyl transferase II (GNT II), which adds one GlcNAc sugar to the terminal α-1,6-mannose arm of the tri-mannose core. Further processing typically involves the attachment of additional GlcNAc, galactose and sialic acid, such as N-acetylneuraminic acid (NANA) moieties (Wildt and Gerngross, 2005). However, a host of additional glycosyltransferases, including fucosyltransferases, GalNAc transferases and GlcNAc phosphotransferases, are known to exist, which further broadens the range of *N*-glycans found on proteins isolated from human sources. In contrast to human N-glycan processing, which involves the removal of mannose followed by the addition of GlcNAc, galactose, fucose and NANA, early *N*-glycan processing in yeast is limited to the addition of mannose and mannosylphosphate sugars(Hamilton et al., 2003; Wildt and Gerngross, 2005). The Golgi apparatus of *S. cerevisiae* contains α-1,2-,α-1,3- and α-1,6-mannosyltransferases as well as mannosylphosphate transferases, which produce N-glycan structures that are mannosylated and hypermannosylated to varying extents. In other yeast such as *K. lactis, H. polymorpha* and *P. pastoris*, a similar set of mannosyltransferases exists, resulting in the production of mostly high-mannose structures that resemble those produced in *S. cerevisiae (Wildt and Gerngross, 2005).*

3.2. Genetic manipulation of *Pichia* glycosylation pathway

The process involved eliminating endogenous yeast glycosylation pathways, and we introduced more than 14 heterologous genes into *Pichia pastoris*, allowing us to replicate the sequential steps of human glycosylation. The enzyme OCH1, an α-1,6 mannosyltransferase adds the first mannose onto the α-1,3 branch of the trimannose core leading to an α-1, 6 extension to initiate the outer-chain, then additional mannoses are transferred into the new created mannose substrate by other mannose transferases and results in hypermannosylation. To knockout of the OCH1 gene in yeast is the critical step to prevent hypermannosylation and then generate a predominately single glycoform Man8GlcNAc2. Besides OCH1 gene, other mannosylphosphate transferases and β-mannosyltransferases can further modify the glycans at secreted proteins into fungal type of glycosylation, and to eliminate these genes is an essential step of engineering the yeast glycosylation pathways into human like. After eliminating yeast enzymes specific for fungal glycosylation modification, an efficient high throughput approach was developed by screening combinatorial libraries of fusing catalytic domain of modification enzymes with localization leader sequence(Choi et al., 2003; Nett et al., 2011). With this approach, α-1,2 mannosidase (MNS1), *N*-acetylglucosaminyltransferase I (GNT1), mannosidase II (MNS2) and *N*-acetylglucosaminyl transferase II (GNT2) were engineered into *Pichia pastoris* and localized from early to later golgi in a sequential order. The engineered *Pichia* strain with these fungal mannosyltransferases knockouts and human or mammalian modification genes knockins is able to express glycoproteins with human type of biantennnary complex glycan with terminal GlcNAc(Gerngross, 2005; Hamilton et al., 2003). Humanization of the glycoengineered *Pichia* glycosylation pathway was continued by introducing a -1,4-galactosyl transferase (GalT) with its selected leader through the library approach mentioned above. But *Pichia pastoris* does not contain activated sugar nucleotide precursor UDP-galactose as the substrate for galatosyl transferase to add galactose to GlcNAc terminated glycans, and the problem was solved by introducing UDP-galactose epimerase. Up to this step, the glycoengineered *Pichia* is able to produce glycoproteins with terminal galactose at its glycans. These strains would be suitable to be used for producing therapeutic monoclonal antibodies (mAbs) since most commercial mAbs produced by mammalian cells with galactose topping on its glycans(Bobrowicz et al., 2004). However, sialic acid containing glycan in therapeutic glycoproteins plays a critical role in its pharmacokinetics and efficacy, to maximal the glycoengineered *Pichia* platform to produce therapeutic biologics, the efforts was advanced to the last stage by further humanization of *Pichia* glycosylation by introducing sialylation pathway. In addition properly localize sialyltransferase at the late golgi, to produce the sugar nucleotide precursor CMP-sialic acid and translocate into the late golgi is a must-have. Four genes including UDP-*N*-acetylglucosamine-2-epimerase/*N*-acetylmannosamine Kinase (GNE), N-acetylneuraminate-9-phosphate synthase (SPS), sialylate-9-P phosphatase (SPP) and CMP-sialic acid synthase (CSS) is used to convert UDP-GlcNAc to CMP-Sialic Acid in the cytosol. CMP-Sialic Acid is transported into the golgi by engineered CMP-sialic acid transporter and then transferred onto the acceptor glycan by sialyltransferase. Sialylated glycoprotein exits the secretory pathway into culture supernatant similar to wild type yeast (Hamilton et al., 2003; Hamilton et al., 2006).

4. Alternative monoclonal antibody production in glycoengineered *Pichia pastoris*

4.1. Selection of glycoengineered *Pichia* expressing mAbs

Conventional mammalian cell lines as expression host secrete glycoprotein usually containing heterogeneous glycans. Mammalian cells, such as Chinese Hamster Ovary (CHO) cells maintain the capability of adding sialic acid at its galatose terminal glycans. However, mAbs expressed in CHO have little or no sialylation at its Fc region due to the steric hindrance of the Fc structure (Nimmerjahn and Ravetch, 2010). On the other hand, glycoengineered *Pichia* provides the possibility of expressing antibody with different glycoforms in relatively more homogenous fashion (Potgieter et al., 2009). Since the engineering of the glycosylation pathway of *Pichia* was a sequential process, which makes it possible to generate a panel of hosts which express glycoprotein with different glycoforms. Each host carrys one major glycan, such as mannose type glycan of Man5GlcNAc2 or hybrid type of glycan (Potgieter et al., 2009), or GlcNAcMan5GlcNAc2(Choi et al., 2003), or complex glycans of GlcNAc2Man5GlcNAc2, Gal1GlcNAcMan3GlcNAc2 or Gal2GlcNAc2Man3GlcNAc2(Hamilton et al., 2003). In addition, hosts with capability of transferring sialic acid onto bi-antennary glycans with terminal galactoses were also available (Bobrowicz et al., 2004; Hamilton et al., 2006). With these hosts bearing different glycosylation, it actually offers a unique tool to produce the same monoclonal antibody or other non-mAb containing various glycoforms at its Fc region, which allows us to study the glycan structure and its activity relationship (SAR) for efficacy, tissue distribution and pharmacokinetics. In consideration of pharmacokinetics (PK), pharmacodynamics (PD) and reducing potential immunogenicity, glycoengineered *Pichia* provides great potential to generate mAbs with more homogeneous glycans and desired efficacy.

4.2. Difference in glycosylation profile of antibodies from CHO and glycoengineered *Pichia*

Mammalian cells, e.g. CHO produced mAbs carry N-linked carbohydrate structures with predominantly core-fucosylated asial-biantennary types with varying degrees of galactosylation. Within a given product, there are lot-to-lot variations even though manufacturing processes are tightly controlled and ensure a high degree of product consistency. Besides complex glycoforms, CHO cells expressed monoclonal antibodies still contains some percentage of Man5 type of glycans. While production of consistent and reproducible mAb glycoform profiles still remains a considerable challenge for CHO cells, variations in cell culture processes play a role in the mAb glycosylation profile. Potential variables in the cell culture physicochemical environment including culture pH, cell culture media composition, raw materials lot-to-lot variations, equipment, and process control differences are just a few examples that can potentially alter glycosylation profiles. In a case study, glycoengineered *Pichia* with the capability of generating human complex glycans has been chosen to express anti-Her2 mAb using amino acid sequence identical to Trastuzumab. The N-glycan composition of anti-HER2 mAb produced in glycoengineered *Pichia* differed from that of CHO produced counterpart primarily in the proportion of GlcNAc2Man3GlcNAc2 (G0), GlcNAc2Man3GlcNAc2Gal(G1) and GlcNAc2Man3GlcNAc2Gal2 (G2) and it was entirely devoid of fucose in the glycan structure. The mAb produced in glycoengineered *Pichia* contains a small number of O-linked single mannose glycan but O-linked glycans were rare in CHO produced Trastuzumab. In another study, using an early stage glycoengineered *Pichia* host, the glycoengineered yeast produced antibody has similar motilities on SDS-PAGE, comparable size exclusion chromatograms and antigen binding affinities compared to its CHO produced comparator but with highly uniform N-linked glycans of that type.

4.3. Bioanalytical characterization of glycoengineered *Pichia* produced antibody

A glycoengineered *Pichia* strain chosen for producing monoclonal antibody is capable of transferring terminal β-1,4 galactose onto biantennary N-linked glycan, which yields antibody entirely devoid of fucose at its glycan structure. MAb was purified through affinity capture using protein A beads and further purified by ion exchange chromatography. Antibody purity by SDS-PAGE and its spectrum from size exclusion chromatography HPLC were compared with CHO produced comparator. Purified mAb was composed of more than 99% fully assembled antibody including double heavy and light chain, and the quality of the antibody profile was comparable to that of mammalian cell expressed antibody. Unlike mammalian cells, glycoengineered *Pichia* derived mAb contains no sialic acid and fucose because *Pichia* inherently lacks these pathways and was not intently engineered in. However, due to the steric hindrance of Fc, the normal IgGs produced in glycoengineered *Pichia* with the capability of transferring terminal β1,4 galactose maintain the glycan heterogeneity. The major glycoform is agalatosylated G0 and various degree of single or two terminal galactosylated glycans. Interestingly when antibody was expressed in another glycoengineered strain, it delivered relatively homogenous type of glycans with more than 90% antibody carrying Man5GlcNAc2 (Potgieter et al., 2009).

4.4. Glycoengineered *Pichia* produced antibody is comparable to CHO derived antibody *in vitro* assays

Glycoengineered *Pichia pastoris* not only provides the capacity of secreted glycoprotein with human like glycans, but it is also capable to assemble heterodimer large molecule like antibody with 16 pairs of disulfide bonds. With the exception of the difference in glycosylation between CHO and glycoengineered *Pichia* produced antibody, they share similar purity based on SDS-PAGE and size exclusion chromatography HPLC analysis. The difference of glycosylation at Fc region has little or no impact on Fab dependant antigen binding in both ELISA based assay and cell based FACS analysis using cell line with antigen expressed on its surface. The antibody and antigen engagement can lead to proper biology function in cell based assays e.g. glycoengineered *Pichia* produced anti-HER2 has demonstrated comparable potencies in receptor inhibition assays *in vitro* compared to CHO derived trastuzumab. They have very similar HER2 and AKT phosphorylation inhibition and therefore both antibodies inhibit tumor cell proliferation at similar levels (Jiang et al., 2011; Liu et al., 2011; Zhang et al., 2011; Potgieter et al., 2009).

4.5. Glycoengineered *Pichia* produced mAb has similar *in vivo* efficacy and serum half-life

As stated before, fungal type hypermannosylated glycans at the antibody Fc region could have detrimental effects on its pharmacokinetics and would result in fast clearance in the blood stream (Liu et al., 2011). These non-human glycans can trigger the human immune response and causes immunogenicity. To ensure antibody produced from glycoengineered *Pichia* is suitable for therapeutic purpose, its PK needs to be monitored and compared with its counterpart from traditional mAb production platform. Comparing to a humanized antibody produced in CHO cell (Trastuzumab) with predominately human complex glycans at its Fc, in both rodent and non human primates, blood time concentration curve of glycoengineered *Pichia* produced mAb (with identical amino acid sequence to Trastuzumab) was almost super-imposable on that of CHO produced Trastuzumab. As a result, the key PK parameters (CL, t1/2, AUC and Vss) were comparable between these two mAbs. Moreover, the same antibody was studied and compared its *in vivo* efficacy in a xenograft nude mice model. With highly expressed receptor antigen on the tumor surface, glycoengineered *Pichia* produced antibody could engage and inhibit the target and therefore prevent the progression of tumor growth. Based on the time course for the average tumor growth, at low, intermediate and high dosage range, glycoengineered *Pichia* produced antibody with humanized glycan shows comparable tumor inhibitory efficacy (Zhang et al., 2011).

4.6. Antibody produced by glycoengineered *Pichia* has better efficacy

Preclinical studies have shown that antibody dependant cell-mediated cytotoxicity (ADCC) is an important part of mechanism of action of therapeutic monoclonal antibodies, especially anti-cancer antibodies, such as Trastumab and Rituximab against tumors (Mori et al., 2007). Some clinical evidence based on genetic analysis of leukocyte receptor (FcγR) polymorphisms of cancer patients treated with anti-CD20 IgG1 Rituximab and anti-HER2 IgG1 Trastuzumab therapies has revealed that ADCC is one of the critical mechanisms responsible for the clinical efficacy of these therapeutic antibodies(Musolino et al., 2008; Kim et al., 2006; Cartron et al., 2002). ADCC enhancement technology is expected to be excised in development of "biobetter" therapeutic antibodies with improved clinical efficacy. A strong correlation with Fc γ receptor affinity and antibody binding to FcγRIIIA in particular has shown to positively correlate with ADCC activity. Trastuzumab produced as Fc engineered or afucosylated mAb showed increased ADCC and improved tumor inhibition in a mouse xenograft model with human immuno-effector cells. A large number of studies with Fc engineered antibodies have firmly established that increased affinity for FcγRIIIA leads to increased NK cell or PBMC-mediated ADCC *in vitro*, and can result in better efficacy *in vivo* in models dependent on immune effector functions. Modulation of the Fc region by utilizing protein and glyco-engineering are two main ways to increase antibody and Fc γ receptor IIIa binding and then to enhance ADCC activity. But conventional therapeutic antibody production cell line such as CHO secrete monoclonal antibody with predominately fucosylated glycans which binds to FcγIIIa with lower affinity compared with afucosylated antibody. There are several technologies capable of generating afucosylated mAbs. This includes Kyowa Hakko/Biowa FUT8 (α-1,6-fucosyltransferase) gene knockout CHO line. Glycart (now part of Roche) utilizes a CHO line with inducible expression of the enzymes β(1,4)-N-acetylglucosaminyltransferase III (GnTIII) and wild-type golgi α-mannosidase II, and this line has been used to produce their GA101 antibody (Heinrich et al., 2009; Friess et al., 2007; Umana et al., 2007), an enhanced anti-CD20 that is currently in late-stage clinical trials for the treatment of Non-Hodgkin lymphoma (NHL). On the other hand, we reconstitute entire human glycosylation pathway in yeast *Pichia de novo* which inherently lacks of fucose transferase and its substrate. We also purposely leave out the fucosylation pathway through engineering, thus glycoengineered *Pichia* naturally produce antibody carrying zero percent fucosylated glycans. Afucosylated antibody from glycoengineered *Pichia* showed 6- and 8-fold increases in affinity for FcγRIIIA polymorphism F158 and V158, respectively compared to the CHO produced counterpart. This increased antibody affinity with Fc gamma receptor IIIa has translated 6-fold increase in NK cell-mediated ADCC activity and 4-fold and 3- fold increase in ADCC with PBMC and monocyte effector cells, respectively. Afucosylated antibodies from glycoengineered *Picha* platform and CHO FUT8 knockout platform have consistently increased affinity for human FcγRIIIa, which results in enhanced ADCC (Zhang et al., 2011). As such glycoengineered *Pichia* can be applied to produce afucosylated antibody where ADCC is part of mechanism of action, this could lead to develop "bio-better" therapeutic antibodies compared to conventional CHO fucosylated platform.

5. One stop shop for antibody development with human glycosylation

5.1. Antibody surface display on glycoengineered *Pichia*

Yeast surface display has been a widely used tool for protein engineering and for antibody engineering in particular (Boder and Wittrup, 1997; Boder and Wittrup, 2000; Wittrup, 2009). However, displaying Fabs or full-length antibodies on the surface of *Pichia pastoris* requires post-translational assembly of the heavy and the light chain. Moreover, the glycans of glycosylphosphatidylinositol (GPI) proteins, most often used as a cell wall anchor for display, play an important role in stabilizing its anchoring on the cell surface. Engineering the glycosylation pathway of *Pichia pastoris* can profoundly modify the glycosylation profile of the endogenous cell surface glycoproteins and recombinant GPI proteins (Bobrowicz et al., 2004; Sethuraman and Stadheim, 2006; Wang et al., 2007; Zhou et al., 2007). Thus the cell wall GPI protein anchors suitable for wild type *Pichia pastoris* strains may not necessarily work well in glycoengineered *Pichia* strains due to the difference of glycosylation profile of cell wall and cell wall anchoring proteins. Using combinatorial approach, we successfully developed a glycoengineered *Pichia* Fab display using a pair of coiled-coil peptides as the linker and *Saccharomyces cerevisiae* Sed1p GPI anchored cell surface protein as an anchoring domain in a host with mammalian mannose type Man5GlcNAc2 N-linked glycans. The system was validated by displaying multiple Fab molecules and sorting mixed Fab-displaying strains based on both expression and affinity. The results demonstrate a high level of concordance in expression and affinity between the displayed Fab and its secreted Fab as well as its full length IgG, which supports the coiled-coil/ScSed1p-based Fab display system as a platform for antibody affinity and expression maturation in glycoengineered *Pichia* strains (Lin et al., 2012).

5.2. Antibody expression platform in glycoengineered *Pichia pastoris*

Glycoengineered *Pichia* strain GFI5.0 was chosen as the expression host for producing mAb. GFI5.0 is capable of transferring terminal β-1,4 galactose onto biantennary complex N-linked glycan, which yields antibody entirely devoid of fucose at its glycan structure(Zhang et al., 2011). DNA sequences encoding the heavy and light chain are under methanol inducible AOX1 promoter. A gene encoding for drug Zeocin® resistance enable selection of transformants and a fragment of a *Pichia pastoris* gene such as the AOX2 promoter and terminator to enable integration of the vector onto the chromosome are part of the expression vector. The expression vector is integrated into *Pichia* chromosome by recombination after linearization followed by electroporation. Transformats are isolated as single colonies from agar plates containing Zeocin® for selection resistance marker (Li et al., 2006). Several strategies could be used through host strain modifications or expression vector optimization could be used to improve the expression level of full length monoclonal antibodies in yeast, for example (a) to increase copy number of heavy and light chain expression cassettes, (b) to improve mRNA stability (c) to optimize codon usage, (d) to screen best signal peptides (e) to engineer folding chaperones, such as immunoglobulin binding protein (BiP) and protein disulfide isomerase (PDI), (f) to delete protease genes and (g) to fuse heavy and light chain with a fusion partner.

5.3. Develop a robust and scalable monoclonal antibody production platform using glycoengineered *Pichia*

Glycoengineered *Pichia pastoris* strain which is capable of producing humanized glycoprotein with terminal galactose for monoclonal antibody production. Like mammalian cell lines, fermentation process plays a critical role to achieve the highest titer as well as maintain good quality glycans. A design of experiments (DoE) approach is often used to optimize the process parameters. In one case study followed by further optimization of the specific methanol feed rate, induction duration, and the initial induction biomass, the resulting process yielded up to 1.6 g/L of for one monoclonal antibody. Even more this process was also scaled-up to 1,200-L scale, and the process profiles, productivity, and product quality were comparable with 30-L scale. The successful scale-up demonstrated that this glycoengineered *Pichia pastoris* fermentation process is a robust and commercially viable process (Berdichevsky et al., 2011). In another case study, an oxygen-limited process was developed and optimized with the use of DoE for the production of monoclonal antibodies in glycoengineered *Pichia pastoris*. The process relied on pulsed feeding of methanol and its productivity was found to depend on biomass concentration and oxygen availability. Oxygen uptake rate was used as a scale-up parameter in demonstrating consistency of the process between the 3 L laboratory scale and the 1200 L pilot manufacturing scale. Scalability and productivity were improved by reducing oxygen consumption and cell growth, allowing extension of the induction phase without the associated mAb fragmentation characteristic of methanol limitation. The final mAb concentration was increased from 1.2 to 1.9 g L^{-1}. Oxygen limitation also improved N-glycan quality in terms of percentage of complex glycans (Ye et al., 2011a).

6. Summary

Glycosylation on therapeutic glycoproteins plays a critical role on its pharmacokinetics, efficacy and tissue targeting. By eliminating pathways responsible for fungal glycosylation and engineering in human glycosylation pathways, glycoengineered *Pichia* expresses glycoprotein with human type of glycans. Glycoengineered *Pichia* was applied as an alternative monoclonal antibody production platform. Comparing glycoengineered *Pichia* produced antibody with traditional mammalian cell CHO produced counterpart, antibody from glycoengineered *Pichia* has comparable *in vitro* functions and *in vivo* efficacy and serum half life. Moreover, by developing antibody yeast surface display in glycoengineered *Pichia* coupled with matured monoclonal antibody production, it becomes possible to integrate antibody discovery and manufacture into a single united platform, and it could increase the probability of success in therapeutic antibody development. Furthermore, glycoengineered *Pichia* antibody producing strain is a scalable and robust antibody production host.

Author details

Dongxing Zha
GlycoFi Inc., A wholly-owned subsidiary of Merck & Co., Inc., Lebanon

7. References

Barbin, K., Stieglmaier, J., Saul, D., Stieglmaier, K., Stockmeyer, B., Pfeiffer, M., Lang, P. and Fey, G.H., 2006. Influence of variable N-glycosylation on the cytolytic potential of chimeric CD19 antibodies. Journal of Immunotherapy 29, 122.

Beck, A., Cochet, O. and Wurch, T., 2010. GlycoFi's technology to control the glycosylation of recombinant therapeutic proteins. Expert Opinion on Drug Discovery 5, 95.

Berdichevsky, M., d'Anjou, M., Mallem, M.R., Shaikh, S.S. and Potgieter, T.I., 2011. Improved production of monoclonal antibodies through oxygen-limited cultivation of glycoengineered yeast. J Biotechnol. 155, 217.

Bergh, M.L.E., Friedman, B., Bamgrover, D., Hayes, M. and Kutzko, J., 1990. Processing and Glycosylation of Human Glucocerebrosidase Expressed in Insect Cells and in Chinese Hamster Ovary Cells. Abstracts of Papers of the American Chemical Society 200, 109-BIOT.

Bobrowicz, P., Davidson, R.C., Li, H., Potgieter, T.I., Nett, J.H., Hamilton, S.R., Stadheim, T.A., Miele, R.G., Bobrowicz, B., Mitchell, T., Rausch, S., Renfer, E. and Wildt, S., 2004. Engineering of an artificial glycosylation pathway blocked in core oligosaccharide assembly in the yeast Pichia pastoris: production of complex humanized glycoproteins with terminal galactose. Glycobiology 14, 757.

Boder, E.T. and Wittrup, K.D., 1997. Yeast surface display for screening combinatorial polypeptide libraries. Nature Biotechnology 15, 553.

Boder, E.T. and Wittrup, K.D., 2000. Yeast surface display for directed evolution of protein expression, affinity, and stability. Applications of Chimeric Genes and Hybrid Proteins, Pt C 328, 430.

Brumshtein, B., Salinas, P., Peterson, B., Chan, V., Silman, I., Sussman, J.L., Savickas, P.J., Robinson, G.S. and Futerman, A.H., 2010. Characterization of gene-activated human acid-beta-glucosidase: Crystal structure, glycan composition, and internalization into macrophages. Glycobiology 20, 24.

Cartron, G., Dacheux, L., Salles, G., Solal-Celigny, P., Bardos, P., Colombat, P. and Watier, H., 2002. Therapeutic activity of humanized anti-CD20 monoclonal antibody and polymorphism in IgG Fc receptor Fc gamma RIIIa gene. Blood 99, 754.

Choi, B.K., Bobrowicz, P., Davidson, R.C., Hamilton, S.R., Kung, D.H., Li, H., Miele, R.G., Nett, J.H., Wildt, S. and Gerngross, T.U., 2003. Use of combinatorial genetic libraries to humanize N-linked glycosylation in the yeast Pichia pastoris. Proc. Natl. Acad. Sci. U. S. A 100, 5022.

Chowdhury, P.S. and Wu, H., 2005. Tailor-made antibody therapeutics. Methods 36, 11.

Correia, I.R., 2010. Stability of IgG isotypes in serum. Mabs 2, 221.

Davidson, R.C., Nett, J.H., Renfer, E., Li, H., Stadheim, T.A., Miller, B.J., Miele, R.G., Hamilton, S.R., Choi, B.K., Mitchell, T.I. and Wildt, S., 2004. Functional analysis of the ALG3 gene encoding the Dol-P-Man: Man(5)GlcNAc(2)-PP-Dol mannosyltransferase enzyme of P-pastoris. Glycobiology 14, 399.

Egrie, J.C. and Browne, J.K., 2001. Development and characterization of novel erythropoiesis stimulating protein (NESP). British Journal of Cancer 84, 3.

Feldhaus, M.J., Siegel, R.W., Opresko, L.K., Coleman, J.R., Feldhaus, J.M.W., Yeung, Y.A., Cochran, J.R., Heinzelman, P., Colby, D., Swers, J., Graff, C., Wiley, H.S. and Wittrup, K.D., 2003. Flow-cytometric isolation of human antibodies from a nonimmune Saccharomyces cerevisiae surface display library. Nature Biotechnology 21, 163.

Ferrara, C., Grau, S., Jager, C., Sondermann, P., Brunker, P., Waldhauer, I., Hennig, M., Ruf, A., Rufer, A.C., Stihle, M., Umana, P. and Benz, J., 2011. Unique carbohydrate-carbohydrate interactions are required for high affinity binding between Fc gamma RIII and antibodies lacking core fucose. Proceedings of the National Academy of Sciences of the United States of America 108, 12669.

Friess, T., Gerdes, C., Nopora, A., Patre, M., Preiss, S., van Puijenbroek, E., Schnell, C., Bauer, S., Umana, P. and Klein, C., 2007. GA101, a novel humanized type IICD20 antibody with glycoengineered Fe and enhanced cell death induction, mediates superior efficacy in a variety of NHL xenograft models in comparison to rituximab. Blood 110, 691A.

Gasdaska, J.R., Sterling, J.D., Regan, J.T., Cox, K.M., Sherwood, S.W. and Dickey, L.F., 2007. Expression of a glyco-optimized anti-CD20 antibody in the aquatic plant lemna with enhanced ADCC activity. Blood 110, 697A.

Gerngross, T., 2005. Production of complex human glycoproteins in yeast. Adv. Exp. Med. Biol. 564, 139.

Hamilton, S.R., Bobrowicz, P., Bobrowicz, B., Davidson, R.C., Li, H.J., Mitchell, T., Nett, J.H., Rausch, S., Stadheim, T.A., Wischnewski, H., Wildt, S. and Gerngross, T.U., 2003. Production of complex human glycoproteins in yeast. Science 301, 1244.

Hamilton, S.R., Davidson, R.C., Sethuraman, N., Nett, J.H., Jiang, Y.W., Rios, S., Bobrowicz, P., Stadheim, T.A., Li, H.J., Choi, B.K., Hopkins, D., Wischnewski, H., Roser, J., Mitchell, T., Strawbridge, R.R., Hoopes, J., Wildt, S. and Gerngross, T.U., 2006. Humanization of yeast to produce complex terminally sialylated glycoproteins. Science 313, 1441.

Hamilton, S.R. and Gerngross, T.U., 2007. Glycosylation engineering in yeast: the advent of fully humanized yeast. Current Opinion in Biotechnology 18, 387.

Heinrich, D.A., Klein, C., Decheva, K., Weinkauf, M., Hutter, G., Zimmermann, Y., Weiglein, T., Hiddemann, W. and Dreyling, M.H., 2009. Anti-CD20 Antibody GA101 Shows Higher Cytotoxicity but Is Competitively Displaced by Rituximab in Mantle Cell Lymphoma. Blood 114, 1059.

Herbst, R., Wang, Y., Gallagher, S., Mittereder, N., Kuta, E., Damschroder, M., Woods, R., Rowe, D.C., Cheng, L., Cook, K., Evans, K., Sims, G.P., Pfarr, D.S., Bowen, M.A., Dall'Acqua, W., Shlomchik, M., Tedder, T.F., Kiener, P., Jallal, B., Wu, H. and Coyle, A.J., 2011. B-Cell Depletion In Vitro and In Vivo with an Afucosylated Anti-CD19 Antibody (vol 335, pg 213, 2010). Journal of Pharmacology and Experimental Therapeutics 336, 294.

Hopkins, D., Gomathinayagam, S., Rittenhour, A.M., Du, M., Hoyt, E., Karaveg, K., Mitchell, T., Nett, J.H., Sharkey, N.J., Stadheim, T.A., Li, H.J. and Hamilton, S.R., 2011. Elimination of beta-mannose glycan structures in Pichia pastoris. Glycobiology 21, 1616.

Jacobs, P.P. and Callewaert, N., 2009a. N-glycosylation Engineering of Biopharmaceutical Expression Systems. Current Molecular Medicine 9, 774.

Jacobs, P.P. and Callewaert, N., 2009b. N-glycosylation Engineering of Biopharmaceutical Expression Systems. Current Molecular Medicine 9, 774.

Jefferis, R., 2009. Recombinant antibody therapeutics: the impact of glycosylation on mechanisms of action. Trends in Pharmacological Sciences 30, 356.

Jiang, Y.W., Li, F., Zha, D.X., Potgieter, T.I., Mitchell, T., Moore, R., Cukan, M., Houston-Cummings, N.R., Nylen, A., Drummond, J.E., McKelvey, T.W., d'Anjou, M., Stadheim, T.A., Sethuraman, N. and Li, H.J., 2011. Purification process development of a recombinant monoclonal antibody expressed in glycoengineered Pichia pastoris. Protein Expression and Purification 76, 7.

Junttila, T.T., Parsons, K., Olsson, C., Lu, Y.M., Xin, Y., Theriault, J., Crocker, L., Pabonan, O., Baginski, T., Meng, G., Totpal, K., Kelley, R.F. and Sliwkowski, M.X., 2010. Superior In vivo Efficacy of Afucosylated Trastuzumab in the Treatment of HER2-Amplified Breast Cancer. Cancer Research 70, 4481.

Kayser, V., Chennamsetty, N., Voynov, V., Forrer, K., Helk, B. and Trout, B.L., 2011. Glycosylation influences on the aggregation propensity of therapeutic monoclonal antibodies. Biotechnology Journal 6, 38.

Kim, D.H., Du Jung, H., Kim, J.G., Lee, J.J., Yang, D.H., Park, Y.H., Do, Y.R., Shin, H.J., Kim, M.K., Hyun, M.S. and Sohn, S.K., 2006. FCGR3A gene polymorphisms may correlate with response to frontline R-CHOP therapy for diffuse large B-cell lymphoma. Blood 108, 2720.

Knappik, A., Ge, L.M., Honegger, A., Pack, P., Fischer, M., Wellnhofer, G., Hoess, A., Wolle, J., Pluckthun, A. and Virnekas, B., 2000. Fully synthetic human combinatorial antibody libraries (HuCAL) based on modular consensus frameworks and CDRs randomized with trinucleotides. Journal of Molecular Biology 296, 57.

Li, H.J. and d'Anjou, M., 2009. Pharmacological significance of glycosylation in therapeutic proteins. Current Opinion in Biotechnology 20, 678.

Li, H.J., Sethuraman, N., Stadheim, T.A., Zha, D.X., Prinz, B., Ballew, N., Bobrowicz, P., Choi, B.K., Cook, W.J., Cukan, M., Houston-Cummings, N.R., Davidson, R., Gong, B., Hamilton, S.R., Hoopes, J.P., Jiang, Y.W., Kim, N., Mansfield, R., Nett, J.H., Rios, S., Strawbridge, R., Wildt, S. and Gerngross, T.U., 2006. Optimization of humanized IgGs in glycoengineered Pichia pastoris. Nature Biotechnology 24, 210.

Li, J. and Zhu, Z.P., 2010. Research and development of next generation of antibody-based therapeutics. Acta Pharmacologica Sinica 31, 1198.

Lin, S., Houston-Cummings, N.R., Prinz, B., Moore, R., Bobrowicz, B., Davidson, R.C., Wildt, S., Stadheim, T.A. and Zha, D.X., 2012. A novel fragment of antigen binding (Fab) surface display platform using glycoengineered Pichia pastoris. Journal of Immunological Methods 375, 159.

Liu, L.M., Stadheim, A., Hamuro, L., Pittman, T., Wang, W.R., Zha, D.X., Hochman, J. and Prueksaritanont, T., 2011. Pharmacokinetics of IgG1 monoclonal antibodies produced in humanized Pichia pastoris with specific glycoforms: A comparative study with CHO produced materials. Biologicals 39, 205.

McCamish, M. and Woollett, G., 2011. Worldwide experience with biosimilar development. Mabs 3, 209.

Mimura, Y., Church, S., Ghirlando, R., Ashton, P.R., Dong, S., Goodall, M., Lund, J. and Jefferis, R., 2000. The influence of glycosylation on the thermal stability and effector function expression of human IgG1-Fc: properties of a series of truncated glycoforms. Molecular Immunology 37, 697.

Misaizu, T., Matsuki, S., Strickland, T.W., Takeuchi, M., Kobata, A. and Takasaki, S., 1995. Role of Antennary Structure of N-Linked Sugar Chains in Renal Handling of Recombinant-Human-Erythropoietin. Blood 86, 4097.

Mori, K., Iida, S., Yamane-Ohnuki, N., Kanda, Y., Kuni-Kamochi, R., Nakano, R., Imai-Nishiya, H., Okazaki, A., Shinkawa, T., Natsume, A., Niwa, R., Shitara, K. and Satoh, M., 2007. Non-fucosylated therapeutic antibodies: the next generation of therapeutic antibodies. Cytotechnology 55, 109.

Musolino, A., Naldi, N., Bortesi, B., Pezzuolo, D., Capelletti, M., Missale, G., Laccabue, D., Zerbini, A., Camisa, R., Bisagni, G., Neri, T.M. and Ardizzoni, A., 2008. Evidence for Linkage Disequilibrium Between Fc gamma RIIIa-V158F and Fc gamma RIIa-H131R Polymorphisms in White Patients, and for an Fc gamma RIIIa-Restricted Influence on the Response to Therapeutic Antibodies IN REPLY. Journal of Clinical Oncology 26, 5491.

Nett, J.H., Gomathinayagam, S., Hamilton, S.R., Gong, B., Davidson, R.C., Du, M., Hopkins, D., Mitchell, T., Mallem, M.R., Nylen, A., Shaikh, S.S., Sharkey, N., Barnard, G.C., Copeland, V., Liu, L.M., Evers, R., Li, Y., Gray, P.M., Lingham, R.B., Visco, D., Forrest, G., DeMartino, J., Linden, T., Potgieter, T.I., Wildt, S., Stadheim, T.A., d'Anjou, M., Li, H.J. and Sethuraman, N., 2012. Optimization of erythropoietin production with controlled glycosylation-PEGylated erythropoietin produced in glycoengineered Pichia pastoris. Journal of Biotechnology 157, 198.

Nett, J.H., Stadheim, T.A., Li, H., Bobrowicz, P., Hamilton, S.R., Davidson, R.C., Choi, B.K., Mitchell, T., Bobrowicz, B., Rittenhour, A., Wildt, S. and Gerngross, T.U., 2011. A combinatorial genetic library approach to target heterologous glycosylation enzymes to the endoplasmic reticulum or the Golgi apparatus of Pichia pastoris. Yeast 28, 237.

Nicolaides, N.C., Sass, P.M. and Grasso, L., 2010. Advances in targeted therapeutic agents. Expert Opinion on Drug Discovery 5, 1123.

Nimmerjahn, F. and Ravetch, J.V., 2007. Fc-receptors as regulators of immunity. Advances in Immunology, Vol 96 96, 179.

Nimmerjahn, F. and Ravetch, J.V., 2008. Fc gamma receptors as regulators of immune responses. Nature Reviews Immunology 8, 34.

Nimmerjahn, F. and Ravetch, J.V., 2010. Antibody-mediated modulation of immune responses. Immunol. Rev. 236, 265.

Oh, D.B., Park, J.S., Kim, M.W., Cheon, S.A., Kim, E.J., Moon, H.Y., Kwon, O., Rhee, S.K. and Kang, H.A., 2008. Glycoengineering of the methylotrophic yeast Hansenula polymorpha for the production of glycoproteins with trimannosyl core N-glycan by blocking core oligosaccharide assembly. Biotechnol. J 3, 659.

Pandhal, J. and Wright, P.C., 2010. N-Linked glycoengineering for human therapeutic proteins in bacteria. Biotechnology Letters 32, 1189.

Pastores, G.M., 2010. Recombinant Glucocerebrosidase (Imiglucerase) as a Therapy for Gaucher Disease. Biodrugs 24, 41.

Potgieter, T.I., Cukan, M., Drummond, J.E., Houston-Cummings, N.R., Jiang, Y.W., Li, F., Lynaugh, H., Mallem, M., McKelvey, T.W., Mitchell, T., Nylen, A., Rittenhour, A., Stadheim, T.A., Zha, D.X. and d'Anjou, M., 2009. Production of monoclonal antibodies by glycoengineered Pichia pastoris. Journal of Biotechnology 139, 318.

Potgieter, T.I., Kersey, S.D., Mallem, M.R., Nylen, A.C. and d'Anjou, M., 2010. Antibody Expression Kinetics in Glycoengineered Pichia Pastoris. Biotechnology and Bioengineering 106, 918.

Projan, S.J., Gill, D., Lu, Z.J. and Herrmann, S.H., 2004. Small molecules for small minds? The case for biologic pharmaceuticals. Expert Opinion on Biological Therapy 4, 1345.

Robak, T. and Robak, E., 2011. New Anti-CD20 Monoclonal Antibodies for the Treatment of B-Cell Lymphoid Malignancies. Biodrugs 25, 13.

Roopenian, D.C. and Akilesh, S., 2007. FcRn: the neonatal Fc receptor comes of age. Nature Reviews Immunology 7, 715.

Sethuraman, N. and Stadheim, T.A., 2006. Challenges in therapeutic glycoprotein production. Curr. Opin. Biotechnol. 17, 341.

Shaaltiel, Y., Bartfeld, D., Hashmueli, S., Baum, G., Brill-Almon, E., Galili, G., Dym, O., Boldin-Adamsky, S.A., Silman, I., Sussman, J.L., Futerman, A.H. and Aviezer, D., 2007. Production of glucocerebrosidase with terminal mannose glycans for enzyme replacement therapy of Gaucher's disease using a plant cell system. Plant Biotechnology Journal 5, 579.

Sinclair, A.M. and Elliott, S., 2005. Glycoengineering: The effect of glycosylation on the properties of therapeutic proteins. Journal of Pharmaceutical Sciences 94, 1626.

Sola, R.J. and Griebenow, K., 2009. Effects of Glycosylation on the Stability of Protein Pharmaceuticals. Journal of Pharmaceutical Sciences 98, 1223.

Takeuchi, M., Inoue, N., Strickland, T.W., Kubota, M., Wada, M., Shimizu, R., Hoshi, S., Kozutsumi, H., Takasaki, S. and Kobata, A., 1989. Relationship Between Sugar Chain Structure and Biological-Activity of Recombinant Human Erythropoietin Produced in Chinese-Hamster Ovary Cells. Proceedings of the National Academy of Sciences of the United States of America 86, 7819.

Tomiya, N., 2009. Humanization of recombinant glycoproteins expressed in insect cells. Trends in Glycoscience and Glycotechnology 21, 71.

Umana, P., Ekkehard, M., Peter, B., Gabriele, K., Ursula, P., Suter, T., Grau, R., Schmidt, C., Herter, S., Gerdes, C., Nopora, A., Patre, M., Moser, S., Sondermann, P., Wheat, L., Dyer, M.J.S., Poppema, S., Bauer, S., Kubbies, M., Strein, P., Fertig, G., Friess, T., Dabbagh, K., DalPorto, J. and Klein, C., 2007. GA101, a novel humanized type IICD20Antibody with glycoengineered Fc and enhanced cell death induction, exhibits superior anti-tumor efficacy and superior tissue B cell depletion in vivo. Blood 110, 694A.

Vaughan, T.J., Williams, A.J., Pritchard, K., Osbourn, J.K., Pope, A.R., Earnshaw, J.C., McCafferty, J., Hodits, R.A., Wilton, J. and Johnson, K.S., 1996. Human antibodies with sub-nanomolar affinities isolated from a large non-immunized phage display library. Nature Biotechnology 14, 309.

von Horsten, H.H., Ogorek, C., Blanchard, V., Demmler, C., Giese, C., Winkler, K., Kaup, M., Berger, M., Jordan, I. and Sandig, V., 2010. Production of non-fucosylated antibodies by co-expression of heterologous GDP-6-deoxy-D-lyxo-4-hexulose reductase. Glycobiology 20, 1607.

Walsh, G., 2010. Post-translational modifications of protein biopharmaceuticals. Drug Discovery Today 15, 773.

Wang, Q.J., Li, L., Chen, M., Qi, Q.Q. and Wang, P.G., 2007. Construction of a novel system for cell surface display of heterologous proteins on Pichia pastoris. Biotechnology Letters 29, 1561.

Ward, E., Mittereder, N., Kuta, E., Sims, G.P., Bowen, M.A., Dall'Acqua, W., Tedder, T., Kiener, P., Coyle, A.J., Wu, H., Jallal, B. and Herbst, R., 2011. A glycoengineered anti-CD19 antibody with potent antibody-dependent cellular cytotoxicity activity in vitro and lymphoma growth inhibition in vivo. British Journal of Haematology 155, 426.

Weaver-Feldhaus, J.M., Lou, J.L., Coleman, J.R., Siegel, R.W., Marks, J.D. and Feldhaus, M.J., 2004. Yeast mating for combinatorial Fab library generation and surface display. Febs Letters 564, 24.

Wildt, S. and Gerngross, T.U., 2005. The humanization of N-glycosylation pathways in yeast. Nature Reviews Microbiology 3, 119.

Wittrup, K., 2009. Protein and Antibody Engineering by Yeast Surface Display. Journal of Biomolecular Screening 14, 883.

Wong, A.W., Baginski, T.K. and Reilly, D.E., 2010. Enhancement of DNA Uptake in FUT8-Deleted CHO Cells for Transient Production of Afucosylated Antibodies. Biotechnology and Bioengineering 106, 751.

Ye, J., Ly, J., Watts, K., Hsu, A., Walker, A., McLaughlin, K., Berdichevsky, M., Prinz, B., Sean, K.D., d'Anjou, M., Pollard, D. and Potgieter, T., 2011a. Optimization of a glycoengineered Pichia pastoris cultivation process for commercial antibody production. Biotechnol. Prog. 27, 1744.

Ye, J.X., Ly, J., Watts, K., Hsu, A., Walker, A., McLaughlin, K., Berdichevsky, M., Prinz, B., Kersey, D.S., d'Anjou, M., Pollard, D. and Potgieter, T., 2011b. Optimization of a glycoengineered Pichia pastoris cultivation process for commercial antibody production. Biotechnology Progress 27, 1744.

Ysebaert, L., Laprevotte, E., Gross, E., Klein, C., Fournie, J., Laurent, G. and Ouillet-Mary, A., 2010. Ga101 Displays Higher Anti-Leukemic Activity Mainly Trough Enhanced Adcc in Chronic Lymphocytic Leukemia. Haematologica-the Hematology Journal 95, 46.

Zhang, N.Y., Liu, L.M., Dumitru, C.D., Cummings, N.R.H., Cukan, M., Jiang, Y.W., Li, Y., Li, F., Mitchell, T., Mallem, M.R., Ou, Y.S., Patel, R.N., Vo, K., Wang, H., Burnina, I., Choi, B.K., Huber, H., Stadheim, T.A. and Zha, D.X., 2011. Glycoengineered Pichia produced anti-HER2 is comparable to trastuzumab in preclinical study. Mabs 3, 289.

Zhou, H.H., 2007. Biologics in the pipeline: Large molecules with high hopes or bigger risks? Journal of Clinical Pharmacology 47, 550.

Zhou, J.G., Zhang, H.C., Liu, X.W., Wang, P.G. and Qi, Q.S., 2007. Influence of N-glycosylation on Saccharomyces cerevisiae morphology: A golgi glycosylation mutant shows cell division defects. Current Microbiology 55, 198.

Chapter 18

Application of Quality by Design Paradigm to the Manufacture of Protein Therapeutics

Ioscani Jimenez del Val, Philip M. Jedrzejewski, Kealan Exley, Si Nga Sou, Sarantos Kyriakopoulos, Karen M. Polizzi and Cleo Kontoravdi

Additional information is available at the end of the chapter

1. Introduction

N-linked glycosylation is of paramount importance for the biopharmaceutical sector given that approximately 40% of all approved therapeutic proteins and eight of the top ten selling biologics of 2010 contain N-linked oligosaccharides [1, 2]. With specific cell productivities in biotechnological processes estimated to reach their theoretical maximum in the near future, and despite anticipated further increases in culture density, the scope for improvement of biotherapeutics is limited in terms of production volume [3]. As a result there is an increased focus on biotherapeutic efficacy in the development and production stages. Modification of the glycoform is a target for drug design as it can enhance efficacy, mode of action and half-life significantly [4]. Simultaneously, N-linked glycosylation plays a key role in the safety of biotherapeutics. Certain N-linked oligosaccharides bound to therapeutic proteins have been found to trigger undesired effects in patients [5-7] thus deeming them a safety concern during process development [8].These elements make N-linked glycosylation a key target for quality control both from the therapeutic efficacy and safety standpoints. A well-defined product may have consistent protein backbones but still a glycoform distribution of more than a hundred detectable isoforms [9]. Narrowing and targeting the glycan profile is expected to improve efficacy and safety significantly. An example of the variety of reported glycoforms found on the crystallisable fragment (Fc) of monoclonal antibodies (mAbs) is presented in panel D of Figure 1. It is worth noting that single monosaccharides (e.g. core Fucose) on the complex glycans present on the crystallisable fragment (Fc) of monoclonal antibodies (mAbs) may alter the therapeutic efficacy of these products.

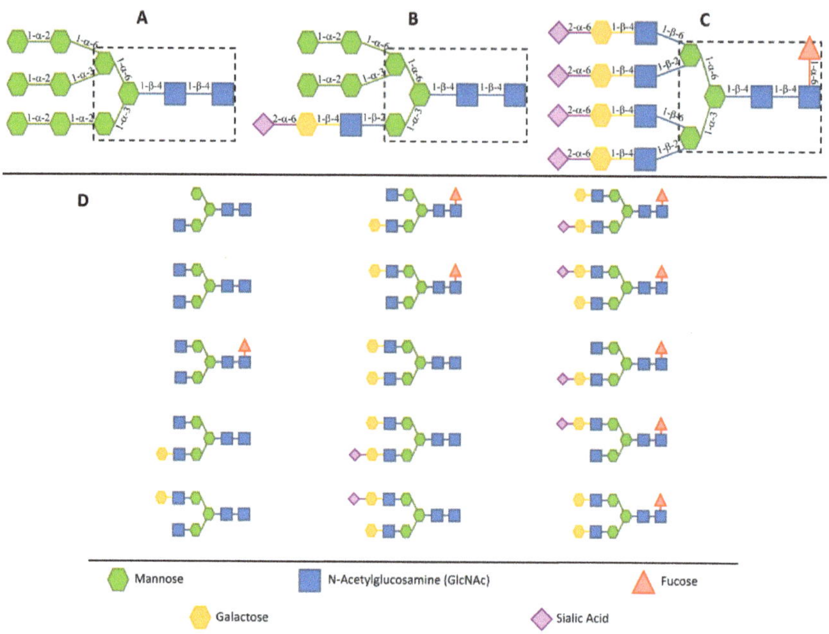

Figure 1. Classification of N-linked oligosaccharides. Panels A, B and C present a high-mannose, a hybrid and a complex tetra-antennary oligosaccharide, respectively. Highlighted by a box is the Man3GlcNAc2 core structure present in all N-linked glycans [10]. Panel D shows the most common glycans observed on the Fc of mAbs [11-13].

1.1. Effect of glycomic profile on therapeutic efficacy

1.1.1. Interferon-β

In 2001, when the human genome was fully sequenced, it was found that it contains fewer genes than expected. However, this might be compensated by the diversity found at the level of protein architecture, structure, transcription patterns, and posttranslational modifications of which glycosylation is the most complex [14]. Glycan moieties have a very diverse number of roles in biological systems, making them relevant for biotherapeutics. Starting with the co-translational attachment of glycan structures, it was suggested by Petrescu et al. that glycan structures act as nucleation sites for remote aromatic parts in an amino acid sequence in order to begin the folding process [15]. Although glycans play an important role in local as well as global folding of proteins it has to be pointed out that they are often not essential to maintain structure [16-18], as is the case for Interferon-β (IFN-β), for example. This naturally occurring glycoprotein is produced in the human body in response to viral infections or other biological threats [19]. Human IFN-β contains a conserved Asn-80 glycan attachment site and as a biotherapeutic, is marketed for the treatment of multiple sclerosis.

There are currently four IFN-β products marketed under the names of AVONEX®, Rebif®, Betaseron® and Extavia®. Crucially, the licensed drugs AVONEX and Rebif® are produced in Chinese hamster ovary (CHO) cells and are therefore fully glycosylated products while Betaseron® and Extavia® are produced in Escherichia coli resulting in an aglycosylated pharmaceutical [20]. AVONEX® and Betaseron® were compared side by side and it was found that the glycosylated form had a 10-fold increased anti-viral assay activity as compared to the *E. coli* product [21]. After deglycosylation the AVONEX® product precipitated due to formation of disulfide bonds between single molecules [22]. Further investigation showed that even when enzymatically deglycosylated, AVONEX® still showed increased heat stability (7°C higher T_m) and anti-viral assay activity (3-fold larger) when compared to Betaseron®, which leads to the conclusion that the glycan structure plays an important role in the folding of the protein.

1.1.2. Monoclonal antibodies

Monoclonal antibodies are currently the highest selling biotechnology products and by 2016 are predicted to make up five of the top ten selling pharmaceutical products worldwide [23]. Currently, there are over 30 FDA approved mAbs which are prescribed for the treatment of cancer and auto-immune disorders [24]. With a total of 205 reported antibodies under trial [25], the total number of approved antibody therapeutics is expected to increase significantly in the foreseeable future. The concept of a "magic bullet" that would go directly to the intended target epitope, yet remain harmless in healthy tissue was introduced in the late 19th century by Paul Ehrlich [26]. Key elements for realising Ehrlich's vision were the development of hybridoma cells by Köhler and Milstein to produce antibodies of predefined specificity [27] and the use of phage display libraries for antibody humanization by Greg Winter and Richard Lerner [28, 29]. Antibodies are generally described as symmetric structures which, in humans, are composed by a pair of heavy chains and an additional pair of light-chains covalently bound together by disulphide bridges. There are five heavy chain isotypes (IgA, IgG, IgE, IgG and IgM) and two light chain isotypes. The spatial arrangement of the four peptide chains results in the Y-shaped structure depicted in 2.

Both the heavy and light chains contain hypervariable segments also referred to as complementary determining regions (CDR). While the CDR is the antigen binding site, located in the antigen binding (Fab) region and responsible for antigen binding, the constant Fc region is responsible for effector functions such as activating the complement cascade or binding to Fc receptors. 80% of all antibodies found in the blood plasma are of the IgG isotype (with a molecular weight of approximately 150kDa) and until today only IgG-type mAb products have been licensed as biopharmaceuticals [1]. As a consequence only the IgG isotype of antibodies and their biological mechanisms will be discussed. Two fundamental modes of biological action of IgG class antibodies are discussed below:

Complement-dependent cellular cytotoxicity (CDC): In this mechanism the antibody binds with high specificity through its CDR to a target cell, and through binding to the Fc region, recruits the C1q component of the complement cascade. The end result of the cascade is the formation of the membrane attack complex, which disrupts the target cell's membrane and eventually leads it to lysis.

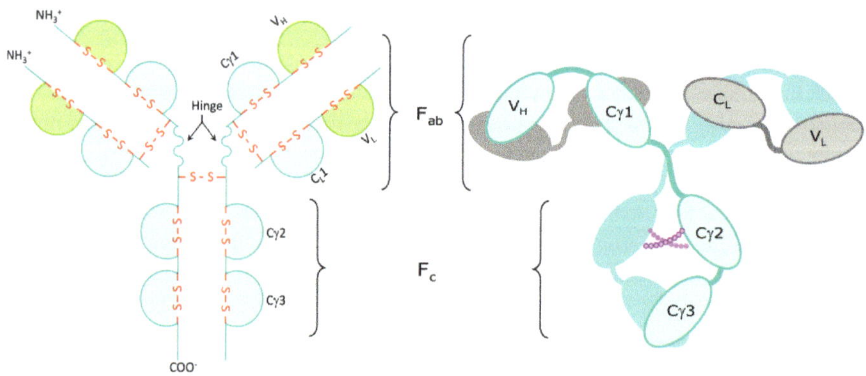

Figure 2. Diagram of immunoglobulin γ (IgG). The structural domains, disulphide bonds and peptide chain orientation are all shown. The antigen-binding (Fab) and the crystallisable (Fc) fragments are also shown along with the consensus Asn-297 glycosylation site on the Cγ2 domains of the heavy chains [1].

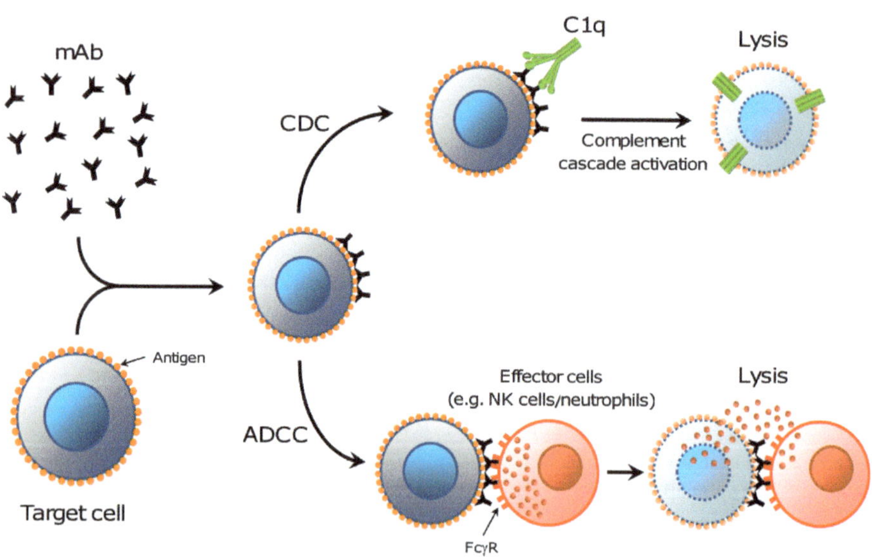

Figure 3. Main effector functions of IgG antibodies.

Antibody-dependent cell-mediated cytotoxicity (ADCC): In this mechanism the antibody Fab binds to a target cell, and through the Fc, binds to Fcγ receptors on the surface of natural killer cells (FcγRIII) and neutrophils (FcγII-A). Through this antibody-mediated interaction, the effector cells release cytokines and cytotoxic granules, which attack the target cell and take it towards apoptosis.

The biological *in vivo* mechanisms that are triggered by antibodies are highly dependent on the glycan moiety on the Cγ2 domains. It has been shown that the Asn297 glycans directly impact the three-dimensional structure of the Fc [30], thus greatly influencing the affinity FcγRs have for the IgG Fc [31, 32]. The close proximity of both Cγ2 domains in the Fc generates considerable steric hindrance, which limits the oligosaccharide structures found there to the complex bi-antennary structures shown in panel D of Figure 1. Depending on the host cell line, the potential carbohydrate residues will have specific features such as bi-secting GlcNAc residues or core fucosylation. Studies into the effect of bisecting structures present in antibody glycan moieties have shown that bisecting GlcNAc sugar residues lead to increased ADCC [33]. However, the presence (or absence) of other monosaccharides has been shown to have a more profound effect on *in vivo* modes of action. Previous reports show that absence of core fucose can increase ADCC up to 50-fold [34, 35]. This finding has been translated into the development of third generation mAbs, which are currently undergoing many promising clinical trials [36]. On the other hand, galactose terminating structures in the Fc fragment of antibodies, (the most abundant terminating saccharides on the Fc glycans of human polyclonal antibodies [37]) increase the affinity of the IgG Fc for the C1q protein substantially. Their removal results in decreased complement activation [31]. Sialic acid terminating glycan moieties can also strongly modulate the efficacy of mAbs. Interestingly, only a small proportion of human IgG is naturally sialylated [38], which can be directly related to the spatial constraints in the Cγ2 domain of IgG antibodies [39]. Sialic acid can be found in either a α2,3- or α2,6-linkage to the galactose moieties of IgGs [40], and although not very abundant, these residues are crucial for immune response modulation (anti-inflammatory activity) *in vivo*, where the α2,6-linkage is preferred over the α2,3 [41]. In fact, anti-inflammatory activity has been abrogated after removal of IgGs containing α2,6-linked sialic acid on their Fc glycan moiety. Similarly, a 10-fold enrichment in sialic acid containing IgGs induced anti-inflammatory response at 10-fold reduced doses. In contrast, IgG with α2,3-linked sialic acid terminating glycan structures failed to induce anti-inflammatory response even at 4-fold higher doses, thus demonstrating the specificity of *in vivo* bioactivity not only to structural but also conformational differences in glycan moieties. In addition to modulating the *in vivo* mechanisms, glycan moieties can also have an effect on pharmacokinetics of biotherapeutic antibodies. It has also been reported that high mannose structures (oligosaccharides with five mannose residues or more) will increase plasma clearance and thus, decrease *in vivo* half-life with a significant negative impact on drug efficacy [42].

1.1.3. Fab region glycans

It is well established that the glycan moieties in the Fc region of therapeutic antibodies have considerable effects on *in vivo* mechanism and pharmacokinetics. However, the Fab region can also be used to enhance characteristics of therapeutic antibodies. It has been estimated that between 20-30% of antibodies also carry a glycan structure in the Fab region, where a potential glycosylation site can be found on both the heavy as well as the light chain [43, 44]. Although the role of variable region carbohydrates is not clear, it is suggested that glycan moieties may influence antigen affinity, specificity, antibody solubility and stability hence, limiting aggregation [45, 46]. A further biological role of Fab region glycans is believed to be *in vivo* half-life modulation. This has been tested by injecting mice with humanized IgG with a variety of glycan moieties at the Asn56 position of the heavy chain variable region [47]. The results indicated that while sialic acid and galactose terminating glycan moieties had very limited effect on clearance, exposed GlcNAc residues showed slightly faster clearance rates. It was proposed that the latter are recognized by Man/GlcNAc receptors and binding interactions may be sufficiently strong to allow for greater clearance. It was later determined that half-life was dependent on the tissue where the antibodies accumulated based on their Fab region glycans, thus providing further scope for improvement in targeting of mAb biotherapeutics. On the other hand, Fab glycosylation has also been reported to produce negative effects in patients. A study has shown that α-1,3 galactose residues on the Fab glycans of the commercial antibody Cetuximab generated anti-α-1,3 galactose IgE-induced anaphylaxis in patients treated with this drug [5]. This effect highlights the relevance of Fab glycans for mAb safety. It has to be noted at this point that the glycoform on the Fab region can have a much more complex structure than the equivalent Fc glycoform including increased sialylation and occurrence of tri-antennary structures [43]. Since the glycan tripeptide sequences will vary within a population of antibodies and its location within the sequence can change it may not come as a surprise that different glycoforms will arise at different positions due to changing accessibility [48]. It was shown that shifting the tripeptide sequence in the variable region of a mAb may change the glycan moiety from a high Man structure to a complex structure. This would suggest that while the glycan precursor is added co-translationally to the protein backbone, the local conformation around the glycan may not allow for further enzymatic processing of the attached high Man structure in the Golgi apparatus. Furthermore changing the position of glycan structure within the polypeptide chain can impact antigen affinity significantly by either contributing to it or by blocking it altogether. Thus, glycan engineering of antibody variable regions represents a strategy towards improving antigen affinity, antibody targeting as well as extending half-life.

1.1.4. Erythropoietin

A further glycoprotein of biopharmaceutical importance is Erythropoietin (EPO), a hormone that binds to the receptors of red blood cell precursors in the bone marrow leading to their survival, proliferation and differentiation, thus increasing the red blood cell count [49-51]. EPO is used in the treatment of anaemia associated with a number of disease states such as chronic renal failure, cancer and HIV infection [52-55]. EPO has a total of four glycan attachment sites, three N-linked sites at amino acid positions 24, 38 and 83,and one O-linked site at amino acid position 126 [56, 57]. The four glycan moieties have been estimated to contribute approximately 40% to the total molecular mass of EPO and probably cover much of the surface of the molecule [58]. The hypothesis of full surface exposure is further supported by an analysis of the glycan moieties, which revealed predominantly fucosylated and tetraantennary complex glycans at the N-linked glycosylation sites [59] and, as such, suggest surface exposure during enzymatic processing. The glycan structures have been shown to play an important role in biological activity. For example, it is known that higher glycan antennarity leads to an increase in EPO's *in vivo* activity [60]. Interestingly, it has been demonstrated that all three N-linked glycan structures are necessary for biological activity *in vivo* while the O-linked glycan structure does not appear to be required for *in vivo* activity [61]. For recombinant human EPO, it has been shown that the N-linked glycan moieties are required for product secretion by the cell as well as ensuring solubility [61, 62].

1.2. Desired characteristics dependent on application

Erythropoietin is also a very good example to demonstrate how desired safety and efficacy characteristics can be enhanced through glycan engineering of existing biotherapeutics. The plasma clearance of EPO is regulated by sialic acid containing carbohydrates [58, 63], where an increase in sialic acid moieties is known to increase glycoprotein half-life and would also explain why higher antennarity is linked to increased plasma half-life. An investigation into EPO glycan structures and biological activity showed that desialylated structures showed increased *in vitro* activity but a much reduced *in vivo* activity, which is a consequence of increased EPO clearance by asialoglycoprotein receptors in the liver [64, 65]. In order to increase protein half-life, increase drug efficacy and thus, decrease dosing rates in patients, a glyco-engineered darbepoetin was created [58]. Darbepoetin features two additional N-linked glycosylation sites that were introduced by changing five amino acid residues through site-directed mutagenesis. The resulting biopharmaceutical commercially marketed as Aranesp® showed three-fold lower plasma clearance rate and results in increased *in vivo* potency over epoetin with three N-linked glycan sites only [66].

2. Current methodology and future application of quality by design initiative

Drug development begins with the discovery of molecules that have shown the biochemical potential to treat illnesses. Based on manufacturability and potential profitability, drug candidates are then selected for optimization and, eventually, for preclinical and clinical trials. To ensure that sufficient material is available for the different phases, manufacturing is concurrently scaled up during this stage in process development. Also throughout these stages, all the data regarding drug safety, efficacy and manufacture is reviewed for approval by the corresponding regulatory agencies. Approval requires that the drug product is produced consistently and that it is safe and efficacious for its indication. Despite decades of advances in drug product manufacturing, pharmaceutical process development and approval is still extremely lengthy, highly expensive and uncertain [67].

In order to reduce their losses, manufacturers have traditionally relied on the so-called quality by testing (QbT) approach for drug development and approval [68]. In QbT, product quality attributes (the ranges for drug substance properties that yield acceptable safety and efficacy) are linked with a specific manufacturing process and its corresponding set of inputs (raw materials and process parameters) during clinical phases of development. The process inputs that empirically show to yield acceptable product quality are defined and are often maintained unchanged after phase II clinical trials so that manufacturers avoid costs associated with additional testing for regulatory compliance [69]. During manufacturing, the process inputs are controlled to remain at their pre-defined set points, and at the end of each batch, the product is tested for compliance with the desired quality [68-71]. This black box approach does not require mechanistic knowledge that relates process inputs with product quality, and because of this, the QbT approach uncouples product end quality from the manufacturing process. The main drawbacks of QbT are [10, 69, 72]:

- Development and approval of pharmaceutical processes under QbT has proven to be extremely time-consuming (between 7 and 10 years [73, 74]) and very expensive (US$1.2 to 1.8 billion per approved drug when the risk of failure is included [67, 75]). This could be attributed, in part, to limited understanding of the relationship between product quality characteristics, therapeutic mechanisms and the effect process inputs have on quality.
- Process control is not established by mechanistic links between inputs and desired product quality. Due to this, the process is susceptible to generate off-spec product, and when this occurs, identifying the source(s) of failure is difficult.
- The range of process conditions approved under QbT is narrow and changes to process conditions outside this range require additional approval, which eventually translates into further delays and expense. This discourages pharmaceutical companies from modifying or optimizing current processes and implementing innovative ones.
- Overall, processes developed and approved under QbT generate a limited amount of knowledge because mechanistic relations between inputs and outputs are not apparent. This greatly restricts transferability of the knowledge gained from one process/product to the next. Moreover, the lack of generated mechanistic information increases the likelihood suboptimal process performance.

Therapeutic proteins currently undergo the same QbT development and approval process as their small molecule counterparts and suffer the same caveats. In the QbT context, bioprocess inputs are empirically defined so that the quality properties (e.g. aggregation, folding, methylation and glycosylation) of the protein lie within ranges that yield acceptable safety and therapeutic efficacy. As occurs with small molecules, QbT disjoins therapeutic protein quality from the bioprocess. This has led to inadequate understanding of the relationship between process inputs and product quality and has greatly limited the potential for bioprocess optimization.

The inefficiencies associated with the QbT approach along with more stringent regulatory requirements have led to manufacturers investing more in drug discovery than process understanding and optimization, which in turn, has translated into a decrease in the cost-effectiveness, number and quality of innovative drugs and pharmaceutical manufacturing processes [67, 69]. In order to overcome these limitations, regulators and industry specialists have proposed the implementation of Quality by Design (QbD) concepts to the manufacture of all new drugs, including therapeutic proteins, in the development pipeline [68, 72, 76].

2.1. Quality by design initiative in pharmaceuticals and how it is expected to affect biopharmaceuticals in the future

Pharmaceutical QbD is a conceptual framework for the development and approval of pharmaceutical manufacturing processes that aims to build quality (particularly with respect to safety and therapeutic efficacy) into the product at every stage of process development [10, 71, 72]. Application of QbD principles to pharmaceutical process development is outlined in the Process Analytical Technology (PAT) guideline "PAT - A Framework for Innovative Pharmaceutical Manufacturing and Quality Assurance" [77] by the US Federal Drug Administration (USFDA) and in the guidance documents "ICH Q8 Pharmaceutical Development" [71], "ICH Q9 Quality Risk Management" [78] and "ICH Q10 Pharmaceutical Quality Systems" [12] from the International Conference on Harmonization (ICH), which is an association constituted by the USFDA, the European Medicines Agency (EMA), the Pharmaceuticals and Medical Devices Agency of Japan (PMDA) and several experts from the pharmaceutical industry.

Implementation of QbD to pharmaceutical manufacturing processes is an information-driven process where all available knowledge on the drug product including, but not limited to, its therapeutic mechanisms, its process of manufacture and potential sources of variability is used to define a range of manufacturing conditions that will ultimately ensure product safety and efficacy when administered to patients. More specifically, the ICH guidelines [12, 71, 78] define the following requirements for the implementation of QbD to pharmaceutical processes:

2.1.1. Definition of the quality target product profile (QTPP)

The QTPP is defined as the set of quality characteristics that would ideally be achieved to ensure that the drug product is safe and efficacious. Considerations to define the QTPP include the route of administration, dosage form, delivery systems, dosage strength, sterility, purity and stability [71].

2.1.2. Identification of the critical quality attributes (CQAs) of the drug product

A CQA is defined in ICH Q8 as "a physical, chemical, biological, or microbiological property or characteristic that should be within an appropriate limit, range, or distribution to ensure the desired product quality" [71]. As the definition implies, identification of CQAs requires thorough physicochemical and biological characterization of the drug product and in-depth knowledge on which of its properties have a higher influence on its safety and efficacy.

2.1.3. Identification of process inputs that affect product CQAs

Once CQAs are identified, it is necessary to determine not only which process inputs (raw materials and process conditions) impact CQAs, but also how these inputs interact to affect the drug product's CQAs. We must note that much of the required knowledge may not be available for certain processes (particularly novel ones) and should be established through the combination of prior knowledge, mechanistic modelling, experimentation and finally, a risk assessment so that the influence of material attributes and process conditions on CQAs is ranked according to likelihood and extent of impact. It is worth mentioning that, under QbD, there is special emphasis on design of experiments (DoE) so that the interaction between individual process inputs and their impact on product CQAs is represented. Crucially, all elements involved in this section directly couple manufacturing process conditions with the CQAs of the drug product in a robust, systematic and information-driven manner.

2.1.4. Selection of the appropriate manufacturing process

With the ranking obtained through the risk assessment mentioned above, a multi-dimensional design space of allowable process input values (and combination thereof) is defined. ICH Q8 defines the design space as "an established multidimensional combination and interaction of material attributes and/or process parameters demonstrated to provide assurance of quality" [71]. Using the design space as a guide, the manufacturing process which is most capable of maintaining process conditions within the ranges that ensure product quality is defined. The selected manufacturing process must be robust such that it minimizes the risk of process conditions falling outside the design space, thus increasing the likelihood of the CQAs being within the range that ensures safety and efficacy.

2.1.5. Definition of a control strategy

With the appropriate manufacturing process in place, a strategy to mitigate the risk of materials and process conditions falling outside acceptable ranges must be established. By applying this risk management strategy, raw material specifications are monitored and controlled such that no impact is observed on the drug product's CQAs. In addition, the process parameters that influence the CQAs are controlled (ideally through online measurements and tight control systems) at every stage of the manufacturing process so that the desired product quality is met.

Conceptually underlying all the elements that constitute QbD implementation is process analytical technology (PAT). PAT is defined by the USFDA and the ICH as "a system for designing, analyzing and controlling manufacturing through timely measurements of CQAs and performance attributes of raw and in-process materials and processes with the goal of ensuring final product quality" [77]. The elements of PAT concerning process design and control have been presented in the description of QbD above. However, more discussion must be provided on the "timely measurement of CQAs and performance attributes" [77] throughout the implementation of QbD to manufacturing process development. It is clear that, from very early stages of QbD-driven process development, methods for accurately identifying and measuring the physical, chemical and biological properties of drug products are essential for defining QTPPs and CQAs. Further along in the implementation of QbD, it is also necessary to accurately measure the material attributes and process parameters that affect product CQAs. Finally, and as the definition of PAT states, once the manufacturing process is selected, appropriate analytical technologies are necessary to monitor it is necessary to measure process parameters in a timely fashion (ideally online) so that the generated data can be used for process control. From this, it is clear that analytical methods constitute a core element throughout process development under the QbD scope [79].

Adoption of the QbD in pharmaceutical process development aims to address all of the limitations described above. The three major regulatory bodies (USFDA, EMA and PMDA) are encouraging implementation of the QbD approach for the development of all new drugs in the pipeline [68, 70, 72, 80]. QbD is expected to reduce process approval time and costs, reduce regulatory intervention and encourage optimization and innovation by building processes around the mechanistic relationships between inputs and product quality. Because these relationships should be based on sound science and engineering principles, process outputs are more predictable and require less regulation which, in turn, would considerably reduce approval time and development costs. In addition, the more ample design space created through QbD would allow inputs to vary more without the need for additional approval. Predictability would also translate into much tighter control systems that would dramatically reduce the likelihood of generating product with unacceptable quality. Finally, the wealth of knowledge generated by the QbD approach along with the broader design space and more flexible regulatory approval characteristics would encourage process optimization and could potentially contribute to the development of novel processes as well as the discovery and design of next generation drugs.

Since the guidelines for QbD were first drafted, the framework has been implemented in the field of small-molecule therapeutics (SMTs) with relative ease. In contrast, the implementation of QbD to protein therapeutics (PTs) has met more resistance. This is likely due to the fact that the physical and chemical processes underlying the manufacture of SMTs is better understood and the mechanisms relating process inputs with SMT quality are easier to define. Conversely, the mechanisms by which living organisms produce PTs are less well understood, and the structural complexity of PTs makes their isolation, separation, purification and overall quality control much more challenging. Despite this, the regulatory agencies and several authors believe that sufficient knowledge is available or can be gained through current experimental and modelling methodologies to elucidate mechanistic relations between bioprocesses and PT quality, thus allowing for implementation of QbD principles to the development of therapeutic protein manufacturing processes in the near future [68, 72].

Implementation of the QbD paradigm in biopharma should increase knowledge on the therapeutic mechanisms of biotherapeutics considerably. This could lead to the improvement of previously existing products and may contribute to the discovery and development of new biologics. It will also generate a rich and systematized knowledge base relating manufacturing conditions with drug products which may lead to more robust process control, process optimization and, potentially, development of novel and efficient platforms for the manufacture of biological therapeutics. Implied in this is the considerable reduction in approval times which would heavily reduce costs of product and process development and eventually translate into lower costs for healthcare providers and patients.

2.2. Other critical quality attributes of protein-based therapeutics

Several drug characteristics other than glycosylation are considered critical quality attributes due to their impact on biological activity, pharmacokinetics or pharmacodynamics, and safety in terms of immunogenicity and toxicity. These can arise from variations in protein structure or from the presence of other adventitious molecules in the product formulation. Protein aggregation is a common protein-related CQA, which can occur at any stage of protein production or processing due to the protein structure, which may leave hydrophilic patches exposed, its concentration in the preparation or the process conditions. Temperature and pH extremes or physical stress can increase a protein's propensity to aggregate. Although protein aggregation is known to affect efficacy, the main concern is the immunogenicity of aggregates [81, 82]. Protein aggregation may also be the result of modification reactions, such as oxidation, which is caused by increased levels of reactive oxygen species. This can additionally affect function, as, for example, in the case of antibodies where oxidation can alter the Fc structure and thus reduce its binding affinity [83]. Conformational changes can also occur at refolding steps of bacterial production platforms and are potentially immunogenic.

Other possible structural changes include protein fragmentation, due to proteolytic enzymes present in the cell culture supernatant or in human plasma, extreme pH or temperature conditions, or because of chemical disruption of peptide bonds, C- and N-terminal truncation and deamidation. The effect of fragmentation is product-dependent, however it is known that it can impact biological activity, serum half life and immunogenicity due to the generation of novel epitopes [84, 85], whereas the remaining aforementioned changes do not appear to adversely affect product potency or safety [86-89], with the exception of deamidation within a complimentarity-determining region, which can affect biological activity [86]. Finally, glycation is another post-translational modification that involves the chemical addition of a monosaccharide on the side chain of a lysine residue. It occurs when a protein is incubated in the presence of reducing sugars, especially fructose and galactose and to a lesser extent glucose, in cell culture [90], and can affect its biological activity [91].

In addition to the product-related critical quality attributes described above, the host cell line, raw materials and process operation can introduce impurities or contaminants with adverse effects on the formulation's suitability for *in vivo* use. Host cell proteins are released mostly at later stages of the cell culture due to cell lysis and can be immunogenic [92], particularly when originating from microbial systems [93]. Additionally, host cell DNA poses considerable risk due to its potential integration and the possibility of a resulting carcinogenic effect. For this reason, the host cell DNA level cannot exceed 10ng per dose [94]. Impurities and contaminants can further be introduced from raw materials and lack of aseptic conditions. The most significant of these in terms of the risk they pose to product integrity are viruses, microbial cells and their products, such as endotoxins, which are highly toxic to humans [95]. Due to the severity of the effects of human injection with such contaminants, sufficient clearance must be demonstrated to regulatory authorities for approval to be gained. A thorough review of the above critical quality attributes and their effects on product safety and efficacy is presented in [96].

2.3. Challenges in implementing QbD in biopharma

Implementation of QbD to bioprocesses first requires thorough characterization of the drug substance (i.e. the therapeutic protein) in order to determine the attributes that define its safety and efficacy. Drug substance characterization, which in many cases is a challenging task, is usually done with liquid chromatography peptide mapping combined with mass spectrometry for amino acid sequencing [97-99], x-ray crystallography for three dimensional structure and different analytical chromatographic or electrophoretic techniques coupled to mass spectrometry for the analysis of N and O glycans which have been extensively reviewed by del Val et al [10]. The propensity of protein aggregation and fragmentation is usually measured through size exclusion chromatography [100-102]. By coupling these techniques to non-clinical (*in vitro* assays), preclinical and clinical trials, the CQAs for protein therapeutics are defined. However, despite the available methods for characterizing proteins and their CQAs, these are not always deemed sufficient to ensure identity. This is evidenced by current US and European legislation concerning Biosimilar products. Both the USFDA and the EMA have required phase I and phase III clinical trials in order to establish that a follow-on product is similar to its brand-named counterpart even when they have been shown to have equivalent CQA attributes according to the available analytical methods [103]. Lack of confidence in the current analytical techniques for drug substance characterization is a critical challenge that must be overcome for appropriate implementation of QbD to biopharmaceutical processes. Development of additional characterization methods are required to compare the product CQAs with the QTPPs defined during the early stages of process development. Furthermore, development of additional non-clinical studies for determining product safety and efficacy is required. *In vitro* assays for product safety and efficacy would dramatically reduce clinical trial associated costs and streamline data acquisition for determination of product CQAs.

The next step under the QbD scope is to define the manufacturing process, the process inputs that affect product CQAs and the mechanisms by which this occurs. In contrast to their small-molecule counterparts which are largely produced *in vitro*, therapeutic proteins are produced by living organisms. Cellular metabolism is extremely complex and several mechanisms by which cells produce therapeutic proteins are yet to be fully described. Moreover, very little information that quantitatively relates process conditions with cell metabolism, protein synthesis and product CQAs is available. According to QbD guidelines, quantitative mechanistic models are ideal tools for relating process inputs with product CQAs. The challenge of relating process inputs with product CQAs may be overcome by additional data generation through an iterative process of DoE-aided experimentation and mechanistic modelling. Through this, sufficient data would be generated such that the effects of process inputs on product CQAs are used to define an enhanced design space that could lead to higher assurance of product safety and efficacy.

After defining the design space, a control strategy must be defined so that process inputs are maintained within the range that ensures product quality. A crucial challenge in defining the control strategy is the ability to measure the process inputs that influence product CQAs. The QbD and PAT guidance documents suggest that control should be established through timely measurements which, ideally, should be performed online. This, again, is not trivial in a biological system. The interior of a bioreactor is a complex environment mainly composed of culture medium, cells, product and co-products of cell culture. The culture medium itself is a complex mixture of nutrients. Many of the common process parameters are readily measurable such as pO_2, pCO_2, temperature cell density and certain metabolite concentrations. However, many of the single components that influence product CQAs are difficult to measure in such a complex mixture. Several analytical methods that have the ability of tracking single components in bioreactors are currently being explored. Some of the most promising technologies are based on infrared spectroscopy, and have been reviewed recently by Landgrebe et al. [104]. Successful implementation of these techniques may lead to absolute online monitoring and process control. In parallel, methods to measure the intracellular concentration of key nutrients and metabolites online are being developed with promising results [105]. On the other hand, mathematical modelling efforts are being developed to describe complex biological processes such as N-linked glycosylation. This work has yielded encouraging results for bioprocess control and optimization, and could potentially aid in cell line development for third generation therapeutic proteins[106].

Almost by definition, QbD is a self-catalytic process because it relies on, and generates, a wealth of information. The more QbD is implemented in bioprocess development and approval, further understanding will be gained and fed into the development of newer processes which will eventually culminate in near-complete description of therapeutic mechanisms, drug product CQAs and therapeutic protein bioprocessing.

3. Overview of production organisms and manufacturing environment

The majority of approved biopharmaceuticals are produced in mammalian cell culture systems, since they are the sole means to deliver proteins with desired glycosylation patterns and thus ensure reduced immunogenicity and higher *in vivo* efficacy and stability [32, 107, 108]. However, mammalian cell culture delivers a heterogeneous mixture of glycan structures which do not all have the same properties. Product half-life and activity is therefore compromised, while higher doses are required for efficacy.

As in April 2012, there are 77 therapeutic glycoproteins out of the total 642 drugs approved by the European Medicines Agency (EMA). Host systems for their production include mammalian cells (65 drugs) and transgenic animals (2 drugs), while several are isolated from the blood plasma of healthy donors (10 drugs), as depicted in Figure 4A. Therapeutic classes of each glycoprotein drug are also presented in Figure 4B and involve mainly: hereditary diseases (Haemophilia A and B, Fabry disease, Gaucher disease, and others; 29% of EMA approved glycoproteins), cancer (leukemia; cancer in ascites; thyroid, stomach, breast, colorectal, etc. cancers or anaemia caused by chronic cancer; 26% of EMA approved glycoproteins), and autoimmune disorders (rheumatoid arthritis, multiple sclerosis, Crohn's disease, Lupus Erythematosus; 18%). Other therapeutic areas for which glycoprotein drugs are prescribed include infertility, acquired injuries/disease (tibial fractures, spondylolisthesis, myocardial infarction), immunosuppresants during transplantations, for hemostasis after surgery, for anaemia caused by chronic kidney disorders, postmenopausal diseases (osteoporosis) and one against the respiratory syncytial virus (containing a monoclonal antibody as an active substance: Synagis®).

The drug categories of current EMA approved glycoproteins (Figure 4C), based also on the classification in [109] include in descending order of approved drugs: polyclonal (mostly human plasma derived drugs) and monoclonal antibodies (mAbs), growth factors (e.g. erythropoietin, stimulating factors, bone morphogenetic proteins), blood factors and coagulants (e.g. factors: VIII, VIIa, IX and fibrinogen & thrombin), therapeutic enzymes (e.g. recombinant β-glucocerebrosidase), hormones (e.g. follitropin, lutropin), fusion proteins (e.g. Enbrel® the extracellular ligand-binding portion of human tumor necrosis factor receptor p75 linked to an analog human Fc portion of human IgG1), and serpins (serine protease inhibitors, e.g. C1 esterase inhibitor). Other classes involve thrombolytics and anticoagulants (e.g. tissue plasminogen activator and protein c, respectively) and cytokines (e.g. interleukins and interferons).

Two other glycoproteins (Leukoscan® and Scintinum®) have been approved by EMA,but are used for radionuclide imaging rather than therapeutics and thus have not been included in the statistics shown in Figure 4A. Furthermore, denosumab a monoclonal antibody produced by Amgen is approved under two brand names Prolia® (for postmenopausal osteoporosis) and as Xgeva® (for bone metastasis in cancer), and hence only one has been included in final list of approved drugs. Two more mAbs have been approved for use in specific EU countries but are not approved by EMA and hence have also not been included (Orthoclone OKT3® and Reopro®). Finally, two drugs that contain the recombinant factor VIII as active ingredient are produced by Bayer Pharma AG using baby hamster kidney (BHK) cells as host cells from an identical fermentation procedure, but since they are purified through slightly different downstream processes, both have been included separately in the statistics shown in Figure 4. Vaccines are also not included in the list of glycoproteins because their production mainly involves the propagation of viruses and not specific glycoprotein production. Moreover, all recombinant vaccines are produced in microbial organisms and hence do not involve glycoprotein production, but amino acid sequences of antigens that are not glycosylated.

CHO (Chinese hamster ovary) cells are the dominant host cells for the production of glycoproteins as far as the approvals in European Union are concerned, with 47 out of the total 77 therapeutic EMA approved glycoprotein drugs using them as host cells. Five drugs produced in CHO cells are biosimilars of recombinant erythropoietin (with reference product being a recombinant epoetin produced from Janssen-Cilag GmbH with brand names Eprex® in UK and Erypo® in Germany). All aforementioned biosimilars were approved in 2007. Biosimilars are defined from the EMA as drugs similar to a biological drug already in the market that are used to treat the same disease and within the same range of doses. Hybridoma cells are the second most frequently used hosts for recombinant glycoprotein production (12 drugs), followed by glycoproteins purified from blood plasma from healthy donors (10 drugs). BHK cells, along with HT-1080 cells, a human sarcoma cell line, with three approved drugs from each cell line are fourth. All approved drugs produced in BHK cells involve blood factors. Regarding HT-1080 cell line, all of the drugs produced from it belong to Shire pharmaceuticals portfolio. The active substances of these drugs involve therapeutic enzymes, while these drugs are also classified as orphans. Orphan drugs, according to the EMA, are prescribed to treat diseases that do not affect more than 5 in 10,000 individuals in the European Union at the time that the drug was submitted for approval. Finally, the host system classification list is completed with transgenic animals. Currently, two drugs are produced from transgenic animals and involve two serpins (serine protease inhibitors): antithrombin III (Atryn®) and C1 esterase inhibitor (conestat alfa; Ruconest®) isolated from goat and rabbit milk, respectively.

3.1. Mammalian cell systems

Mouse myeloma cell lines SP2/0 and NS0 are frequently used for research purposes [110], while human embryonic kidney 293 (HEK293) cells are often utilized to produce material for pre-clinical trials or research [111]. CHO cells are the workhorse of industrial production because of their ability to do gene amplification, which increases the level of product specificity and the ability to grow in serum-free suspension conditions [112]. Similarly, the myeloma cell-derived rodent NS0 cells have a high efficiency in producing recombinant immunoglobulin proteins and they can be cultured in both serum or serum/protein free suspension which reduces manufacturing costs during large-scale protein production [113]. Since CHO and NS0 originate from different mammalian species, the amount and type of glycosidases and transferases they have varies. Studies have reported that NS0 cells can produce glycoproteins that are highly immunogenic to human [114].

Human cells, on the other hand, are alternative hosts. The fact that human cells generate recombinant products with PTMs similar to the native counterparts secreted inside the human body gives them an advantage over CHO and NS0, which lack some essential glycosylation enzymes, e.g. bisecting N-acetylglucosamine transferase and α2-6 sialyltransferase. HEK 293 cells is the most-widely used human cell line for research into recombinant protein production and studies show that HEK 293 cells are capable of manufacturing Xigris (activated protein C) with proper γ-carboxylation and propeptide digestion at its glutamate amino acid residues, which CHO cells fail to offer [115]. In addition, the main advantage of protein expression in human cells is the low level of immunogenic reactions, with limited expression of N-glycolylneuraminic acid (Neu5Gc)-bearing erythropoietin (EPO) in human HT1080 fibrosarcoma cell line than the CHO-derived ones [116]. Some new human-based expression cell lines have been developed and are increasingly adopted by industry. For example, the Per.C6 cell line derived from the human retina cells have shown to give a protein titre of more than 2 g/L in simple fed-batch culture [117]. Unlike CHO cells, which often require product selection, the Per.C6 cell line does not require any gene amplification or selection strategy, nor does it need a large gene copy number for stable protein expression [118]. Per.C6 derived-EPO is free of Neu5Gc thus limiting immunogenicity [119].

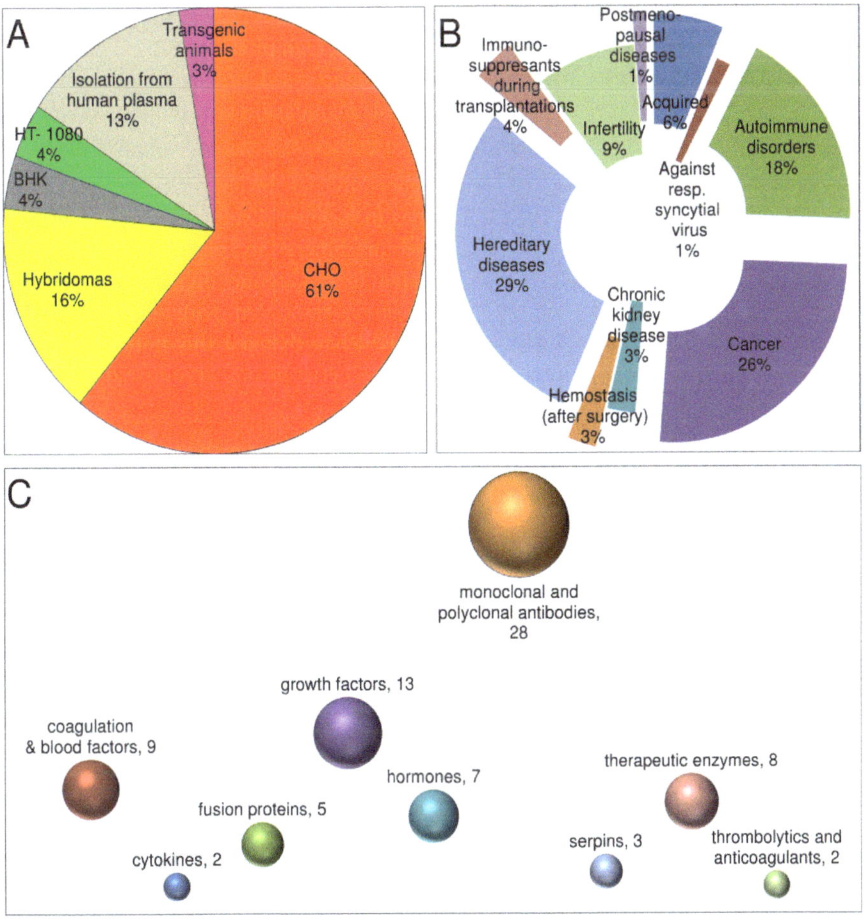

Figure 4. The 77* EMA approved therapeutic glycoprotein drugs (up to April 2012) sorted by: A. Host system used for their production. B. Therapeutic area†. C. Drug category of the active substance based mainly on the classification in [109].

* Actual total number of drugs is 80, but two drugs are used for radionuclide imaging and hence do not consist therapeutic proteins and one active substance has been approved two times, for details see main text.

† Hereditary over autoimmune diseases at some cases are difficult to be distinguished, herein as hereditary diseases those that involve being born with the symptoms of the disease have been accounted for. In this graph, also the drugs Prolia and Xgeva that contain the same active substance but are authorised to treat different diseases have been individually taken into consideration and hence, the number of EMA glycoprotein therapeutic approved drugs for this case are 78.

3.2. Microbial cell systems

Microbial cell systems from both eukaryotic and prokaryotic origin are commonly used for the expression of heterologous proteins. They have long been established for use in the production of recombinant proteins. This is due to their ease of use, the vast wealth of genetic and biochemical knowledge which has amassed, high protein yield and inexpensive production costs. The large availability of well-characterised host strains coupled with extensive customized expression vectors can provide manufacturers with functional protein of high yields.

However, these microbial expression systems generally lack the ability to perform human-like PTMs. Prokaryotic systems have difficulties in forming disulphide bonds and many species, such as *Escherichia coli,* lack the machinery to perform glycosylation. Glycosylation was always felt to be exclusive to eukaryotic domain, however the discovery that the pathogenic *Campylobacter jejuni,* associated with gastroenteritis in humans, can perform N-linked glycosylation [120] has changed this notion. However, there are several notable and important differences between prokaryotic and eukaryotic N-glycosylation (Table 1).

Waker and collaborators identified the native glycosylation capability in *Campylobacter jejuni* and demonstrated that their N-linked glycosylation pathway was also functional when transferred to *Escherichia coli* [121]. The *C. jejuni* glycosylation machinery is encoded by the gene locus *pgl,* consisting of 12 genes encoding for various glycosyltransferases (GTase), enzymes involved in sugar biosynthesis and an oligosaccharyltransferase (OT), all of which share similarity to their eukaryotic counterparts [122]. As in eukaryotes lipid-bound sugar chain is transfer *en bloc* onto an asparagine amino acid in a specific sequence. Their work shows that it is perhaps possible to clone a universal N-linked glycosylation cassette in *E. coli.*

A research team from Cornell University have used a bottom-up engineering approach to assemble a synthetic glycosylation pathway within *E. coli* to produce human-like glycoproteins. The challenge was to introduce a pathway which would lead to an Asn-GlcNAc linkage, then couple this with the mammalian glycosylation pathway which would generate an acceptable sugar moiety, but this was achieved using metabolic engineering ([123] [124] [125]. The *E. coli* was then further engineered to synthesis a mannose3-N-acetylglucosamine2 (Man3GlcNAc2) glycan chain, a common core structure shared in all eukaryotes [126]. However, many challenges still remain. Most notably, the yield of glycosylated protein in this system is extremely low (~1%). In addition, the currently identified bacterial oligosaccharide transferases and known homologs do not to recognise the triplet eukaryotic consensus amino acid sequence, but rather a longer sequence which means protein engineering of glycoproteins will be required in order to use a microbial system. Finally, further sugar addition reactions are required to the core structure Man3GlcNAc2 to produce viable glycoproteins and each of these must be engineered separately into the host system.

	C. jejuni	Eukaryotes
N-glycosylation site consensus sequence	Asp/Glu-Yaa-Asn-Xaa-Ser/Thr (where Yaa, Xaa are not Pro [127]	Asn-Xaa-Ser/Thr
Sugar Linkage	Bac-Asn	GlcNAcβ1-Asn
Sugar Moiety transferred to Asn	(GalNAc5GlcBac)	Glc3Man9GlcNAc2
Further sugar moiety processing after transfer	Not Observed	Yes
Lipid carrier	Undecaprenyl pyrophosphate (Und-PP)	Dolichol Phosphate (Dol-P)
oligosaccharyltransferase	One protein (subunit)	Greater than eight different subunits
Sugar Moiety Transfer Location	Periplasma	ER
Protein State on Transfer	Folded	Unfolded

Table 1. Differences between bacterial and eukaryotic N-linked glycosylation.

Genetic engineering of *E. coli* is an ambitious task because the species does not naturally contain any N-glycosylation machinery, meaning everything must be inserted from the ground up. Non-mammalian eukaryotes are also seen as attractive expression systems and possess an N-glycosylation pathway with a common core structure, although the end-product is remarkably different. Although yeast can perform many PTMs, similar to higher eukaryotic cells, they produce a non-human N-glycosylation profile of high mannose content, which can elicit an immune response in humans. Therefore microbial expression systems are only employed in the production of therapeutic proteins if the protein is functional without PTM, such as human insulin which is produced in *E. coli* [128], or where the PTM is required for folding and stability but does not affect drug efficacy, such as in several vaccines which are produced in *Pichia pastoris* [129, 130] (examples are presented in Table 2).

Recently steps have been taken to genetically engineer humanized N-glycosylation pathways in non-mammalian eukaryotic expression systems such as yeast (*Pichia Pastoris*) and insect cell line Sf9. A synthetic glycosylation pathway has been established within the methyltrophic yeast species *Pichia pastoris* [131-133]. The challenge has been to remove hypermannosylation and replace this with glycosylation machinery to produce a more human-like glycan profile. This has been achieved by the removal (knock-out) of α-1-6-mannosyltransferase enzyme (OCH1 in *P. pastoris*) and replacing this with various mammalian GTases which will synthesise a human-like glycan chain.

The glyco-engineered *P. pastoris* developed by GlycoFi (a subsidiary of Merck & Co. Inc.) is at a more advanced stage than the *E. coli* expression systems. The humanized *P. pastoris* contains all the machinery to produce an N-Glycan chain of complex type, including the gene for the most complex step of human N-glycosylation, terminal sialylation. In total the engineered *P. pastoris* strain contains a set of 14 mammalian genes integrated into its genome. These include glycosyltransferases, enzymes involved in sugar biosynthesis and sugar transporters. The result has been the successful production of a glycoprotein with and an oligosaccharide of the human complex type a highly homogeneous oligosaccharide of the human complex type [131-133].

The N-glycosylation pathway in insects is more similar to mammalian eukaryotic species than *P. pastoris*. Insect cell lines do not hypermannosylate but trim N-glycan to the core structure before adding GlcNAc to a mannose sugar, therefore halting N-glycosylation maturation before mammalian cells, and not gaining a complex or hybrid glycan chain [134]. The glycoengineering of insect cell lines does not require all the mammalian genes added in *P. pastoris* to reach an oligosaccharide of the human complex type.

There are several examples of glycoengineered lepidopteran insect cell lines [135-138]. Specifically, a research group from University of Wyoming have transformed Spodoptera frugiperda (Sf9) cells with higher eukaryotic genes encoding for N-glycosylation pathway including gene products to enable terminal sialylation [136, 139]. Unlike previous glycoengineered species, which have incorporated genes into the genome, the glycosylation genes are placed under inducible plasmids in order to reduce metabolic overload, negative impact on growth rate, and long-term instability [139]. The mammalian glycosylation genes are present on three vectors encoding six mammalian genes in total. These plasmids were transformed into Sf9 cell line which resulted in the sialylated N-glycosylation of a glycoprotein.

Product Name	Manufacturer	Description	Expression System	Eurkaryotic or Prokaryotic
GlucaGen®	Novo-Nordisk	Glucagon for acute hypoglycaemia	Saccharomyces cerevisiae	Eurkaryotic
HEPLISAV™	Dynavax Technologies	Hepatitis B vaccine#	Hansenula polymorpha	Eurkaryotic
Engerix-B	GlaxoSmithKline	Hepatitis B vaccine	Saccharomyces cerevisiae	Eurkaryotic
KALBITOR®	Dyax Corp.	Inhibitor of the protein kallikrein in treatment of hereditary angioedema	Pichia Pastoris	Eurkaryotic
Humulin	Genentech/ EliLilly	Human Insulin	Escherichia Coli	Prokaryotic

Table 2. Examples of commercialized recombinant biopharmaceuticals and the expression systems employed for their production. Products marked with # are currently in clinical trials.

3.3. Manufacturing conditions

From the discussion relating glycan moieties of IFN-β, EPO and both the antibodies' Fc as well as the Fab region to *in vivo* function, it becomes apparent that exerting control over the glycoform of a biotherapeutics would be highly desirable. We know that many process conditions will affect the glycoform such as high CO_2 concentration, which will lead to increased osmolality and can limit growth, antibody production and also affect the glycoform [140]. Similarly culture modes, growth phase, and temperature among other factors will have significant effects on the quality of biotechnology products. This has been extensively summarised by del Val et al. [10] and an overview is presented in Table 3.

Process Condition	Product/Cell Line: Effect		Proposed cause(s)
Nutrient availability	mAb/human hybridoma: change in microheterogeneity when glucose is substituted by fructose, mannose and galactose		Differential efficiency with which the cells consume the different carbon sources [141]
	IFNγ/CHO ↓ site occupancy on at glucose<0.3mM		Low biosynthesis of UTP and glucosamine 6- phosphate [142, 143]
	IFNγ/CHO ↓ sialylation and ↑ high mannose glycans at 0.35mM <glucose<0.7mM and glutamine at 0.1mM		Decline of UDP-GlcNAc synthesis [144]
Dissolved O_2	Effect depends on cell line and product	EPO-Fc/CHO: no effect between 10-90% DOT, ↓ sialylation at 100% DOT	Mechanism not defined [145]
		hFSH/CHO: maximum sialylation at 100%DOT	Mechanism not defined [146]
		IL-2 glycovariant/BHK-21: no effect at 0% DOT	Mechanism not defined [147]
		IgG1 mAb/murine hybridoma CC9C10: ↑galactosylation with ↑ DOT	UDP-Gal increases with DOT due to increased flux through glycolysis [148]
Ammonia and pH	The effect between NH_4^+ and pH is synergic	mPL-I/murine: ↓ site occupancy on at 9mM NH_4^+	Mechanism not defined [149]
		TNF-IgG/CHO: 40% ↓ in galactosylation and sialylation	↑pH inhibits GalT and SiaT activities. ↑pH causes GnTI, ManII, GalT and SiaT to mislocalize in Golgi [150-152]
		IL-2/BHK-21: ↑ branching in at 15mM NH_4^+	↑Concentration of UDP-GlcNAc and UDP-GalNAc reduces the concentration of UDP-Ga l[153]
		EPO/CHO: ↓ sialylation of at 10mM NH_4^+	↑Concentration of UDP-GlcNAc decreases transport of CMP-Neu5Ac into Golgi [154, 155]
		mPL-I/murine: 28%↓ sialylation at 6.8<pH<7.8	↑pH inhibits GalT and SiaT activities. ↑pH causes GnTI, ManII,

Process Condition	Product/Cell Line: Effect	Proposed cause(s)
		GalT and SiaT to mislocalize in Golgi [149]
	Highest levels of IgG3 mAb/murine hybridoma galactosylation at pH=7.4, lowest levels of sialylation at pH=6.9	↑pH inhibits GalT and SiaT activities. ↑pH causes GnTI, ManII, GalT and SiaT to mislocalize in Golgi [156]
Viability, growth phase and temperature	EPO/CHO: ↓ branching at 32°C<T<37°C	↓ temperature decreases UDP-GlcNAc and UDP-GalNAc concentration [157]
Serum and lipid supplements	IFNγ/CHO: ↑ branching with lipid supplementation	Mechanism not understood because serum composition is unknown [158]
	IgG1/CHO: ↑ sialylation with serum-free medium	Mechanism not understood because serum composition is unknown [159]
	IL2/BHK-21: ↑ fucosylation and ↑ sialylation with serum-free medium	Mechanism not understood because serum composition is unknown [147]
	mAb/murine hybridomas: broader glycan distribution and ↑ galactosylation with BSA supplementation	Mechanism not understood because serum composition is unknown [160]
pCO_2 and osmolality	IgG2/SP2/0: unaffected up to 250mmHg pCO_2 ↑ galactosylation at 320mOsm/kg and high pCO_2 25% ↓ in galactosylation at 195mmHg pCO_2 and 435mOsm/kg	osmotic stress induces ↑pH which inhibits GalT and SiaT activities and causes them to mislocalize [161]
Stirring speed	tPA/CHO: 72% ↓ site occupancy at 40<S<200rpm	Shear stress increases protein syhtnesis wich reduces ER retention time [162]
Culture modes	I IFNγ/CHO: low dilution rates yield low site occupancy 0.5<D<0.8day^{-1}.	Concentration of metabolites inhibits GalT and SiaT activity and mislocalizes these compounds. [143]

Table 3. Effect of bioprocess conditions on therapeutic protein glycosylation. The upward arrow denotes increase and the downward arrow represents decrease.

From Table 3, it is evident that many process parameters influence protein N-linked glycosylation. These effects could initially be seen as potential sources for variability. However, if their underlying mechanisms were to be fully understood quantitatively, they could serve as variables for the modulation and control of glycosylation-associated quality attributes of therapeutic proteins.

3.4. Medium formulation

When formulating media for mammalian cell culture for the production of secreted glycoproteins, the carbon source is of paramount importance. The rate at which glycosyltransferases can process glycans is subject to changes in substrate concentrations. The substrates of the glycosyltransferase reactions are the antibodies as well as nucleotide sugar donors (NSD). In the case of biotherapeutics, the important NSDs are UDP-GlcNAc, UDP-Gal, GDP-Fuc CMP-NeuAc and CMP-NeuGc as well as UDP-GalNAc when murine cell lines are employed over CHO cells. High levels of CMP-NeuGc terminating carbohydrate moieties are unfavorable in a therapeutic context as they are oncofetal and potentially immunogenic [163, 164]. The metabolism pathways and biosynthesis of sugars in mammalian cells are well known and are graphically represented in Figure 5 where the transport into the Golgi is also indicated. Expanding on this, the glycoform can be controlled and the concept of feeding strategies is based on the hypothesis that addition of specific metabolic intermediates of the nucleotide-sugar biosynthesis to culture medium will drive metabolic flux towards the desired NSD and eventually influence the glycoform through the increase of desired rates of reaction [165]. Attention must be paid though to a number of inhibitory mechanisms which naturally regulate the depicted metabolic network.

In contrast to supplementing the culture medium with glycosylation substrates to increase certain reaction rates, other strategies have consisted of adding glycosylation reaction inhibitors to achieve desired glycoforms. More specifically, non-reactive fucose analogues have been added to mammalian culture medium to avoid core fucosylation of mAb Fc glycans [166]. Others have added mannosidase inhibitors to prevent mAb Fc oligosaccharides form reaching more processed states [167].

3.5. In-process analysis of glycoproteins

In general, the complete analysis of oligosaccharide structures and linkages requires a series of steps; for example, enzymatic fragmentation, chromatographic separation steps and either mass spectrometry (MS) or nuclear magnetic resonance spectroscopy (NMR) to determine the chemical structure of the fractions [168]. The technology for these offline analyses has improved in terms of the amount of material required, the sensitivity of the method, and the speed of analysis, but there is still room for improvement in miniaturization, speed, and throughput when it comes to analytics for bioprocessing. Particularly if analyses are going to be used to inform manufacturing operations in real-time, then faster methods with lower sample requirements will be required. Preferably, these should also be automatable to reduce the requirement for human expertise.

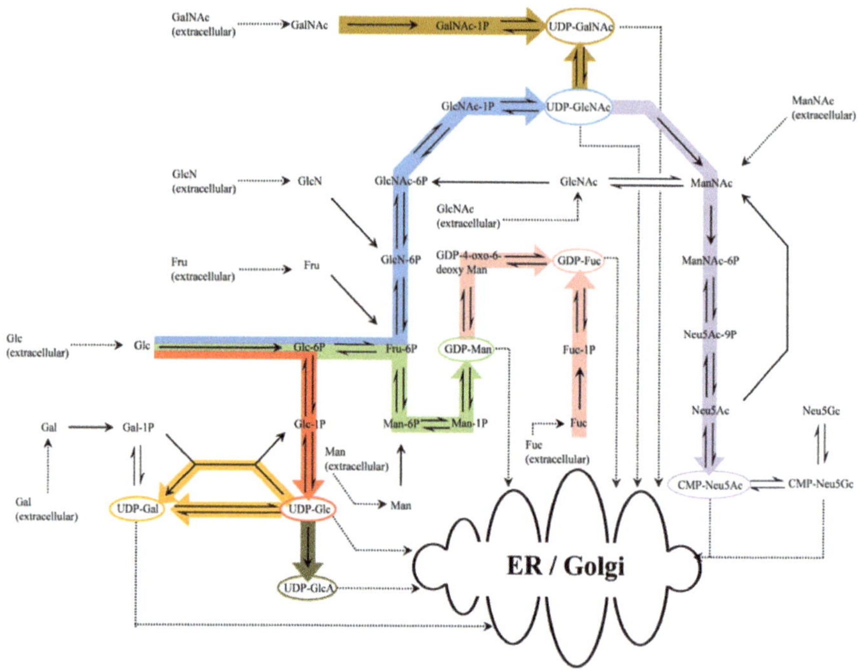

Figure 5. The metabolic pathway of mammalian nucleotide sugar metabolism.

Early attempts at the high throughput analysis of post-translational modifications included a microtitre plate-based assay, where a series of steps including capture, desalting and reduction on beads, followed by elution, tryptic digestion, fractionation, and isolation of N-linked glycopeptides were performed in low volume before MALDI-TOF MS was used to identify structures [169]. While this method had the advantage of small sample volume, it still was a lengthy procedure and while technically automatable, would be difficult to implement in-process.

One of the most important advances is the introduction of online methods for product analysis which will eventually enable real-time or quasi-real time control over the bioreactor environment in order to influence the product quality. Towards this end a number of new systems have been developed usually based on automated sampling followed by desalting via HPLC and analyses that have significantly shorter timescales from sampling to information.

For example,a two-dimensional system capable of analyzing up to 6 fractions from a separations operation was recently reported. Using a single HPLC system capable of running two columns simultaneously with independent gradients and switching between columns coupled to ESI-MS, fractions from size exclusion or ion exchange chromatography steps were analyzed for charge heterogeneity and size variation. An online concentration step was used to analyse dilute fractions to elucidate minor size variants. While unable to give complete peptide mapping of glycoforms, this represents a first step towards monitoring some of the important QbD properties online including glycoform heterogeneity and N-cyclisation [170]. In another recent report, a method for the rapid detection and differentiation of sialic acids by HPLC was developed that is capable of analysing the content of *N*-acetylneuraminic acid versus *N*-glycolylneuraminic acid in about five minutes [171] simply by using a shorter column.

In a further example, Mittermayr et al, made a serendipitous discovery which might lead to more rapid sample processing times when analysing glycans from the Fc region of mAbs. They initially sought to compare the analysis capabilities of a newly developed hydrophilic interaction chromatography method with capillary electrophoresis coupled to laser induced fluorescence analysis, but found that the techniques were actually highly complementary in their resolving power. Thus, using a combination of the two in a 2D analysis, separation time for each could be reduced to 20 minutes. However, some very similar structures could not be resolved from each other, making this method incomplete [172].

Since measuring glycoform profile in real-time will remain a challenge in the near future, another approach is to measure surrogate markers that correlate with the glycan structure (and ultimately make a link using metabolic modelling). This idea is very similar to finding biomarkers for diagnostic indicators for disease—in essence, one or more 'biomarkers' for particular glycan structures of interest (e.g. high mannose, highly branched, or high levels of sialic acid endcapping) should exist. Once these are identified a fluorescent, *in vivo* biosensor can be designed to monitor each 'biomarker' non-invasively in real-time. A variety of fluorescence monitoring equipment is available to accurately determine fluorescence levels in small volumes and in high throughput and this can then be exploited for process design, medium formulation, and cell line engineering experiments. We have demonstrated the utility of FRET-based biosensors for monitoring essential metabolites such as glucose and glutamine [173]. These can then be paired with metabolic models to predict the trajectory of glycoforms given the current nutrient availability.

Also interesting are a suite of tools developed initially for glycomic analysis of whole cell samples (often in the context of disease investigations), but which might be adaptable to monitoring proteins secreted during bioprocessing.

One of the most promising of these techniques is the lectin microarray. Lectins are naturally occurring carbohydrate binding proteins which show some degree of specificity and have been shown to detect glycoproteins in less than 1 pg amounts [174]. Previous work has used lectins conjugated to chromatography resin (e.g [175]) or to magnetic beads for recovering glycan containing proteins from a complex mixture, for example serum [176]. The process can be done in small volumes using microtitre plates and the glycoproteins can later be analysed by MS in a process that can be automated by the use of liquid handling apparati [177].

Lectins immobilised to beads and fluorescence quenching of quantum dots have also been used to quantify specific types of glycoforms, suggesting a proof-of-principle for lectin based analysis [178]. However, the main issue with lectin-based detection as a standard is that the specificity of individual lectins is broad, and therefore, specific detection of individual glycoforms is currently not possible. However, the possibility of using protein engineering or synthetic biology techniques to evolve panels of lectins with precise and varying specificity exists and could result in a platform amenable to high throughput in-process detection [179].

3.6. Experimental strategies for cell line modification

Another strategy to reduce glycan heterogeneity is genetic modification of cell lines. This can be through the overexpression or knockout of the genes encoding for glycosyl transferase enzymes. Such attempts aim to produce mAbs with specific biological functions or to avoid the expression of mAbs with glycan structures that are potentially immunogenic to humans. Umaña's group overexpressed *N*-acetylglucosylaminlytransferase III and V (GnT III & V) in a tetracyclin-regulated manner, with an aim to introduce bisecting GlcNAc and tri-antennary structures respectively in CHO-DUKX cells. Despite the successful production of mAb with desired glycan structure, overexpression of GnT III & V enzymes greatly impeded cell growth [180]. In the same year, Weikert and collaborators suggested the possibility of genetically engineering CHO cells for terminal galactose or sialic acid addition in order to encourage CDC and modulate inflammation. This group showed that overexpression of human β-1,4 galactosyltransferase (GT) or α-2,3 sialyltransferase (ST) genes reduced the level of terminal GlcNAc and more than 90 % of IgG Fc-oligosaccharides were sialylated [181]. The genetic knockout of α-1,6 fucosyltransferase (FUT8) gene achieved via constitutive expression of small interfering RNA (siRNA) produced mAbs where around 60% of them were defucosylated. This increased ADCC activity up to 100-fold in in vitro assays [182]. Genetic engineering is therefore a potential approach to produce mAbs under the QbD strategy, but optimisation is indeed required to minimise possible side-effects.

Chaperone engineering can be another key parameter in boosting productivity. ER chaperones are responsible for correct protein folding and co-overexpression of chaperone genes and other regulatory elements (e.g. ERp57, calnexin/calreticulin, and/or protein disulfide isomerise) exhibited position effects in Productivity. Disulfide isomerase in particular increased specific mAb productivity by 55% in transient gene expression, but there were no/negative effects in stable gene expression [33]. In addition, targeting the unfolded protein response (UPR) pathway is another approach to enhancing recombinant protein yield. The spliced form of X-box binding protein 1 (XBP-1s) is the spliced form of the parental XBP-1 protein and only exists upon the induction of ER-stress. Studies showed that overexpressing XBP-1s in mAb expressing-CHO-T cells under hypothermic condition increased total mAb concentration by 36 % [26].

4. *In silico* protein glycosylation studies

Mathematical modeling is a powerful tool for *in silico* studies of complex phenomena. A high-fidelity model of protein production and glycosylation would be useful for bioprocess design, culture media formulation, and the design of genetic engineering strategies. Early computational studies concerning intracellular glycosylation highlighted particular aspects of the biological machinery. Monica et al. investigated the role of diffusion in the trans-Golgi network on limited sialylation [183]. The model assumed an isotropic compartment with respect to both substrate and enzyme concentration and concluded that diffusion limitations are not significant with respect to the sialyltransferase-catalysed reactions. This finding is particularly significant with respect to other glycosyltransferase reactions as sialic acid is known to be at the lowest abundance compared to other nucleotide sugar species. Shelikoff et al. presented a first approach towards the mathematical modelling of macroheterogeneity in glycoproteins [184]. The work focused on the attachment of the glycan precursor to the Asp-X-Ser/Thr tripeptide sequence, which takes place in the ER. While site occupancy currently receives little attention in the production of mAbs, this aspect of glycosylation may be of greater significance in the near future as cellular antibody productivity keeps increasing and thus, placing more strain on the ER, which may lead to increased macroheterogeneity.

The first mathematical investigation of glycosylation microheterogeneity was carried out by Umaña et al. in 1997 as part of a study into the effect of glycosyltransferase overexpression in a mammalian cell line in order to exert control over the glycoform [185]. As part of this study, a *Central Reaction Network* (CRN) to monitor a total of 33 species comprising mannosidases, GlcNAc-transferases (GnTs) and terminating upon the addition of the first galactose, which prevents further processing through GnTs or ManII of the glycan structure [186], was proposed. The calculations of species abundance were based on enzyme concentrations as well as distribution, kinetic constants of reaction, protein half-life in Golgi, the Golgi volume and finally the specific glycoprotein productivity. The underlying mode of operation for the Golgi is assumed to be the vesicular transport model, which states that vesicles will bud off their respective Golgi compartment at their bulk concentration and fuse with the next compartment in series. This mode of operation can be idealised and viewed as four continuously stirred tank reactors (CSTRs) in series, representing the cis-, medial-, trans-Golgi cisternae and the trans-Golgi network. The work paid particular attention to GnTIII, which catalyses the transfer of a bi-secting GlcNAc to an agalactosylated glycan moiety upon which no further GnTs can act and thus, capping antennarity. The authors of the study examined the overexpression of the particular glycosyltransferase in a number of *in silico* experiments and confirmed their hypothesis that antennarity was reduced and hybrid glycan content increased. This study provided an important first insight into the power of mathematical modelling as an approach towards glycan engineering.

Krambeck and Betenbaugh extended the above described work through the inclusion of further glycosyltransferases and thus, expanding the number of structures from 33 to 7,565 structures resulting from a total of 22,871 reactions, which accounts for core fucosylation, galactosylation and sialylation of carbohydrate moieties [187]. The model was also not specific to a single glycan site and viewed the Golgi compartments as four CSTRs in series. Building on previous work, enzyme dissociation constants from experimental investigations were employed for each glycosyltransferase and competitive product inhibition was taken into account. Furthermore, the model was evaluated and fitted against experimental data, where the glycoform data was obtained from recombinant human thrombopoietin (TPO) in which an average of 5.4 occupied N-glycan sites have been reported per molecule [188]. Model optimisation was based on an averaged TPO glycan site, where enzyme concentrations were altered to give a closest fit to experimental data. Krambeck and Betenbaugh argue that while dissociation constants and kinetic data for the glycosyltransferases exist in literature, the enzyme concentration in the Golgi is cell line-dependent and will be subject to change based on culture conditions. Golgi-resident enzyme concentrations were changed to match data and resulted in improved model simulation results.

Krambeck et al. extended the model further in a follow up study, where the model was tailored to analyse glycoforms from mass spectrometric data [189]. Mass spectrometry is based on mass to charge ratio of species and, thus, will not be able to distinguish between different structures of same molecular mass. The presented model attempts to resolve the issue through the prediction of alternative structures content for the same mass spectrometric data peak and eventually screen for glycan disease markers in humans. The model was extended through the addition of further glycosylation enzymes to makeup a total of 19 glycosyltransferases and subsequently enzyme activities were adjusted to match normal and malignant human monocyte N-glycan mass spectra. Through application of limiting conditions based on prior knowledge, glycans of probable negligible abundance can be omitted resulting in a total of 10,000 – 20,000 structures. The model gives valuable insight into changes in enzyme activity as a result of different diseases, but is rather limited in its application to the development and production of biotherapeutics, where large specificity with respect to the glycan structure and accessibility to the protein backbone is required for a highly accurate predictive model.

Hossler et al. attempted to improve the predictive ability of glycan distribution models through the variation of reaction-related variables [190]. This model was the first to discriminate between reaction mechanisms for different glycosylation enzymes. ManI and ManII were modelled assuming Michaelis-Menten kinetics with substrate competition; the remaining transferases were modelled using a rapid equilibrium, random, Bi-Bi mechanism. While all previous models assumed the vesicular transport regime, which is modelled as a series of CSTRs, this study explored the hypothesis of the Golgi maturation model, which states that each compartment undergoes a maturation process to transform from early cisternae to late cisternae. In an idealised case this can be described by a plug flow reactor (PFR) and, thus, travelling through a tubular reactor, representing the Golgi cisternae. Modelling the Golgi apparatus as a single reactor of constant enzyme concentration showed that a long enough residence time will lead to highly processed glycan structures and changes in glycosyltransferase concentration can lead to a targeted glycoform for most glycan species. However, the authors of the study argued that a total of four reactors will be required to accurately simulate changes in the enzyme concentration along the length of the Golgi apparatus. The results show a greater appearance of under-processed glycan structures in the final product and generally less deviation from the data obtained from the CSTR-in-series model. A decrease in protein residence time was shown to have a much larger impact on the four PFR model than the CSTR model, with many more under-processed glycans were observed for the PFR case. It was further shown that modifications of the enzyme concentrations for the PFR-in-series model could lead to the most targeted glycoform and thus, demonstrating enzyme localization to be a very potent approach in glycan engineering. Hossler et al. concluded that while the actual biological mechanism will be less idealised than assumed in the study, the PFR-in-series model will give a more true distribution as demonstrated by comparison with experimental data.

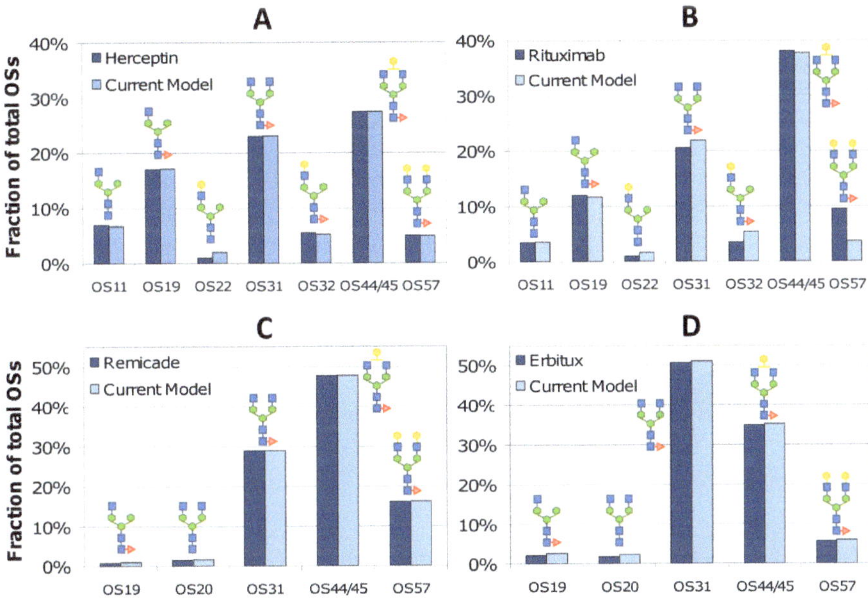

Figure 6. Calculated vs. experimental oligosaccharide profiles with the Jimenez del Val model [106]. The dark bars correspond to the experimental oligosaccharide profiles, and the light bars represent those calculated with the model. Comparison between Herceptin (A), Rituxan (B), Remicade (C), and Erbitux (D) are shown.

Recently, Jimenez del Val et al. developed a model specifically tailored to the glycan found on the Asn297 position of an IgG antibody constant region [191]. The model includes a cisternal maturation approach and expands on rate expressions for various enzymes to include Michaelis-Menten, sequential Bi-Bi and random order Bi-Bi kinetics for specific glycosyltransferases as reported in previous literature for each enzyme. The model considers the Golgi apparatus to be a single PFR of constant diameter, no axial dispersion within the compartment, constant flow and no mass transfer limitation where enzyme recycling along the biological reactor length leads to changes in glycosyltransferase concentrations. As a result it proposes a novel representation of enzyme concentrations along the length of a PFR as normal distribution functions. The unknowns of the three parameter normal functions for the spatial distribution of the enzymes where found through optimization-based methods, where the minimum amount of total enzyme necessary to achieve terminal oligosaccharide processing, including 50% sialylation, was sought. A further extension of previous mathematical models was the incorporation of protein-mediated nucleotide sugar donor transport into the Golgi cisternae. Again, due to the assumption of Golgi-resident protein recycling, a distribution of transport proteins is expected along the length of the PFR, which was estimated using an optimisation-based method. The optimisation was based on the assumption that the rate of by-product dephosphorylation is much faster than nucleotide sugar donor accumulation and, therefore, parameter values for minimum transport protein concentration were determined. Further, it was argued that while the above distributions should not change significantly, the dissociation constants will differ for individual glycoproteins as well as glycan sites within a glycoprotein. This accounts for steric hindrance and much reduced sialylation in the antibody Fc region and, hence, by taking dissociation constants for various commercial mAbs from literature, a glycoform was obtained. A comparison with experimental data, the Krambeck and Bettenbaugh, as well as the Hossler et al. model showed that the hereby obtained mathematical tool presented the closest fit to experimental data for most glycan species analysed as shown in Figure 6. Furthermore the model was demonstrated to show good fit to experimental data under gene silencing as presented in Figure 7.

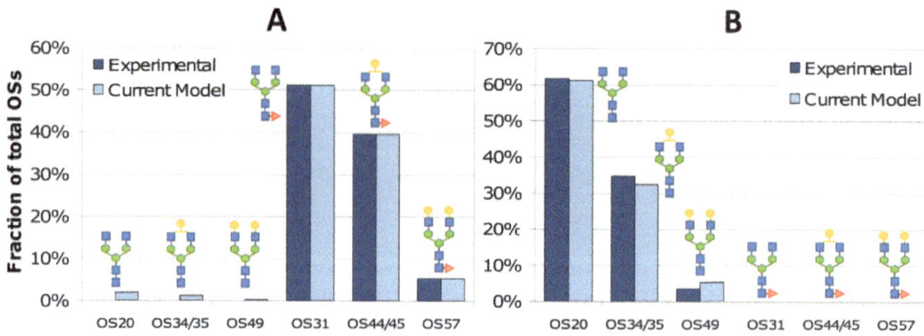

Figure 7. A: Compares the reported oligosaccharide profiles (dark bars) with the ones calculated with the Jimenez del Val model (light bars), before FucT gene silencing. B: Compares reported and calculated oligosaccharide profiles after FucT silencing. Experimental results from [192].

5. Conclusions and outlook

The main goal of QbD is to ensure product quality by building it into the manufacturing process. The initiative provides practical incentives, such as shorter approval times and higher flexibility towards changes in manufacturing process conditions, for industrial production. Despite the volume of data required for QbD, the benefits far outweigh the effort. The first goal of QbD for biopharmaceuticals should be to narrow the glycomic profile of glycoprotein-based drugs based on existing knowledge of the desired structures for the application at hand and of the effect of manufacturing conditions and media/feed formulation on the availability of nucleotide sugars and the resulting glycan profile. This can be achieved more effectively through the combination of fundamental biological techniques for cell engineering, methodologies allowing rapid glycan analysis (in particular *in vivo* biosensors for 'biomarkers' of the desired glycan structure), and rational engineering design of manufacturing conditions.

Rapid analytical tools will allow us to examine more samples in-process with the aim of controlling and potentially optimising conditions in real time. An enabling tool is mathematical modelling, which, given its tremendous progress in successfully simulating the modification of complex glycoproteins, could in the future allow us to collect in process information about a fermentation run arising from various analyses at line and use it to infer the current state of the system and design improved operation strategies in terms of supplementation of nutrients or precursors, or adjustment of DOT, culture pH or other key conditions.

At the same time, while CHO cells clearly remain the dominant host system, research on other, more prominent cell lines that provide better glycosylation, or higher yield remains active. Yeast, insect, plant cells, and transgenic animals are amongst the more likely host systems to replace CHO cells. However, a significant amount of research effort regarding further bioprocess optimisation, as well as how they can be engineered to produce complex therapeutic molecules, such as heparin is still ongoing for CHO cells. Given the continued investment in CHO research, the number of existing production platforms, and pending patents as well as the stringency of the pharmaceutical regulatory agencies, they will likely remain a relevant industrial production host for at least another few decades.

That being said, the ideal expression system would achieve a high level expression of recombinant protein at a low cost. This implies that microbial hosts would lead to a significant cost advantage. Advances in humanising the PTM machinery in microbial hosts, *P. pastoris* in particular, offer significant promise. However, the aforementioned research developments have been achieved in lab-scale fermentation. To reap their benefits, the glycosylation profile needs to be demonstrated to be consistently homogeneous in large-scale fermentations. Overall, it is clear that the QbD initiative dictates a unified engineering and scientific approach to potentiate control over the glycomic profile of cell culture-derived protein-based drugs.

Author details

Ioscani Jimenez del Val, Sarantos Kyriakopoulos and Cleo Kontoravdi
Centre for Process Systems Engineering, Department of Chemical Engineering,
Imperial College London, London, UK

Philip M. Jedrzejewski, Kealan Exley and Si Nga Sou
Centre for Process Systems Engineering, Department of Chemical Engineering,
Imperial College London, London, UK
Division of Molecular Biosciences, Department of Life Sciences,
Imperial College London, London, UK
Centre for Synthetic Biology and Innovation, Imperial College London, London, UK

M. Polizzi
Division of Molecular Biosciences, Department of Life Sciences,
Imperial College London, London, UK
Centre for Synthetic Biology and Innovation, Imperial College London, London, UK

6. References

[1] EvaluatePharma, World Preview 2016 "Beyond the Patent Cliff". 2011.
[2] Walsh, G., Biopharmaceutical benchmarks 2010. Nature Biotechnology, 2010. 28(9): p. 917-924.
[3] De Jesus, M. and F.M. Wurm, Manufacturing recombinant proteins in kg-ton quantities using animal cells in bioreactors. European Journal of Pharmaceutics and Biopharmaceutics, 2011. 78(2): p. 184-188.
[4] Kawasaki, N., et al., The Significance of Glycosylation Analysis in Development of Biopharmaceuticals. Biological & Pharmaceutical Bulletin, 2009. 32(5): p. 796-800.
[5] Chung, C.H., et al., Cetuximab-induced anaphylaxis and IgE specific for galactose-alpha-1,3-galactose. N Engl J Med, 2008. 358(11): p. 1109-1117.
[6] Purcell, R.T. and R.F. Lockey, Immunologic Responses to Therapeutic Biologic Agents. Journal of Investigational Allergology and Clinical Immunology, 2008. 18(5): p. 335-342.
[7] Sethuraman, N. and T.A. Stadheim, Challenges in therapeutic glycoprotein production. Current Opinion in Biotechnology, 2006. 17(4): p. 341-346.
[8] EMA Guideline on immunogenicity assessment for biotechnology-derived therapeutic proteins. 2007.
[9] Balaquer, E. and C. Neususs, Intact glycoform characterization of erythropoietin-alpha and erythropoietin-beta by CZE-ESI-TOF-MS. Chromatographia, 2006. 64(5-6): p. 351-357.
[10] del Val, I.J., C. Kontoravdi, and J.M. Nagy, Towards the Implementation of Quality by Design to the Production of Therapeutic Monoclonal Antibodies with Desired Glycosylation Patterns. Biotechnology Progress, 2010. 26(6): p. 1505-1527.
[11] Kamoda, S., R. Ishikawa, and K. Kakehi, Capillary electrophoresis with laser-induced fluorescence detection for detailed studies on N-linked oligosaccharide profile of therapeutic recombinant monoclonal antibodies. J Chromatogr A, 2006. 1133(1-2): p. 332-339.
[12] Stadlmann, J., et al., Analysis of immunoglobulin glycosylation by LC-ESI-MS of glycopeptides and oligosaccharides. Proteomics, 2008. 8(14): p. 2858-2871.
[13] Nakano, M., et al., Capillary electrophoresis-electrospray ionization mass spectrometry for rapid and sensitive N-glycan analysis of glycoproteins as 9-fluorenylmethyl derivatives. Glycobiology, 2009. 19(2): p. 135-143.
[14] Venter, J.C., et al., The sequence of the human genome. Science, 2001. 291(5507): p. 1304-+.
[15] Petrescu, A.J., et al., Statistical analysis of the protein environment of N-glycosylation sites: implications for occupancy, structure, and folding. Glycobiology, 2004. 14(2): p. 103-114.

[16] Helenius, A., How N-Linked Oligosaccharides Affect Glycoprotein Folding in the Endoplasmic-Reticulum. Molecular Biology of the Cell, 1994. 5(3): p. 253-265.
[17] Imperiali, B. and S.E. O'Connor, Effect of N-linked glycosylation on glycopeptide and glycoprotein structure. Current Opinion in Chemical Biology, 1999. 3(6): p. 643-649.
[18] Olden, K., J.B. Parent, and S.L. White, Carbohydrate Moieties of Glycoproteins - a Re-Evaluation of Their Function. Biochimica Et Biophysica Acta, 1982. 650(4): p. 209-232.
[19] Sen, G.C. and P. Lengyel, The Interferon System - a Birds-Eye-View of Its Biochemistry. Journal of Biological Chemistry, 1992. 267(8): p. 5017-5020.
[20] Taniguchi, T., et al., Expression of the Human Fibroblast Interferon Gene in Escherichia-Coli. Proceedings of the National Academy of Sciences of the United States of America-Biological Sciences, 1980. 77(9): p. 5230-5233.
[21] Runkel, L., et al., Structural and functional differences between glycosylated and non-glycosylated forms of human interferon-beta (IFN-beta). Pharmaceutical Research, 1998. 15(4): p. 641-649.
[22] Sola, R.J. and K. Griebenow, Effects of Glycosylation on the Stability of Protein Pharmaceuticals. Journal of Pharmaceutical Sciences, 2009. 98(4): p. 1223-1245.
[23] EvaluatePharma, Beyond the Patent Cliff, in World Preview to 20162011, EvaluatePharma Ltd. p. 34.
[24] Reichert, E.D.J.M., World Bispecific Antibody Summit, September 27–28, 2011, Boston, MA, in World Bispecific Antibody Summit, L. Bioscience, Editor 2012, mAbs: Boston, MA. p. 13.
[25] Rekhert, J.M. and J.B. Wenger, Development trends for new cancer therapeutics and vaccines. Drug Discovery Today, 2008. 13(1-2): p. 30-37.
[26] Strebhardt, K. and A. Ullrich, Paul Ehrlich's magic bullet concept: 100 years of progress. Nature Reviews Cancer, 2008. 8(6): p. 473-480.
[27] Kohler, G. and C. Milstein, Continuous Cultures of Fused Cells Secreting Antibody of Predefined Specificity. Nature, 1975. 256(5517): p. 495-497.
[28] McCafferty, J., et al., Phage antibodies: filamentous phage displaying antibody variable domains. Nature, 1990. 348(6301): p. 552-4.
[29] Barbas, C.F., 3rd, et al., Assembly of combinatorial antibody libraries on phage surfaces: the gene III site. Proc Natl Acad Sci U S A, 1991. 88(18): p. 7978-82.
[30] Krapp, S., et al., Structural analysis of human IgG-Fc glycoforms reveals a correlation between glycosylation and structural integrity. Journal of Molecular Biology, 2003. 325(5): p. 979-989.
[31] Hodoniczky, J., Y.Z. Zheng, and D.C. James, Control of recombinant monoclonal antibody effector functions by Fc N-glycan remodeling in vitro. Biotechnology Progress, 2005. 21(6): p. 1644-1652.
[32] Raju, T.S., Terminal sugars of Fc glycans influence antibody effector functions of IgGs. Current Opinion in Immunology, 2008. 20(4): p. 471-478.
[33] Zou, G.Z., et al., Chemoenzymatic Synthesis and Fc gamma Receptor Binding of Homogeneous Glycoforms of Antibody Fc Domain. Presence of a Bisecting Sugar

Moiety Enhances the Affinity of Fc to Fc gamma IIIa Receptor. Journal of the American Chemical Society, 2011. 133(46): p. 18975-18991.

[34] Shields, R.L., et al., Lack of fucose on human IgG1 N-linked oligosaccharide improves binding to human Fc gamma RIII and antibody-dependent cellular toxicity. Journal of Biological Chemistry, 2002. 277(30): p. 26733-26740.

[35] Matsumiya, S., et al., Structural comparison of fucosylated and nonfucosylated Fc fragments of human immunoglobulin G1. Journal of Molecular Biology, 2007. 368(3): p. 767-779.

[36] Friess T, H.F., Bauer S, et al., GA101, a novel therapeutic type II CD20 antibody with outstanding efficacy and clear superiority in subcutaneous non-Hodgkin lymphoma xenograft models as well as superior B cell depletion. , in American Association for Cancer Research2008: San Diego, CA. .

[37] Jefferis, R., et al., A comparative study of the N-linked oligosaccharide structures of human IgG subclass proteins. Biochem J, 1990. 268(3): p. 529-537.

[38] Parekh, R.B., et al., Association of Rheumatoid-Arthritis and Primary Osteo-Arthritis with Changes in the Glycosylation Pattern of Total Serum Igg. Nature, 1985. 316(6027): p. 452-457.

[39] Raju, T.S., et al., Glycoengineering of therapeutic glycoproteins: In vitro galactosylation and sialylation of glycoproteins with terminal N-acetylglucosamine and galactose residues. Biochemistry, 2001. 40(30): p. 8868-8876.

[40] Takahashi, N., et al., Comparative Structural Study of the N-Linked Oligosaccharides of Human Normal and Pathological Immunoglobulin-G. Biochemistry, 1987. 26(4): p. 1137-1144.

[41] Anthony, R.M., et al., Recapitulation of IVIG anti-inflammatory activity with a recombinant IgG fc. Science, 2008. 320(5874): p. 373-376.

[42] Goetze, A.M., et al., High-mannose glycans on the Fc region of therapeutic IgG antibodies increase serum clearance in humans. Glycobiology, 2011. 21(7): p. 949-959.

[43] Youings, A., et al., Site-specific glycosylation of human immunoglobulin G is altered in four rheumatoid arthritis patients. Biochemical Journal, 1996. 314: p. 621-630.

[44] Abel, C.A., Spiegelberg, H. L., and Grey, H. M., The carbohydrate content of fragments and polypeptide chains of human >G-myeloma protein of different heavy chain subclasses. Biochemistry and Cell Biology-Biochimie Et Biologie Cellulaire, 1968. 7: p. 1271-1278.

[45] Coloma, M.J., et al., Position effects of variable region carbohydrate on the affinity and in vivo behavior of an anti-(1 -> 6) dextran antibody. Journal of Immunology, 1999. 162(4): p. 2162-2170.

[46] Holland, M., et al., Differential glycosylation of polyclonal IgG, IgG-Fc and IgG-Fab isolated from the sera of patients with ANCA-associated systemic vasculitis. Biochimica Et Biophysica Acta-General Subjects, 2006. 1760(4): p. 669-677.

[47] Huang, L.H., et al., Impact of variable domain glycosylation on antibody clearance: An LC/MS characterization. Analytical Biochemistry, 2006. 349(2): p. 197-207.

[48] Wright, A., et al., Antibody Variable Region Glycosylation - Position Effects on Antigen-Binding and Carbohydrate Structure. Embo Journal, 1991. 10(10): p. 2717-2723.

[49] Dandrea, A.D., H.F. Lodish, and G.G. Wong, Expression Cloning of the Murine Erythropoietin Receptor. Cell, 1989. 57(2): p. 277-285.

[50] Lodish, H.F., et al., The erythropoietin receptor: Biogenesis, dimerization, and intracellular signal transduction. Cold Spring Harbor Symposia on Quantitative Biology, 1995. 60: p. 93-104.

[51] Constantinescu, S.N., S. Ghaffari, and H.F. Lodish, The erythropoietin receptor: Structure, activation and intracellular signal transduction. Trends in Endocrinology and Metabolism, 1999. 10(1): p. 18-23.

[52] Winearls, C.G., et al., Effect of Human Erythropoietin Derived from Recombinant-DNA on the Anemia of Patients Maintained by Chronic-Hemodialysis. Lancet, 1986. 2(8517): p. 1175-1178.

[53] Eschbach, J.W., et al., Recombinant Human Erythropoietin in Anemic Patients with End-Stage Renal-Disease - Results of a Phase-Iii Multicenter Clinical-Trial. Annals of Internal Medicine, 1989. 111(12): p. 992-1000.

[54] Markham, A. and H.M. Bryson, Epoetin Alfa - a Review of Its Pharmacodynamic and Pharmacokinetic Properties and Therapeutic Use in Nonrenal Applications. Drugs, 1995. 49(2): p. 232-254.

[55] Cazzola, M., F. Mercuriali, and C. Brugnara, Use of recombinant human erythropoietin outside the setting of uremia. Blood, 1997. 89(12): p. 4248-4267.

[56] Browne, J.K., et al., Erythropoietin - Gene Cloning, Protein-Structure, and Biological Properties. Cold Spring Harbor Symposia on Quantitative Biology, 1986. 51: p. 693-702.

[57] Egrie, J.C., et al., Characterization and Biological Effects of Recombinant-Human-Erythropoietin. Immunobiology, 1986. 172(3-5): p. 213-224.

[58] Egrie, J.C. and J.K. Browne, Development and characterization of novel erythropoiesis stimulating protein (NESP). British Journal of Cancer, 2001. 84: p. 3-10.

[59] Sasaki, H., et al., Site-Specific Glycosylation of Human Recombinant Erythropoietin - Analysis of Glycopeptides or Peptides at Each Glycosylation Site by Fast Atom Bombardment Mass-Spectrometry. Biochemistry, 1988. 27(23): p. 8618-8626.

[60] Misaizu, T., et al., Role of Antennary Structure of N-Linked Sugar Chains in Renal Handling of Recombinant-Human-Erythropoietin. Blood, 1995. 86(11): p. 4097-4104.

[61] Delorme, E., et al., Role of Glycosylation on the Secretion and Biological-Activity of Erythropoietin. Biochemistry, 1992. 31(41): p. 9871-9876.

[62] Dube, S., J.W. Fisher, and J.S. Powell, Glycosylation at Specific Sites of Erythropoietin Is Essential for Biosynthesis, Secretion, and Biological Function. Journal of Biological Chemistry, 1988. 263(33): p. 17516-17521.

[63] Egrie, J.C., et al., The Role of Carbohydrate on the Biological-Activity of Erythropoietin. Glycoconjugate Journal, 1993. 10(4): p. 263-263.

[64] Fukuda, M.N., et al., Survival of Recombinant Erythropoietin in the Circulation - the Role of Carbohydrates. Blood, 1989. 73(1): p. 84-89.

[65] Spivak, J.L. and B.B. Hogans, The Invivo Metabolism of Recombinant Human Erythropoietin in the Rat. Blood, 1989. 73(1): p. 90-99.
[66] Egrie, J.C., et al., Darbepoetin alfa has a longer circulating half-life and greater in vivo potency than recombinant human erythropoietin. Experimental Hematology, 2003. 31(4): p. 290-299.
[67] Paul, S.M., et al., How to improve R&D productivity: the pharmaceutical industry's grand challenge. Nat Rev Drug Discov, 2010. 9(3): p. 203-14.
[68] Yu, L.X., Pharmaceutical quality by design: Product and process development, understanding, and control. Pharmaceut Res, 2008. 25(4): p. 781-791.
[69] Hinz, D.C., Process analytical technologies in the pharmaceutical industry: the FDA's PAT initiative. Analytical and Bioanalytical Chemistry, 2006. 384(5): p. 1036-1042.
[70] Kenett, R.S. and D.A. Kenett, Quality by Design applications in biosimilar pharmaceutical products. Accredit Qual Assur, 2008. 13(12): p. 681-690.
[71] ICH Harmonized Tripartite Guideline on Pharmaceutical Development Q8(R2). 2009.
[72] Rathore, A.S. and H. Winkle, Quality by design for biopharmaceuticals. Nat Biotechnol, 2009. 27(1): p. 26-34.
[73] Kaitin, K.I., Deconstructing the drug development process: the new face of innovation. Clin Pharmacol Ther, 2010. 87(3): p. 356-61.
[74] Reichert, J.M., Trends in development and approval times for new therapeutics in the United States. Nat Rev Drug Discov, 2003. 2(9): p. 695-702.
[75] DiMasi, J. and H. Grabowski, The Cost of Biopharmaceutical R&D: Is Biotech Different? Manage Decis Econ, 2007. 28(4-5).
[76] Kenett, R. and D. Kenett, Quality by Design applications in biosimilar pharmaceutical products. Accreditation and Quality Assurance, 2008. 13(12): p. 681-690.
[77] USFDA, Guidance for industry: PAT -- A framework for innovative pharmaceutical development, manufacturing, and quality assurance, 2004, US Food and Drug Administration: Rockville, MD.
[78] ICH Harmonized Tripartite Guideline on Quality Risk Management Q9. 2005.
[79] Vogt, F.G. and A.S. Kord, Development of Quality-By-Design Analytical Methods. Journal of Pharmaceutical Sciences, 2011. 100(3): p. 797-812.
[80] Rathore, A.S., et al., Quality by Design: Industrial Case Studies on Defining and Implementing Design Space for Pharmaceutical Processes Part 2. BioPharm International, 2009. 22(1): p. 36+.
[81] Rosenberg, A., Effects of protein aggregates: An immunologic perspective. The AAPS Journal, 2006. 8(3): p. E501-E507.
[82] Hermeling, S., et al., Antibody response to aggregated human interferon alpha2b in wild-type and transgenic immune tolerant mice depends on type and level of aggregation. Journal of Pharmaceutical Sciences, 2006. 95(5): p. 1084-1096.
[83] Liu, D., et al., Structure and stability changes of human IgG1 Fc as a consequence of methionine oxidation. Biochemistry, 2008. 47(18): p. 5088-5100.

[84] Vlasak, J. and R. Ionescu, Fragmentation of monoclonal antibodies. Mabs, 2011. 3(3): p. 253-263.
[85] Page, M., et al., Fragmentation of Therapeutic Human-Immunoglobulin Preparations. Vox Sanguinis, 1995. 69(3): p. 183-194.
[86] Vlasak, J. and R. Ionescu, Heterogeneity of Monoclonal Antibodies Revealed by Charge-Sensitive Methods. Current Pharmaceutical Biotechnology, 2008. 9(6): p. 468-481.
[87] Keck, R., et al., Characterization of a complex glycoprotein whose variable metabolic clearance in humans is dependent on terminal N-acetylglucosamine content. Biologicals, 2008. 36(1): p. 49-60.
[88] Harris, R.J., Processing of C-Terminal Lysine and Arginine Residues of Proteins Isolated from Mammalian-Cell Culture. Journal of Chromatography A, 1995. 705(1): p. 129-134.
[89] Goetze, A.M., M.R. Schenauer, and G.C. Flynn, Assessing monoclonal antibody product quality attribute criticality through clinical studies. Mabs, 2010. 2(5): p. 500-507.
[90] Quan, C., et al., A study in glycation of a therapeutic recombinant humanized monoclonal antibody: Where it is, how it got there, and how it affects charge-based behavior. Analytical Biochemistry, 2008. 373(2): p. 179-191.
[91] Nakajou, K., et al., The effect of glycation on the structure, function and biological fate of human serum albumin as revealed by recombinant mutants. Biochimica Et Biophysica Acta-General Subjects, 2003. 1623(2-3): p. 88-97.
[92] Sharma, B., Immunogenicity of therapeutic proteins. Part 3: Impact of manufacturing changes. Biotechnology Advances, 2007. 25(3): p. 325-331.
[93] Wadhwa, M., et al., Immunogenicity of GM-CSF products in cancer patients following immunostimulatory therapy with GM-CSF. Developments in biologicals, 2003. 112: p. 61-7.
[94] Organization, W.H., Requirements for the use of animal cells as in vitro substrates for the production of biologicals, in WHO Expert Committee on Biological Standardization1998: Geneva.
[95] Shepherd, A.J., N.J. Wilson, and K.T. Smith, Characterisation of endogenous retrovirus in rodent cell lines used for production of biologicals. Biologicals, 2003. 31(4): p. 251-260.
[96] Eon-Duval, A., H. Broly, and R. Gleixner, Quality attributes of recombinant therapeutic proteins: An assessment of impact on safety and efficacy as part of a quality by design development approach. Biotechnology Progress, 2012: p. n/a-n/a.
[97] Garnick, R.L., Peptide-Mapping for Detecting Variants in Protein Products. International Symposium on Continuous Cell Lines - an International Workshop on Current Issues, 1992. 76: p. 117-130.
[98] Gennaro, L.A., O. Salas-Solano, and S. Ma, Capillary electrophoresis-mass spectrometry as a characterization tool for therapeutic proteins. Analytical Biochemistry, 2006. 355(2): p. 249-258.
[99] Chin, E.T. and D.I. Papac, The use of a porous graphitic carbon column for desalting hydrophilic peptides prior to matrix-assisted laser desorption/ionization time-of-flight mass spectrometry. Analytical Biochemistry, 1999. 273(2): p. 179-185.

[100] Wei, W., Instability, stabilization, and formulation of liquid protein pharmaceuticals. International Journal of Pharmaceutics, 1999. 185(2): p. 129-188.

[101] Jones, A.J.S., Analysis of Polypeptides and Proteins. Advanced Drug Delivery Reviews, 1993. 10(1): p. 29-90.

[102] Ahrer, K., et al., Analysis of aggregates of human immunoglobulin G using size-exclusion chromatography, static and dynamic light scattering. Journal of Chromatography A, 2003. 1009(1-2): p. 89-96.

[103] EMA Guideline on similar biological medicinal products containing biotechnologically-derived proteins as active substance: non-clinical and clinical issues. 2006.

[104] Landgrebe, D., et al., On-line infrared spectroscopy for bioprocess monitoring. Applied Microbiology and Biotechnology, 2010. 88(1): p. 11-22.

[105] Behjousiar, A., C. Kontoravdi, and K.M. Polizzi, In Situ Monitoring of Intracellular Glucose and Glutamine in CHO Cell Culture. Plos One, 2012. 7(4): p. e34512.

[106] del Val, I.J., J.M. Nagy, and C. Kontoravdi, A dynamic mathematical model for monoclonal antibody N-linked glycosylation and nucleotide sugar donor transport within a maturing Golgi apparatus. Biotechnology Progress, 2011. 27(6): p. 1730-1743.

[107] Jefferis, R., Glycosylation as a strategy to improve antibody-based therapeutics. Nat Rev Drug Discov, 2009. 8(3): p. 226-234.

[108] Wright, A. and S.L. Morrison, Effect of glycosylation on antibody function: implications for genetic engineering. Trends in Biotechnology, 1997. 15(1): p. 26-32.

[109] Aggarwal, S., What's fueling the biotech engine--2010 to 2011. Nat Biotechnol, 2011. 29(12): p. 1083-9.

[110] Chu, L. and D.K. Robinson, Industrial choices for protein production by large-scale cell culture. Curr Opin Biotechnol, 2001. 12(2): p. 180-7.

[111] Bollin, F., V. Dechavanne, and L. Chevalet, Design of Experiment in CHO and HEK transient transfection condition optimization. Protein Expr Purif, 2011. 78(1): p. 61-8.

[112] Kim, J.Y., Y.G. Kim, and G.M. Lee, CHO cells in biotechnology for production of recombinant proteins: current state and further potential. Appl Microbiol Biotechnol, 2012. 93(3): p. 917-30.

[113] Trill, J.J., A.R. Shatzman, and S. Ganguly, Production of monoclonal antibodies in COS and CHO cells. Curr Opin Biotechnol, 1995. 6(5): p. 553-60.

[114] Jenkins, N., R.B. Parekh, and D.C. James, Getting the glycosylation right: implications for the biotechnology industry. Nat Biotechnol, 1996. 14(8): p. 975-81.

[115] Suttie, J.W., Report of Workshop on expression of vitamin K-dependent proteins in bacterial and mammalian cells, Madison, Wisconsin, USA, April 1986. Thromb Res, 1986. 44(1): p. 129-34.

[116] Llop, E., et al., Structural analysis of the glycosylation of gene-activated erythropoietin (epoetin delta, Dynepo). Anal Biochem, 2008. 383(2): p. 243-54.

[117] Zhang, J., Mammalian cell culture for biopharmaceutical production, in Manual of Industrial Microbiology and Biotechnology, R.H. Baltz, J.E. Davies, and A.L. Demain, Editors. 2010, ASM Press: Washington, DC. p. 145-156.
[118] Zhu, J., Mammalian cell protein expression for biopharmaceutical production. Biotechnol Adv, 2011.
[119] Diaz, S.L., et al., Sensitive and specific detection of the non-human sialic Acid N-glycolylneuraminic acid in human tissues and biotherapeutic products. PLoS One, 2009. 4(1): p. e4241.
[120] Young, N.M., et al., Structure of the N-linked glycan present on multiple glycoproteins in the Gram-negative bacterium, Campylobacter jejuni. Journal of Biological Chemistry, 2002. 277(45): p. 42530-9.
[121] Wacker, M., et al., N-Linked Glycosylation in Campylobacter jejuni and Its Functional Transfer into E. coli. Science, 2002. 298(5599): p. 1790-1793.
[122] Szymanski, C.M., et al., Evidence for a system of general protein glycosylation in Campylobacter jejuni. Molecular Microbiology, 1999. 32(5): p. 1022-1030.
[123] Wacker, M., et al., Substrate specificity of bacterial oligosaccharyltransferase suggests a common transfer mechanism for the bacterial and eukaryotic systems. Proc Natl Acad Sci U S A, 2006. 103(18): p. 7088-93.
[124] Schwarz, F., et al., A combined method for producing homogeneous glycoproteins with eukaryotic N-glycosylation. Nat Chem Biol, 2010. 6(4): p. 264-6.
[125] Lehrer, J., et al., Functional characterization and membrane topology of Escherichia coli WecA, a sugar-phosphate transferase initiating the biosynthesis of enterobacterial common antigen and O-antigen lipopolysaccharide. J Bacteriol, 2007. 189(7): p. 2618-28.
[126] Valderrama-Rincon, J.D., et al., An engineered eukaryotic protein glycosylation pathway in Escherichia coli. Nat Chem Biol, 2012. 8(5): p. 434-6.
[127] Kowarik, M., et al., Definition of the bacterial N-glycosylation site consensus sequence. EMBO J, 2006. 25(9): p. 1957-1966.
[128] Sorensen, H.P. and K.K. Mortensen, Advanced genetic strategies for recombinant protein expression in Escherichia coli. J Biotechnol, 2005. 115(2): p. 113-28.
[129] Hardy, E., et al., Large-scale production of recombinant hepatitis B surface antigen from Pichia pastoris. J Biotechnol, 2000. 77(2-3): p. 157-67.
[130] Ebrahimi, S.M., et al., Cloning, expression and purification of the influenza A (H9N2) virus M2e antigen and truncated Mycobacterium tuberculosis HSP70 as a fusion protein in Pichia pastoris. Protein Expression and Purification, 2010. 70(1): p. 7-12.
[131] Hamilton, S.R. and T.U. Gerngross, Glycosylation engineering in yeast: the advent of fully humanized yeast. Current Opinion in Biotechnology, 2007. 18(5): p. 387-392.
[132] Hamilton, S.R., et al., Humanization of Yeast to Produce Complex Terminally Sialylated Glycoproteins. Science, 2006. 313(5792): p. 1441-1443.
[133] Nett, J.H., et al., Optimization of erythropoietin production with controlled glycosylation-PEGylated erythropoietin produced in glycoengineered Pichia pastoris. J Biotechnol, 2012. 157(1): p. 198-206.

[134] Geisler, C. and D.L. Jarvis, Substrate Specificities and Intracellular Distributions of Three N-glycan Processing Enzymes Functioning at a Key Branch Point in the Insect N-Glycosylation Pathway. Journal of Biological Chemistry, 2012. 287(10): p. 7084-7097.

[135] Hollister, J., et al., Engineering the protein N-glycosylation pathway in insect cells for production of biantennary, complex N-glycans. Biochemistry, 2002. 41(50): p. 15093-15104.

[136] Aumiller, J.J., J.R. Hollister, and D.L. Jarvis, A transgenic insect cell line engineered to produce CMP-sialic acid and sialylated glycoproteins. Glycobiology, 2003. 13(6): p. 497-507.

[137] Yun, E.Y., et al., Galatosylation and sialylation of mammalian glycoproteins produced by baculovirus-madiated gene expression in insect cells. Biotechnol Lett, 2005. 27(14): p. 1035-9.

[138] Okada, T., et al., N-Glycosylation engineering of lepidopteran insect cells by the introduction of the beta 1,4-N-acetylglucosaminyltransferase III gene. Glycobiology, 2010. 20(9): p. 1147-1159.

[139] Aumiller, J.J., et al., A new glycoengineered insect cell line with an inducibly mammalianized protein N-glycosylation pathway. Glycobiology, 2012. 22(3): p. 417-28.

[140] Takuma, S., C. Hirashima, and J.M. Piret, Dependence on glucose limitation of the pCO(2) influences on CHO cell growth, metabolism and IgG production. Biotechnology and Bioengineering, 2007. 97(6): p. 1479-1488.

[141] Tachibana, H., et al., Changes of monosaccharide availability of human hybridoma lead to alteration of biological properties of human monoclonal antibody. Cytotechnology, 1994. 16(3): p. 151-157.

[142] Hayter, P.M., et al., Glucose-limited chemostat culture of chinese hamster ovary cells producing recombinant human interferon-gamma. Biotechnol Bioeng, 1992. 39(3): p. 327-335.

[143] Nyberg, G.B., et al., Metabolic effects on recombinant interferon-gamma glycosylation in continuous culture of Chinese hamster ovary cells. Biotechnol Bioeng, 1999. 62(3): p. 336-347.

[144] Wong, D.C.F., et al., Impact of dynamic online fed-batch strategies on metabolism, productivity and N-glycosylation quality in CHO cell cultures. Biotechnol Bioeng, 2005. 89(2): p. 164-177.

[145] Trummer, E., et al., Process parameter shifting: Part I. Effect of DOT, pH, and temperature on the performance of Epo-Fc expressing CHO cells cultivated in controlled batch bioreactors. Biotechnol Bioeng, 2006. 94(6): p. 1033-1044.

[146] Chotigeat, W., et al., Role of environmental conditions on the expression levels, glycoform pattern and levels of sialyltransferase for hFSH produced by recombinant CHO cells. Cytotechnology, 1994. 15(1-3): p. 217-221.

[147] Gawlitzek, M., H.S. Conradt, and R. Wagner, Effect of different cell culture conditions on the polypeptide integrity and N-glycosylation of a recombinant model glycoprotein. Biotechnol Bioeng, 1995. 46(6): p. 536-544.

[148] Kunkel, J.P., et al., Dissolved oxygen concentration in serum-free continuous culture affects N-linked glycosylation of a monoclonal antibody. J Biotechnol, 1998. 62(1): p. 55-71.
[149] Borys, M.C., D.I. Linzer, and E.T. Papoutsakis, Culture pH affects expression rates and glycosylation of recombinant mouse placental lactogen proteins by Chinese hamster ovary (CHO) cells. Bio/Technology, 1993. 11(6): p. 720-724.
[150] Gawlitzek, M., et al., Ammonium alters N-glycan structures of recombinant TNFR-IgG: degradative versus biosynthetic mechanisms. Biotechnol Bioeng, 2000. 68(6): p. 637-646.
[151] Axelsson, M.A., et al., Neutralization of pH in the Golgi apparatus causes redistribution of glycosyltransferases and changes in the O-glycosylation of mucins. Glycobiology, 2001. 11(8): p. 633-644.
[152] Rivinoja, A., et al., Elevated Golgi pH impairs terminal N-glycosylation by inducing mislocalization of Golgi glycosyltransferases. J Cell Physiol, 2009. 220(1): p. 144-154.
[153] Gawlitzek, M., U. Valley, and R. Wagner, Ammonium ion and glucosamine dependent increases of oligosaccharide complexity in recombinant glycoproteins secreted from cultivated BHK-21 cells. Biotechnol Bioeng, 1998. 57(5): p. 518-528.
[154] Yang, M. and M. Butler, Effects of ammonia on CHO cell growth, erythropoietin production, and glycosylation. Biotechnol Bioeng, 2000. 68(4): p. 370-380.
[155] Rijcken, W.R.P., et al., The Effect of Increasing Nucleotide Sugar Concentrations on the Incorporation of Sugars into Glycoconjugates in Rat Hepatocytes. Biochemical Journal, 1995. 305: p. 865-870.
[156] Müthing, J., et al., Effects of buffering conditions and culture pH on production rates and glycosylation of clinical phase I anti-melanoma mouse IgG3 monoclonal antibody R24. Biotechnol Bioeng, 2003. 83(3): p. 321-334.
[157] Ahn, W.S., et al., Effect of culture temperature on erythropoietin production and glycosylation in a perfusion culture of recombinant CHO cells. Biotechnol Bioeng, 2008. 101(6): p. 1234-1244.
[158] Jenkins, N., et al., Effect of lipid supplements on the production and glycosylation of recombinant interferon-gamma expressed in CHO cells. Cytotechnology, 1994. 15(1-3): p. 209-215.
[159] Lifely, M.R., et al., Glycosylation and biological activity of CAMPATH-1H expressed in different cell lines and grown under different culture conditions. Glycobiology, 1995. 5(8): p. 813-822.
[160] Serrato, J.A., et al., Differences in the glycosylation profile of a monoclonal antibody produced by hybridomas cultured in serum-supplemented, serum-free or chemically defined media. Biotechnol Appl Biochem, 2007. 47(Pt 2): p. 113-124.
[161] Schmelzer, A.E. and W.M. Miller, Hyperosmotic stress and elevated pCO2 alter monoclonal antibody charge distribution and monosaccharide content. Biotechnol Prog, 2002. 18(2): p. 346-353.

[162] Senger, R.S. and M.N. Karim, Effect of shear stress on intrinsic CHO culture state and glycosylation of recombinant tissue-type plasminogen activator protein. Biotechnol Prog, 2003. 19(4): p. 1199-1209.

[163] Lavecchio, J.A., A.D. Dunne, and A.S.B. Edge, Enzymatic Removal of Alpha-Galactosyl Epitopes from Porcine Endothelial-Cells Diminishes the Cytotoxic Effect of Natural Antibodies. Transplantation, 1995. 60(8): p. 841-847.

[164] Sanai, Y., M. Yamasaki, and Y. Nagai, Monoclonal-Antibody Directed to a Hanganutziu-Deicher Active Ganglioside, Gm2 (Neugc). Biochimica Et Biophysica Acta, 1988. 958(3): p. 368-374.

[165] Hills, A.E., et al., Metabolic control of recombinant monoclonal antibody N-glycosylation in GS-NS0 cells. Biotechnology and Bioengineering, 2001. 75(2): p. 239-251.

[166] Alley, S.C., et al. SEA technology: a novel strategy for enhancing antibody effector function. in Proceedings of the American Association for Cancer Research Annual Meeting. 2010.

[167] Siadak, T., et al. Enhancing Biological Activity of Immunoglycoproteins by a Convenient Method of Generating Preferred Glycovariants. in Cell Culture Engineering XI. 2008.

[168] Brooks, S.A., Strategies for Analysis of the Glycosylation of Proteins: Current Status and Future Perspectives. Molecular Biotechnology, 2009. 43(1): p. 76-88.

[169] Bailey, M.J., et al., A platform for high-throughput molecular characterization of recombinant monoclonal antibodies. Journal of Chromatography B-Analytical Technologies in the Biomedical and Life Sciences, 2005. 826(1-2): p. 177-187.

[170] Alvarez, M., et al., On-line characterization of monoclonal antibody variants by liquid chromatography-mass spectrometry operating in a two-dimensional format. Analytical Biochemistry, 2011. 419(1): p. 17-25.

[171] Hurum, D.C. and J.S. Rohrer, Five-minute glycoprotein sialic acid determination by high-performance anion exchange chromatography with pulsed amperometric detection. Analytical Biochemistry, 2011. 419(1): p. 67-69.

[172] Mittermayr, S., et al., Multiplexed Analytical Glycomics: Rapid and Confident IgG N-Glycan Structural Elucidation. Journal of Proteome Research, 2011. 10(8): p. 3820-3829.

[173] Behjousiar, A., C. Kontoravdi, and K.M. Polizzi, <italic>In Situ</italic> Monitoring of Intracellular Glucose and Glutamine in CHO Cell Culture. Plos One, 2012. 7(4): p. e34512.

[174] Chen, P., et al., Identification of N-glycan of alpha-fetoprotein by lectin affinity microarray. Journal of Cancer Research and Clinical Oncology, 2008. 134(8): p. 851-860.

[175] Wang, Y.H., S.L. Wu, and W.S. Hancock, Monitoring of glycoprotein products in cell culture lysates using lectin affinity chromatography and capillary HPLC coupled to electrospray linear ion trap-Fourier transform mass spectrometry (LTQ/FTMS). Biotechnology Progress, 2006. 22(3): p. 873-880.

[176] Loo, D., A. Jones, and M.M. Hill, Lectin Magnetic Bead Array for Biomarker Discovery. Journal of Proteome Research, 2010. 9(10): p. 5496-5500.

[177] Choi, E., et al., High-throughput lectin magnetic bead array-coupled tandem mass spectrometry for glycoprotein biomarker discovery. Electrophoresis, 2011. 32(24): p. 3564-3575.
[178] Kim, Y.P., et al., On-chip detection of protein glycosylation based on energy transfer between nanoparticles. Biosensors & Bioelectronics, 2009. 24(5): p. 1189-1194.
[179] Gerlach, J.Q., et al., Glycobiomimics and glycobiosensors. Biochemical Society Transactions, 2010. 38: p. 1333-1336.
[180] Umana, P., et al., Engineered glycoforms of an antineuroblastoma IgG1 with optimized antibody-dependent cellular cytotoxic activity. Nature Biotechnology, 1999. 17(2): p. 176-180.
[181] Weikert, S., et al., Engineering Chinese hamster ovary cells to maximize sialic acid content of recombinant glycoproteins. Nat Biotechnol, 1999. 17(11): p. 1116-1121.
[182] Imai-Nishiya, H., et al., Double knockdown of alpha1,6-fucosyltransferase (FUT8) and GDP-mannose 4,6-dehydratase (GMD) in antibody-producing cells: a new strategy for generating fully non-fucosylated therapeutic antibodies with enhanced ADCC. BMC Biotechnol, 2007. 7: p. 84.
[183] Monica, T.J., D.C. Andersen, and C.F. Goochee, A mathematical model of sialylation of N-linked oligosaccharides in the trans-Golgi network. Glycobiology, 1997. 7(4): p. 515-521.
[184] Shelikoff, M., A.J. Sinskey, and G. Stephanopoulos, A modeling framework for the study of protein glycosylation. Biotechnology and Bioengineering, 1996. 50(1): p. 73-90.
[185] Umana, P. and J.E. Bailey, A mathematical model of N-linked glycoform biosynthesis. Biotechnology and Bioengineering, 1997. 55(6): p. 890-908.
[186] Schachter, H., Biosynthetic Controls that Determine the Branching and Microheterogeneity of Protein-Bound Oligosaccharides. Biochemistry and Cell Biology-Biochimie Et Biologie Cellulaire, 1986. 64(3): p. 163-181.
[187] Krambeck, F.J. and M.J. Betenbaugh, A mathematical model of N-linked glycosylation. Biotechnology and Bioengineering, 2005. 92(6): p. 711-728.
[188] Inoue, N., et al., Asn-linked sugar chain structures of recombinant human thrombopoietin produced in Chinese hamster ovary cells. Glycoconjugate Journal, 1999. 16(11): p. 707-718.
[189] Krambeck, F.J., et al., A mathematical model to derive N-glycan structures and cellular enzyme activities from mass spectrometric data. Glycobiology, 2009. 19(11): p. 1163-1175.
[190] Hossler, P., B.C. Mulukutla, and W.S. Hu, Systems Analysis of N-Glycan Processing in Mammalian Cells. Plos One, 2007. 2(8).
[191] Jimenez del Val, I., J.M. Nagy, and C. Kontoravdi, A dynamic mathematical model for monoclonal antibody N-linked glycosylation and nucleotide sugar donor transport within a maturing Golgi apparatus. Biotechnology Progress, 2011. 27(6): p. 1730-43.

[192] Imai-Nishiya, H., et al., Double knockdown of alpha 1,6-fucosyltransferase (FUT8) and GDP-mannose 4,6-dehydratase (GMD) in antibody-producing cells: a new strategy for generating fully non-fucosylated therapeutic antibodies with enhanced ADCC. Bmc Biotechnology, 2007. 7.

Permissions

The contributors of this book come from diverse backgrounds, making this book a truly international effort. This book will bring forth new frontiers with its revolutionizing research information and detailed analysis of the nascent developments around the world.

We would like to thank Sefana Petrescu, Ph.D, for lending her expertise to make the book truly unique. She has played a crucial role in the development of this book. Without her invaluable contribution this book wouldn't have been possible. She has made vital efforts to compile up to date information on the varied aspects of this subject to make this book a valuable addition to the collection of many professionals and students.

This book was conceptualized with the vision of imparting up-to-date information and advanced data in this field. To ensure the same, a matchless editorial board was set up. Every individual on the board went through rigorous rounds of assessment to prove their worth. After which they invested a large part of their time researching and compiling the most relevant data for our readers. Conferences and sessions were held from time to time between the editorial board and the contributing authors to present the data in the most comprehensible form. The editorial team has worked tirelessly to provide valuable and valid information to help people across the globe.

Every chapter published in this book has been scrutinized by our experts. Their significance has been extensively debated. The topics covered herein carry significant findings which will fuel the growth of the discipline. They may even be implemented as practical applications or may be referred to as a beginning point for another development. Chapters in this book were first published by InTech; hereby published with permission under the Creative Commons Attribution License or equivalent.

The editorial board has been involved in producing this book since its inception. They have spent rigorous hours researching and exploring the diverse topics which have resulted in the successful publishing of this book. They have passed on their knowledge of decades through this book. To expedite this challenging task, the publisher supported the team at every step. A small team of assistant editors was also appointed to further simplify the editing procedure and attain best results for the readers.

Our editorial team has been hand-picked from every corner of the world. Their multi-ethnicity adds dynamic inputs to the discussions which result in innovative outcomes. These outcomes are then further discussed with the researchers and contributors who give their valuable feedback and opinion regarding the same. The feedback is then collaborated with the researches and they are edited in a comprehensive manner to aid the understanding of the subject.

Apart from the editorial board, the designing team has also invested a significant amount of their time in understanding the subject and creating the most relevant covers. They scrutinized every image to scout for the most suitable representation of the subject and create an appropriate cover for the book.

The publishing team has been involved in this book since its early stages. They were actively engaged in every process, be it collecting the data, connecting with the contributors or procuring relevant information. The team has been an ardent support to the editorial, designing and production team. Their endless efforts to recruit the best for this project, has resulted in the accomplishment of this book. They are a veteran in the field of academics and their pool of knowledge is as vast as their experience in printing. Their expertise and guidance has proved useful at every step. Their uncompromising quality standards have made this book an exceptional effort. Their encouragement from time to time has been an inspiration for everyone.

The publisher and the editorial board hope that this book will prove to be a valuable piece of knowledge for researchers, students, practitioners and scholars across the globe.

List of Contributors

Benjamin Luke Schulz
The University of Queensland, School of Chemistry and Molecular Biosciences, St Lucia, QLD, Australia

Joanne E. Nettleship
OPPF-UK, Research Complex at Harwell, R92 Rutherford Appleton Laboratories, Harwell Oxford, Didcot, UK
Division of Structural Biology, Oxford University, Welcome Trust Centre for Human Genetics, Roosevelt Drive, Oxford, UK

Marius D. Surleac, Laurentiu N. Spiridon, Robi Tacutu, Adina L. Milac, Stefana M. Petrescu and Andrei-J Petrescu
Institute of Biochemistry of the Romanian Academy, Bucharest, Romania

Vitor H. Pomin
Program of Glycobiology, Institute of Medical Biochemistry, University Hospital Clementino Fraga Filho, Federal University of Rio de Janeiro, Rio de Janeiro, Brazil

Nagatoshi Fujiwara
Department of Bacteriology, Osaka City University Graduate School of Medicine, Japan

Jumpei Hayakawa and Morio Ishizuka
Department of Applied Chemistry, Faculty of Science and Engineering, Chuo University, Tokyo, Japan

Masahiro Nagahama, Masataka Oda and Keiko Kobayashi
Tokushima Bunri University, Japan

Oshrat Levy-Ontman
Department of Chemical Engineering, Sami Shamoon College of Engineering, Beer Sheva, Israel

Mariela Odjakova, Eva Popova and Merilin Al Sharif
Department of Biochemistry, Faculty of Biology, Sofia University, Bulgaria

Roumyana Mironova
Roumen Tsanev Institute of Molecular Biology at the Bulgarian Academy of Sciences, Bulgaria

Kevin D. Smith and Jennifer Behan
School of Life, Sport and Social Sciences, Edinburgh Napier University, Sighthill Campus, Edinburgh, EH11 4BN, Scotland

Gerardine Matthews-Smith
School of Nursing, Midwifery and Social Care, Edinburgh Napier University, Sighthill Campus, Edinburgh, EH11 4BN, Scotland

Anthony M. Magliocco
Department Chair Anatomic Pathology, Esoteric Laboratory Services, H Lee Moffitt Cancer, Tampa, Florida, USA

Anna Janik, Mateusz Juchimiuk, Joanna Kruszewska, Jacek Orłowski, Monika Pasikowska and Grażyna Palamarczyk
Institute of Biochemistry and Biophysics, Polish Academy of Sciences, Laboratory of Fungal Glycobiology, Warsaw, Poland

Paula Santoyo-Ramos, María Cristina Castañeda-Patlán and Martha Robles-Flores
Department of Biochemistry, Faculty of Medicine, Universidad Nacional Autónoma de México (UNAM), Mexico City, Mexico

Kristian Prydz, Gro Live Fagereng and Heidi Tveit
Department of Molecular Biosciences, University of Oslo, Oslo, Norway

Brian J. Arey
Bristol-Myers Squibb Co, USA

Julio C. Valencia and Vincent J. Hearing
Pigment Cell Biology Section, Laboratory of Cell Biology, National Cancer Institute, Bethesda, USA

Céline Raymond and Yves Durocher
Département de Biochimie, Université de Montréal, Montréal, Canada

Céline Raymond, Erika Lattová, Hélène Perreault and Yves Durocher
MabNet NSERC Network for the Manufacturing of Single-Type Glycoform Monoclonal Antibody, Canada

Anna Robotham, John Kelly and Yves Durocher
Human Health Therapeutic Portfolio of the National Research Council Canada, Montreal and Ottawa, Canada

Erika Lattová and Hélène Perreault
Chemistry Department, University of Manitoba, Winnipeg, Canada

Dongxing Zha
GlycoFi Inc., A wholly-owned subsidiary of Merck & Co., Inc., Lebanon

Ioscani Jimenez del Val, Sarantos Kyriakopoulos and Cleo Kontoravdi
Centre for Process Systems Engineering, Department of Chemical Engineering, Imperial College London, London, UK

Philip M. Jedrzejewski, Kealan Exley and Si Nga Sou
Centre for Process Systems Engineering, Department of Chemical Engineering, Imperial College London, London, UK
Division of Molecular Biosciences, Department of Life Sciences, Imperial College London, London, UK
Centre for Synthetic Biology and Innovation, Imperial College London, London, UK

M. Polizzi
Division of Molecular Biosciences, Department of Life Sciences, Imperial College London, London, UK
Centre for Synthetic Biology and Innovation, Imperial College London, London, UK